AF810747

Biological Magnetic Resonance

Volume 24

Biomedical EPR, Part B: Methodology, Instrumentation, and Dynamics

A Continuation Order Plan is available for this series. A continuation order will bring delivery of each new volume immediately upon publication. Volumes are billed only upon actual shipment. For further information please contact the publisher.

KAP ARCHIEF

Biological Magnetic Resonance

Volume 24

Biomedical EPR, Part B: Methodology, Instrumentation, and Dynamics

Edited by

Sandra R. Eaton
University of Denver
Denver, Colorado

Gareth R. Eaton
University of Denver
Denver, Colorado

and

Lawrence J. Berliner
University of Denver
Denver, Colorado

KLUWER ACADEMIC / PLENUM PUBLISHERS
NEW YORK, BOSTON, DORDRECHT, LONDON, MOSCOW

Library of Congress Cataloging-in-Publication Data

Biomedical EPR/edited by Sanda S. Eaton, Gareth R. Eaton, Lawrence J. Berliner.
p. cm. — (Biological magnetic resonance, ISSN 0192-6020; v. 23–24)
Includes bibliographical references and index.
ISBN 0-306-48506-0 (v. 23) — ISBN 0-306-48556-7 (v. 23 e-book) — ISBN 0-306-48532-X (v. 24) — ISBN 0-306-48533-8 (v. 24 e-book)
1. Electron paramagnetic resonance spectroscopy. I. Eaton, Sandra S. II. Eaton, Gareth R. III. Berliner, Lawrence J. IV. Series.
QH324.9.M28 B56 vol. 23, etc.
[QP519.9.E433]
572′.36—dc22

2004042156

ISBN: 0-306-48532-X
eISBN: 0-306-48533-8

© 2005 Kluwer Academic / Plenum Publishers, New York
233 Spring Street, New York, New York 10013

http://www.wkap/nl

10 9 8 7 6 5 4 3 2 1

A C.I.P. record for this book is available from the Library of Congress

All rights reserved

No part of this book may be reproduced, stored in a retrieval system, or transmitted in any form or by any means, electronic, mechanical, photocopying, microfilming, recording, or otherwise, without written permission from the Publisher, with the exception of any material supplied specifically for the purpose of being entered and executed on a computer system, for exclusive use by the purchaser of the work.

Permissions for books published in Europe: *permissions@wkap.nl*
Permissions for books published in the United States of America: *permissions@wkap.com*

Printed in the United States of America

Dedication:

To the students whom we hope to stimulate to become the next generation of biomedical EPR researchers.

Contributors

Albert H. Beth • Department of Molecular Physiology and Biophysics, Vanderbilt University, Nashville, TN 37232

Theodore G. Camenisch • Department of Biophysics, Medical College of Wisconsin, Milwaukee, WI 53226

Gareth R. Eaton • Department of Chemistry and Biochemistry, University of Denver, Denver, Colorado 80208

Sandra S. Eaton • Department of Chemistry and Biochemistry, University of Denver, Denver, Colorado 80208

Jimmy B. Feix • Department of Biophysics, Medical College of Wisconsin, Milwaukee, WI 53226

Jack H. Freed • Department of Chemistry, and Chemical Biology, Baker Laborator, Cornell University, Ithaca, New York 14853-1301

Wojciech Froncisz • Jagiellonian University, Krakow, Poland

Fabian Gerson • Department of Chemistry, University of Basel, Klingelbergstrasse 80, CH-4056 Basel, Switzerland

Georg Gescheidt • Department of Chemistry, University of Basel, Klingelbergstrasse 80, CH-4056 Basel, Switzerland

László I. Horváth • Institute of Biophysics, Biological Research Centre, 6701 Szeged, Hungary

Eric J. Hustedt • Department of Molecular Physiology and Biophysics, Vanderbilt University, Nashville, TN 37232

James S. Hyde • Department of Biophysics, Medical College of Wisconsin, Milwaukee, WI 53226

Lowell Kispert • Chemistry Department, The University of Alabama, Box 870336, Tuscaloosa, Al 35487

Candice Klug • Department of Biophysics, Medical College of Wisconsin, Milwaukee, WI 53226

Vsevolod A. Livshits • Centre of Photochemistry, Russian Academy of Sciences, 117421 Moscow Russia

Marvin Makinen • Department of Biochemistry and Molecular Biology, The University of Chicago, Cummings Life Science Center, 920 East 58th Street, Chicago, IL 60637

Derek Marsh • Max-Planck-Institut für Biophysikalische Chemie, Abteilung Spektroskopie, 37070 Göttingen, Germany

Devkumar Mustafi • Department of Biochemistry and Molecular Biology, The University of Chicago, Cummings Life Science Center, 920 East 58th Street, Chicago, IL 60637

Tibor Páli • Institute of Biophysics, Biological Research Centre, 6701 Szeged, Hungary

Joseph J. Ratke • Department of Biophysics, Medical College of Wisconsin, Milwaukee, WI 53226

George A. Rinard • Department of Engineering and University of Denver, Denver, Colorado 80208

Charles P. Scholes • Department of Chemistry, University at Albany – State University of New York, Albany, NY 12222

Robert A. Strangeway • Milwaukee School of Engineering, Milwaukee, WI 53226

PREFACE

There has not been an attempt to cover the full scope of biological EPR in a single volume since *Biological Applications of Electron Spin Resonance* edited by Swartz, Bolton, and Borg in 1972. In three decades there have been enormous changes in the field. Our original plan for one volume expanded into two. A stimulus for an updated book at this time was the 70th birthday of James S. Hyde (May 20, 2002), one of the leaders in the development of EPR instrumentation and methodology applied to biological problems. To symbolically tie this book to Jim Hyde's efforts, we choose the title "Biomedical EPR", which is the name of the NIH-funded National Biomedical EPR Center founded by Harold Swartz and James Hyde at the Medical College of Wisconsin in 1975. This Center has been funded continuously since then, and has been a focal point of new developments and applications in biomedical research. Many of the authors of chapters in this book have been close associates of Jim Hyde, and several have been long-term members of the Advisory Committee of the Center.

There is a long history underlying most of the topics in these books. Some of this history was surveyed in *Foundations of Modern EPR*, edited by Eaton, Eaton, and Salikhov (1998). It is helpful to keep in mind that theoretical and experimental studies of spin relaxation preceded the development of EPR and NMR. The early work of Waller and of Gorter, for example, focused on spin relaxation (see *Foundations of Modern EPR*). Long development periods, and indirect paths from initial concept to biomedical application are the norm. Even new instrumentation or methodology developments, with few exceptions, require of the order of 10 to 15 years from "invention" to general application. No one could have predicted that the attempt to make a better measurement of the deuterium magnetic moment would lead to functional magnetic resonance imaging (fMRI), and if such a prediction had been made, it would have been dismissed as ridiculous. Those who sponsor research, and nurture researchers, enrich humanity by not demanding proof of relevance. We each pursue goals that inspire us, and hope that they will be of benefit. This book is part of a story as it unfolds.

Contributors were asked to make this book more "pedagogical" than "review." The goal is a multi-author introduction to biomedical EPR with up-to-date examples, explanations, and applications, pointing toward the future. Thus, the book is aimed not just at readers who are EPR experts, but at biomedical researchers seeking to learn whether EPR technology and methodology will be useful to solve their biomedical problems. The derivation and explanation of the underlying theory and methodology for many of the topics presented would require separate books. The authors were asked to keep the background and theory to a minimum, referring whenever possible to other texts and reviews to lead the reader to additional information. The referencing in most chapters is thus to be tutorial and helpful, rather than to be comprehensive or to reflect priority of discovery. There is a focus on papers with a biological orientation. Thus, for example, although the fact that oxygen in solution broadens CW EPR spectra has been known since 1959 (see the chapter by Hauser and Brunner in *Foundations of Modern EPR*), the citations in the oxymetry chapter in this book to biologically relevant literature about oxygen broadening start about twenty years later. The perspective in each chapter is presented from the viewpoint of people involved in cutting-edge research.

Chapters, including our own, were peer-reviewed, usually by at least two referees in addition to the editors. We thank the referees for their assistance in improving the pedagogy of the chapters. The editors have added cross references between chapters.

In these volumes, we did not include some topics that had been reviewed recently. Spin Labeling I (1976) and II (1979), and the two volumes in this series that are successors to these, volumes 8 (1989) and 14 (1998), emphasize nitroxyl radicals. Volume 13 (1993) emphasizes paramagnetic metals, especially in enzymes, and transient EPR and spin trapping. Volume 18 (2004) describes *in vivo* EPR. Volume 19 (2000) is about measuring distances between unpaired electrons. Volume 21 of the Biological Magnetic Resonance series includes chapters on instrumentation (Bender), sensitivity (Rinard, Quine, Eaton, and Eaton), and a survey of low-frequency spectrometers (Eaton and Eaton). Other chapters of interest can be found in the list of contents of related prior volumes, at the end of each of these volumes. Some volumes in the series Metal Ions in Biological Systems, edited by Sigel focus on EPR. See, for example, Volume 22 (*ENDOR, EPR, and Electron Spin Echo for Probing Coordination Spheres*, 1987).

Although the focus of this book is on biomedical applications of EPR, and the examples used in this book therefore are largely from the biomedical field, an analogous treatise could focus on materials science, traditional small-molecule chemistry, or solid state physics. There are, of course, unifying theoretical, instrumental, and experimental methodologies that cross disciplinary applications. EPR has the great power of specificity for unpaired electron spins, and as Jim has said more than once, "there are spins everywhere."

Biological applications of EPR encompass measuring metal ion environments in proteins at liquid helium temperature and measuring NO production in living animals. The variety of technologies and methodologies required is so wide that a researcher who is expert in one may be almost unaware of another. The landscape is rich and the horizons extend as far as we can see. These two volumes, which should be read as a single treatise, have the goal of helping biomedical researchers see a little further.

Some potential users will need a more extensive basic introduction to EPR. The reader unfamiliar with EPR may want to start with the Introduction to the chapter by Subramanian and Krishna in Part B (Volume 24), which includes a concise survey of the basic principles of EPR. The Swartz, Bolton and Borg book (1972) mentioned above also is a good place to start. Among the several complete texts on EPR, those by Carrington and McLachlan (1967), by Weil, Bolton and Wertz (1994), and by Atherton (1993) are particularly appropriate for beginners who have a good physical chemistry background. Eaton and Eaton (1997) present an introduction to CW and pulsed EPR, with an emphasis on practical experimental aspects for the novice. Experimental and instrumental aspects of EPR are treated in Fraenkel (1959) and Reiger (1972), but the two major and most highly recommended sources are Alger (1968) and Poole (1967, 1983). Jim Hyde also wrote a brief summary of instrumental aspects of EPR (1995). It is hoped that some readers will enjoy learning some of the historical background of the field. Some of the chapters in this book provide a glimpse, and Foundations of Modern EPR (1998) captures the thinking of pioneers in the field on the occasion of the 50th anniversary of the discovery.

Pictures of experimental EPR spectra beyond those in these books may help the reader's understanding. Many spectra are reproduced in the texts cited above, and in Yen (1969), McGarvey (1966), Goodman and Raynor (1970), Drago (1992), Gerson (1970), and Gerson and Huber (2003). Some early reviews of spin labeling remain very useful introductions to the fundamentals of CW EPR of nitroxyl radical line shapes (Griffith and Wagoner, 1969; Jost, Wagoner, and Griffith, 1971; Jost and Griffith, 1972; Gaffney, 1974).

There is not enough space in these two volumes to teach the underlying principles of pulsed EPR in depth, nor to illustrate the wide range of applications. Readers are directed to several other books for more on these topics: Kevan and Swartz (1979), Keijzers et al. (1989), Hoff (1989), Kevan and Bowman (1990), Dikanov and Tsvetkov (1992), Schweiger and Jeschke (2000), and Berliner, Eaton, and Eaton, (2000) (volume 19 in this series).

For those readers unfamiliar with the practical methodology of EPR, it is reasonable to ask "how long will it take to run an EPR spectrum?" The answer depends strongly on what one wants to learn from the sample, and can range from a few minutes to many weeks. Even the simple question, are there any unpaired electrons present, may take quite a bit of effort to answer, unless one already knows a lot about the sample. Column fractions of a nitroxyl-spin-labeled polymer can be monitored for radicals about as fast as the samples can be put in the spectrometer. This is an example of an application that could be automated. On the other hand, the spins may have relaxation times so long that they are difficult to observe without saturation or so short that they cannot be observed except at very low temperature where the relaxation times become long enough (e.g., Co(II) in many environments). If one wants to know the concentration of Co(II) in a sample, need for quantitative sample preparation, accurate cryogenic temperature control, careful background subtraction, and skillful setting of instrument parameters lead to a rather time-consuming measurement.

Other reasonable questions include "how much will this cost?" and "how/where can I do this?" EPR measurements require a significant investment in instrumentation, but spectrometer systems are available from several vendors. The largest manufacturers, Bruker BioSpin EPR Division, and JEOL, market general-purpose spectrometers intended to fulfil most analytical needs. The focus is on X-band (ca. 9-10 GHz) continuous wave (CW) spectrometers, with a wide variety of resonators to provide for many types of samples. Accessories facilitate control of the sample temperature from <4K to ca. 700 K. Magnets commonly range from 6-inch to 12-inch pole face diameters. Smaller, table-top spectrometers are available from Bruker, JEOL, and Resonance Instruments. Some of these have permanent magnets and sweep coils for applications that focus on spectra near g = 2, and others have electromagnets permitting wide field sweeps. Bruker makes one small system optimized for quantitation of organic radicals and defect centers, such as for dosimetry. Bruker and JEOL market pulsed, time-domain spectrometers as well as CW spectrometers. Bruker and JEOL market spectrometers for frequencies lower than X-band, which are useful for study of lossy samples. Bruker markets high-frequency (95 GHz), high-field EPR spectrometers that require superconducting magnets, not electromagnets.

Volume 23 begins with an appreciation of the contributions that Jim Hyde made to biomedical EPR, with some historical perspective by Helmut Beinert and Harold Swartz of the mutual stimulation of Jim, the NIH Research Resource "Center" funding program, and the collaborations it spawned.

Among the common analytical tools available to those who study the properties of matter, whether biological or non-biological, ESR has the special feature that it is very sensitive to the anisotropy of the environment of the unpaired electron. The CW EPR spectral line shape is strongly influence by motions that are of the order of the anisotropies in hyperfine couplings and in g-values. Electron spin relaxation times are also sensitive to molecular motions. These effects give rise to the ability to measure rates and anisotropies of molecular motions, and stimulate the extensive field of spin labeling. One of the first physical parameters of spin labels to be exploited, the incomplete averaging of anisotropic g and hyperfine values, remains central to many uses of nitroxyl spin probes and spin labels. Freed (volume 24 chapter 9) explores the motions reported in great detail by nitroxyl EPR spectra. The saturation transfer technique developed by Jim Hyde and Larry Dalton (1979) is crucial to learning about the dynamics of biological membranes (Marsh et al., volume 24 chapter 11, and Beth and Husted, chapter 12). Beth and Husted show the sensitivity of Q-band (ca. 35 GHz) and W-band (ca. 95 GHz) EPR for analyzing complex anisotropic rotational dynamics, and emphasize the utility of global analysis of spectra obtained at two or more microwave frequencies. Basosi in volume 23 chapter 13 illustrates the kinds of information that can be learned about motions of metal ions.

There are contributions to the CW lines shape and some relaxation properties from electron-nuclear and electron-electron couplings. The dipolar part of the interaction is the basis for distance measurements. Electron-electron distance measurements were the topic of Volume 19 in this series (Berliner, Eaton, and Eaton, 2000), and the Eatons have presented a concise summary of this topic in chapter 8, and in Eaton and Eaton (2002). Because the electron dipole is larger than the nuclear dipoles, EPR measures distances that are larger than the distances measured by NMR. Multiple resonance techniques provide more detail about the spin environment than do "normal" EPR techniques. ENDOR is a very important tool for resolving hyperfine structure. ENDOR of species in frozen glassy solutions is described by Mustafi and Makinen in volume 24 chapter 4 and ENDOR of radicals in fluid solution is described by Gerson and Gescheidt in chapter 5. Next, Lowell Kispert (chapter 6) describes CW, pulsed, and multiquantum ELDOR as ways of probing electron-electron spin-spin interactions.

Many fundamental studies are directly relevant to biomedical science, but the goal of it all is to understand function and malfunction of living systems. It is important to perceive the relevance to human studies of early explorations in plants, for example. The chapter on free radicals and medicine (volume 23 chapter 3) surveys many of the motivations for investigating free radical phenomena. We thank Hal Swartz for coordinating the several contributors to this chapter, and for writing the introduction that give his overall perspective on this important area of science. How far we have come toward studies of animals and humans is reflected by several chapters. Hal Swartz and Nadeem Khan in volume 23 chapter 9 discuss the achievements to date and future possibilities in EPR spectroscopy of function *in vivo*. Depending on your point of view, Hal's perspective could be described as realistic or pessimistic. Maybe some reader will be stimulated to demonstrate clinical importance of tools that Hal says are unlikely to have major application. The major focus of research for *in vivo* EPR is the development of methodology to measure O_2 concentrations for medical purposes. Modern instrumentation facilitates a new focus on measurement of relaxation times, and use of relaxation properties to measure O_2, distances, etc. (see, for example, volume 24 chapters 1 and 8 by Eaton and Eaton). Benjamin Williams and Howard Halpern in volume 23 chapter 11 and Sankaran Subramanian and Murali Krishna in chapter 12 describe the fundamentals *in vivo* EPR spectroscopy and imaging by CW and pulsed low frequency EPR, respectively. Both of these chapters relate to measurement of O_2 in vivo. Oximetry is also the topic of the very detailed chapter 10 by Subczynski and Swartz.

Reactive free radicals, including superoxide (see volume 23 chapter 4 by Vásquez-Vivar, Martásek, and Kalyanaraman), are studied by the spin-trapping technique. In chapter 5 Ron Mason and Maria Kadiiska describe trapping of reactive radicals *in vivo*, with *ex vivo* detection of the EPR signal. Then, in chapter 6 Keszler and Hogg demonstrate linear regression analysis of multiple spin-trapped spectra to obtain kinetic information.

Application of EPR to understanding complex biological systems is illustrated by the examples of melanin (Sarna and Plonka, volume 23, chapter 7) and photosynthesis (Tikhanov and Subczynski, chapter 8).

As is emphasized in the introductory perspectives by Beinert and Swartz, many of Jim Hyde's contributions were innovations in instrumentation. The tight coupling to biomedical applications, first via visitors to Varian Associates and then with his colleagues at the Medical College and visitors to the National Biomedical ESR Center, focused Jim's instrumentation and methodology development on biomedical problems. Saturation recovery (Eaton and Eaton, volume 24 chapter 1) is applied in several biological investigations, including the oximetry measurements mentioned above. Loop-gap resonators (volume 24 chapter 2 by Rinard and Eaton) were the enabling technology for some of the recent developments in stopped-flow and rapid mixing EPR (Scholes, volume 24, chapter 3) and for the ability to study small spin-labeled protein samples, which gave the impetus to rapid application of site-directed spin labeling (Feix and Klug, Volume 24 chapter 10). In volume 24, chapter 13, Jim presents a perspective on the role of instrumentation in biomedical research. One of the trends foreseen, increased use of computer capability for fast digitization and post-processing, is illustrated by volume 24 chapter 7 (Hyde et al.). Having all frequencies in a magnetic resonance spectrometer phase-locked to a single master oscillator, then using fast digital detection and time-locked subsampling permits, for example, study of multiple harmonics of the field modulated signal.

We thank the authors for contributing to this book, and we also thank the many anonymous referees whose attention to both large and small matters helped improve the chapters. Beverly Ventura, Biophysics Research Institute, Medical College of Wisconsin, helped several authors with grammar and layout of their chapters. The final formatting of all chapters was done by one of the editors (SSE). We add special thanks to Hal Swartz, whose recruitment of Jim Hyde from Varian to the Medical College of Wisconsin set the stage for much of what is presented in these books, and who somehow could not say no to repeated entreaties to write on yet another topic. He ended up writing the introductory chapter on the background of Jim Hyde and the ESR Center in Milwaukee, and three chapters covering *in vivo* spectroscopy, oximetry, and free radicals in medicine. The last of these was a major effort, since we persuaded him to take contributions from many co-authors and assemble them into the overall chapter, which was probably more work than just writing it all himself!

Overall, even though we introduced these two books as a successor to the Swartz, Bolton, and Borg 1972 book, these books are still just a preface to the future of biomedical applications of EPR.

S. S. Eaton
G. R. Eaton
L. J. Berliner
Denver, Colorado

REFERENCES

Alger, R. S. (1968) *Electron Paramagnetic Resonance: Techniques and Applications*, Wiley-Interscience.

Atherton, N. M. (1993) *Principles of Electron Spin Resonance*. Ellis Horwood PTR Prentice Hall, London. This book has a strong emphasis on hyperfine interactions and ENDOR.

Carrington, A. and McLachlan, A. D. (1967) *Introduction to Magnetic Resonance*, Harper and Row, 1967. This book provides a parallel treatment of NMR and EPR.

Dikanov, S. A., and Tsvetkov, Yu. D. (1992) *Electron Spin Echo Envelope Modulation (ESEEM) Spectroscopy*. CRC Press, Boca Raton, FL.

Drago, R. S. (1992) *Physical Methods for Chemists*, 2nd ed. Saunders, Orlando, FL.

Eaton, S. S., and Eaton, G. R., (1997) Electron Paramagnetic Resonance, in *Analytical Instrumentation Handbook*, G. W. Ewing, ed., Marcel Dekker, 2nd ed., 767-862. The third edition will be published in 2004.

Eaton, G. R., Eaton, S. S., and Salikhov, K., eds., (1998) *Foundations of Modern EPR*, World Scientific Publishing, Singapore.

Eaton S. S. and G. R. Eaton, Electron Paramagnetic Resonance Techniques for Measuring Distances in Proteins, in *Structures and Mechanisms: from Ashes to Enzymes*, ACS Symposium Series 827, American Chemical Society, Washington, D. C., 2002, 321-339.

Fraenkel, G. K., (1959) Paramagnetic Resonance Absorption, in *Technique of Organic Chemistry*, Vol. I, Part IV, 3rd. ed., Chapter XLII, pp. 2801-2872. This is an old article, but it contains much valuable information not readily available elsewhere.

Gaffney, B. J. (1974) Spin Label Measurements in Membranes. *Methods Enzymol.* **32**, 161-198.

Gerson, F. (1970) *High Resolution E.S.R. Spectroscopy*. Verlag Chemie, Weinheim.

Gerson, F. and Huber, W. (2003) *Electron Spin Resonance Spectroscopy of Organic Radicals*. Wiley-VCH, Weinheim.

Goodman, B. A. and Raynor, J. B. Raynor, Electron Spin Resonance of Transition Metal Complexes, in *Advances in Inorganic Chemistry and Radiochemistry*, vol 13, 1970.

Griffith, O. H. and Waggoner, A. S. (1969) Nitroxide free Radicals: Spin Labels for Probing Biomolecular Structure. *Acct. Chem. Res.* **2**, 17-24.

Hausser, K. H. and Brunner, H. (1998) The Effect of Concentration and Oxygen in EPR Chapter H.2 in *Foundations of Modern EPR*, edited by Eaton, G. R., Eaton, S. S., and Salikhov, K. M., World Scientific, Singapore.

Hoff , A. J., ed., (1989) *Advanced EPR : Applications in Biology and Biochemistry*. Elsevier, Amsterdam.

Hyde, J. S., and Dalton, L. R. (1979) Saturation-Transfer Spectroscopy, in *Spin Labeling II: Theory and Applications*. L. J. Berliner, ed., Academic Press, New York.

Hyde, J. S. (1995) Electron Paramagnetic Resonance. Chapter 13 in *Handbook of Microwave Technology*, Volume 2, Ishii, T. K., ed., Academic Press.

Jost, P., Waggoner, A. S., and Griffith, O. H. (1971) Spin Labeling and Membrane Structure, in *Structure and Function of Biological Membranes*, Rothfield, L., ed., Academic Press, New York.

Jost, P. and Griffith, O. H. (1972) Electron Spin Resonance and the Spin Labeling Method. *Methods in Pharmacology* **2**, 223-276.

Keijzers, C. P., Reijerse, E. J., and Schmidt, J. (1989) *Pulsed EPR: A New Field of Applications*. North Holland, Amsterdam.

Kevan, L. and Schwartz, R. N. (1979) *Time Domain Electron Spin Resonance*. Wiley, New York.

Kevan, L. and Bowman, M. K. (1990) *Modern Pulsed and Continuous-Wave Electron Spin Resonance*. Wiley, New York.

McGarvey, B. R. (1966), Electron Spin Resonance of Transition Metal Complexes, in *Transition Metal Chemistry*, vol. 3, R. L. Carlin, ed., Marcel Dekker.

Poole, C. P., Jr. (1967, 1983) *Electron Spin Resonance: A Comprehensive Treatise on Experimental Techniques*, Wiley-Interscience, 1967. Second edition, Wiley, 1983. Note that both editions should be consulted. Especially for the topics of microwave components the coverage is more extensive in the first edition than in the second edition.

Rieger, P. H. (1972) Electron Spin Resonance, in *Techniques of Chemistry*, Vol. 1, Part III A, pp. 499-598.

Schweiger, A., and Jeschke, G. (2001) *Principles of Pulse Electron Paramagnetic Resonance*. Oxford University Press, Oxford.

Sigel, A., and Sigel, H., eds. (1999) *Interrelations between Free Radicals and Metal Ions in Life Processes*. Marcel Dekker, New York.

Swartz, H. M., Bolton, J. R., and Borg, D. C. (1972) *Biological Applications of Electron Spin Resonance*. Wiley

Weil, J. A., Bolton, J. R., and Wertz, J. E. (1994) *Electron Paramagnetic Resonance: Elementary Theory and Practical Applications*. Wiley, New York.

Yen, T. F. (1969) *Electron Spin Resonance of Metal Complexes*, Plenum.

Contents

Section I. Instrumentation and Methodology

Chapter 1

Saturation Recovery EPR

Sandra S. Eaton and Gareth R. Eaton

Chapter 2

Loop-Gap Resonators

George A. Rinard and Gareth R. Eaton

Chapter 5

Solution-ENDOR of Some Biologically Interesting Radical Ions

Fabian Gerson and Georg Gescheidt

Chapter 6

Electron-Electron Double Resonance

Lowell D. Kispert

Chapter 7

Digital Detection by Time-Locked Sampling in EPR

James S. Hyde, Theodore G. Camenisch, Joseph J. Ratke, Robert A. Strangeway, Wojciech Froncisz

Chapter 8

Measurement of Distances Between Electron Spins Using Pulsed EPR

Sandra S. Eaton and Gareth R. Eaton

Section II. Motion, Proteins, and Membranes

Chapter 9

ESR and Molecular Dynamics

Jack H. Freed

Chapter 10

SDSL: A Survey of Biological Applications

Candice S. Klug and Jimmy B. Feix

Chapter 11

Saturation Transfer Spectroscopy of Biological Membranes

Derek Marsh, László I. Horváth, Tibor Páli And Vsevolod A. Livshits

Chapter 12

Saturation Transfer EPR: Rotational Dynamics of Membrane Proteins

Albert H. Beth and Eric J. Hustedt

Chapter 13

Trends in EPR Technology

James S. Hyde

Chapter 14

Prognosis

Sandra S. Eaton and Gareth R. Eaton

Contents of Previous Volumes

Index

I

Instrumentation and Methodology

Chapter 1

Saturation Recovery EPR

Sandra S. Eaton and Gareth R. Eaton
Department of Chemistry and Biochemistry, University of Denver, Denver, Colorado 80208

Abstract: Saturation recovery EPR measures electron spin relaxation times, T_{1e}. Measurement techniques and applications to relaxation mechanisms, oximetry, Heisenberg exchange and spin-spin distance measurements are discussed.

1. MOTIVATION

From early times, it has been noticed that the CW EPR spectra of some spin systems saturate at lower microwave powers than do other spin systems. There developed a qualitative and semi-quantitative understanding of ways to use these observations to characterize radicals, and in some cases to identify that a normally slowly-relaxing radical was in proximity to a faster-relaxing radical. The general understanding of the effects of a rapidly relaxing metal ion predates the EPR measurements. Recall, for example, that transition metal ions were added to some of the earliest NMR samples to shorten the proton relaxation times. Also, by the late 1950s it was understood that O_2 in solution broadened the lines of CW EPR spectra (see review by Hausser and Brunner, 1998).

A qualitative understanding of relaxation times is essential for selection of parameters for CW EPR and ENDOR experiments and for prediction of feasibility of pulsed EPR experiments. Quantitatively measured relaxation times provide insight into electronic structure, motion, and other processes that contribute to relaxation.

Early estimates of relaxation times were obtained from power saturation curves, however analysis of these data is complicated by the dependence of saturation on the T_1T_2 product rather than on either T_1 or T_2 individually, and on spectral diffusion (Eaton and Eaton, 2000a). Most EPR signals in biomolecules are inhomogeneously broadened due to unresolved hyperfine structure. There is inherently more information in the relaxation times of individual spin packets and the rates of energy transfer among them than in the envelope of the CW line shape. Consequently, time-domain methods are of particular importance for the study of biomolecular systems. This chapter describes the saturation recovery (SR) method of measuring electron spin lattice relaxation (T_{1e}) and examples of applications to biological samples. Some applications of T_{1e} measurements are discussed together with spin echo measurements of T_2, and related relaxation times, in the chapter on distance measurements (ch. 8)

2. BRIEF HISTORY

The measurement of T_1 by CW SR was first demonstrated for nuclear spins by Bloembergen (1949). Subsequently there have been many applications to electron spins. Bloembergen and Wang (1954) measured the change in the z magnetization following a microwave pulse by using a pickup coil outside the resonant cavity. They mentioned, but did not apply, a technique in which recovery would be monitored in a low-level microwave field after the end of the saturating microwave pulse. This method became known as saturation recovery. Weissman and coworkers (1957) manually stepped the microwave amplitude in a measurement of a 20-minute T_{1e} of triphenylmethyl radical at 1.2 K. The first SR spectrometers capable of measurements faster than those that used manual switching were described in 1958 (Davis et al., 1958; Giordmaine et al., 1958), 1959 (Bowers and Mims, 1959) and 1960 (Pastor, et al., 1960). Collins et al. (1959) used a bimodal resonator and Scott and Jeffries (1962) demonstrated the use of a transmission cavity for SR. Early applications of SR to organic radicals were published by Venkataraman and coworkers (Rengen et al., 1972; 1974a,b, 1979; Lingam et al., 1972; Fessenden et al., 1981; Venkataraman, 1982). Theory of relaxation, applied to SR experiments, and illustrated with the case of nitroxyl radicals in fluid solution, was detailed by Freed (1974). The modern development of the field is largely due to Hyde and coworkers (Huisjen and Hyde, 1974a,b; Percival and Hyde, 1975, 1976; Hyde, 1979). Detailed descriptions of SR spectrometers have been published by Huisjen and Hyde (1974b), Percival and Hyde (1975), Mailer et al. (1985), Beck et al. (1991), and Quine et al. (1992, 1996). The development of SR at Varian Instruments has been described by Hyde (1998). Summaries of the early literature can be found in Standley and Vaughan (1969).

3. INFORMATION CONTENT OF SATURATION RECOVERY CURVES

It is important to distinguish between the experimentally determined response to a perturbation of the spin populations, which might be termed an "effective T_{1e}," and the "true T_{1e}," that characterizes transitions between particular electron spin energy levels. In a short-pulse SR experiment there frequently will be contributions to the recovery curve from spectral diffusion processes including molecular tumbling (in fluid solution), nuclear spin relaxation or cross relaxation (Yin and Hyde, 1987a). In these cases a "T_{1e} measurement" produces a recovery curve that is "the sum of all possible relaxation pathways rather than the relaxation between only the observed levels" (Hyde, 1979, page 27). The spectral diffusion processes can be characterized using electron-electron double resonance (ELDOR). The effects of these spectral diffusion processes on the SR response often can be mitigated in a long-pulse SR experiment (Hyde, 1979). However, we have observed that in some samples that contain methyl groups that are rotating at a rate that is comparable to the electron Larmor frequency, even long-pulse SR curves are not single exponentials. ELDOR curves indicate that electron spin cross relaxation is rapid for these samples (Harbridge et al., 2002), which may cause the deviation from single exponential behavior.

4. PRACTICAL ASPECTS OF EXPERIMENTAL METHODOLOGY

4.1 Description of the Continuous Wave SR Experiment

The pulse sequence for CW SR is sketched in Figure 1a. A higher-power pulse is applied to saturate the EPR transition. The length of this pulse defines whether the experiment is viewed as long- or short-pulse SR. As soon as possible after this pump pulse, the EPR signal is detected with lower-power continuous wave microwaves. The minimum time between the end of the pulse and the beginning of observation is determined by switching transients and by the ring-down time of the resonator. In principle, there should be no signal when the magnetic field is set off resonance. However, in reality, there are switching transients and responses of the resonator to heating that result in a "background" signal that is observed even when the field is set off resonance. Correction for these artifacts is performed by taking the difference between on- and off-resonance responses. The spectrometers in the Hyde laboratory employ dual-channel boxcar detection with low-frequency (e.g., 28 Hz) modulation of the magnetic field to alternately collect on- and off-resonance responses at each point in the recovery curve (Percival and Hyde, 1975; Hyde, 1979). The field stepping also eliminates noise at frequencies lower than the pulse repetition rate and the field-stepping rate. Modulation of the magnetic field is not an effective way to correct for non-resonance artifacts if the spectral width exceeds the field modulation range. In our laboratory we signal average blocks of SR responses on- and off- resonance and take the difference between the two sets of data (Quine et al., 1992), which permits measurements to be performed for broad transition metal signals.

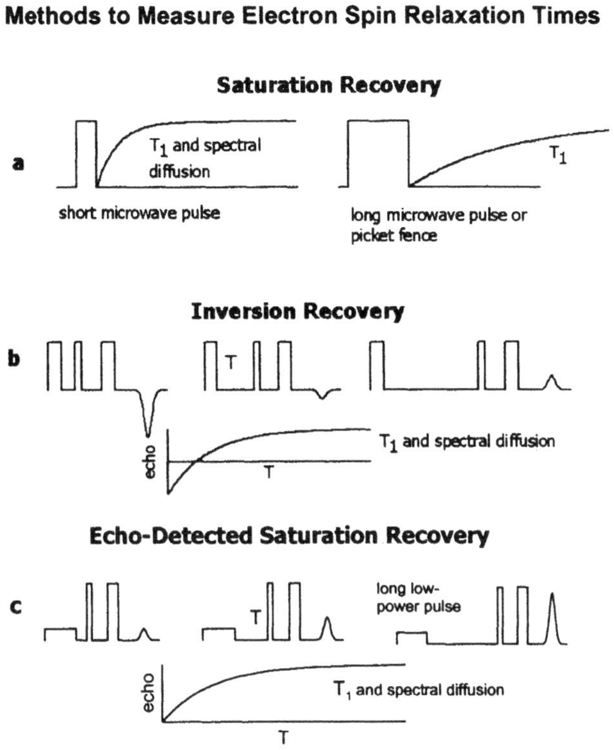

Figure 1. Pulse sequences for measurement of T_1, as discussed in the text. The sketches emphasize that in many cases the time constant that is measured is an "effective" T_1 that includes spectral diffusion contributions.

4.2 Contributions from Spectral Diffusion

In a CW-detected SR experiment (just as in CW EPR) the resonator is critically coupled and the resonator Q typically is relatively high. For example, at X-band the rectangular TE_{102} resonator has Q about 3600 where $Q = \omega/\Delta\omega$. At 9.5 GHz this value of Q corresponds to a half-power bandwidth, $\Delta\omega$, of 2.6 MHz. At g = 2 there is about 2.8 MHz/G so 2.6 MHz is about 1 Gauss (0.1 mT) at g = 2. Thus, this bandwidth means that any process that moves the resonant field by about a gauss on the time scale of the experiment can appear to be a relaxation process. Typical Q-values for X-band loop-gap resonators are roughly 1000 (depending on sample size and solvent lossiness), which means that the bandwidth for the LGR is about a factor of four greater than for the TE_{102} cavity, although the bandwidth still is small compared with the widths of EPR spectra, even for many radicals.

If it is observed that the experimental recovery time constant depends upon the length of the saturating pulse, then it can be concluded that spectral diffusion processes are contributing to the recovery. The length of the saturating pulse is then increased until a limiting value of the apparent relaxation time constant is observed. This limiting time constant is the best approximation to T_{1e} that can be obtained by SR. Cases in which the limiting value of the time constant still is not a "true T_{1e}" were documented in a series of papers by Manenkov and coworkers (Manenkov et al., 1962; Manenkov and Prokhorov, 1962; Manenkov and Pol'skii, 1964). They showed that although long pulses often suppress the effects of spectral diffusion on the recovery curve, it is possible to achieve conditions where the observed recovery is independent of the pulse width, but there is a steady state where the effect of spectral or spin diffusion is roughly balancing T_{1e}. In this case the recovery following the pump pulse exhibits contributions from spectral or spin diffusion. Manenkov et al. focused specifically on the importance of two paths, spin-lattice relaxation and cross relaxation to other states of neighboring multi-level paramagnetic centers, such as Cr(III), Fe(III), Nd(III) and Gd(III). In the Cr(III) systems studied, at various concentrations and at 1.7 and 4.2 K, cross relaxation, T_{x1}, was always shorter than T_{1e} and therefore contributed to the recovery curve. However, if $T_{x1} > T_{1e}$ cross relaxation would not contribute to the long-pulse SR curves. Daraseliya and Manenkov (1970) showed that one could "quench" the effect of cross relaxation by sweeping rapidly through the line during the saturating pulse, which saturated all sublevels, and the recovery was a true T_{1e}. This paper also pointed out that in an inhomogeneously broadened line there could be "a spectrum of cross-relaxation times T_{x1}." These ideas were applied to Fe(III) in $Al(NO_3)_3{\cdot}9H_2O$ crystals (Manenkov and Milyaev, 1970) and to Nd(III) (Daraseliya et al., 1970). The latter paper presented a theoretical basis for the prior papers, and showed that for many sets of assumptions the recovery curve will not be a simple exponential in T_{1e}. For Nd(III) the experimental data reveal longer T_{1e} when the method of rapid scan through the line during saturation is used than when the recovery is deconvoluted into T_{x1} and T_{1e}. Unfortunately, the quantitative basis for this analysis was not presented.

In our studies of radicals trapped in irradiated solids (Harbridge et al., 2002) we observed that long pump pulses removed the contributions from rapid nuclear spin relaxation. However when $T_{x1} < T_{1e}$, residual contributions from T_{x1} were present in the SR curves even after long pump pulses.

4.3 Selection of Pump and Observe Powers

The saturation factor, S, for an EPR signal is given by equation (1).

$$S = \frac{1}{1 + \gamma^2 B_1^2 T_{1e} T_{2e}} \tag{1}$$

where $\gamma = 1.7608 \times 10^7$ rad s^{-1} G^{-1}, and B_1 is the microwave magnetic field at the sample $= \sqrt{QP}$, P is the power incident on the resonator in watts and Q is the resonator quality factor. The pump power is selected to make S small. Typical values of P that are currently used in SR spectrometers are 10's to a few hundred mW. It might seem that higher pump power would always be better. However, higher pump powers cause greater resonator heating, which can cause baseline and frequency drift so there often is an optimum pump power for a particular sample and spectrometer configuration.

If S is small during the pump pulse, the amplitude of the SR signal changes from almost zero immediately after the pump pulse, back to the normal unsaturated signal amplitude, which is proportional to $\chi''\eta Q\sqrt{P}$. χ'' is the imaginary component of the RF susceptibility, which is proportional to the number of unpaired spins in the sample. η is the filling factor of the sample in the resonator. Q is the quality factor for the resonator. χ'', η, and Q are fixed by the sample and resonator properties. Thus, the dynamic range of the SR signal is linearly proportional to the square root of the observe power. To obtain an accurate value of T_{1e}, it is important to keep the observe power low enough that the saturation factor, S (eq. (1)), is approximately equal to 1 (Huisjen and Hyde, 1974b; Mailer et al., 1985; Fajer et al., 1986). For example, if the resonator has an efficiency of about 1 $G/\sqrt{W}$ where W is the power incident on the resonator in watts, and if $T_{2e} = 2 \times 10^{-6}$ s, $T_{1e} = 100 \times 10^{-6}$ s, then the observe power should be kept well below 10 μW. These are reasonable parameters for organic radicals near liquid nitrogen temperature. At lower temperature, T_{1e} increases by at least the inverse square of the temperature (Eaton and Eaton, 2000a), and the observe power should be reduced proportionately. At room temperature, T_{1e} and T_{2e} are much shorter than at lower temperatures and higher observe powers can be used.

There is an inherent trade off between S/N and signal distortion. In practice, SR curves are initially recorded at higher observe power to facilitate signal detection, and then the observe power is decreased until no further increase in the recovery time constant can be detected. Yin and Hyde (1989) showed that the rates of bimolecular collisions between nitroxyl radicals can be measured in SR experiments that use observing power so high that the "effective T_{1e}" is altered. The advantage of using higher observe power is higher signal-to-noise.

4.4 Comparison of CW-Saturation Recovery with Inversion Recovery or Echo-Detected Saturation Recovery

Inversion recovery and echo-detected SR are two other pulse techniques that have been used to measure T_{1e} (Figure 1b,c). In an inversion recovery experiment the first pulse is a 180° pulse that inverts the spin magnetization. The time between the inverting pulse and the two-pulse spin echo sequence that is used to monitor the net magnetization is varied. The disadvantage of the inversion recovery sequence is that spectral diffusion processes with time constants that are shorter than T_{1e} contribute to the recovery curve. In echo-detected SR (Figure 1c) a longer lower-power pulse is used to saturate the EPR transition and the net magnetization as a function of time is monitored with two-pulse spin echoes. In principle, the recovery curve obtained by echo-detected SR method should have the same time constant as obtained by CW-SR, if the lengths and powers of the saturating pulses are comparable.

An advantage of CW-SR is that the saturating microwave pulse is produced by a CW source, so it can, in principle, be made as long as is required to saturate the spectral diffusion processes. By contrast most sources used in spin echo and inversion recovery experiments are pulsed travelling wave tube (TWT) amplifiers for which the maximum length of a pulse is about 10 μs. The limitations on lengths of pulses can be partially overcome by using a series of pulses (a picket fence of pulses), but when this is done the length of time between the pulses in the pulse train must be short relative to the time constants for the spectral diffusion processes. Another advantage of CW detection is that it can be used even when T_2 is so short, or echo envelope modulation is so deep, that it would be difficult to observe an echo. Short T_2's can arise in fluid solutions of nitroxyl radicals where T_2 is dominated by incomplete motional averaging of g and A anisotropy and T_{1e} typically is substantially longer than T_{2e}. The relative benefits of long and short saturating pulses to obtain various kinds of relaxation information in CW-SR experiments have been described by Hyde and coworkers (Hyde,

1979; Yin and Hyde, 1987b) and some examples are included in the following section.

Provided that T_2 is long enough to permit spin-echo detection, the echo-detected SR frequently is more sensitive than CW-SR. The detection system in the CW-SR experiment does not use magnetic field modulation, because the modulation could function as a relaxation process or cause passage effects. Thus, an SR spectrometer does not benefit from the noise-rejection achieved by phase-sensitive detection at the modulation frequency that is used in most CW spectrometers. Also, it is necessary to use very low observe powers to avoid perturbing the system in a CW-SR experiment. However, the detector filter time constant, chosen appropriate to the signal recovery time constant, usually is much longer for CW detection than for echo detection. The longer filter time constant improves the signal-to-noise proportional to the square root of the time constant. An advantage of inversion recovery or echo-detected SR is that high power pulses are used to form the spin echoes. These pulses typically are "hard" pulses that fully invert the majority of the spins over a field range of several gauss. By contrast, in the CW-SR experiment the low observe power creates a "soft" pulse, which is only a small perturbation of the magnetization for the spins within a narrower field range. As a result, the CW-SR experiment detects a much smaller change in magnetization than the inversion recovery or spin echo experiment.

5. APPLICATIONS

The following paragraphs give brief descriptions of four of the more widely used applications of SR. The discussions are intended to show the range of information that can be obtained by SR and are not intended to be comprehensive reviews.

5.1 Characterization of Electron Spin Relaxation Processes and Mechanisms

Values of T_{1e} measured by SR have been crucial in elucidating processes and mechanisms of electron spin relaxation. Many of these experiments and their interpretation are discussed in a chapter in a prior volume of this series (Eaton and Eaton, 2000a). Prabhananda and Hyde (1986) noted the potential utility of multifrequency SR measurements in distinguishing between possible mechanisms of spin lattice relaxation, although that capability was not available at that time. Spin lattice relaxation rates at 95 GHz (W-band),

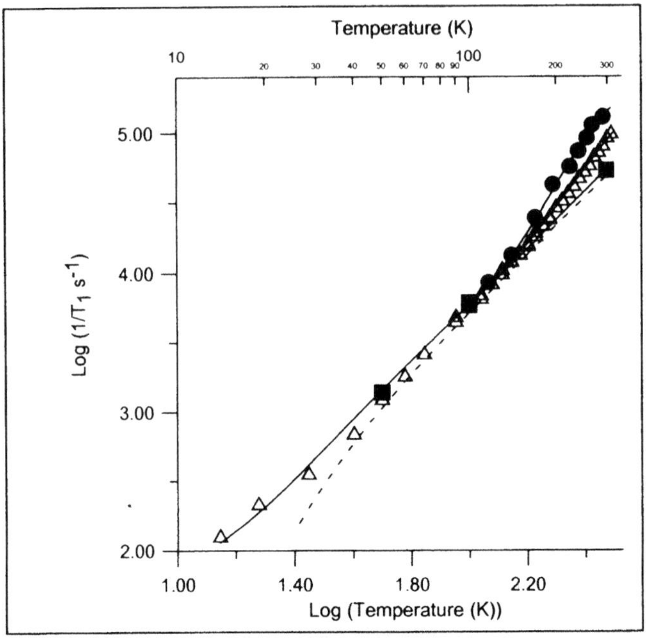

Figure 2. Temperature dependence of spin-lattice relaxation rates at W-band (■), X-band (Δ), and S-band (•) in the perpendicular region of the spectrum for tempol doped 1:500 into 4-OH-2,2,6,6-tetramethyl-piperidine. The solid lines through the data are fit functions as described in Eaton et al. (2001), including a frequency-dependent contribution from an activated process that is attributed to methyl group rotation. The dashed line is the contribution to the fit function from the Raman process.

9.5 GHz (X-band), and S-band (3.2 GHz) as a function of temperature for 4-hydroxy-2,2,6,6-tetramethylpiperidinol-oxy (Tempol) doped into its diamagnetic analog are shown in Figure 2 (Eaton et al., 2001). The relaxation rate between 40 and 100 K is dominated by the Raman process and is independent of microwave frequency. Above about 130 K there is an additional contribution to the relaxation. If only data at X-band had been available, it would not have been possible to determine whether this additional contribution was due to a local vibrational mode or a thermally activated process. However, the multifrequency data show that the relaxation rate decreases in the order S-band > X-band > W-band. This frequency dependence is characteristic of a contribution from a thermally-activated process, which can be described by a spectral density function, and inconsistent with a local mode (Eaton et al., 2001). The fit lines in Figure 2 were obtained with an activation energy of 1100 K (2.2 kcal/mole), which is in good agreement with that for rotation of nitroxyl methyl groups. E_a = 1100 K for methyl rotation was obtained by ENDOR for a nitroxyl in a solid host (Barbon et al., 1999) and E_a = 2.0 to 2.3 kcal/mole (1030 to 1150 K) was obtained from the temperature dependence of spin echo dephasing for nitroxyls in glassy solutions (Tsvetkov and Dzuba, 1990; Shushkakov et al., 1989; Nakagawa et al., 1992). Thus, the multifrequency data support the proposal that rotation of the nitroxyl methyl groups at rates comparable to the electron Larmor frequency makes significant contributions to T_{1e} at S-band and X-band in the motional regime that is present in the doped solid near room temperature. Work is currently underway to determine the importance of this contribution to T_{1e} for nitroxyl radicals that are slowly tumbling in fluid solution.

5.2 Oximetry – the Effect of Molecular Oxygen on T_{1e}

EPR oximetry is the measurement of oxygen concentration via the effect of oxygen on the EPR signal from a probe molecule. Many of these studies are based on linewidth changes or changes in CW power saturation curves. However, changes in T_{1e} are inherently more sensitive than CW lineshape to collisions with oxygen, and therefore SR is the preferred oximetry method when measurements of small changes are required (Hyde et al., 1990).

For nitroxyl spin labels in fluid solution, the time constant obtained in a long-pulse SR experiment is T_{1e} (Kusumi et al., 1982). In the absence of oxygen T_{1e} was between about 1.5 and 5 μs for a range of spin labels and tumbling correlation times in membranes between 0 and 36 °C. The reciprocal of T_{1e}, T_{1e}^{-1}, for long-chain doxyl spin labels in lipid bilayers increased linearly with the partial pressure of oxygen when the gas surrounding a TPX capillary was changed from nitrogen to 100% oxygen at temperatures between 4 and 37° C. The oxygen transport parameter, W, was defined as $W = T_{1e}^{-1}(\text{air}) - T_{1e}^{-1}(N_2)$ and was proposed as a monitor of the bimolecular collision rate between oxygen and the spin label (Kusumi et al., 1982). The oxygen transport parameter has been used to characterize the oxygen permeability of phosphatidylcholine-cholesterol membranes by varying the position of the spin label relative to the polar head groups (Subczynski et al., 1989). For the liquid-crystalline phase it was found that addition of cholesterol decreased oxygen transport in the membrane. In the absence of cholesterol, the incorporation of a double bond at the C9-C10 position of the alkyl chain in the lipid was found to decrease the oxygen transport parameter at all positions along the chain in phosphatidylcholine membranes (Subczynski et al., 1991). Values of W for a spin-labeled retinal analog incorporated into rhodopsin were 10 to 60 times smaller than in water and 1.1 to 20 times smaller than in model membranes, which indicates that membrane proteins create significant resistance to the transport of molecular oxygen inside and across the membrane (Subczynski et al., 1992a). Studies of Chinese hamster ovary plasma membrane also indicated lower oxygen permeability for the membrane than for a comparable layer of water (Subczynski et al., 1992b). However, it was concluded that oxygen permeability of cell plasma membranes is not the rate-limiting step for cellular respiration.

In a series of measurements of oximetry, Hyde and coworkers explored the relative information content of three variants of the SR measurement, "short pulse," "long pulse," and "high observe power." The recovery signal after a short saturating pulse is multi-exponential, and in favorable circumstances can distinguish several relaxation times: the electron T_{1e}, the nitrogen nuclear T_{1n}, the spectral diffusion time, and/or the Heisenberg exchange rate (Yin et al., 1987). After a long saturating pulse, the various electron spin energy levels tend to be equalized, and the recovery tends to a single exponential, which is T_{1e}. Short and long are defined relative to the relevant relaxation times and motional regime. For nitroxyl radicals in fluid solution, relaxation times are fast enough that a 5 μs pulse (Kusumi et al., 1982) or 20-40 μs pulse (Fajer et al., 1986) is "long." Similarly, high observe power is the power at which the second term in the expression ($T_1^{-1} + \gamma^2 B_1^2 T_1 T_2$) cannot be ignored, i.e. the saturation factor (eq. (1)) is significantly less than 1. Usually, one performs SR measurements with very low observing power to avoid artificial shortening of the measured T_{1e}. Lower observing power usually results in poorer signal-to-noise. However, Hyde et al. (1990) showed that if the goal is to measure the oxygen transport parameter, high observing powers can be used to measure an effective T_{1e} at two oxygen concentrations.

In a short-pulse SR experiment the recovery time constant for nitroxyl spin labels in fluid solution has contributions from (i) Heisenberg exchange due to collisions between spin labels and collisions between oxygen and spin labels, (ii) nitrogen nuclear spin relaxation and (iii) the intrinsic T_{1e} for the spin label in the absence of oxygen (Yin and Hyde, 1987a). The effects of the multiple contributions to the recovery curves can be interpreted in terms of a model based on population differences between electron spin energy levels. Experimental SR curves can be analyzed in terms of multiple exponentials (Yin and Hyde, 1987a,b). In 1-palmitoyl-2-oleoylphosphatidylcholine bilayers containing a transmembrane α-helical peptide the oxygen transport parameter was substantially higher in the middle of the bilayer than on the edges (Subczynski et al., 1998). Analysis of short-pulse SR data has been used to identify regions of a bacteriorhodopsin-rich reconstituted membrane in which oxygen transport is a factor of 5 slower than in the bulk (Ashikawa et al., 1994). It was speculated that these domains consist of lipids in contact with two proteins and/or in contact with protein and boundary lipids. In influenza viral membrane the oxygen transport rate in the slow-oxygen-transport domains is a factor of two smaller than in purple membrane, where bacteriorhodopsin is aggregated, which suggests that the slow oxygen transport regime in the viral membrane is a cholesterol-rich raft domain (Kawasaki et al., 2001). The contributions to the recovery curve from nitroxyl-oxygen collisions were distinguished from nitroxyl-nitroxyl interactions by varying oxygen and nitroxyl concentrations.

5.3 Heisenberg Exchange due to Nitroxyl-nitroxyl Collisions

The decrease in T_{1e} (increase in $1/T_{1e}$) due to nitroxyl-nitroxyl collisions has been used to characterize lipid dynamics using stearic acids spin-labeled at various positions along the chain. The nitroxyl Heisenberg exchange rates determined by analysis of short-pulse SR were shown to be proportional to spin label concentration, and were judged to be valid indicators of bimolecular collision rates (Yin et al., 1987). The use of ^{14}N and ^{15}N labels combined with short-pulse SR overcomes spectral overlap problems that are encountered with CW methods of measuring Heisenberg exchange (Yin et al., 1988). SR studies of ^{14}N and ^{15}N-labeled labeled stearic acid pairs in model membranes prepared from lipids with varying chain lengths showed little dependence of lateral diffusion on chain length (Feix et al., 1987). For doxylstearates labeled at the 5-, 12-, or 16- positions, collision frequencies were higher in dimyristoyl-phosphatidylcholine (DMPC) relative to that in dielaidoylphosphatidylcholine (DEPC), consistent with CW studies that showed increased order in DEPC than in DMPC (Yin et al., 1990). Addition of cytochrome c to cardiolipin bilayers caused an increase in T_{1e} that was attributed to changes in the molecular dynamics in the vicinity of double bonds in the acyl chains of the bilayer (Pinheiro et al., 1993). The effects of lutein and cholesterol on collision frequencies among doxyl-stearic acids in lipid bilayers confirmed the occurrence of vertical fluctuations of alky chain ends near the bilayer surface (Yin and Subczynski, 1996).

5.4 Distance Between a Radical and a More Rapidly Relaxing Metal Ion

The enhancement in the spin lattice relaxation rate for a slowly relaxing spin due to dipolar interaction with a more rapidly relaxing metal ion can be analyzed to determine the distance between the two paramagnetic centers. This major application of SR was recently reviewed by Eaton and Eaton (2000b) and applications in photosynthesis were reviewed by Lakshmi and

Brudvig (2000). An example of the use of SR to determine the distance between the Fe(III) in metmyoglobin and a spin label is discussed in ch. 8 of this volume.

6. PROGNOSIS

Until recently SR experiments could be performed only on locally-constructed spectrometers. The recent addition of an option for SR to the Bruker Elexsys bridge will make the technique more widely available. We expect that SR will be much more widely used in the future due to the availability of commercial instrumentation.

7. ACKNOWLEDGMENT

Our work in this area is supported by National Institutes of Health Grant EB002807 (formerly GM21156).

8. REFERENCES

Ashikawa, I., Yin, J.-J., Subczynski, W. K., Kouyama, T., Hyde, J. S., and Kusumi, A. (1994). Molecular organization and dynamics in bacteriorhodopsin-rich reconstituted membranes: discrimination of lipid environments by the oxygen transport parameter using pulse ESR spin-labeling technique. *Biochemistry* **33**, 4947-4952.

Barbon, A., Brustolon, M., Maniero, A. L., Romanelli, M., Brunel, L.-C. (1999). Dynamics and spin relaxation of tempone in a host crystal. An ENDOR, high-field EPR, and electron spin echo study. *Phys. Chem. Chem. Phys.* **1**, 4015-4023.

Beck, W. F., Innes, J. B., Lynch, J. B., and Brudvig, G. W. (1991). Electron spin-lattice relaxation and spectral diffusion measurements on tyrosine radicals in proteins. *J. Magn. Reson.* **91**, 12-29.

Bloembergen, N. (1949) On the interaction of nuclear spin in a crystalline lattice. *Physica* **15**, 386-426.

Bloembergen, N. and Wang, S. (1954) Relaxation effects in para- and ferromagnetic resonance. *Phys. Rev.* **93**, 72-83.

Bowers, K. D., and Mims, W. B. (1959) Paramagnetic relaxation in nickel fluosilicate. *Phys. Rev.* **115**, 285-295.

Collins, S. A., Kyhl, R. L., and Strandberg, M. P. W. (1959) Spin lattice relaxation from state of negative susceptibility. *Phys. Rev. Lett.* **2**, 88-90.

Daraseliya, D. M. and Manenkov, A. A. (1970) Quenching of cross correlation in inhomogeneously broadened epr lines. *JETP Letters* **11**, 224-226 (337-339 in Russian)

Daraseliya, D. M., Epifanov, A. S., and Manenkov, A. A. (1970) Relaxation in inhomogeneously broadened EPR lines. *Soviet Physics JETP* **32**, 244-249 (**59**, 445-456 in Russian).

Davis, C. F. Jr., Strandberg, M. W. P., and Kyhl, R. L. (1958) Direct measurement of electron spin-lattice relaxation times. *Phys. Rev.* **111**, 1268-1272.

Eaton, S. S. and Eaton, G. R. (2000a). Relaxation times of organic radicals and transition metal ions, *Biol. Magn. Reson.* **19**, 29-154.

Eaton, S. S., and Eaton, G. R. (2000b). Determination of distances based on T_1 and T_m, *Biol. Magn. Reson.* **19**, 347-381

Eaton, S. S., Harbridge, J., Rinard, G. A., Eaton, G. R., and Weber, R. T. (2001) Frequency dependence of electron spin relaxation for three S=1/2 species doped into diamagnetic solid hosts, *Appl. Magn. Reson.* **20**, 151-157.

Fajer, P., Thomas, D. D., Feix, J. B., and Hyde, J. S. (1986). Measurement of rotational molecular motion by time-resolved saturation transfer electron paramagnetic resonance, *Biophys. J.* **50**, 1195-1202.

Feix, J. B., Yin, J. J., and Hyde, J. S. (1987) Interactions of nitrogen-14:nitrogen-15 stearic acid spin-label pairs: effects of host lipid alkyl chain length and unsaturation.

Fessenden, R. W., Hornak, J. P., and Venkataraman, B. (1981) Electron spin-lattice relaxation of transient free radicals. *J. Phys. Chem.* **74**, 3694-3704.

Freed, J. H. (1974) Theory of saturation and double resonance in electron spin resonance spectra. VI. Saturation recovery. *J. Phys. Chem.* **78**, 1156-1167.

Giordmaine, J. A., Alsop, L. E., Nash, F. R., and Townes, C. H. (1958) Paramagnetic relaxation at very low temperature. *Phys. Rev.* **109**, 302-311.

Harbridge, J. R., Eaton, S. S., and Eaton, G. R. (2002) Electron spin lattice relaxation in radicals containing two methyl groups, generated by γ-irradiation of polycrystalline solids, *J. Magn. Reson.* **159**, 195-206.

Hausser, K. H. and Brunner, H. (1998) The Effect of concentration and oxygen in EPR, in *Foundations of Modern EPR*, G. R. Eaton, S. S. Eaton, and K. Salikhov, eds., World Scientific, Singapore, 469-481.

Huisjen, M. and Hyde, J. S. (1974a) Saturation recovery measurements of electron spin-lattice relaxation times of free radicals in solution. *J. Chem. Phys.* **60**, 1682-1683.

Huisjen, M. and Hyde, J. S. (1974b) A pulsed EPR spectrometer. *Rev. Sci. Instrum.* **45**, 669-675.

Hyde, J. S. (1979) Saturation recovery methodology, in Kevan, L., and Schwartz, R. N., eds., *Time Domain Electron Spin Resonance*, Wiley-Interscience, New York.

Hyde, J. S., Yin, J.-J., Feix, J. B., and Hubbell, W. L. (1990) Advances in spin label oximetry. *Pure & Appl. Chem.* **62**, 255-260.

Hyde, J. S. (1998) Saturation recovery in *Foundations of Modern EPR*, G. R. Eaton, S. S. Eaton, and K. Salikhov, eds., World Scientific, Singapore, 607-618.

Kawasaki, K., Yin, J.-J., Subczynski, W. K., Hyde, J. S., and Kusumi, A. (2001). Pulse EPR detection of lipid exchange between protein-rich raft and bulk domain in the membrane: methodology develoment and it application to studies of influenza viral membrane. *Biophys. J.* **80**, 738-748.

Kusumi, A., Subczynski, W. K., and Hyde, J. S. (1982) Oxygen transport parameter in membranes deduced by saturation recovery measurements of spin-lattice relaxation times of spin labels. *Proc. Natl. Acad. Sci. USA* **79**, 1854-1858.

Lakshmi, K. V., and Brudvig, G. W. (2000) Electron paramagnetic resonance distance measurements in photosynthetic reaction centers, *Biol. Magn. Reson.* **19**, 513-567.

Lingam, K. V., Nair, P. G., and Venkataraman, B. (1972) Spin lattice relaxation studies of semiquinone ions, *Proc. Indian Acad. Sci. A* **76**, 207- 220.

Mailer, C., Danielson, J. D. S., and Robinson, B. H. (1985) Computer-controlled pulsed electron-paramagnetic-resonance spectrometer. *Rev. Sci. Instrum.* **56**, 1917-1925.

Manenkov, A. A. and Milyaev, V. A. (1970) Paramagnetic relaxation processes in $Al(NO_3)_2 9H_2O:Fe^{3+}$ single crystals at helium temperatures. *Soviet Physics JETP* **31**, 427-428 (**58**, 796-799 in Russian)

Manenkov, A. A. and Pol'skii, Yu. E. (1964) Relaxation processes in the paramagnetic resonance of Gd^{3+} in CaF_2. *Soviet Physics JETP* **45**, 1425-1429 (1963) (**18**, 985-987 (1964) in transl.)

Manenkov, A. A. and Prokhorov, A. M. (1962) SpinlLattice relaxation and cross-relaxation interactions in chromium corundum. *Soviet Physics JETP* **42**, 75-83 (54-59 in translation).

Manenkov, A. A., Milyaev, V. A. and Prokhorov, A. M. (1962) The relaxation time of Cr^{3+} and Fe^{3+} ions in single crystals of rutile. *Soviet Physics Solid State* **4**, 388-391 (280-283 in translation).

Nakagawa, K., Candelaria, M. B., Chik, W. W. C., Eaton, S. S., and Eaton, G. R. (1992). Electron spin relaxation times of chromium(V). *J. Magn. Reson.* **98**, 81-91.

Pace, J. H., Sampson, D. F., and Thorp, J. S. (1960). *Phys. Rev. Lett.* **4**, 18-19.

Pastor, R. C. and Hoskins, R. H. (1960) Paramagnetic resonance in charred dextrose. *J. Chem. Phys.* **32**, 264-269.

Percival, P. W. and Hyde, J. S. (1975) Pulsed EPR spectrometer II. *Rev. Sci. Instrum.* **46**, 1522-1529.

Percival, P. W. and Hyde, J. S. (1976) Saturation-recovery measurements of the spin-lattice relaxation times of some nitroxides in solution. *J. Magn. Reson.* **23**, 249-257.

Pinheiro, T. J. T., Bratt, P. J., Davis, I. H., Doetschman, D. C., and Watts, A. (1993). Spin-lattice relaxation times of phospholipid aminoxyl spin labels in cardiolipin-cytochrome c bilayers: a pulse saturation recovery study, J. C. S. Perkin Trans 2, 2113-2117.

Prabhananda, B. S., and Hyde, J. S. (1986). Study of molecular motion in liquids by electron spin relaxation: halogenated *p*-semiquinone anions in alcohols, *J. Chem. Phys.* **85**, 6705-6712.

Quine, R. W., Eaton, S. S. and Eaton, G. R. (1992) A saturation recovery electron paramagnetic resonance spectrometer. *Rev. Sci. Instrum.* **63**, 4251-4262.

Quine, R. W., Rinard, G. A., Ghim, B. T., Eaton, S. S., and Eaton, G. R. (1996) A 1-2 GHz pulsed and continuous wave electron paramagnetic resonance spectrometer. *Rev. Sci. Instrum.* **67**, 2514-2527.

Rengen, S. K., Khakhar, M. P., Prabhananda, B. S., and Venkataraman, B. (1972) Electron spin-lattice relaxation in organic free radicals in solution. *Pure & Appl. Chem.* **32**, 287-305.

Rengen, S. K., Khakhar, M. P., Prabhananda, B. S., and Venkataraman, B. (1974a) Study of molecular motions in liquids by electron spin lattice relaxation measurements. I. Semiquinone ions in hydrogen bonding solvents. *Pramana* **3**, 95-121.

Rengen, S. K., Khakhar, M. P., Prabhananda, B. S., and Venkataraman, B. (1974b) Study of molecular motions in liquids by electron spin lattice relaxation measurements. II. 2,5-Di-tert-butylsemiquinone Ion in Acetonitrile and Tetrahydrofuran. *J. Magn. Reson.* **16**, 35-43.

Rengen, S. K., Bhagat, V. R. Sastry, V. S. S., and Venkataraman, B. (1979). Magnetic field-pulsed ELDOR spectroscopy, *J. Magn. Reson.* **33**, 227-240.

Scott, P. L., and Jeffries, C. D. (1962). Spin-lattice relaxation in some rare-earth salts at helium temperature; observation of the phonon bottleneck, *Phys. Rev.* **127**, 32-51.

Shushkakov, O. A., and Dzuba, S. A. (1989). Barriers for internal rotation of methyl groups screening nitroxide paramagnetic fragments as determined by an electron spin-echo technique, *J. Struct. Chem.* **30**, 75-80 (1989).

Standley, K. J., and Vaughan, R. A. (1969) *Electron Spin Relaxation Phenomena in Solids*, Plenum Press, N. Y.

Subczynski, W. K., Hyde, J. S., and Kusumi, A. (1989). Oxygen permeability of phosphtidylcholine-cholesteral membranes, *Proc. Natl. Acad. Sci. (U.S.)* **86**, 4474-4478.

Subczynski, W. K., Hyde, J. S., and Kusumi, A. (1991). Effect of alkyl chain unsaturation and cholesterol intercalation on oxygen transport in membranes: a pulsed ESR spin labelling study, *Biochemistry* **20**, 8578-8590.

Subczynski, W. K., Renk, G. E., Crouch, R. K., Hyde, J. S., and Kusumi, A. (1992a). Oxygen diffusion-concentration product in rhodopsin as observed by a pulse ESR spin labelling method, *Biophys. J.* **63**, 573-577.

Subczynski, W. K., Hopwood, L. E., and Hyde, J. S. (1992b) Is the mammalian cell plasma membrane a barrier to oxygen transport? *J. Gen. Physiol.* **100**, 69-87.

Subczynski, W. K., Lewis, R. N. A. H., McElhaney, R. N., Hodges, R. S., Hyde, J. S., and Kusumi, A. (1998) Molecular organization and dynamics of a 1-palmitoyl-2-oleoylphosphatidylcholine bilayer containing a transmembrane α-helical peptide, *Biochemistry* **37**, 3156-3164.

Tsvetkov, Yu. D., and Dzuba, S. A. (1990). Pulsed EPR and molecular motion, *Appl. Magn. Reson.* **1**, 179-194.

Venkataraman, B. (1982) Time resolved electron spin resonance spectroscopy. *Current Science* **51**, 397-400.

Weissman, S. I., Feher, G., and Gere, E. A. (1957) The spin relaxation time of triphenylmethyl at low temperatures. *J. Am. Chem. Soc.* **79**, 5584-5585.

Yin, J.-J., and Hyde, J. S. (1987a) Spin label saturation-recovery electron spin resonance measurements of oxygen transport in membranes. *Z. Physikal. Chem. Neue Folge* **153**, 57-65.

Yin, J.-J., and Hyde, J. S. (1987b) Application of rate equations to ELDOR and saturation recovery experiments on ^{14}N:^{15}N spin-label pairs, *J. Magn. Reson.* **74**, 82-93.

Yin, J.-J., and Hyde, J. S. (1989) Use of high observing power in electron spin resonance saturation-recovery experiments in spin-labeled membranes. *J. Chem. Phys.* **91**, 6029-6035.

Yin, J.-J., Feix, J. B., and Hyde, J. S. (1988) Solution of the nitroxide spin-label spectral overlap problem using pulse electron spin resonance. *Biophys. J.* **53**, 525-531.

Yin, J.-J., Feix, J. B., and Hyde, J. S. (1990) Mapping of collision frequencies for stearic acid spin-labels by saturation-recovery electron paramagnetic resonance. *Biophys. J.* **58**, 713-720.

Yin, J.-J., Pasenkiewicz-Gierula, M., and Hyde, J. S. (1987) Lateral diffusion of lipids in membranes by pulse saturation recovery electron spin resonance. *Proc. Natl. Acad. Sci. USA* **84**, 964-968.

Yin, J.-J., and Subczynski, W. K. (1996). Effects of lutein and cholesterol on alkyl chain bending in lipid bilayers: a pulse electron spin resonance spin labeling study. *Biophys. J.* **71**, 832-839.

Chapter 2

Loop-Gap Resonators

George A. Rinard and Gareth R. Eaton
Department of Engineering and Department of Chemistry and Biochemistry, University of Denver, Denver, Colorado 80208

Abstract: Lumped-circuit resonators known as loop-gap resonators (LGR) have many advantages for EPR measurements for frequencies at or below X-band. This chapter introduces principles that underlie the design and use of LGR's and provides practical guidance concerning their use.

1. INTRODUCTION

Over the years, quite a number of resonant and traveling-wave microwave and RF structures have been used in various EPR experiments (Poole, 1967; Poole and Farach, 1999). However, along with the adoption of X-band (ca. 9-10 GHz) as the "normal" frequency for EPR studies, the resonant cavity, and especially the TE_{102} rectangular cavity, became the most common resonator structure for EPR. Indeed, the Varian E-231 TE_{102} cavity was the subject of a large number of papers that provided the researcher with good characterization of the interaction of microwaves with the sample, and of the modulation field distribution over the sample (see references cited in Eaton and Eaton, 1980; Dalal et al., 1981; More et al., 1984). In recent years, Mazúr and coworkers have made measurements on Bruker TE_{102} X-band cavity resonators in support of efforts at spin quantitation (Mazúr et al., 1997, 2000). However, for pulsed EPR at X-band, and for most EPR measurements at frequencies lower than X-band, researchers have found it preferable to use a dielectric or lumped-element resonator. Lumped-element resonators in which the inductor consists of a loop and the capacitor is a gap have become known as loop-gap resonators. The early experiments by Zavoisky used a resonant structure that we would now call a loop-gap resonator. In this chapter, we introduce some of the principles that underlie the design and use of loop-gap resonators, and give some practical guidance to their use.

The simplest topology for a loop-gap resonator is the cylinder with a slot (gap) cut in one side (Fig. 1), as pictured in the papers by Hardy and Whitehead (1981), Hyde and Froncisz (1981) and Froncisz and Hyde (1982). A loop, coaxial with the cylindrical LGR, at the end of the transmission line inductively couples the microwaves from the transmission line to the resonator (Figures 1, 2, 3). The loop connects the center conductor of the coaxial transmission line to the shield. Capacitive coupling, with the center conductor forming an antenna instead of a shorted loop, can also be used (Figure 2).

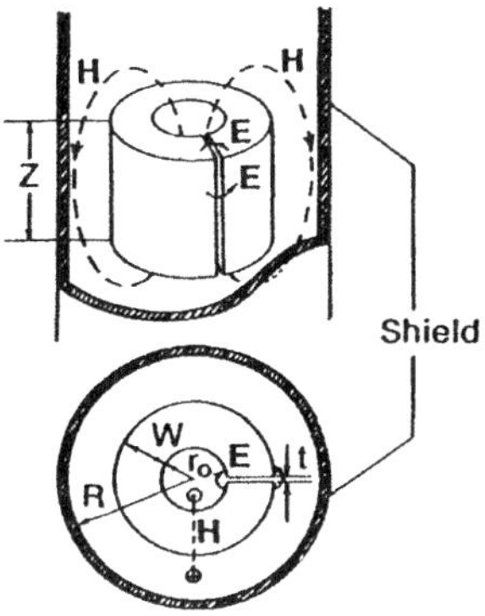

Figure 1. Sketch of a loop gap resonator. The electrical field of the microwaves is mostly confined to the capacitive gap, and the magnetic field to the inductor, the open central cylinder where the sample tube is placed, and in the return flux path outside the resonator body. The resonator usually is surrounded by a shield made of a good conductor such as a metal or metallic paint on plastic or quartz. The shield helps confine the microwaves in the resonator structure, so that it does not radiate. Without the shield the resonator is a dipole radiator, resulting in low Q and baseline instability when objects, such as the spectrometer operator, move near the resonator (this is colloquially called "handwaving effects"). Not shown in this diagram are the means of coupling to the transmission line and the means of mechanically supporting the resonator. For further details, see Figures 2 and 3. (Mehdizadeh et al., 1983).

2. HISTORY

As noted above, the history of loop-gap resonators goes back to the beginning of EPR; indeed, it is the fundamental resonant structure in the magnetron that graces the covers of the Radiation Laboratory series of books. In 1965 a UHF LGR was used in a dynamic nuclear polarization experiment on metallic sodium (Reichert and Townsend). The modern use of lumped-circuit resonators in magnetic resonance dates from the publication by Schneider and Dullenkopf (1977) of the slotted tube resonator for NMR, by Hardy and Whitehead (1981) of the split-ring resonator for NMR, and the publication in 1981-82 of the loop-gap resonator for EPR (Hyde and Froncisz, 1981; Froncisz and Hyde, 1982). Since then, there have been many implementations of various forms of loop-gap resonators (LGR) in EPR, and in NMR. A central theme in these applications is the considerable design flexibility presented by the LGR, permitting a resonator to be designed for a sample, rather than selecting a sample size and shape that can be used in a pre-existing cavity resonator.

An explanation of the concept of cavity resonators, in the context of the design of a klystron, shows a picture of a "loop and a circular plate condenser" (Harrison, 1947) that is similar to a LGR used for EPR. A textbook introduces the concept of a cavity resonator by starting with a discrete circuit of an inductor (L) and capacitor (C), and then putting many inductors in parallel connected to the capacitor, and finally merging these into an enclosed cavity (Squires, 1963). Recently, Hyde has found value in making the reverse transformation, starting with cavity resonators, and pushing walls together in various ways to form capacitive elements, and then converting this into a design for a loop-gap resonator (personal communication, 2002). These concepts are important in understanding how to optimize resonators for EPR measurements.

There is an extensive, parallel, development of resonators for NMR, including such topologies as the crossed coil, birdcage, Alderman-Grant, and various helix and saddle coil designs. For an entrée to this literature, see the Encyclopedia of NMR (e.g., Hill, 1996; Hayes, 1996). Some of the NMR resonators are conceptually loop-gap resonators. The coaxial NMR cavity described by Kan et al. (1973), the slotted tube resonator (Schneider and Dullenkopf, 1977), the decoupling coil described by Alderman and Grant (1979) that has become known as the Alderman-Grant resonator, and the split-ring resonator (Hardy and Whitehead, 1981) all are conceptual forerunners of the class of resonators that are collectively called loop-gap resonators in EPR. Grist and Hyde (1985) used a LGR for ^{31}P NMR at 1.5 T. Lurie et al. (2002) used an Alderman-Grant resonator for EPR at 564 MHz and a solenoid for NMR at 856 KHz in a proton-electron double resonance imaging (PEDRI) measurement.

3. WHY SHOULD ONE USE LOOP-GAP RESONATORS?

Loop-gap resonators are advantageous for CW EPR measurements at frequencies below X-band, where cavity resonators are inconveniently large, and for measurements of limited size samples at X-band, where a LGR can have a higher filling factor than a cavity resonator. The lower Q of a LGR relative to a cavity results in less demodulation of source noise, and consequently better signal-to-noise (S/N) in dispersion spectra (Hyde et al. 1982b).

The advantages of LGRs for pulsed EPR include:

- Large filling factor
- Good S/N for small samples
- Reasonable physical size at low frequencies
- Large B_1 per square root watt
- Use of lower power results in less detector overload (and lower cost)
- Fairly uniform B_1 over the sample
- Easy to achieve low Q for short ringdown time
- Cooling the resonator along with the sample may decrease thermal background noise
- Larger bandwidth (lower Q) allows two or more simultaneous frequencies
- Ability to rotate an entire EPR spectrum with a pulse, and hence do FT EPR
- Facilitates EPR at low frequencies where cavity resonators would be impractical.

The corresponding disadvantages include:

- The critically-coupled Q is lower than for a cavity resonator, but the impact of this on reducing sensitivity is partially offset by higher filling factor for a LGR than for a cavity resonator.
- Small capacitive gaps may lead to arcing at high pulse power
- Careful sample positioning is required to take advantage of B_1 uniformity, especially if the length of the LGR is small.
- LGR heating occurs if the thermal mass is small
- Background signals from the materials of construction.
- Microphonics can be a problem, but LGRs may not be inherently worse than cavities; this merits further study
- Large frequency shift as the coupling is changed, resulting in difficulty in tuning and maintaining field/frequency relationship as temperature is varied and when samples are changed

Reviews of loop gap resonator design and application have been published by Hyde and Froncisz (1986, 1989). A fairly comprehensive survey of LGRs is available from the National Biomedical ESR Center. This chapter provides an introduction to the basic concepts of loop gap resonators and analogous resonators, with the equations needed to understand them. A few selected examples of applications are included to inspire future use of LGRs, but no attempt has been made at comprehensiveness in citation of the literature, either of LGRs or their application.

4. BASICS

In EPR one is interested in measuring the effect of electron spin on the sample's magnetic susceptibility. The role of the resonator is to concentrate the RF magnetic field, B_1, in the sample and make the signal produced by a change in magnetic susceptibility at resonance as large as possible. By "lumped-element" resonator, is meant a structure in which the region of high electric field and high magnetic field are readily identifiable and spatially separated. The sample is placed in the inductive element, where the B field is large and the E field is small.

In the following, SI units are used in the equations. However, we follow the convention in EPR of expressing magnetic flux density, B_1, in Gauss (G) rather than Tesla (T). $1\ G = 10^{-4}\ T$.

The EPR signal for CW is given by Eq. (1) Similar expressions can be found in most introductions to EPR; we use the notation of Rinard et al. (1999a).

$$V_{EPR} = \chi'' \eta Q_L \sqrt{Z_0 P} \tag{1}$$

Where: χ'' = Imaginary part of paramagnetic susceptibility
η = Resonator filling factor
Q_L = Resonator loaded Q
Z_0 = Transmission line characteristic impedance
P = Spectrometer power delivered to resonator

The Q_L of a LGR is in general lower than that for a cavity resonator at the same frequency. Despite this fact, the filling factor in a LGR can be many times that for a high Q cavity and the $\eta\ Q_L$ product is often comparable to that for a cavity. A LGR may also be desirable over a cavity for lossy samples, because of its lower Q and the fact that the E and B fields usually are separated better in a LGR than in a cavity resonator. In addition, cavity resonators are not practical for frequencies below about 1 GHz.

The filling factor, η, is a measure of how effective the sample is in affecting the resonator and is proportional to the ratio of B_1^2 integrated over sample to B_1^2 integrated over the entire resonator (Poole, 1967). It has a maximum value of one. In practice, the filling factor is usually much less than 1. A sample in a standard 4 mm o.d. tube (ca. 3 mm i.d.) in a TE_{102} X-band resonator has a filling factor of roughly 0.01, while that for a LGR can be more than an order of magnitude larger. A LGR inherently has a larger filling factor than does a cavity resonator at the same frequency, and the filling factor can be increased by making the return flux cross section large relative to the sample loop cross section. The filling factor of a LGR is limited by the wall thickness of the sample tube. For example, if the loop is 4.2 mm diameter to hold a 4 mm o.d. tube and the sample is 3 mm od, and long relative to the resonator so that the length can be ignored, the filling factor is reduced relative to that for a completely full loop by a factor of $(3/4.2)^2 = 0.5$.

The parameters η and Q_L depend on resonator geometry. In general, for a LGR, η depends on the ratio of the area of the resonator loop to the area of the path for the reentant magnetic flux outside of the resonator. This reentrant loop should have an area about an order of magnitude larger than the resonator loop. The filling factor will be higher if the sample is long enough to include the fringing flux outside the loop. However, this results in a non-uniform B_1 over the sample. Increasing the length of the loop, provided it is filled with sample, will increase the filling factor. For the best filling factor in limited sample applications, the loop should be designed, as much as practical, to match the sample size.

The inductance of the loop is proportional to the square of its diameter and, for a given frequency, the smaller the loop the larger the capacitance of the gap must be (see Eqs. 2, 4). The capacitance of the gap can be increased by increasing its area and by decreasing the gap spacing (see Eq. (3)). For room temperature operation it may be necessary to fill the gap with a low loss dielectric such as Teflon, not only to increase the capacitance, but also to prevent arcing for high power pulse applications. In general, such dielectrics should be avoided for cryogenic operation because of the high temperature coefficients of their dielectric constant and of their dimensions.

The resistive loss in a resonator can be due to the materials of construction (e.g., aluminum has higher resistance than copper, and hence an aluminum resonator has lower Q than an otherwise identical copper resonator), or due to the sample itself. The word "lossy" characterizing a sample, or the solvent in which a sample is dissolved, means that power is "lost" in the sample by conversion of electromagnetic radiation to heat by interaction with the molecular dipoles or ionic conduction in the substance. Thus, water reduces the Q of the resonator, and saline solution reduces it even more.

It is not possible to completely separate the E and B fields in the LGR and the fields are less separable at a given frequency the larger the loop, or at higher frequencies for a loop of a given size. For these cases the gap spacing will in general be larger and there will be more fringing of the E field into the loop. The larger the loop, the closer the resonator becomes to a cavity where the fields are well mixed. The E field fringing from the edges of the capacitive element is often the limiting feature that determines the size and placement of a lossy sample in a LGR.

5. TOPOLOGIES OF LOOP GAP RESONATORS

The descriptor LGR virtually speaks for itself in terms of resonator geometry. However, the LGR can take on a number of different forms depending on the frequency, resonator shield, field modulation provision, and support structure. One particularly simple embodiment of the LGR can be constructed by wrapping thin copper around a sample tube (Lin et al., 1985; LoBrutto et al., 1986). The copper can be self-supporting. It is also possible to make resonators from thin-film, copper coated Teflon etched to create desired patterns using circuit board techniques (Ghim et al., 1996). The resonator can be supported by the sample tube itself. This approach complicates changing sample tubes, but shows just how simple it can be to implement a LGR.

To help localize the electric field in the LGR, it is convenient to shield the gap with another conductor, creating what have become known as bridged LGRs. The first report was by Ono and co-workers (Ono et al., 1986a,b; Ogata et al., 1986; Hirata and Ono, 1996). Bridged LGRs were extensively developed by Schweiger and coworkers (Pfenninger et al., 1988; Forrer et al., 1990). Rotating the shield relative to the gaps makes a frequency-tunable LGR (Symons, 1995).

The basic field distribution of a LGR is rather like a dipole pattern, extending into space unless confined by a conducting shield. There usually is a hole in the shield for convenient sample placement. The size and conductivity of the shield affect the resonant frequency, η, Q_L, and B_1 per square root watt. If the resonator is not properly shielded, there will be "hand waving effects" due to motion near the resonator. As described above, the shield should be at least 3 to 4 times the diameter of the LGR to maintain a good filling factor. The shield may be a simple conducting cylinder concentric with the resonator (Froncisz and Hyde, 1982) or may consist of a reentrant loop, which may be considered a part of the resonator itself (Sotgiu, 1985; Sotgiu and Gualtieri, 1985; Quine et al., 1996).

When a separate shield is used, some means of supporting the LGR is required. The original Froncisz and Hyde paper (1982) depicted the LGR as free in space. Obviously, some means for supporting the LGR is required. The trick is to support the LGR with non-lossy dielectric material such as Teflon, Rexolite 1422, or polystyrene that is compatible with the experimental environment (temperature, etc.) and that does not have an interfering EPR signal. Differential temperature coefficients of expansion, and cracking upon thermal cycling, limit use of some plastic materials, and impurity signals prevent use of most ceramic or oxide materials, especially for cryogenic operation. The Bruker "split ring" implementation of the LGR solves this problem by incorporating the return flux region and the sample region into one structure that can be supported by the outside rim. This is analogous to a 3-loop-2-gap LGR, in one mode of which two of the loops provide the return flux path for the third loop (Wood et al., 1984). For some reentrant LGR designs no separate support is required (Sotgiu, 1985; Sotgiu and Gualtieri, 1985; Quine et al., 1996).

Hyde and Froncisz (1989) reviewed several resonator topologies, pointing out designs intended for applications such as ENDOR and ELDOR. Several examples are sketched in Figure 2. Additional examples are sketched in some of the patents on LGRs, USA patents 4,435,680, 4,446,429, 4,453,147, 4,480,239, and 4,504,788. In a multi-purpose resonator, whose resonant element is sketched in Fig. 2d, the sample goes into the larger, center, loop, and the smaller loops are for the return flux. A different choice was made for a 2-loop-1-gap resonator for Q-band (Froncisz et al., 1986) and an X-band LGR designed for continuous and stopped flow studies of small amounts of samples (Hubbell et al., 1987). Putting the sample in the smaller loop increases the B_1 at the sample for a fixed incident microwave power, since the integral of the power over the cross section must be equal in the two loops. The 2-loop-1-gap resonator has been found to have advantages over the 1-loop-1-gap resonator in terms of thermal and mechanical stability (Hubbell et al., 1987).

The resonator component that contains the loops and gaps has to be shielded, coupled to the transmission line, and mechanically supported in an assembly that mates to the rest of the spectrometer. One example of a full assembly is shown in Fig. 3. Another full assembly is shown in Hyde et al. (1985).

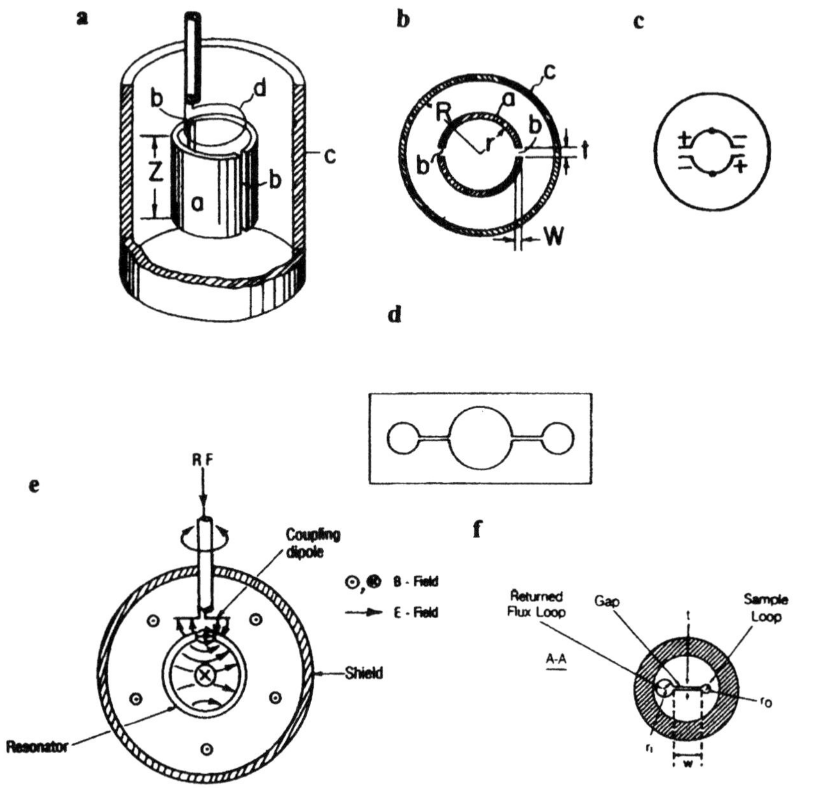

Figure 2. Some LGR topologies that have been proposed. This figure is derived from several figures that show even more possible arrangements of loops, gaps, and locations of wires for RF coils for ENDOR. a, b, and c are three views of a 1-loop-2-gap resonator (Froncisz and Hyde, 1984; Hyde and Froncisz, 1989). The charges near the gaps denote regions of high electric field, and the large black dots in c label the points of minimum electric field. These are locations at which ENDOR or modulation coils can be placed with minimal effect on the microwave distribution. Sketch d is a 3-loop-2-gap resonator of the type (Wood et al., 1984) used in a resonator designed to be the analog of the "multi-purpose" TE_{102} rectangular cavity resonator (Hyde et al., 1989). Inductive coupling is shown in a, and capacitive coupling is shown in e. Resonator f is a 2-loop-1-gap LGR for very small samples, which are placed in the small loop (Hubbell et al., 1987; Froncisz et al., 1986).

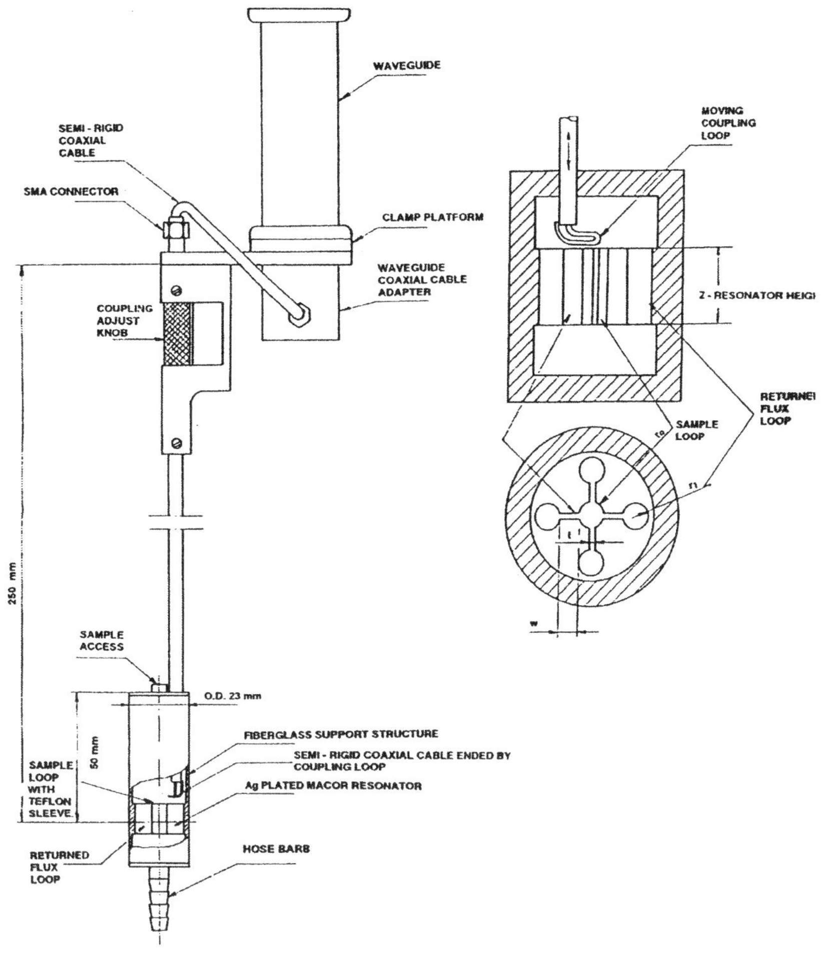

Figure 3. The "rising sun" LGR topology used in this resonator illustrates some practical features of LGR design and operation. The sample tube goes into the central loop. The transmission line coupling loop inductively couples to one of the four outer loops, which jointly serve as the return flux loops for the center loop. The position of the coupling loop can be changed, and hence the impedance match to the resonator changed, by moving the loop up and down by turning the knurled knob near the coax-to-waveguide transition. The resonator itself is supported by the fibreglass support structure. There is a Teflon sleeve in the center loop to protect the metal plating from abrasion by the sample tube. This reduces the filling factor, but is a practical necessity to ensure long life. In some other resonators additional support for the LGR is needed for geometric reasons, and usually is provided by a low-loss plastic such as Rexolite, although this support structure is rarely shown in published papers. This figure is similar to one published in Rinard et al., 1994.

It is often desirable to have the LGR designed for a standard 4 mm sample tube. However, when there is unlimited sample better S/N can be obtained, particularly for lossless samples, by using a larger LGR at low frequencies (Froncisz et al., 1989; Halpern et al., 1989; Rinard et al., 2002a). A 4 mm loop is somewhat large for X-band, and the capacitance gap becomes narrow and the gap large. To alleviate this, multiple gaps and reentrant loops can be used as shown in Figure 3 (Wood et al., 1984; Rinard et al., 1994). Since the gaps are effectively in series, each can be reduced in size nearly proportional to the number of gaps used. The thin-film LGRs can also be made with multiple gaps (Ghim et al., 1996).

The standard TE_{102} rectangular cavity has many properties favorable as a "multipurpose" resonator, and Hyde et al. (1989) created a 3-loop-2-gap LGR having a form factor similar to the multipurpose cavity, so that it could use the accessories designed for the Varian E-231 cavity. For example, the central loop is the same diameter (11 mm) as the sample access stack of the E231 cavity resonator so standard EPR Dewars and flat cells fit it, and the coupling to the waveguide is similar to the Varian coupler.

It is also possible to place a loop-gap-type resonator inside a standard cavity resonator to increase the B_1 at the sample relative to that in the cavity without the LGR inserted (Anderson et al., 1985; Britt and Klein, 1987).

6. COUPLING TO RESONATORS

The LGR can be designed as a series L–C circuit. The inductor, L, will have a series resistance on the order of milliohms, which will be the impedance of the circuit at resonance. The LGR can also be designed as a parallel L–C circuit, and at resonance, the low resistance in series with the inductor is transformed into a high resistance on the order of kiloohms or greater. For the most efficient coupling of power, and to prevent power reflection, the impedance of the resonator circuit must be equal to the characteristic impedance of the line (typically 50 ohms). Therefore, a coupling circuit is required between the resonator and the line to transform the very low or very high resistance into 50 ohms (Figure 4). Coupling (matching) a resonator to the microwave/RF transmission line is an important design and construction feature. Various types of transformers and electric probes have been used. The most common features have been moveable inductive loops or adjustable capacitors at the end of a semi-rigid coaxial cable, but Gordon couplers have also been used (Britt and Klein, 1987; Ichikawa et al., 1989; Oles et al., 1989).

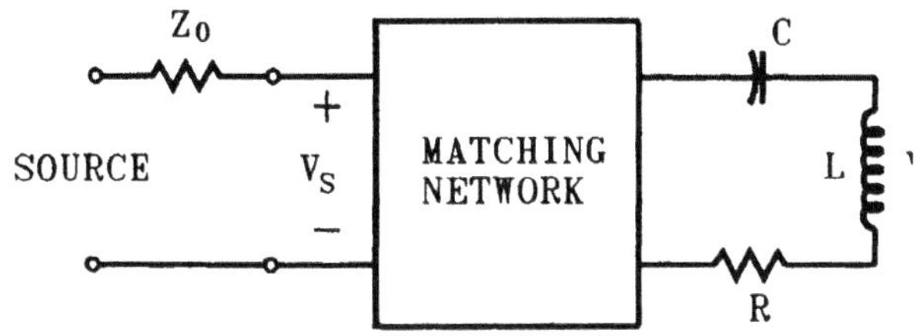

Figure 4. A resonator (R, L, C) is coupled to a microwave source via a matching network, where V_E is the spin system voltage and V_S is the magnetic resonance signal voltage sensed at the detector side of the impedance matching network.

The coupling circuit should be as lossless as possible, and therefore is usually composed of low loss inductors and/or capacitors, which must be non-magnetic if they are close to the resonator. A thorough review of the various coupling techniques, complete with design equations is given in (Rinard et al., 1993). Inductive coupling was also analysed in the context of NMR coils used for imaging and spectroscopy (Froncisz et al. 1986). The most common coupling methods are series capacitive, and mutual inductive.

Imperfect electrical contact, mechanical instabilities, and/or backlash in the adjustable element can be the cause of noise or irreproducibility in the EPR spectra. Experience in many labs (unpublished) has been that wear of the adjustable elements limits the useful life of some resonators.

In series capacitive coupling, a series capacitor couples the parallel resonant L-C circuit to the transmission line. The frequency is adjusted until the L-C circuit is almost resonant but still inductive. The reactance of the series capacitor cancels this inductive component and the circuit is purely resistive. By adjusting the frequency and the size of the coupling capacitor, the value of the resultant resistive component can be varied over a wide range. Critical coupling is obtained when this resistive component is equal in value to the characteristic impedance of the line.

Mutual inductive coupling is obtained by means of a small loop on the end of the transmission line coupled to the inductance (loop) of the series L-C circuit. In this case, the coupling loop couples with the inductance of the resonator in a way to transform the very low resistance of the resonator loop into a resistance on the order of that of the characteristic impedance of the transmission line. When the impedances are equal, the resonator is critically coupled and the power reflected from the resonator/coupling circuit is a minimum (Poole, 1967, p.38; Rinard et al., 1993).

When series capacitance coupling is used, the resonant frequency of the coupled LGR system will be slightly lower than the resonant frequency of the isolated LGR. With mutual inductive coupling the resultant resonant frequency will be slightly higher than that of the LGR itself. In either case the frequency will be closer to the LGR resonant frequency the higher the Q of the circuit, and the frequency difference will become larger if the resonator is overcoupled to reduce its Q for pulse application. This shift in frequency is of little consequence as far as the operation of the resonator is concerned, and for high Q resonators may be less than that caused by the sample.

Other types of coupling are possible, but in general they use the same types of reactive impedance transformations. The type of coupling used may depend more on the physical geometry of the LGR than electrical considerations. However, the LGR in general is a balanced structure and the coupling method should be balanced, particularly if there are significant openings in the shield. The common coaxial transmission line is not balanced. With insufficient shielding, RF currents can exit the LGR shield and flow on the outside of the coaxial line causing, at the least, hand waving effects. A balun (Balanis, 1982) is a device for converting from a balanced to an unbalanced configuration. Mutual inductive coupling is like a two winding transformer and acts somewhat like a balun, and may be preferable in some cases to capacitive coupling, but it is not perfectly balanced and for critical applications a balun may still be desirable.

7. DESIGN EQUATIONS

Design equations for LGRs and coupling are presented in (Froncisz and Hyde, 1982; Mehdizadeh et al., 1983; Wood et al., 1984; Froncisz et al., 1986; Rinard et al., 1993, 1994, 1999). Some of the more important equations are given below.

The LGR parameters are as follows: r = inside radius of loop, z = length of loop and gap, (distances are in meters) σ = conductivity of loop, w = width of gap, n = number of gaps, $\varepsilon_0 = 8.854\mathrm{x}10^{-12}$ F m^{-1} is the permittivity of free space, ε_r (dimensionless) is the dielectric constant in the gap, and t = thickness of gaps. $\mu_0 = 4\ \pi\ \mathrm{X}\ 10^{-7}$ Hm^{-1} For mutual coupling, L_1 and M_1 are the self and mutual inductance of the coupling loop, C_1 is the capacitance for series capacitance coupling, and Z_0 is the characteristic impedance of the transmission line.

Inductance (in Henries) (Terman, 1943): $L = \dfrac{\mu_0 \pi r^2}{z + 0.9r}$ (2)

Capacitance (in Farads): $C = \varepsilon_r \varepsilon_0 \frac{(w+t)(z+t)}{nt}$ (3)

This formula is derived from the definition that the capacitance of a pair of parallel plates, each of area A and distance d apart is $C = \varepsilon_r \varepsilon_0 \frac{A}{d}$, where the linear dimensions of the plate are increased by the plate spacing to account for electric field fringing.

Resonant frequency of LGR (in Hz): $\nu_0 = \frac{1}{2\pi\sqrt{LC}}$ (4)

Resistance (in ohms) (Rinard et al., 1999a): $R = 2\sqrt{\frac{\mu_0 \pi \nu_0}{\sigma}}\left(\frac{\pi r + \frac{w}{3}}{z}\right)$ (5)

Eq. (5) is the a.c. resistance due to skin effect. Eq. (5) is a modification of Rinard et al., 1999a, Eq. (29), as described in Rinard et al., 1999c, Eq. (17), to account for the linear distribution of the current in the gap.

Critically coupled loaded $Q_L = \frac{\pi \nu_0 L}{R}$ (6)

Froncisz and Hyde (1982) provide equations for resonator frequency and for Q_L from Hardy and Whitehead (1981) that accounts for the effects of the shield by including the dimensions of the resonator and shield.

Critical coupling mutual inductance (Rinard et al., 1993):

$$M_1 \approx \sqrt{\frac{R}{Z_0}} L_1 \quad (7)$$

Where M_1 is the mutual inductance between the input loop and the resonator loop, and L_1 is the inductance of the input loop.

Critical coupling series capacitance (Rinard et al., 1993):

$$C_1 \approx \sqrt{\frac{R}{Z_0}} C \quad (8)$$

Where C_1 is the capacitance between the input line and the parallel loop and gap.

As a practical matter, characterization of a resonator in the laboratory will exploit the easiest measurements and use these to calculate other parameters of interest. One would measure the resonant frequency, ν, and Q, either in a spectrometer or with a network analyzer on a test bench. In the spectrometer, Q could be estimated by measuring the half-power bandwidth (Dalal et al., 1981) or the ring down time constant following a pulse. Of the calculations above, that of C is probably the most accurate, since the as-built dimensions will be known better than will the actual surface resistance after machining. However, the calculation of L would be the most accurate if the gap is very small or difficult to determine. Using the measured resonant frequency and the calculated C (or L), L (or C) can be determined from Eq. (4). The calculated inductance (capacitance), together with the measured Q and ν, can be used to calculate the resistance R. The resistance will be of the order of a few hundredths of an ohm for typical small LGRs. Examples are given in Rinard et al., 1993, 1994 and 1999b.

Once one has calculated the resistance, R, and the inductance, L, then it is possible to estimate the B_1 per square root watt dissipated in the resonator. From first principles, (Rinard et al., 1999a)

$$B_{1t} = \frac{\phi}{A} \tag{9}$$

ϕ is the total magnetic flux in the loop of area A ($A=\pi r^2$), and B_{1t} is the linearly polarized (total) component of magnetic field.

From the definition of inductance for a coil with N turns,

$$\phi = \frac{LI}{N} \tag{10}$$

where I is the current in the walls of the resonator that generate the magnetic flux, and N is the number of turns.

Often, as in a simple one-loop LGR, N = 1. The current I is given by

$$I = \sqrt{\frac{P}{R}} \tag{11}$$

Combining these formulae,

$$\frac{B_{1t}}{\sqrt{P}} = \frac{L}{AN\sqrt{R}} \tag{12a}$$

Because it is derived from power, this B_{1t} is the RMS value. For EPR, we are interested in the magnitude (amplitude) of the circularly polarized (rotating) component of the magnetic field, B_1. The ratio of magnitude to RMS is $\sqrt{2}$ and the magnitude of the rotating component is ½ that of the total field. Therefore,

$$\frac{B_1}{\sqrt{P}} = \frac{L}{AN\sqrt{2R}} \tag{12b}$$

which now is in terms of parameters either measured or reliably estimated.

The best experimental estimate of B_1 is by pulsed methods. The power at which a maximum FID or echo is observed is approximately that corresponding to the B_1 for a $\theta = \pi/2$ pulse, if the pulse repetition time is long relative to T_1. A more precise measurement of the B_1 for a $\pi/2$ pulse is the null of the "T echo" in a three-pulse echo sequence, $\pi/2$ - τ - $\pi/2$ - T - $\pi/2$ - T (echo) (Perman et al., 1989). If τ is smaller than T, the T echo is the 3rd echo of the 4 generated by this sequence (this is one of the echoes Mims called "unwanted").

$$\theta = \gamma B_1 t_p \tag{13}$$

where θ is the pulse turning angle and t_p is the length of the pulse. The magnetogyric ratio, γ, for the electron is 1.7608×10^7 rad $s^{-1}G^{-1}$.

The use of these equations is illustrated by calculations using parameters for a typical resonator. Consider a single-turn S-band LGR made of copper with: loop radius 2.03 mm, length 10.16 mm, gap width 3.05 mm and gap thickness 0.18 mm. The parameters calculated from Eqs. (2), (3), and (4) are L = 1.34 nH, C = 1.66 pF, ν_0 = 3.37 GHz. The measured Q was 590. Calculating R from Eq. (6) gives R = 0.024 Ω. From Eq. (12b),

$$\frac{B_1}{\sqrt{P}} = 4.7\frac{\text{PeakGauss}}{\sqrt{\text{Watt}}} \tag{14}$$

8. MAGNETIC FIELD MODULATION

For most CW EPR measurements, it is important to modulate the magnetic field and use phase-sensitive detection (Poole, 1967) at the modulation frequency. In order to get the modulation field to the sample, either the metal of the shield and resonator has to be thin relative to the penetration (skin depth) at the modulation frequency (see Poole, 1967, pages 73-76), or there have to be breaks in the conductor. For example, thinly plated ceramic resonators can be used with magnetic field modulation, but resonators made of solid metal have to be either small in one dimension, as is the Bruker split-ring resonator, or have slots cut in them to allow modulation field penetration, as in several of the resonators described in our papers. The depth of penetration for electromagnetic energy is represented by the skin depth (Poole 1967, pp. 73-75)

$$\delta = \sqrt{\frac{1}{\pi \nu \mu \sigma}} \tag{15}$$

In general, the resonator should be several skin depths thick at its resonant frequency in order to have a high Q and a fraction of a skin depth thick at the field modulation frequency for good modulation field penetration. For the resonator whose dimensions are given above, the metallic layer of the resonator should be on the order of 0.01 mm thick for a field modulation of up to ca. 100 KHz.

The purpose of field modulation is to encode the EPR signal with a reasonably high frequency modulation of the magnetic field, B_0, and reduce the effect of low frequency noise (Anderson, 1960; Poole, 1967). If the penetration of the resonator is low, eddy currents may be induced into the resonator structure that can interact with the static field and produce extraneous signals at the same frequency as the encoded EPR signal. Therefore, the penetration of the resonator to field modulation should be as good as possible while still maintaining a high Q. Resonators made of plated quartz or plated Macor ceramic usually have very little attenuation of the modulation field, but resonators made of solid metal with a few slots cut through the metal usually significantly attenuate the modulation field. We have found attenuations of ca. 30 to 40% in several resonators made of slotted solid metal.

A 300 MHz LGR with multilayer construction has been shown to yield greater modulation field penetration than a bridged loop-gap resonator, by suppression of eddy currents, while having similar Q values (Sato et al., 1999).

One might find mechanical resonances excited by the magnetic field modulation at particular modulation frequencies, which would then have to be avoided in operation. The eddy-current-induced noise depends on the amplitude of the magnetic field modulation, and may be the major contributor to noise in the EPR spectrum at high modulation amplitudes. Even with broad CW EPR spectra, the modulation-induced noise may be large enough that modulation larger than a few G might not be practical.

9. LGR FOR TIME DOMAIN EPR

Most of the parameters for a good CW LGR also apply for pulse applications. A critically coupled LGR is used for saturation recovery (SR), but it is necessary to be able to overcouple the resonator to low Q for FID and ESE measurements. Magnetic field modulation is not required for pulse applications. For CW, the EPR signal is in a very narrow band and centered about the spectrometry frequency. The spectrum is generated by sweeping the field at a frequency that is orders of magnitude lower than the RF frequency. In CW spectroscopy, noise is reduced by narrow band, low frequency filtering. In time domain EPR, the entire spectrum is recorded in a very short time, so the detector bandwidth has to be much larger that for CW EPR. The consequence of the large detector bandwidth needed for pulse EPR is that to obtain good S/N it is necessary to co-add many measurements of the response of the spin system.

The frequency excitation bandwidth of a pulse is also an important experimental parameter, affecting the selection of resonator Q. Mims (1965a,b) established a criterion for choice of Q in a pulse experiment by pointing out that for any shaped pulse, the Q must be equal to or less than that required to match the half power bandwidth of the resonator to the half power bandwidth of the pulse.

$$\Delta\nu = \frac{\nu}{Q} = \text{half power width of Fourier transform of pulse of length } t_p \qquad (16)$$

Other authors have expressed somewhat similar ideas in terms of other criteria, such as the width of the EPR signal excited. See, for example, Hornak and Freed (1986), Bowman (1990), Saalmuetter et al. (1995), and the discussion in Eaton and Eaton (2002). Since the focus here is on the resonator, the Mims criterion will be used in this Chapter.

The bandwidth of the resonator and spectrometer must be high enough (i.e., the Q low enough) to pass all of the frequency components of the pulse and EPR spectrum. The loaded Q of the resonator can be expressed in terms of its bandwidth as,

$$Q_L = \frac{\nu_0}{\Delta\nu} \tag{17}$$

Where $\Delta\nu$ = half power resonator bandwidth. Therefore, large bandwidth requires a low Q_L. The resonator will ring (exponentially damped oscillation at its resonance frequency) after the end of an RF pulse. The time constant of the power ring down is,

$$TC = Q_L / 2\pi\nu \tag{18}$$

(when measuring signal voltage, the time constant will be twice this value) and there will be a dead time of as many as ca. 20 time constants, depending on signal strength, before the pulse power has decreased to the level that the much weaker EPR signal can be recorded. Although some researchers have constructed lossy resonators to obtain low Q_L at critical coupling, the most efficient way to reduce the Q is by over coupling the resonator (Rinard et al., 1994). There is a complicated tradeoff between dead time (longer, the higher the Q) and EPR signal (higher, the higher the Q). The highest Q meeting the Mims criterion and consistent with experimental relaxation times will give the strongest EPR signal. If the experimental goal can accept a finite dead time, then the best EPR signal occurs with a Q that is larger than the minimum achievable by overcoupling. However, the experimental goal might be to observe a rapidly-decaying signal, or to observe echo envelope modulation, which require the minimum feasible dead time. These are experimental tradeoffs that will have to be optimized by each researcher. However, the Mims criterion yields a low enough Q to be a good starting point for such optimization.

For both pulse and CW applications an important parameter is resonator efficiency, defined by Eq. (19) as (Hyde and Froncisz, 1989),

$$\Lambda = \frac{B_1}{\sqrt{P}} \tag{19}$$

where P is the power into the resonator, and B_1 is the magnitude of the circularly polarized component of the microwave magnetic field.

Note that if the resonator is overcoupled, some power is reflected (see Rinard et al., 1994, and the worked example below), and the reflected power must be subtracted from the incident power to obtain the value of P used in Eq. (19). An important spectrometer system design criterion is the pulse power required to generate the desired B_1 in the resonator. From basic considerations it can be shown (Rinard et al., 1999a) that for a given frequency and resonator conductivity,

$$Q_L \propto d \tag{20}$$

and that

$$\Lambda \propto \frac{1}{\sqrt{dz}} \tag{21}$$

More extensive discussions of the scaling of B_1, power required, etc., on the size and frequency of a resonator are in (Eaton et al., 1998; Rinard et al., 1999a, 1999b, 1999c). For convenience, we cite here useful equations from our prior papers, with notation specific to loop-gap-resonators (LGR), for Q and B_1.

$$Q = \left(\frac{\sqrt{2}}{8} \sqrt{\mu_0} \sqrt{\sigma} \right) d\omega^{1/2} \tag{22}$$

where d is the diameter of the LGR, μ_0 is the permeability in a vacuum, σ is the conductivity of the surface of the resonator and ω is the EPR frequency.

$$B_1 = \frac{\sigma^{1/4} \mu_0^{3/4} \sqrt{P}}{2^{1/4} \omega^{1/4} \sqrt{\pi dz}} \tag{23}$$

where z is the length of the LGR.

If the resonator is overcoupled, then some of the power incident on the resonator is reflected from the resonator (Rinard et al., 1994). The following equations relate the reflected power, via the coupling coefficient, β, to the resonator Q, which is often the most convenient measure of the degree of overcoupling, and thence to the power that gets to the sample in the resonator.

We define the coupling coefficient, β, such that measurement of the critically coupled loaded Q, Q_H, and the overcoupled loaded Q, which we simply call Q, allows calculation of β with the formula

$$Q = \frac{2Q_H}{1+\beta}. \text{ Hence, } \beta = \frac{2Q_H}{Q} - 1. \tag{24}$$

For a critically coupled resonator, $\beta = 1$, and for an overcoupled resonator $\beta > 1$.

The fraction of incident power transmitted to the sample in the resonator is given by

$$T_{P\beta} = \frac{4\beta}{(1+\beta)^2} \tag{25}$$

For example, using typical values for a TE_{102} cavity overcoupled from $Q_H = 3350$ to $Q = 1408$, so $\beta = 3.8$, then 0.66 of the incident power enters the resonator. Therefore, $B_1 = \sqrt{0.66}$ times the B_1 for the critically coupled cavity. The power, P_β, required to produce a given B_1 in an overcoupled resonator relative to that required for a critically coupled resonator is given by

$$p_\beta = \frac{Q_H}{Q}\left(\frac{1}{2-\left(\frac{Q}{Q_H}\right)}\right) \tag{26}$$

When the quality factor is reduced from Q_H by adding loss to the resonator the power required to produce a given B_1 will increase by the factor $p_L = \frac{Q_H}{Q}$ (27)

Thus, for large β there is a potential for using only half as much power in the overcoupled resonator case relative to that required for the same Q in the inherently low Q resonator. The pulsed EPR signal for a given B_1 and Q is higher if the Q was achieved by overcoupling from Q_H than if the Q was achieved by higher loss in the resonator (due to different materials of construction, or a lossy sample). The relative signal amplitudes for the overcoupled and inherently low-Q cases are given by

$$V_\beta = \sqrt{\frac{Q}{Q_H}\left(2-\frac{Q}{Q_H}\right)} \tag{28}$$

and

$$V_L = \sqrt{\frac{Q}{Q_H}} \tag{29}$$

respectively. In the limit, the overcoupled resonator yields $\sqrt{2}$ times the signal voltage of the inherently low-Q resonator. Experimental confirmation of these predictions is in (Rinard et al., 1994).

10. SELECTION OF THE Q OF A LGR

As discussed above, it is usually better to construct the LGR with the highest possible Q for a given desired resonator size, which means using high-conductivity materials, and overcoupling it to achieve low Q for FID and ESE measurements, rather than purposefully constructing a LGR with low Q. However, the reflected power due to overcoupling might cause a problem in some spectrometer systems. Also if the spectrometer is power-limited, there may be an incentive to maximize the resonator Q. This may determine whether or not to use overcoupling and what degree of overcoupling would be needed. However, it is usually desirable to design the spectrometer to use an overcoupled resonator (Rinard et al., 1994). In general, one needs to reduce the dead time by decreasing Q, especially for samples with short T_2. However, the largest Q should be used consistent with the bandwidth, the dead time, and S/N. There will be an optimum Q that gives the best S/N, but this may not be the best Q for obtaining desired information in a particular experiment, as discussed above.

Another consideration of the Q to be selected is the pulse length. Mims (1965, 1972) matched the half-power bandwidth of the resonator with the half-power width of the Fourier transform of a pulse, yielding

$$Q \le \frac{2\pi \nu t_p}{\alpha} \tag{30}$$

where t_p is the width of the pulse between half amplitude points.

For a Gaussian time pulse, the width of which is measured at the half amplitude points, $\alpha = 3.9$. For an isosceles triangular pulse, measured at half amplitude points, $\alpha = 4$. Today's technology allows generation of very rectangular pulses on the order of tens of nanoseconds long. For a rectangular pulse, $\alpha = 5.6$ (Mims reported 6.6, which must have been a typo). For rough estimates to guide setting up an experiment, use of a value of 4 for short pulses and 6 for long pulses is a reasonable approximation.

Using $\alpha = 5.6$, Mims' criterion for rectangular pulses becomes,

$$Q \le \frac{2\pi \nu t_p}{5.6} \le 1.12 \nu t_p \tag{31}$$

This result specifies the Q necessary to pass the frequency spectrum of the pulse. The frequency spectrum for a rectangular pulse has half power when,

$$\frac{\sin(\frac{\pi\Delta\nu t_p}{2})}{\frac{\pi\Delta\nu t_p}{2}} = 0.707\text{, or } \pi\Delta\nu t_p \leq 2.8 \text{ or } t_p \leq \frac{2.8}{\pi\Delta\nu} = \frac{0.88}{\Delta\nu} \quad (32)$$

where $\Delta\nu$ is the bandwidth of the EPR spectra in Hertz.

For example, if it is required to excite one gauss of spectra (2.8 MHz), the resonator bandwidth should be at least $\Delta\nu = 2.8$ MHz, (larger for more uniform excitation). If $\nu = 9$ GHz, from Eq. (17) $Q \leq 3200$. Using these values in Eq. (31) yields $t_p \approx 320\text{ns}$. If the pulse length, t_p, were 16 ns, at 9 GHz, the Q would have to be < 160 to meet the criterion of Eq. (30). The lower the frequency, the lower the Q required for a given t_p. For example, using t_p = 320 ns, as in the above X-band example, at 250 MHz the Q would have to be reduced to less than about 90, and a 16 ns pulse would require Q < 5. The pulse length for a given bandwidth may be any value less than that given by Eq. (32). Likewise, the Q may be any value less than that specified by Eq. (31) using the equality sign in Eq. (32). However from a practical standpoint, the dead time of the spectrometer as determined by the ring down time of the resonator will be shortest for the longest practical pulse length (lowest power pulse) and the lowest Q. There is an optimum Q above or below which the S/N will be lower, if the full spectrum is excited.

Other criteria for the choice of resonator Q include the magnitude of B_1 needed over the frequency bandwidth, and the width of the spectrum to be studied. A pulse length short enough to provide a broad frequency spectrum to excite all the spins in the sample may require a B_1 for a 90-degree pulse that is unattainable in the spectrometer. This is particularly important for non-uniformly broadened samples or for imaging using magnetic field gradients. The relaxation time is also important. The pulse length, t_p, as calculated from setting $\Delta\nu$ = sample spectral width, may be close to the spin relaxation time T_2. One should make $t_p \leq 0.1 \times T_2$ to prevent the spins from relaxing excessively during excitation. Q is more important than t_p for uniformly broadened samples, since the single line spectra can be excited by a pulse that is longer that that given by Eq. (32), but the Q must be low enough to faithfully reproduce the EPR spectra. It can be shown that the spectra will be very good, even if the Q is somewhat higher then that given by Eq. (31), but the comments about the optimum Q still apply. If some means for killing the Q can be utilized to shorten the resonator ring down time following the pulse, then return to the higher Q to record the EPR signal, higher S/N may be achieved (Rinard et al., 2002a). If there were a single, T_2-determined line, these criteria on Q and t_p would result in an excitation bandwidth many times the line width. This is an interesting, but rather special case. More commonly, the spectrum is inhomogeneously broadened.

11. MEASURING B_1 IN THE LGR

Measurement of the absolute value of B_1 in a resonator is one of the more difficult tasks in EPR. Calculations, based on the details of resonator size and materials of construction, are as good as measurements in a well-characterized resonator. We start by discussing B_1 in the well-known TE_{102} rectangular cavity resonators. A large literature, cited in (Dalal et al., 1981; More et al., 1984; Poole, 1967 pp. 394-420; Eaton and Eaton, 1985; Bales and Kevan, 1970), provides confidence that the B_1 is given by Eq. (33), where κ = ca. 0.02 for a Varian TE_{102} E-231 cavity resonator.

$$B_1 = \kappa\sqrt{QP}\ \text{Gauss} \tag{33}$$

where P is the incident power in watts and Q is the loaded Q.

Eaton and Eaton (1985) found $\kappa = 0.0159$ as the coefficient in Eq. (33) for a Varian E231 cavity without a dewar insert. Dalal et al. (1981) discussed formulae for Q of a cavity and presented experimental data for the effect of a Dewar insert on the B_1. Note that this is strongly dependent on the wall thickness of the Dewar. The use of this relation to estimate B_1 for the purpose of determining T_1 and T_2 from continuous saturation was discussed by More et al. (1984).

As a convenient rule of thumb, use $\kappa = 0.02$, and with $Q = 3000$, then B_1 = 1 G per square root watt. Bruker cites a value of 1.4 G per square root watt for their TE_{102} cavity, which has a slightly higher frequency than the Varian cavity. Using the notation of Eq. (19), $\Lambda \approx 1$ for an X-band TE_{102} cavity resonator. X-band LGRs have been made with Λ as high as 10 (Hyde and Froncisz, 1989).

One can also use the method of perturbing spheres to measure B_1 with similar results. Note that in the formulae given in the book by Ginzton (1957), H_1 is what we would call $2B_1$, since the perturbing spheres method measures the linearly polarized component of B_1 at the position of measurement.

The CW continuous saturation of a well-characterized sample such as Fremy's salt can also be used to estimate B_1 (More et al., 1984).

When pulsed EPR is available, it is convenient to measure B_1 in the resonator, for the exact experimental conditions of interest, by measuring the turning angle for the pulse, using Eq. (13).For example, a 40 ns pulse gives B_1 = 2.2 G (assuming bandwidth is adequate). If the Q is low enough and the relaxation time is long enough (e.g., the E' center in irradiated quartz (Eaton and Eaton, 1993), the turning angle can be used to measure B_1 in a critically coupled resonator also. Bimodal LGR and Crossed-loop Resonator (CLR)

Two LGRs can be constructed so that they are orthogonal and intersect at the sample (Hyde and Froncisz, 1989; Tsapin et al., 1992). An alternative approach to orthogonality was described by Piasecki et al. (1996) for an S-band bimodal resonator. This approach uses two resonators coaxial with one another, but arranged such that the integral over all space of the fields from the two are equal to zero, instead of making the resonators locally orthogonal. The orthogonality of the two resonators has advantages for both CW and time domain EPR. For CW EPR, the source and detector are very well isolated and phase noise from the source is essentially eliminated. For time domain EPR the isolation reduces the dead time of the spectrometer. Following a high-power pulse, the power in the resonator has to decay to less than the signal power in order to observe the signal. Depending on the signal strength, this may require a change in power by ca. 140 dB or about 16 time constants (Mims 1965, page 324), or even more. A commonly cited value for NMR is 21 time constants (Gerstein and Dybowski, 1985). Isolation provided by the CLR between the pulse power and the signal resonator decreases the dead time by the amount of the isolation. Let the dead time, dt, be n power decay time constants, tc, for the resonator, dt = ntc. Then,

$$dB = 20\log\left(e^{-\frac{dt}{tc}}\right) = 20(-n)\log(e) = -8.7n \tag{34}$$

$$\text{or } n = \frac{dB \text{ isolation}}{8.7} \tag{35}$$

Thus, with 60 dB isolation in the CLR, the dead time is shortened by ca. 7 resonator time constants. In addition, each LGR can be optimized and Q-switched (Rinard et al., 2002a) to improve S/N and further reduce the dead time.

It has proven difficult to extend the Piasecki et al. (1996) approach to X-band, because of the small size of critical components. Success has been achieved with the CLR design at frequencies at 250 MHz (Rinard et al.,

2002a), L-band (Rinard et al., 2000), S-band (Rinard et al., 1996a,b) and X-band (unpublished).

Preliminary tests simulating animal movement also indicates that artifact from animal movement can be greatly reduced by using a CLR.

12. VARIABLE TEMPERATURE

The commercial X-band cavity resonators typically are equipped with a Dewar that inserts into the cavity. The temperature of the sample is changed while the resonator is kept at ambient temperature. Because LGRs are smaller than cavities at the same frequency, and because the LGR is generally designed to have a large filling factor, there is not sufficient space in the sample loop for the traditional flow-through Dewar. Consequently, the entire resonator and sample assembly is cooled. In the limit, the spectrometer system performance is limited by the thermal noise of the resonator. With a cross loop resonator, it is also possible to cool the resonator and the first stage microwave amplifier to reduce thermal noise (Rinard et al., 1999b,d). In addition, in a well-designed cryostat there is less temperature gradient over the sample when the entire assembly is cooled. However, it is much more complicated to adjust the tuning and coupling of a resonator that is a large distance (down in the cryostat) from the control devices. A disadvantage of low-temperature EPR with LGRs is that EPR signals from the resonator itself and from the shield and other structures are more important in LGRs than in cavity resonators. Usually, when performing low-temperature EPR with a cavity resonator, the sample is cooled in a Dewar that fits within the cavity, and the cavity itself remains at room temperature, but a LGR is usually cooled along with the sample, and signals from species such as Cu(II) in the LGR and its shield become more important as the resonator assembly is cooled.

13. MECHANICAL CONSIDERATIONS

Perhaps the best material for constructing the LGR is pure silver since it has the lowest electrical resistively of all metals. However, pure copper is nearly as good electrically. Both pure silver and pure copper are very soft and difficult to machine. A good substitute is a tellurium-copper alloy number 145. The tellurium content is low and the electrical resistively of this alloy is nearly as good as copper. Its main advantage is that it is very machinable albeit carbide-cutting tools are required since the alloy is somewhat abrasive.

A researcher can build a functional LGR using fairly readily available tools - drill press, small lathe, small milling machine, etc., starting with a piece of Macor that will be plated after machining, or with a piece of copper. The support structure can be made from polystyrene and Teflon, also using simple tools. The hardest part is making the capacitive gap small enough to achieve the desired frequency. Slotting saws are readily available down to ca. 0.15 mm (0.006 inch size in the USA). The actual cut will be a bit wider. Some machine shops can cut the slot with wire using electrical discharge machining (EDM). Some shops can achieve about 0.05 mm (0.002 in) (Hyde, personal communication, 2003) with this technique, but many shops are not able to achieve less than 0.25 mm (0.010in), which places a practical limit on actual construction of LGRs. We have found that it is practical to make the reentrant resonators, such as are used in the CLR described in (Rinard et al., 2002b) by machining them in two pieces and assembling them. If the surfaces are very smooth and clean, and the brass screws are fastened very tightly, the Q is not reduced very much by the finite resistance of the joints. This method of construction facilitates fine adjustment of the capacitive gap.

For cryogenic applications requiring a cryostat, remote tuning and coupling adjustments are required and can require some ingenuity.

14. COMMERCIAL RESONATORS

Some of the LGRs designed by Hyde and Froncisz have been produced in Poland and marketed in the USA by Medical Advances (Milwaukee) and, as of 2002, by Molecular Specialties (Milwaukee).

The Bruker "split-ring" resonator has features in common with the LGR, including a sample loop about the same size as the sample tube, capacitive gaps that define the resonant frequency, and a confined return flux region. These resonators (called probeheads by Bruker) are available in sizes for several tube diameters and at several microwave frequencies.

15. APPLICATIONS OF LUMPED-CIRCUIT RESONATORS

Moving from the standard rectangular TE_{102} microwave cavity to the LGR freed the minds of investigators to adapt the resonator to the experiment, rather than try to fit the experiment to the cavity, however "general-purpose" the commercial designs may be. The advantages of LGRs listed above clearly indicate that LGR-type resonators will be dominant for CW EPR at frequencies below X-band, and for pulsed EPR at X-band and below. The possibility of high filling factor for very small samples, in conjunction with the development of site-directed spin labeling created a renaissance in spin labeling of biological molecules (Hubbell et al., 1998, and see ch. 10 by Klug and Feix). The high filling factor for small samples was exploited, soon after the development of the LGR, for saturation-transfer of spin labeled muscle fibers (Thomas et al., 1983). The good separation of E and H fields facilitates the use of LGRs for in vivo studies (Halpern et al., 1989), and other studies of lossy samples, such as in electrochemistry (Allendoerfer et al., 1988) and stopped flow EPR (Hubbell et al., 1987; Jiang et al., 1993; Sienkiewicz et al., 1994, and see the chapter by Scholes). EPR at frequencies below X-band has been facilitated by use of the LGR and related lumped-circuit resonators. Examples include oximetry in a mouse in a 25 mm diameter LGR at 1 GHz (Subczynski et al., 1986) and at 250 MHz (Halpern et al., 1989). The LGR topology also lends itself to orienting the B_1 field parallel to the B_0 field to observe forbidden transitions (Rothenberger et al., 1986). Essentially all pulsed EPR studies using the Bruker Elexsys series spectrometers use either a dielectric resonator or a split-ring resonator.

The small amount of sample required when a LGR is used at X-band (Hubbell et al., 1987) is an enabling technology for the combination of site-directed spin labelling (SDSL) and EPR (see ch. 10 by Klug and Feix for examples of applications of SDSL). Hubbell et al. (1987) used a 2-loop, 1-gap LGR with the sample in the small (1 mm diameter) loop and the larger (4 mm diameter) loop used for the return flux. This LGR had an active volume of 1.2 μL and a sensitivity about 50 times that of a TE_{102} resonator for the same number of spins in an aqueous sample. The small dimensions resulted in small volumes of sample required for rapid flow and stopped flow measurements of short-lived radicals. Subsequently, this X-band LGR found use for measuring small volumes of spin-labeled proteins (Cornish et al., 1994), and for time-resolved studies of spin-labeled mutants (Shin et al., 1993). The studies of structure and motion (Hubbell et al., 1996) of T4-lysozyme, rhodopsin, and other proteins by Hubbell and others was made possible by the ability to measure the EPR spectra of very small amounts of spin-labeled protein.

All pulsed EPR measurements of distance between electron spins have used resonators that are either of the loop-gap type or were stimulated by the LGR. The large B_1, low Q (and related high bandwidth), and high filling factor all are features essential to the successful pulsed ELDOR and related measurements (see Biol. Magn. Reson. 19).

At frequencies below X-band, cavity resonators become awkwardly large, and below ca. 3 GHz it would be difficult to put a cavity resonator in a practical magnet. Furthermore, to have a filling factor even as large as in an X-band cavity (ca. 1%), enormous amounts of sample would be required. The amount of sample required would make impossible most studies of proteins at frequencies below X-band. Thus, the LGR has opened up the spectral range below X-band for EPR. This is especially evident for in vivo EPR (see ch. 12 of vol. 23 by Subramanian and Krishna, and ch. 11 of vol. 23 by Williams and Halpern; He et al., 2002).

16. FURTHER INFORMATION

In addition to the published literature cited, two booklets reviewing "Resonators for Pulsed EPR," compiled by Dr. Gareth Eaton, and a booklet containing a collection of reprints of LGR papers by Hyde and coworkers, are available from the National Biomedical ESR Center, Medical College of Wisconsin, Milwaukee, Wisconsin, USA.

17. ACKNOWLEDGMENTS

Our work in resonator design, construction, and application has been funded by NSF and NIH, and is currently supported by NIH grants P41 RR12257 (Professor H. J. Halpern, University of Chicago, and GRE) and EB002807 (formerly GM21156), and by NSF grant DBI-9986942. Early in this work, GRE benefited from discussions with James S. Hyde and Wojciech Froncisz. Richard W. Quine and Sandra S. Eaton have helped with the testing of LGRs and CLRs in our laboratory.

18. REFERENCES

Alderman, D. W. and Grant, D. M. (1979) An Efficient Decoupler Coil Design which Reduces Heating in Conductive Samples in Superconducting Spectrometers. *J. Magn. Reson.* **36**, 447-451.

Allendoerfer, R. D., Froncisz, W., Felix, C. C., and Hyde, J. S. (1988) Electrochemical Generation of Free Radicals in an EPR Loop-Gap Resonator. *J. Magn. Reson.* **76**, 100-105.

Anderson, J. R., Venters, R. A., Bowman, M. K., True, A. E., and Hoffman, B. M. (1985) ESR and ENDOR Applications of Loop-Gap Resonators with Distributed Circuit Coupling. *J. Magn. Reson.* **65**, 165-168.

Anderson, W. A. (1960) Magnetic Field Modulation for High Resolution NMR, in "NMR and EPR Spectroscopy," by the NMR-EPR Staff of Varian Associates, Pergamon, New York, pp.180-184.

Balanis, C. A. (1982) Antenna Theory, Analysis and Design. Harper and Row, New York, pp. 365-368.

Bales, B. L. and Kevan, L. (1970) Paramagnetic Relaxation of Silver Species in γ-Irradiated Frozen Aqueous Solutions. *J. Phys. Chem.* **74**, 4644-4653.

Bowman, M. K. (1990) Fourier Transform Electron Spin Resonance, in *Modern Pulsed and Continuous Wave Electron Spin Resonance*, L. Kevan and M. K. Bowman, eds., Wiley, N. Y., 1-42, especially p. 28-29.

Britt, R. D. and Klein, M. P. (1987) A Versatile Loop-Gap Probe for Low-Temperature Electron Spin-Echo Studies. *J. Magn. Reson.* **74**, 535-540.

Christidies, T., Froncisz, W., Oles, T., and Hyde, J. S. (1994) Probehead with Interchangeable Loop-Gap Resonators and rf Coils for Multifrequency EPR/ENDOR. *Rev. Sci. Instrum.* **65**, 63-67.

Cornish, V. W., Benson, D. R., Altenbach, C. A., Hideg, K., Hubbell, W. L., and Schultz, P. G. (1994) Site-Specific Incorporation Of Biophysical Probes Into Proteins. *Proc. Natl. Acad. Sci. USA* **91**, 2910-2914.

Dalal, D. P., Eaton, S. S. and Eaton, G. R. (1981) The Effects of Lossy Solvents on Quantitative EPR Studies, *J. Magn. Reson.* **44**, 415-428.

Eaton, S. S., and Eaton, G. R. (1980) Signal Area Measurements in EPR. *Bull. Magn. Reson.* **1**, 130-138.

Eaton, G. R. and Eaton, S. S. (1985) Relaxation Times for the Organic Radical Signal in the EPR Spectra of Oil Shale, Shale Oil, and Spent Shale, *J. Magn. Reson.* **61**, 81-89.

Eaton, S. S. and Eaton, G. R. (1993) Irradiated Fused Quartz Standard Sample for Time Domain EPR. *J. Magn. Reson.* **A102**, 354-356.

Eaton, G. R., Eaton, S. S. and Rinard, G. A. (1998) Frequency Dependence of EPR Sensitivity, in *Spatially Resolved Magnetic Resonance: Methods, Materials, Medicine, Biology, Rheology, Geology, Ecology, Hardware*, P. Blümler, B. Blümich, R. Botto, and E. Fukushima, eds., Wiley-VCH, Weinham, p. 65-74.

Eaton, S. S. and Eaton, G. R. (2000) Relaxation Times of Organic Radicals and Transition Metal Ions, in *Distance Measurements in Biological Systems by EPR*, G. R. Eaton, S. S. Eaton, and L. J. Berliner, eds., *Biol. Magn. Reson.* **19**, 29-154.

Forrer, J., Pfenninger, S., Eisenegger, J., and Schweiger, A. (1990) A Pulsed ENDOR Probehead With The Bridged Loop-Gap Resonator: Construction And Performance. *Rev. Sci. Instrum.* **61**, 3360-3367.

Froncisz, W., and Hyde, J. S. (1982). The Loop-Gap Resonator: A New Microwave Lumped Circuit ESR Sample Structure. *J. Magn. Reson.* **47**, 515-521.

Froncisz, W., Oles, T., and Hyde, J. S (1986a) Q-Band Loop-Gap Resonator, *Rev. Sci. Instrum.* **57**, 1095-1099.

Froncisz, W., Jesmanowicz, A., and Hyde, J. S. (1986b) Inductive (Flux Linkage) Coupling to Local Coils in Magnetic Resonance Imaging and Spectroscopy. *J. Magn. Reson.* **66**, 135-143.

Froncisz, W., Oles, T., and Hyde, J. S (1989) Murine L-Band Loop-Gap Resonator. *J. Magn. Reson.* **82**, 109-114.

Gallay, R. and van der Klink, J. J. (1986) Resonator and Coupling Structure for Spin-Echo ESR. *J. Phys. E. Sci. Instrum.* **19**, 226-230

Gerstein, B. C. and Dybowski, C. R. (1995) Transient Techniques in NMR of Solids. Academic Press, page 221.

Ghim, B. T., Rinard, G. A., Quine, R. W., Eaton, S. S. and Eaton, G. R. (1996) Design and Fabrication of Copper-Film Loop-Gap Resonators. *J. Magn. Reson.* **A120**, 72-76.

Ginzton, E. L. (1957) Microwave Measurements. McGraw-Hill Book Co, New York, p.448.

Grist, T. M. and Hyde, J. S. (1985) Resonators for in Vivo ^{31}P NMR at 1.5 T. *J. Magn. Reson.* **61**, 571-578.

Halpern, H. J., Spenser, D. P., van Polen, J., Bowman, M. K., Nelson, A. C., Dowey, E. M., and Teicher, B. A. (1989) Imaging Radio Frequency Electron-Spin-Resonance Spectrometer With High Resolution And Sensitivity For In Vivo Measurements. *Rev. Sci. Instrum.* **60**, 1040-1050.

Hardy, W. N. and Whitehead, L. A. (1981) Split-Ring Resonator For Use In Magnetic Resonance From 200-2000 Mhz. *Rev. Sci. Instrum.* **52**, 213-216.

Harrison, A. E. (1947) Klystron Tubes. McGraw-Hill Book Company, New York, pp. 12-13.

Hayes, C. E. (1996) Birdcage and Other High Homogeneity Radiofrequency Coils for Whole Body Magnetic Resonance, in Encyclopedia of Nuclear Magnetic Resonance, Grant, D. M., and Harris, R. K., eds., John Wiley & Sons, Chichester, pp. 968-974.

He, G., Samouilov, A., Kuppusamy, P., and Zweier, J. L. (2002) In Vivo Imaging Of Free Radicals: Applications From Mouse To Man. *Molecular and Cellular Biochem.* **234/235** 359-367.

Hill, H. D. W. (1996) Probes for High Resolution, in Encyclopedia of Nuclear Magnetic Resonance, Grant, D. M., and Harris, R. K., eds., John Wiley & Sons, Chichester, pp. 3762-3771.

Hirata, H. and Ono, M. (1996) Resonance Frequency Estimation Of A Bridged Loop-Gap Resonator Used For Magnetic Resonance Measurements. *Rev. Sci. Instrum.* **67**, 73-78.

Hornak, J. and Freed, J. H. (1986) Spectral Rotation in Pulsed ESR Spectroscopy. *J. Magn. Reson.* **67**, 501-518.

Hubbell, W. L., Froncisz, W., and Hyde, J. S. (1987) Continuous and Stopped Flow EPR Spectrometer Based on a Loop Gap Resonator, *Rev. Sci. Instrum.* **58**, 1879-1886.

Hubbell, W. L., Mchaourab, H. S., Altenbach, C., and Lietzow, M. A. (1996) Watching proteins move using site-directed spin labelling. *Structure (London)* **4**, 779-783.

Hubbell, W., L., Fross, A., Langren, R., and Lietzow, M. A. (1998) Recent advances in site-directed spin labeling of proteins. *Curr. Opinion Struct. Biol.* **8**, 649-656.

Hyde, J. S. and Froncisz, W. (1981) ESR S-Band Microwave Spectrometer For Process Control. *National Electronics Conference Proceedings* **35**, 602-606.

Hyde, J. S., Froncisz, W. and Kusumi, A. (1982) Dispersion Electron Spin Resonance With The Loop-Gap Resonator. *Rev. Sci. Instrum.* **53**, 1934-1937.

Hyde, J. S., Yin, J.-J., Froncisz, W., and Feix, J. (1985) Electron-Electron Double Resonance (ELDOR) with a Loop-Gap Resonator. *J. Magn. Reson.* **63**, 142-150.

Hyde, J. S. and Froncisz, W. (1986) Loop-Gap Resonators, in "Electron Spin Resonance" Symons, M. C. R., ed., Specialist Periodical Reports, Royal Society of Chemistry, London, Vol. 10, pp. 175-184.

Hyde, J. S. and Froncisz, W. (1989) Loop Gap Resonators. in "Advanced EPR: Applications in Biology and Biochemistry," Hoff, A. J., ed., Elsevier, Amsterdam, pp. 277-306.

Hyde, J. S., Froncisz, W. and Oles, T. (1989) Multipurpose Loop-Gap Resonator. *J. Magn. Reson.* **82**, 223-230.

Hyde, J. S., Yin, J.-J., Froncisz, W. and Feix, J. B. (1985) Electron-Electron Double Resonance (ELDOR) with a Loop-Gap Resonator. *J. Magn. Reson.* **63**, 142-150.

Ichikawa, T., Yoshida, H. and Westerling, J. (1989) Coupling Structure for the Loop-Gap Resonator. *J. Magn. Reson.* **85**, 132-136.

Jiang, J. Bank, J. F., and Scholes, C. P. (1993) Subsecond Time-Resolved Spin Trapping Followed by Stopped-Flow EPR of Fenton Reaction Products. *J. Am. Chem. Soc.* **115**, 4742-4746.

Kan, S., Gonord, P., Duret, C., Salset, J., and Vibet, C. (1973) A Versatile And Inexpensive Electronic System For A High Resolution NMR Spectrometer. *Rev. Sci. Instrum.* **44**, 1725-1733.

Lin, C. P., Bowman, M. K., and Norris, J. R. (1985) A Folded Half-Wave Resonator for ESR Spectroscopy. *J. Magn. Reson.* **65**, 369-374.

LoBrutto, R., Smithers, G. W., Reed, G. H., Orme-Johnson, W. H., Tan, S. L., and Leigh, J. S. Jr. (1986) Observation of Manganese(II)-Ligand Superhyperfine Couplings in Complexes with Proteins by Electron Spin-Echo Spectroscopy. *Biochem.* **25**, 5654-5660.

Lurie, D. J., Li, H., Petryakov, S., and Zweier, J. L. (2002) Development of a REDRI Free-Radical Imager Using a 0.38 T Clinical MRI System. *Magn. Reson. Med.* **47**, 181-186.

Mazúr, M. Morris, H., and Valko, M. (1997). Analysis of the Movement of Line-like Samples of variable Length along the X-Axis of a Double TE_{104} and a Single TE_{102} Rectangular Cavity. *J. Magn. Reson.* **129**, 188-200.

Mazúr, M., Valko, M., and Morris, H.,. (2000). Analysis of the Radial and Longitudinal Effect in a Double TE_{104} and a Single TE_{102} Rectangular Cavity. *J. Magn. Reson.* **142**, 37-56.

Mehdizadeh, M., Ishii, T. K., Hyde, J. S., and Froncisz, W. (1983) Loop-Gap Resonator: A Lumped Mode Microwave Resonant Structure. *IEEE Trans. Microwave Theory Tech.* **MTT31**, 1059-1064.

Mims, W. B. (1965a) Electron Echo Methods in Spin Resonance Spectrometry. *Rev. Sci. Instrum.* **36**, 1472-1479.

Mims, W. B. (1965b) Electron Spin Echoes, in Electron Paramagnetic Resonance, S. Geschwind, ed., Plenum, New York.

More, K. M., Eaton, G. R. and Eaton, S. S. (1984) Determination of T_1 and T_2 by Simulation of EPR Power Saturation Curves and Saturated Spectra: Application to Spin-Labeled Iron Porphyrins, *J. Magn. Reson.* **60**, 54-65.

Oles, T., Hyde, J. S. and Froncisz, W. (1989) Gordon Coupler for K-Band Loop Gap Resonator. *Rev. Sci. Instrum.* **60**, 389-391.

Ono, M., Ogata, T., Hsieh, K.-C., Suzuki, M., Yoshida, E., and Kamada, H. (1986a) Development of Lop-Gap Resonator and a Surface Coil With an Electric Shield. 28th Rocky Mountain Conference, Abstract 105, Denver, CO, USA, August 1986.

Ono, M., Ogata, T., Hsieh, K.-C., Suzuki, M., Yoshida, E., and Kamada, H. (1986b) L-Band ESR Spectrometer Using a Loop-Gap Resonator for In Vivo Analysis. *Chem. Lett.* 491-494.

Ogata, T., Ono, M., Yoshida, E., and Kamada, H. (1986) Application of L-band ESR Techniques to In Vivo Analysis. 28th Rocky Mountain Conference, Abstract 103, Denver, CO, USA, August 1986.

Perman, W. H., Bernstein, M. A., and Sandstrom, J. C. (1989) A Method for Correctly Setting the rf Flip Angle. *Magn. Reson. Med.* **9**, 16-24.

Pfenninger, S., Schweiger, A., Forrer, J., and Ernst, R. R. (1988) Echo-Detected ESR Spectroscopy with Magnetic Field Vector Jumps: A Novel Approach for Improving the Spectral Resolution in Disordered Systems. *Chem. Phys. Lett.* **151**, 199-204.

Pfenninger, S., Froncisz, W., Forrer, J., Luglio, J., and Hyde, J. S. (1995) General Method for Adjusting the Quality Factor of EPR Resonators. *Rev. Sci. Instrum.* **66**, 4857-4965.

Piasecki, W., Froncisz, W., and Hyde, J. S. (1996) Bimodal Loop-Gap Resonator. *Rev. Sci. Instrum.* **67**, 1896-1904.

Poole, C. P. Jr. (1967) Electron Spin Resonance: A Comprehensive Treatise on Experimental Techniques. Wiley-Interscience, New York, page 291.

Poole, C. P. Jr. and Farach, H. A. (1999) Resonators, Chapter III in Handbook of Electron Spin resonance, Vol. 2, Poole, C. P. Jr. and Farach, H. A., eds., AIP Press Springer-Verlag, New York.

Quine, R. W., Rinard, G. A., Ghim, B. T., Eaton, S. S. and Eaton, G. R. (1996) A 1-2 GHz Pulsed and Continuous Wave Electron Paramagnetic Resonance Spectrometer. *Rev. Sci. Instrum.* **67**, 2514-2527.

Reichert, J. F., and Townsend, J. (1965). Dynamic Nuclear Enhancement in Metallic Sodium, *Phys. Rev.* **137**, 476-482.

Rinard, G. A., Quine, R. W., Eaton, S. S. and Eaton, G. R. (1993) Microwave Coupling Structures for Spectroscopy, *J. Magn. Reson.* **A105**, 134-144.

Rinard, G. A., Quine, R. W., Eaton, S. S., Eaton, G. R., and Froncisz, W. (1994) Relative Benefits of Overcoupled Resonators vs. Inherently Low-Q Resonators for Pulsed Magnetic Resonance. *J. Magn. Reson.* **A108**, 71-81.

Rinard, G. A., Quine, R. W., Ghim, B. T., Eaton, S. S. and Eaton, G. R. (1996a) Easily Tunable Crossed-Loop (Bimodal) EPR Resonator. *J. Magn. Reson.* **A122**, 50-57.

Rinard, G. A., Quine, R. W., Ghim, B. T., Eaton, S. S. and Eaton, G. R. (1996b) Dispersion and Superheterodyne EPR Using a Bimodal Resonator. *J. Magn. Reson.* **A122**, 58-63.

Rinard, G. A., Eaton, S. S., Eaton, G. R., Poole, C. P. Jr., and Farach, H. A. (1999a) Sensitivity in EPR Measurements, *Handbook of Electron Spin Resonance*, C. P. Poole, Jr., and H. A. Farach, eds, vol 2, AIP Press, 1-23

Rinard, G. A., Quine, R. W., Song, R., Eaton, G. R. and Eaton, S. S. (1999b) Absolute EPR Spin Echo and Noise Intensities, *J. Magn. Reson.* **140**, 69-83

Rinard, G. A., Quine, R. W., Harbridge, J. R., Song, R., Eaton, G. R. and Eaton, S. S. (1999c) Frequency Dependence of EPR Signal-to-Noise, *J. Magn. Reson.* **140**, 218-227.

Rinard, G. A., Quine, R. W., and Eaton, G. R. (1999d) A Cryogenically Coolable Microwave Limiter. *J. Magn. Reson.* **136**, 207-210.

Rinard, G. A., Quine, R. W., and Eaton, G. R. (2000) An L-band Crossed-Loop (Bimodal) Resonator, *J. Magn. Reson.* **144**, 85-88.

Rinard, G. A., Quine, R. W., Eaton, G. R. and Eaton, S. S. (2002a) 250 MHz Crossed Loop Resonator for Pulsed Electron Paramagnetic Resonance, *Magn. Reson. Engineer.* **15**, 37-46.

Rinard, G. A., Quine, R. W., Eaton, S. S., and Eaton, G. R. (2002b) Frequency Dependence of EPR Signal Intensity, 248 MHz to 1.4 GHz, *J. Magn. Reson.* **154**, 80-84.

Rothenberger, K. S., Nilges, M. J., Altman, T. E., Glab, K., Belford, R. L., Froncisz, W., and Hyde, J. S. (1986) L-Band Parallel Mode EPR. Measurement of Quadrupole Coupling Through Direct Observation of Secondary Transitions. *Chem. Phys. Lett.* **124**, 295-298.

Saalmueller, J. W., Long, H. W., Maresch, G. G., and Spiess, H. W. (1995) Two-Dimensional Field-Step ELDOR. A Method for Characterizing the Motion of Spin Probes and Spin Labels in Glassy Solids. *J. Magn. Reson. A* **117**, 193-208.

Schneider, H. J. and Dullenkopf, P. (1977). Slotted Tube Resonator: A New NMR Probe Head At High Operating Frequencies. *Rev. Sci. Instrum.* **48**, 68-73.

Shin, Y.-K., Levinthal, C., Levinthal, F. and Hubbell, W. L. (1993) Colicin E1 Binding to Membranes: Time-Resolved Studies of Spin-Labeled Mutants. *Science* **259**, 960-963.

Sienkiewicz, A., Qu, K., and Scholes, C. P. (1994) Dielectric Resonator-Based Stopped-Flow Electron Paramagnetic Resonance. *Rev. Sci. Instrum.* **65**, 68-74.

Squires, T. L. (1963) *Introduction to Microwave Spectroscopy.* George Newnes Ltd., London, p. 53-54.

Sotgiu, A. (1985) Resonator Design for In Vivo ESR Spectroscopy. *J. Magn. Reson.* **65**, 206-214.

Sotgiu, A. and Gualtieri (1985) Cavity Resonator For In Vivo ESR Spectroscopy. *J. Phys. E Sci. Instrum.* **18**, 899-901.

Subczynski, W. K., Lukiewicz, S., and Hyde, J. S. (1986) Murine In Vivo L-Band ESR Spin-Label Oximetry with a Loop-Gap Resonator. *Magn. Reson. Med.* **3**, 747-754.

Symons, M. C. R. (1995) Whole Body Electron Spin Resonance Imaging Spectrometer. in "Bioradicals Detected by ESR Spectroscopy." Ohya-Nishiguchi, H. and Packer, L. eds. Birkhäuser Verlag, Basel, Switzerland.

Terman, F. E. (1943) Radio Engineers Handbook, McGraw-Hill, New York, page 55.

Thomas, D. D., Wendt, C. H., Froncisz, W., and Hyde, J. S. (1983) Saturation Transfer EPR Spectroscopy on Spin-Labeled Muscle Fibers Using a Loop-Gap Resonator. *Biophys. J.* **43**, 131-135.

Tsapin, A. I., Hyde, J. S., and Froncisz, W. (1992) Bimodal Loop-Gap Resonator. *J. Magn. Reson.* **100**, 484-490.

Wood, R. L., Froncisz, W., and Hyde, J. S. (1984) The Loop-Gap Resonator. II. Controlled Return Flux Three-Loop, Two-Gap Microwave Resonators for ENDOR and ESR Spectroscopy. *J. Magn. Reson.* **58**, 243-253.

Chapter 3

EPR Interfaced To Rapid Mixing

Charles P. Scholes
Department of Chemistry, University at Albany –State University of New York, Albany, NY 12222

Abstract: We describe the development of flow and stopped-flow EPR starting with large mixing devices designed for standard metallized EPR cavities and proceeding to micro mixing systems intimately attached to micro resonant structures. The characteristics of resonant structures dedicated to close attachment of liquid mixing systems with low dead volumes, short mixing times, and minimal microwave microphonic perturbation are outlined. The application of such systems to study of radical systems, either spin labelled, naturally occurring, or trapped, is reviewed, and recent experiments to probe rapid folding kinetics of site-directed spin labeled protein are summarized.

1. INTRODUCTION

1.1 Technical Background for Flow and Stopped-Flow EPR

As early as 1940 Britton Chance had provided insightful, groundbreaking technical designs for flow, accelerated flow, and stopped-flow apparatus to follow rapid chemical reactions (Chance, 1940a; Chance, 1940b). Limitations from dead time, quality of mixing, and signal-to-noise in UV-Vis detection were described. As compared with the continuous flow method, the stopped-flow method, in which observations starting soon after mixing were made on a stationary but still reacting system, provided an important saving in reactant.

1.2 Flow and Stopped-Flow EPR Using Conventional Metallized EPR Cavities

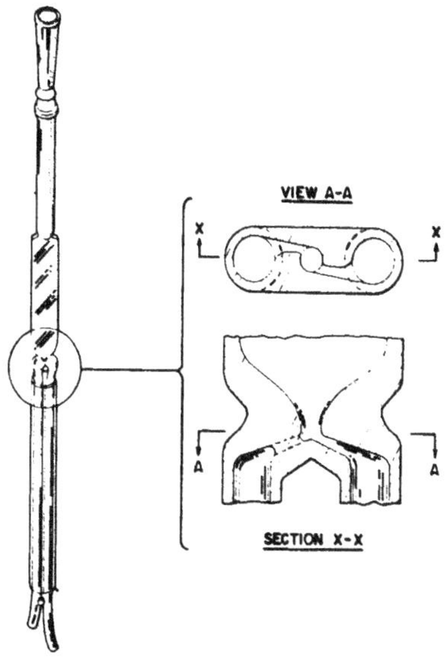

Figure 1. Mix-and-flow cell is shown. This had inner dimensions of 45 X 9 X 0.25 mm for the flat chamber and 1.1 and 1.5 mm for the diameters of the offset inlet jets and of the mixing chambers. The dead volume from mixer to center of the EPR-active zone was 50 μL. Level X-X exhibits a top view through the mixing chamber, and A-A exhibits a cross-sectional view through the mixing chamber. Figure used with permission of D. Borg.

By the late 1950's EPR flow flat cells of a type still marketed by Wilmad had appeared. These cells were designed for insertion into the electric field-free region of a TE_{102} standard cavity. One such flat cell was reported to have a volume between mixer and center of the EPR active zone of 140 μL (Yamazaki et al., 1960). The flow was driven by compressed gas, and with a reported flow rate of 12 mL/s, spectra of radicals of age greater than ~15 ms could be obtained. The spectra of p-semiquinone and ascorbate radicals following enzymatic oxidation, respectively of quinol and ascorbate by peroxide, were observed during flow and collected by rapid recorder sweep (Yamazaki et al., 1960). The reactant usage for an overall experiment of the type reported by Yamazaki et al. (1960) was ~ 1 L (Borg and Elmore, 1967), thus limiting experiments to readily available, inexpensive compounds. Improved EPR cells of the mix-and-flow type shown in Figure 1 with smaller dead volumes and mixers placed as close as possible to the EPR cavity were reported by Borg (Borg, 1964a; Borg, 1964b). With a ~ 50 μL dead volume from mixer to the center of the EPR-active zone, Borg's fastest device allowed a dead time of several milliseconds. Detailed multi hyperfine line EPR spectra of short-lived tyrosyl radical intermediates arising from oxidation of a 10^{-2} M tyrosine solution were recorded using a rapid field sweep during a flow which lasted ~3 s. Even here the reactant usage was >25 mL in 3 s. Although electrically driven solenoids were used for stopping flow, their stopping times were limited, and they provided electrical pick-up interference. It would appear that positive displacement devices for providing stopped-flow shots were avoided because of their potential for creating high pressures leading to breakage of glassware and to stopping/starting microphonic transients. It was recognized in principle that a Q-band cavity could provide considerably higher EPR sensitivity with considerable smaller usage of sample if equipped with a sample tube of ~0.2 mm diameter (Borg and Elmore, 1967).

2. THE LOOP GAP RESONATOR BASED STOPPED-FLOW SYSTEM

In the 1980's the technology of small resonant structures, in contrast to large TE_{102} microwave cavities, had emerged. The loop-gap resonator (LGR) (see Chapter on Loop Gap Resonators) was notably developed at the National ESR Center (Froncisz and Hyde, 1982). Compared to a standard X-band cavity, the LGR had a greater filling factor, much smaller size, low Q, and relative insensitivity to dielectric losses. For small point samples the LGR could have a 50-fold sensitivity improvement over a standard rectangular cavity. With considerable isolation between magnetic and electric parts of the microwave field, the LGR was particularly useful for small static liquid samples, especially when T_1-sensitive dispersion methods were used that took advantage of their low Q (Froncisz et al., 1985). In 1987 Hubbell et al. (1987) reported a novel design for continuous and stopped-flow EPR based on a loop-gap resonator (LGR) shown in Figure 2. Compared to conventional metallized cavity resonators, this design had a very high filling factor, relatively low resonator quality factor Q, and good signal-to-noise ratio for small microliter (μL) sample volumes. The small size of the resonator made the very close location of the mixer to the resonator possible. The close location led to a small ~6 μL dead volume (including mixer volume plus volume to center of EPR active zone) and ~4 ms dead time at a total flow rate of 1.5 mL/s. The liquid delivery system (manufactured by Update Instruments, Madison, WI) used a positive displacement syringe ram with a programmable dynamic braking motor. This delivery system fed the highly efficient Wiskind grid mixer, whose purpose was to assure that the mixing was completed at the mixer so that time-zero for a chemical reaction started at the mixer. Constant flow velocity with high efficiency mixing is important for obtaining EPR signals at precisely controlled times after mixing and for inferring the existence of faster unresolved kinetics during the dead time. For kinetics at a single field approximately 30 μL of each reactant was used per shot, and for the 100 ms period of a rapid field scan the amount of each reactant consumed during flow with a 2.4 ms dead time was ~130 μL. In concept the LGR was well adapted for kinetic studies of biological materials, which are characterized by a high dielectric loss and are often limited in supply. Integration of the LGR-based stopped-flow system into an overall EPR spectrometer was similar to that shown below in Figure 8 for the DR-based system. This stopped-flow device was used by Shin et al. (1993) to follow the interaction of the channel-forming fragment of toxin colicin E1 with membranes. The fragment interacted with the membrane in two distinct steps which were monitored by site-directed spin labels at separate, distinct positions: (i) rapid adsorption to the surface on the time scale of 5 s; and (ii) slow, rate-limiting insertion of the hydrophobic central helices into the membrane interior on the time scale of 100 s.

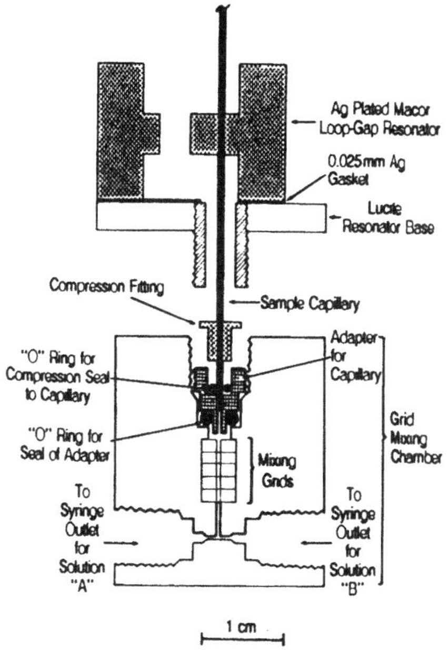

Figure 2. Figure showing the adaptation of the grid mixer to the loop gap resonator. The Lucite four-grid Wiskind mixer is commercially available from Update Instruments, Madison, WI. The sample capillary is sealed in place by O-ring fittings which are compressed as the mixer is screwed into the threaded extension from the resonator base. Figure reprinted from Hubbell et al. (1987) by permission of J. S. Hyde.

2.1 Application of LGR-Based Stopped-Flow EPR to Time-Resolved Spin Probe Oximetry

Our impetus for developing time-resolved spin probe oximetry was rapidly to measure oxygen consumption by cytochrome *c* oxidase, the enzyme which consumes the majority of oxygen for all aerobic organisms (Jiang et al., 1992). Cytochrome *c* oxidase takes four electrons from reduced ferrocytochrome *c* (cyt c^{++}), combines them in a sequential fashion with oxygen and protons to provide H_2O plus ferricytochrome *c* (cyt c^{+++}), and couples the resultant free energy derived from this reaction to creating a transmembrane proton chemical gradient.

$$4\text{cyt } c^{++} + 4H^{+} + O_2 \text{--->} 2H_2O + 4 \text{ cyt } c^{+++}$$

In probing cytochrome *c* oxidase kinetics under those limited turnover conditions where finite ratios of substrate (usually the cyt c^{++} reductant) to enzyme hold, researchers had generally followed the transient spectrophotometric changes of cyt c^{++}. We wanted to measure oxygen consumption in a similar time regime to the time regime in which cyt c^{++} consumption was measured. This would be done at a limited ratio of cyt c^{++} to cytochrome oxidase enzyme and also at a low, limited mole ratio of oxygen consumed to cytochrome oxidase enzyme.

The relaxation times of freely tumbling nitroxide spin probes such as CTPO (2,2,5,5-Tetramethyl-3-pyrrolin-1-oxyl-3-carboxamide; see Figure 3) are sensitive to oxygen concentration. Molecular collisions between spin probe and paramagnetic di-oxygen modulate the Heisenberg exchange between probe spin and the triplet oxygen molecule (Windrem and Plachy, 1980), thereby enhancing the magnetic T_1 and T_2 relaxation rates of the spin probe. In pilot time-resolved spin probe oximetry following the development of the low-Q X-band LGR, Froncisz et al. (1985) used T_1-sensitive dispersion (χ') spin probe oximetry to measure oxygen consumption as shown in Figure 3 by microliter-size volumes containing $\sim 10^3$ respiring cells. The oxygen-induced change in the field modulated adiabatic rapid-passage dispersion signal from CTPO was detected 90 ° out-of-phase with respect to 100 KHz field modulation. The 90 ° out-of-phase dispersion signal diminished with diminishing oxygen concentration, and the signal change was proportional to oxygen concentration at oxygen concentrations below 50 μM, i.e., the solution oxygen concentration in equilibrium with 20 % Air (Froncisz et al., 1985; Jiang et al., 1992).

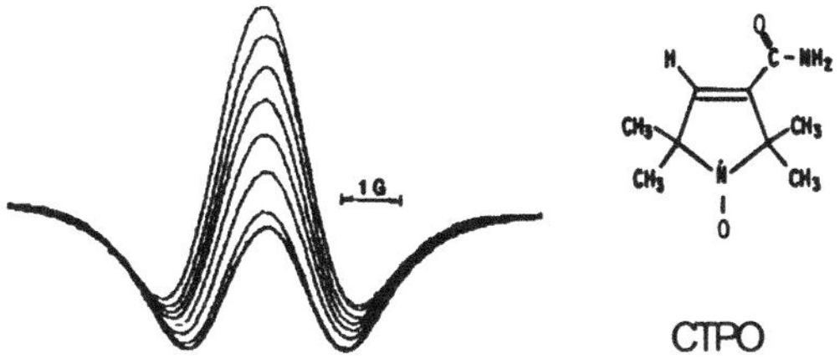

CTPO

Figure 3. Spectra showing the time course of the 90 ° out-of-phase dispersion signal from the central $m_I = 0$ nitrogen hyperfine line of CTPO over the course of O_2 consumption by Chinese hamster ovary cells. Scans were taken every two minutes, and the central feature diminished as the oxygen was consumed. Figure reprinted from Froncisz et al. (1985) by permission of J. S. Hyde.

The emergence of spin probe oximetry methodology (Lai et al., 1982; Froncisz et al., 1985), availability of stopped-flow EPR apparatus capable of rapid time resolution on small volumes (Hubbell et al., 1987), and our interest in structure and function of cytochrome oxidase (van Camp et al., 1978; Stevens et al., 1982; Mascarenhas et al., 1983; Fan et al., 1988) led us to develop time-resolved spin probe oximetry. The technique was applied to microliter volumes of purified mammalian cytochrome *c* oxidase enzyme at micromolar concentrations of enzyme. Continuously in time, micromolar oxygen concentration changes were followed where the amount of oxygen consumed became comparable with or ultimately even less than the amount of enzyme. The time-resolved spin probe oximetry method (Jiang et al., 1992) contrasted sharply with oxygen measuring methods such as polarography (Davies, 1962) or oxygen-dependent phosphorescence, both of which used mL (vs μL-sized) samples with time resolution limited to >100 ms by stirring (Vanderkooi et al., 1987; Wilson et al., 1988). In most polarographic oxygen monitoring experiments the amount of oxygen consumed was orders of magnitude larger than the amount of cytochrome *c* oxidase, unlike the case presented here.

In the kinetic case which is shown here, oxygen consumption was limited by the very availability of oxygen itself (i.e., the system became anaerobic) while cyt c^{++} remained in excess. Zeroth-order kinetic rate behavior (i.e., a constant rate of oxygen consumption) was observed until almost all the oxygen was exhausted as shown in Figure 4. (Zeroth order kinetics mean that another processes besides the binding of oxygen, such as internal electron transfer within the cytochrome oxidase, limited the rate of oxygen consumption.) The oxygen concentration where the rate diminished to half its maximal value was about 0.3 μM, and at that point the rate of oxygen consumption went exponentially to zero. Time-resolved spin-probe oximetry thus presented a novel method for studying oxygen-limited kinetics in the submicromolar oxygen regime where rapid kinetic methods had not been previously applied.

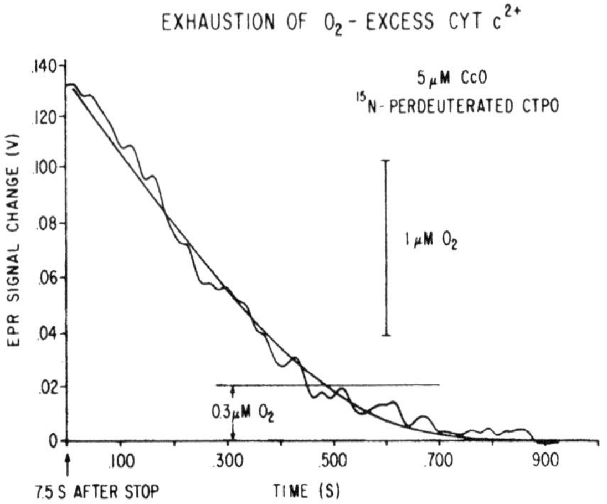

Figure 4. This trace provides the details of oxygen consumption by 5 μM cytochrome c oxidase in the presence of excess ferrocytochrome reductant after the point in time, which was about 7.5 s after mixing, where oxygen became rate limiting; concentration of CTPO was 200 μM. The line 0.3 μM above the baseline is the estimate of the point where the slope went to half its initial value and where oxygen itself became rate limiting. (Figure reprinted with permission from Jiang et al., 1992; Copyright 1992 American Chemical Society.)

2.2 Application of LGR-Based Stopped-Flow EPR to Time-Resolved Spin Trapping of Hydroxyl Radicals

Hydrogen peroxide in the presence of ferrous ion had long been known as a strong oxidizing agent (Fenton, 1894). Fenton's reagent (H_2O_2-Fe^{+2}) generated highly reactive oxidizing hydroxyl radicals (Buettner, 1987), and toxicity from oxidative stress has been attributed to the highly reactive hydroxyl radical (Halliwell and Gutteridge, 1989). The overall reaction: had been proposed as a major source of reactive, short-lived, toxic radicals, especially the $HO^{\bullet}$ that damage membranes and DNA (Tullius, 1988). $HO^{\bullet}$ radical can be identified by spin trapping since its spin-trap adduct shown in Figure 5 give a characteristic, stable EPR signal. The upper spectrum in Figure 5A is that of the spin-trapped DMPO-OH adduct following its reaction with $HO^{\bullet}$.

$$Fe^{2+} + H_2O_2 \text{ ----> } Fe^{3+} + HO^{-} + HO^{\bullet}$$

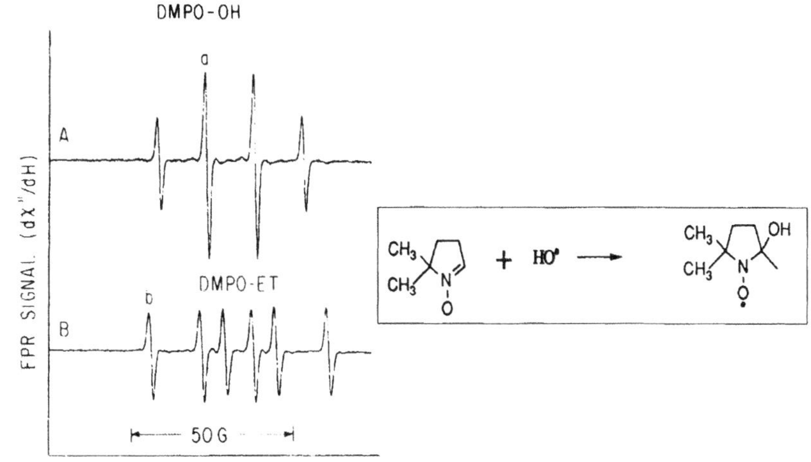

Figure 5. This figure provides traces of the spin trapped DMPO-OH and DMPO-ET, which are respectively the spin adducts of HO• and the ethanol a-radical as taken approximately 1 min after mixing reactants. Spectrum **A** resulted from 1:1 mixing of a solution containing 40 mM DMPO, 400 μM EDTA, 1.2 mM H_2O_2, in 40 mM pH 7.4 phosphate buffer, 150 mM KCl with a second solution containing 200 μM ferrous ammonium sulfate in 150 mM KCl. This spectrum contained about 25 μM DMPO-OH spin adduct. Spectrum **B** is from the spin trapped DMPO-ET adduct; reactants to create DMPO-ET were the same as those of DMPO-OH except that the solution after mixing contained 1.65 M ethanol. This spectrum contained 28 μM DMPO-ET. (Figure reprinted with permission from Jiang et al., 1993; Copyright 1993 American Chemical Society.)

When they are trapped, other reactive secondary radicals also instigated by Fenton chemistry give their own characteristic signatures as shown for DMPO-ET in Figure 5B, where DMPO-ET is the DMPO spin trapped adduct of the ethanol α-radical (Yamazaki and Piette, 1990). Fe(II)-EDTA in the presence of H_2O_2 provided a source of hydroxyl radicals whose rapid kinetic creation ended in less than a second [Jiang, 1993 #12]. These kinetic changes were followed by observing the change of the derivative peak of the appropriate DMPO spin adduct. Thus time resolved spin trapping using the LGR system provided evidence for rapid trapping of Fenton radicals such as shown in Figure 6A. With ability to directly detect kinetics of radical creation in a time regime not previously studied, we measured the initial rates of radical production and avoided the complications of later chain and side reactions (with which Fenton chemistry is replete). The initial rate of trapped $HO^•$ radical production was linearly dependent on both the initial Fe(II)-EDTA concentration and on the initial H_2O_2 concentration so that a second-order rate constant for production of trapped $HO^•$ was easily determined (Jiang et al., 1993). A comparison of the DMPO-OH trapping (trace 6A on a 1 s time scale) with DMPO-ET (trace 6B on a 3.5 s time scale) showed that the build-up of DMPO-ET took longer than that of DMPO-OH and did not simply parallel the production of trapped hydroxyl radical. In summary, stopped-flow EPR carried out with the LGR-based stopped-flow system gave time resolution to follow in well under 100 ms the initial kinetics of spin trapped adducts at micromolar concentrations and in microliter samples.

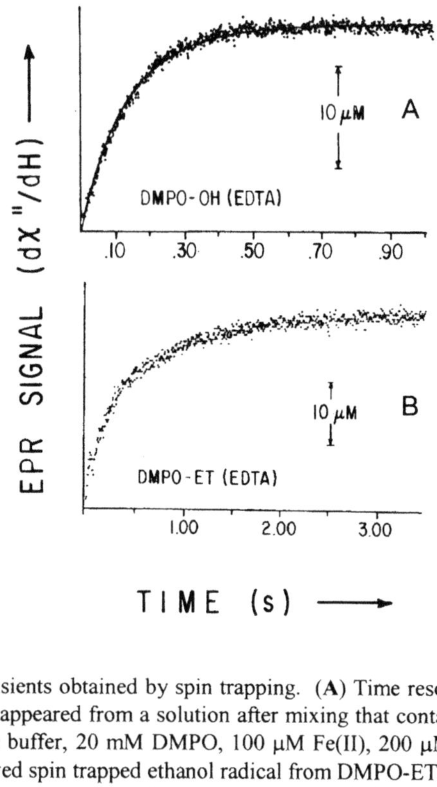

Figure 6. Kinetic transients obtained by spin trapping. (**A**) Time resolved spin trapped $HO^{\bullet}$ from DMPO-OH as it appeared from a solution after mixing that contained 150 mM KCl, 20 mM pH 7.4 phosphate buffer, 20 mM DMPO, 100 μM Fe(II), 200 μM EDTA, and 600 μM H_2O_2. (**B**) Time resolved spin trapped ethanol radical from DMPO-ET as it appeared over 3.5 s. Concentrations of reactants were as in A except that the solution contained 1.65 M ethanol. (Figure reprinted with permission from Jiang et al., 1993; Copyright 1993 American Chemical Society.)

3. DIELECTRIC RESONATOR-BASED STOPPED-FLOW EPR

3.1 Technical Development

Richtmyer (Richtmyer, 1939) first pointed out that unmetallized dielectric objects can function as resonators. By the late 1970's low loss barium tetra-titanate materials with high temperature stability in resonance frequency were becoming available (Cohn, 1968; Plurde et al., 1975; Plurde and Ren, 1981) A disk-shaped dielectric resonator (DR) operating in the $TE_{01\delta}$ mode will have a vertical microwave magnetic field (B_1) extending up through its center where the sample is located and will confine the electric field (E_1) within the dielectric disk. Several different versions of DR-based EPR resonators were reported (Dykstra and Markham, 1986; Walsh and Rupp, 1986; Bromberg and Chan, 1992), most notably the EPR application of Dykstra and Markham (1986) which, when applied to small point and liquid samples, yielded more than two orders of magnitude higher EPR signals than a standard TE_{102} EPR cavity.

The DRs which formed the heart of the microwave resonant probe, developed by (Sienkiewicz et al., 1994) and shown in Figure 7, were loss-free, high dielectric ceramics made of zirconium titanate ($\varepsilon = 30$; MuRata Inc., DRT series). Incorporating small size and a high sample filling factor, they were toroids, of 6 mm o.d. and 2.5 mm height with a 1-2 mm sample hole. The DRs were enclosed in a cylindrical shield consisting of silvered Rexolite 1422 plastic (C-LEC Plastic, Inc., Beverly, NJ) with top and bottom lids of copper foil and a top and bottom sample tube guidance provided by brass inserts. (We have recently discovered that the Q of the resonant structure can be increased several fold by using high conductivity silver foil as an outer shield rather than silver paint; with this procedure adequate field modulation is admitted via fine slits around the circumference of the shield.) Our primary resonator consisted of two stacked ceramic DRs separated by a (nonsilvered) spacer of low dielectric Rexolite; varying the size of the spacer varied the resonator frequency and provided a larger EPR active zone. We called this DR configuration a double-stacked DR. A critical feature of any resonant structure for EPR is its coupling to the microwave transmission line. Stopped-flow experiments involve rapid sample movement through the sample capillary and concomitant vibrations; thus the coupling and matching structures needed to be remote from the sample capillary to avoid microphonic disturbance of them. To couple magnetically to the $TE_{01\delta}$ mode of the DR, we used an antenna loop oriented with its plane perpendicular to axis of the cylindrical dielectric shield. The coupling loop was connected by a short section of semi-rigid coax to a modified SMA connector, which in turn formed the center of a T-shaped structure whose three arms were the incoming microwave transmission line, the coupling loop, and our coaxial tuner. The latter tuner, based on a smoothly adjustable short, modified the coupling between resonator and microwave line. The resonant structure and the tuning drive were rigidly fastened by a plastic (Delrin) support to standard X-band wave guide so that the probe could directly replace a typical microwave cavity in a conventional EPR spectrometer. Separate 100 KHz external modulation coils (Medical Advances) were initially used, and more recently, 100 KHz modulation coils were incorporated into the Delrin support structure.

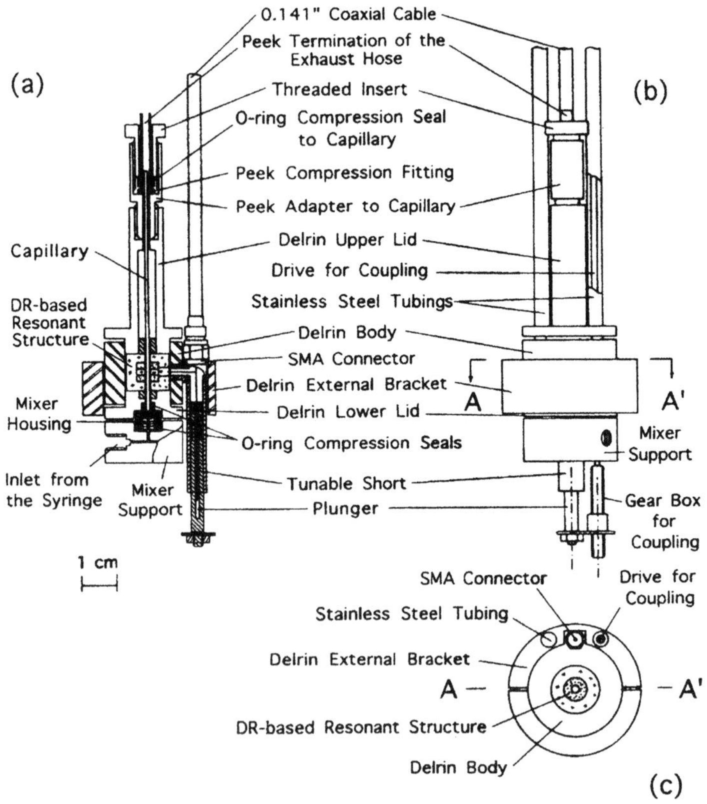

Figure 7. DR-based EPR probe for stopped-flow showing: (a) The cross-sectional view of the Delrin body and of the DR-based resonant structure including the coupler. (b) Front view of the EPR probe showing the supporting stainless steel tubing, the microwave line (0.141 in. coaxial cable), the Delrin bracket, and the gearbox for tuning. (c) Horizontal cross section (A - A') with the location of the supporting stainless steel tubings and of the central part of the SMA connector. Figure reprinted with permission from Sienkiewicz et al. (1994).

Our initial EPR probe contained a commercially available Wiskind grid mixer and mixer support (Update Instrument, Madison, WI,) similar in design, dead volume, and location to that used with the LGR system of Hubbell et al. (1987). The liquid mixer and mixer support system were attached to the resonator body by brass screws, and in later versions, coils for coolant were integrated into the mixer support. The overall schematic of the system which also includes rapid field scan coils is shown in Figure 8.

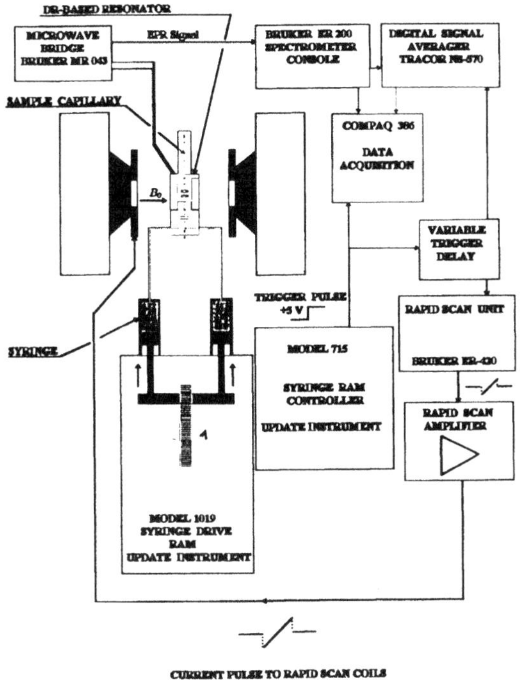

Figure 8. Schematic of the flow and stopped-flow probe interfaced to a Bruker ER-200 D-SRC EPR spectrometer and to a data collection system based on Scientific Software Services Systems software. Figure reprinted from Sienkiewicz et al. (1999) with permission from Elsevier Science Publishers.

The unloaded Q of the DR probe was ~1800 and the loaded Q with an aqueous 300 μM TEMPO sample (0.6 mm i.d. 0.84 mm o.d. sample tube, EPR active volume ~2 μL) was ~1000. The EPR signal from the double stacked DR was compared with that from a standard Bruker TE_{102} cavity with the same sample, at the same incident non-saturating microwave power, and with the same field modulation (Jaworski et al., 1997). For a point DPPH sample the signal from the DR was fifty-fold larger and for a TEMPO line sample 3 cm long the signal from the DR was thirty-fold larger.

Calculations of Jaworski et al. (1997) showed that the filling factor for the DR was two orders of magnitude larger than for the TE_{102} cavity because the DR highly concentrated the microwave magnetic field, B_1, on the sample. In a comparison of saturation behavior of TEMPO in DR devices and the LGR as well to that in the TE_{102} cavity (Sienkiewicz et al., 1994), the TEMPO EPR signal reached its maximum amplitude at about 5 mW in the DR and LGR and at about 100 mW in the TE_{102}. The implication is that the B_1 field in the former two devices was 4-5 times larger than in the TE_{102} at the same microwave power. The maximum amplitude saturable TEMPO signal at ~5 mW from the DR was about 1.7 times bigger than the maximum amplitude signal from the TE_{102} at ~100 mW. Because of its higher Q and comparable filling factor, the signal-to-noise in the absorption mode from the DR was ~25-50 % higher than that from the LGR. (See Table I of Sienkiewicz et al. (1994).) Under such absorption conditions, detector noise is the limiting noise source. On the other hand the signal-to-noise for high power dispersion ($d\chi'/dH$) EPR was better for the LGR whose low Q makes it a less sensitive demodulator of klystron FM noise.

For small liquid samples of the size and shape used in stopped-flow measurements, the EPR sensitivity and ease of DR operation in the commonly used AFC-locked absorption mode represented an advantage for the DR. The DR isolated the sample from the electric part of the microwave field and concentrated the microwave magnetic field on the sample. Compared to the LGR, which has fringing electric fields near its gap, such a DR offered a better separation of E_1 away from the sample position. The resonator frequency shift for the DR upon insertion of the water-filled EPR tube was only 3 MHz vs >100 MHz for the LGR; this was practical proof of the isolation of the microwave electric field from the sample. This isolation contributed to the insensitivity of the DR to flow and stopped-flow-induced sample tube movements. We consider the LGR-based stopped-flow system of Hubbell et al. (1987) as the seminal step in mini resonator design for flow and stopped-flow, but in relation to our aim of elucidating kinetics on the sub-10 ms time scale, we concluded that that the DR-based system was less sensitive to flow-induced transients. (See Figure 6 of Sienkiewicz et al. (1994).)

In summary, the major benefits of the DR system were: 1) It incorporated a small, high sensitivity resonator system that was insensitive to stopped-flow induced noisy transients. 2) The DR system was cheap, robust, and easily assembled. 3) It contained a microwave coupling scheme based on an adjustable short and coupling loop that provided finesse in tuning and freedom from microphonics. It is this system which we subsequently used in our first application to protein folding (Qu et al., 1997), whose frequency and loss characteristics we have modeled (Jaworski et al., 1997), and which we have modified with rapid field scan in order to obtain more complete information on overall spectra (Sienkiewicz et al., 1999).

3.2 Application of DR-Based Stopped-Flow EPR to the Folding of Cysteine-Specifically Labeled Cytochrome *c*

Cytochrome *c* is a globular heme protein of 12.5 kDa molecular weight which has been a paradigm subject for folding. Yeast iso-1-cytochrome *c* (iso-1-cyt *c*), which is the explicit system we study, has a known crystal structure, shown by the ribbon diagram of Figure 9.

There are three helices, termed the N-terminal, C-terminal, and 60's helices (Louie and Brayer, 1990), and four omega loops, A, B, C, and D (Leszczynski and Rose, 1986; Fetrow et al., 1989). Using stopped-flow methods, the folding of cytochromes was first followed from optical changes at the heme and from heme-induced tryptophan fluorescence quenching emanating from the sole Trp59. Fluorescence quenching reported overall compacting of the cytochrome structure (Tsong, 1976; Nall and Landers, 1981; Brems and Stellwagen, 1983).

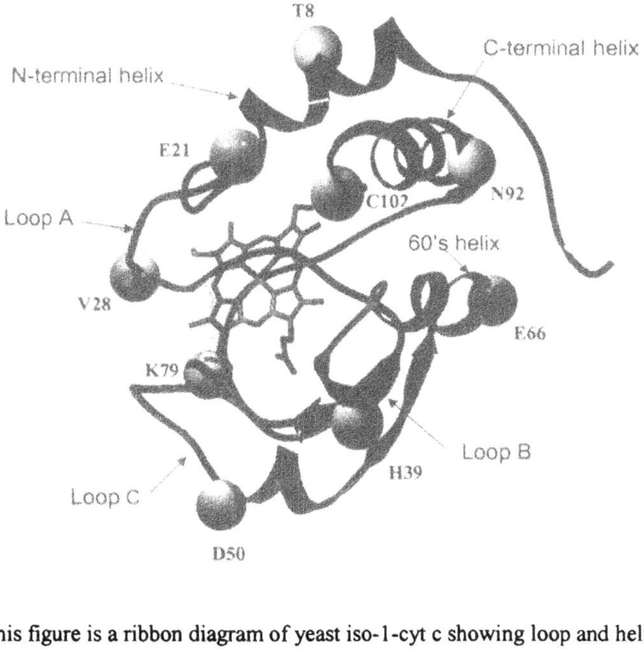

Figure 9. This figure is a ribbon diagram of yeast iso-1-cyt c showing loop and helical regions and the positions where we have recently (DeWeerd et al., 2001) created cysteine-directed labeling positions. (Note loop D happens to be hidden in this presentation.) C102 is the naturally occurring cysteine that was initially labeled. Figure reprinted with permission from DeWeerd et al. (2001); Copyright 2001 American Chemical Society.

NMR-monitored deuterium-hydrogen exchange following quenched-flow provided evidence for folding-induced formation of stable secondary helical structures (Roder et al., 1988). These stopped-flow and quenched flow techniques using standard rapid mixers routinely had millisecond or longer dead times. There were at least three exponential phases observable by all these techniques (Nall and Landers, 1981; Brems and Stellwagen, 1983; Colón et al., 1996): a fast phase in the 5-100 ms range, an intermediate phase in the 0.5-1 s range, and a minor slow phase in the 5-25 s range. NMR-monitored deuterium-hydrogen exchange of amide protons subsequent to a quenched-flow procedure indicated that the 5-100 ms phase represented nucleation-collapse of the N-C terminal helices (Roder et al., 1988; Elöve et al., 1992; Elöve et al., 1994; Englander et al., 1998). The intermediate 0.5-1 s phase was attributed to states trapped with improper heme-histidine ligation that convert into properly folded protein following dissociation of the non-native histidine iron ligands (Sosnick et al., 1994; Colón et al., 1997). The slowest 5-25 s phase is most likely due to cis/trans isomerization of prolyl peptide bonds (Osterhout and Nall, 1985). These previous findings have in many cases not provided information on the particular sites where we have recently attached labels at position in Figure 9 above and Table 1 below (DeWeerd et al., 2001), and the information which we have obtained from these labelling sites goes beyond just amplifying pre-existing information.

There also were changes occurring within the dead time of a standard rapid mixer, too early explicitly to be measured. What was initially called a "burst" stage (≤5 ms) of folding was suggested to be formation of a compact state with incompletely defined helical structure (Creighton, 1994; Colón et al., 1996; Sosnick et al., 1997). The explicit elucidation of very early submillisecond folding/compaction has been the subject of intense investigation (Winkler and Gray, 1998). Folding/compaction may start as early as 50 μs after initiation of folding, although specifically where and how is a subject for our own research. A question fundamental to the protein folding process that we aim to answer by probing at numerous spin labeled sites is whether early folding is simply a response of a polymer to change in solvent quality (Englander et al., 1998) or whether it represents early folding events with specific energy barriers, pathways, and activation energies (Shastry and Roder, 1998; Shastry et al., 1998).

3.3 Iso-1-Cyt *c* Labeled at its Naturally Occurring Cys102 – Characterization and Stopped-Flow

Following the development of dielectric resonator-based stopped flow EPR (Sienkiewicz et al., 1994), we monitored the kinetics of protein folding as the folding altered the mobility of a cysteine-specific spin label, methanethiosulfonate spin label, MTSSL in Figure 10 (Qu et al., 1997). The spin label for this initial study was attached at the sole cysteine (Cys102) naturally occurring within the C-terminal helix of yeast iso-1-cyt *c*. We called the spin labeled wild type iso-1-cyt *c* C102-SL. The spin label attached to unfolded protein demonstrated sub-nanosecond mobility and

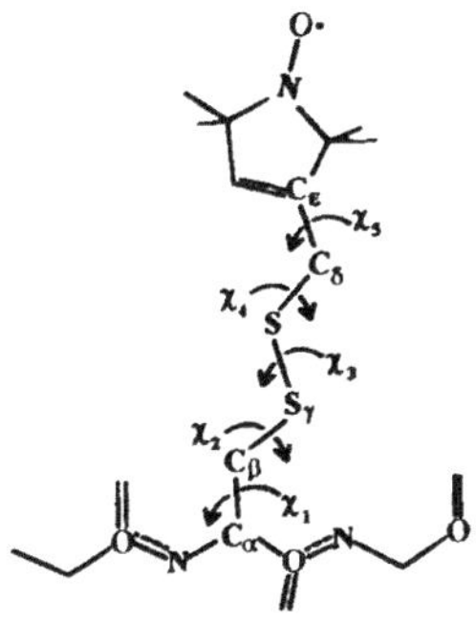

Figure 10. This figure shows the MTSSL spin-label attached to the cysteine sulfur via a disulfide linkage. The angles χ are angles about which rotation may occur. In the folded protein it appears that free motion only occurs about χ_4 and χ_5.

sharp, intense derivative EPR features. The label on the folded protein was more encumbered by its environment and showed broader and less intense derivative EPR features indicating nanosecond or longer tumbling times. (The tumbling time for cytochrome *c* itself in aqueous buffer at room temperature is ~3 ns.) Figure 11 illustrates the sharpening of the spin label derivative EPR signal under denaturing conditions and compares the signals of folded and chemically or thermally denatured C102-SL.

Stopped-flow EPR of C102-SL revealed a mono-exponential, guanidinium-induced unfolding process; a ~20 ms unfolding time occurred in the presence of 2 M guanidinium (GdnHCl) denaturant, as shown in Figure 12. When this unfolding was compared by stopped-flow EPR of the spin label reporter and by stopped-flow UV-Vis of the heme, a nearly identical single-exponential time dependence was measured by both spin label and UV-Vis spectroscopy (Qu et al., 1997). Unfolding has generally been found to exhibit this global, single exponential unfolding, reported similarly by all types of spectroscopy monitoring different sites.

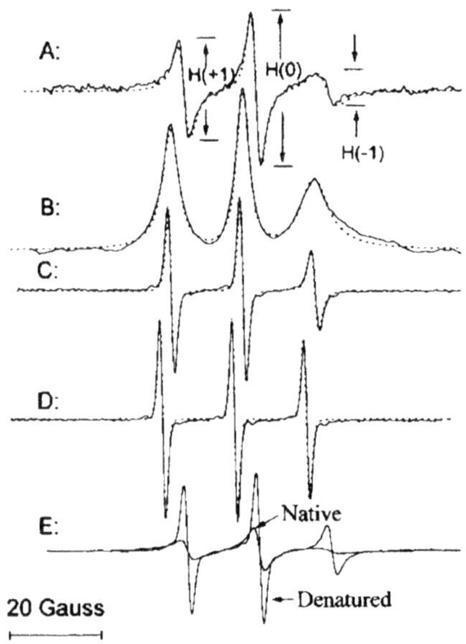

Figure 11. The EPR spectra of spin labeled yeast iso-1-cytochrome c in 0.1 M pH 5 sodium acetate buffer. (**A**) Folded protein at 15 °C. (**B**) First integral of A. (**C**) Denatured protein in 2 M GdnHCl at 15 °C. (**D**) C102-SL at 75 °C. (**E**) Comparison of native and GdnHCl-denatured C102-SL at the same system gain. Figure reprinted with permission from Qu et al. (1997); Copyright 1997 American Chemical Society.

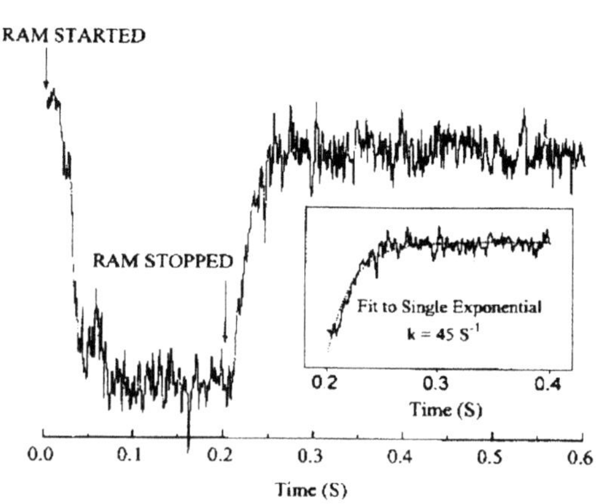

Figure 12. The kinetic unfolding transient EPR signal following 1:1 mixing of 80 μM C102-SL in 0.1 M phosphate buffer with 4 M GdnHCl denaturant pH 6.5, 2 °C. The central H(0) EPR feature was monitored. The data presented is the result of 2 summed transients. Inset shows fit of kinetic transient to single exponential with unfolding rate constant $k = 45\ s^{-1}$. This experiment required less than 150 μL of 80 μM spin labeled protein. Figure reprinted with permission from Qu et al. (1997); Copyright 1997 American Chemical Society.

The more complex refolding kinetics of our labeled cytochrome were studied by stopped-flow EPR (Figure 13A at pH 6.5). The spin probe signal at pH 6.5 (and 5.0) showed a fast kinetic process (~100 ms times or faster at 2 °C), compatible with the time range over which hydrogen/deuterium exchange sensed early helix formation of the C & N helices (Roder et al., 1988; Elöve et al., 1994; Sosnick et al., 1994). At pH 6.5 an additional slower kinetic phase (~0.5 s) was reported by the spin probe attached at C102. Heme-ligation-sensitive UV-Vis absorption spectroscopy exclusively indicated this slower (0.5-1 s) folding (Figure 13B), a folding attributed to recovery from the kinetic trap of pH-dependent heme-histidine mis-ligation (Sosnick et al., 1994; Colón et al., 1997).

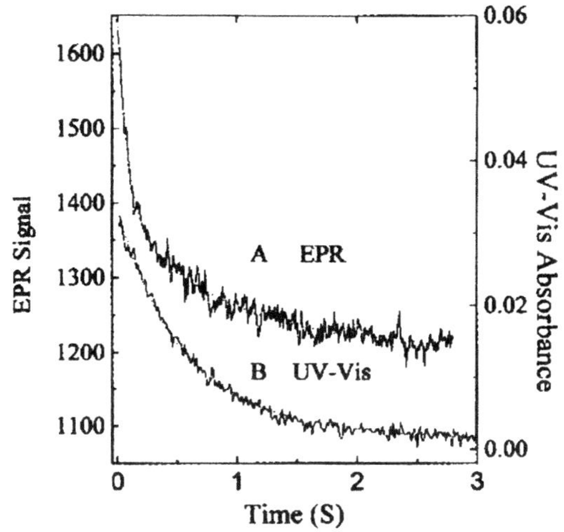

Figure 13. This figures provides a comparison of the EPR and UV-Vis stopped-flow kinetics of protein refolding where initially 80 μM unfolded C102-SL in pH 6.5, 1.8 M GdnHCl was mixed 1:1 with pH 6.5, 0.1 M phosphate buffer at 2 °C. (**A**) The EPR trace exhibiting biphasic behavior with rates of 12 and 1.2 s^{-1}. (**B**) The UV-Vis kinetic trace obtained at 407 nm exhibiting a slower biphasic behavior with rates of 1.8 and 0.2 s^{-1}. Figure reprinted with permission from Qu et al. (1997); Copyright 1997 American Chemical Society.

The work of Qu et al. (1997) showed evidence of a faster probe immobilization and an incipient "burst" of folding that occurred in a time less than the ~7 ms dead time of the 1997-vintage grid-mixer stopped-flow apparatus. To resolve such a faster refolding process, a micro ball-mixer with submicroliter dead volume was subsequently integrated with the dielectric resonator (Grigoryants et al., 2000). Thermal melting characterization made us aware that spin labeling the C102 position had led to a 20 °C lower melting temperature than that of unlabeled wild type protein (Qu et al., 1997). The location of the C102 sulfur prior to labeling is a region of hydrophobic packing (Louie and Brayer, 1990), and so labeling C102 from within that region had destabilized the protein, probably by distorting the hydrophobic packing.

We had noted that the utility of spin labeling had been markedly improved through cysteine-specific spin labeling via methanethiosulfonate (Hubbell and Altenbach, 1994; Hubbell et al., 1998; Hubbell et al., 2000), and our next goal was to attach MTSSL in new cysteine-mutated sites. The new labeling sites (See Figure 9 and Table 1) were selected as external, hydrophilic, mutation-tolerant sites. These sites, when mutated to cysteine and spin labeled, were less perturbing to folding than the naturally occurring C102. Furthermore, labeling at numerous non-perturbing locations presented the possibility of following folding at numerous locations, not just C102. In summary, the initial kinetic work on C102-SL (Qu et al., 1997) put the folding study of spin labeled protein into the framework of other recent folding measurements. It pointed toward a systematic folding study with cysteine-directed mutants, and it pointed toward technical improvements in mixer-resonator design.

Table 1. Summary of amino acids that have been mutated to cysteine

Residue[a]	Labeled Derivative	2°Struct[b]	Substitutions[c]
Thr8	T8C-SL	N-Helix	KTNRDS
Glu21	E21C-SL	Loop A	EDQGNAK
Val28	V28C-SL	Loop A	KTVIQAE
His39	H39C-SL	Loop B	KQTHAV
Asp50	D50C-SL	Loop C	TDEANK
Glu66	E66C-SL	60's helix	EADNGQKS
Lys79	K79C-SL	Loop D	TK
Asn92	D92-CL	C-helix	ANQGEVKDTL

[a]Name and number of the residue that was replaced in the C102S protein, where the wild-type Cys102 is replaced by Ser; C102S serves as wild type both here and in yeast, and *it cannot be labeled at C102.* [b]The secondary structure in which the residue is found. [c]These are the single letter codes for other amino acids found at this position in 106 homologous eukaryotic cytochromes *c*. Table reprinted with permission from DeWeerd et al. (2001); Copyright 2001 American Chemical Society.

3.3.1 Variable Velocity Liquid Flow EPR Applied to Submillisecond Protein Folding

A schematic of Grigoryants' elegant capillary mixer intimately connected to a mini EPR probe is provided in Figure 14. Liquid to be mixed flowed from two inlet tubes, through a narrow annular slit, past a platinum ball-mixer, and into the outlet tube which immediately entered the DR. With this device we followed the submillisecond refolding kinetics of iso-1-cyt *c* labeled at its at naturally occurring C102.

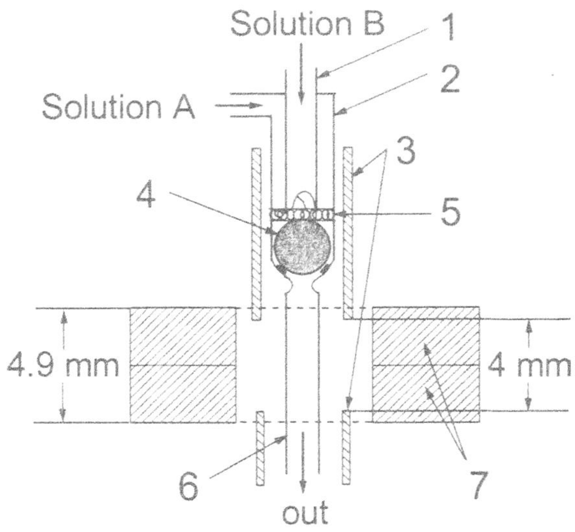

Figure 14. Schematic of the prototype mini ball mixer integrated to a 9.5 GHz dielectric resonator. The system has a dead volume ~0.5 μL and ~150 μs delivery time to center of the observation zone with a flow velocity of 3300 μL/s. Solutions A and B are two solutions to be mixed. Parts are as follows: (1) 0.3 mm i.d. inlet capillary; (2) 0.6 mm i.d. inlet capillary; (3) 0.9 mm i.d. silver shield tubes; (4) 0.57 mm Pt sphere/ball mixer; (5) 0.1 mm frit made from 0.076 mm Pt wire; (6) 0.4 mm i.d. outlet capillary; (7) two ceramic toroids of the dielectric resonator. The height of the dielectric resonator structure is 4.9 mm, and the height of the EPR-active zone is 4 mm. Figure reprinted from Grigoryants et al. (2000) with permission from the Biophysical Society.

By changing the programmed flow velocity, we changed the delivering time of mixed liquids from the mixing point at the ball mixer to the zone of observation (i.e., the dielectric resonator). It was necessary first to calibrate the time resolution of the instrument; the calibration process also led to an empirical estimate of the dead volume and gave insight into the mixing efficiency. The timing calibration of the instrument was performed by using the rapid destruction of TEMPO (2,2,6,6-tetramethyl-4-piperidinol) by sodium dithionite, a reaction which was initially determined by stopped-flow EPR to exhibit fast linear decay. The fast linear decay time meant that during flow there was a meaningful and detectable change in the EPR signal as a function of the flow velocity. For the TEMPO-dithionite reaction the decay of signal intensity depended linearly on the delivery (dead) time between mixing and the center of the EPR-active zone and inversely upon the flow velocity. In Figure 15 we present a plot of the EPR signal amplitude obtained during flow versus the inverse of the flow velocity. This plot was also made versus the delivery time which was the age of the sample computed from the linearly decaying amplitude of the EPR signal (and not from a pre-conceived concept of dead volume). At all but the slowest velocity the signal amplitude scaled linearly with the inverse of flow velocity. The implication of this linear scaling was that the actual time for turbulent mixing in the narrow annular region near the ball (Regenfuss et al., 1985) was much less than the dead time for the sample to flow from mixer to the EPR-active volume, at all but the slowest velocity. This meant that the time-zero for the start of folding is set at the mixer. Thus the subsequent time for flow from the mixer to the center of the EPR-active zone is the time for folding kinetics and not for a complex compendium of solution mixing and folding kinetics. The dead volume obtained from the decay rate of the calibrating chemical reaction and the known fluid flow rate was about 0.45 μL (Grigoryants et al., 2000), a volume which was virtually identical to the liquid volume from the middle of the ball to the middle of the DR. At the slowest velocity (0.2 μL/ms) and longest delivery times (>2 ms) there was evidence for inefficient mixing since the signal during flow deviated in a positive direction from linear decay. That deviation implied that at low fluid velocity the time for turbulent mixing was not much less than the time for the sample to flow from the mixing element to the EPR-active volume. Fortunately, kinetics in the >2 ms time regime can be independently obtained by the stopped-flow method with the velocity before stopping >0.2 μL/ms. The limitation on the fastest flow was the high pressure needed to achieve that flow combined with the stress tolerance of our thick-walled (presently 4.6 mm i.d., 30 mm o.d.) glass driver syringes; we refrained from flowing at a faster rate than 3 mL/s.

We have so far used solutions whose viscosity is about that of water (0.01 poise). Conceivably more viscous solvents than the solutions used here could lead to a more inefficient, less turbulent mixing and to the need for higher flow velocity to induce efficient, more turbulent mixing. A Reynolds number approach to turbulence (see Ch. 41 in Feynman et al. (1964)) indicates that turbulence increases with flow velocity and decreases with viscosity. With higher flow velocity would come the higher RAM pressures which could endanger the glass driver syringes.

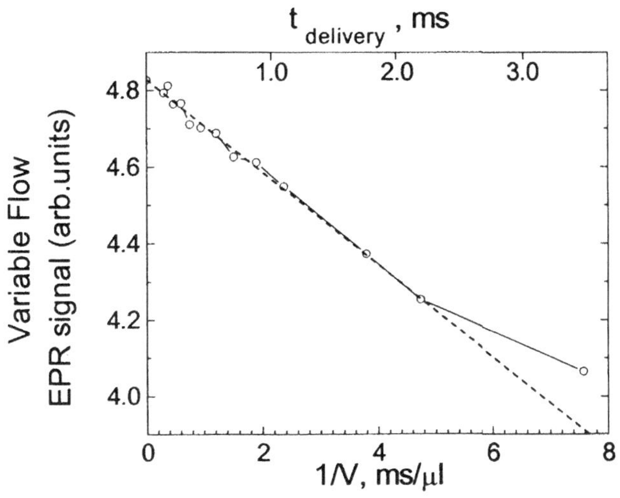

Figure 15. Presents the velocity flow/time calibration curve for our micro ball mixer. We show the dependence of the intensity of the continuously flowing EPR signal upon the reciprocal of the flow velocity for the mixing of 2 mM TEMPO with 30 mM Na-dithionite contained in 0.1 M KPO_4 buffer, and we show the time scale that we calibrated from the linear decay of the dithionite-TEMPO reaction for delivery of solutions from mixing to the center of the EPR-active zone. Evidence for inefficient mixing was observed at the slowest flow rate (i.e., flow velocity <0.20 μL/ms or reciprocal velocity >5 ms/μL). Figure reprinted from Grigoryants et al. (2000) with permission from the Biophysical Society.

This "Variation of Flow Velocity" method here allowed us initially to reach a kinetic resolution time of 150 μs (and most recently, 65 μs), two orders faster than the former stopped-flow method (Qu et al., 1997). Direct evidence was provided (Figure 16) for a burst of folding or protein compacting at C102 that occurred within 150 μs of mixing at 20 °C and within 500 μs of mixing at 7 °C. There clearly was a temperature dependence to this burst, implying an activation energy to the early submillisecond folding process at C102-SL. The reactant usage for the overall set of experiments which involved separate measurements at different flow velocities was 2 mL of 400 μM spin labeled C102SL. The ball-mixer device also served nicely as a highly efficient stopped-flow probe with total reactant usage of about 20 μL/shot. Replacing the grid mixer enabled us in stopped-flow mode to obtain folding times at room temperature below 10 ms as shown in Figure 17. The ball mixer device is the first ever created to give submillisecond time resolution with well-defined dead times in EPR flow mode measurements, and the technology opens a new area to biophysics for the kinetic study of radicals. Development to decrease the dead volume further, to computerize the system so that the velocity may be continuously swept, and to obtain overall rapidly scanned spectra during flow is underway.

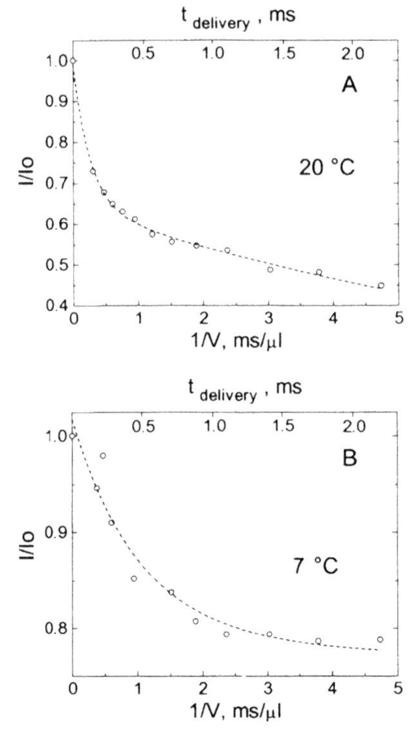

Figure 16. (**A**) shows the decreasing amplitude of the EPR signal at 20 °C represented by the fractional signal intensity I/I_o (shown with open circles) of the central EPR feature. 0.4 mM spin labeled C102-SL in 0.8 M GdnHCl was mixed 1:1 with dilute 0.1 M Na acetate buffer to obtain a 0.4 M GdnHCl concentration in which the protein was ~70 % refolded. I_o was the initial unfolded derivative EPR signal amplitude, and I was the time-dependent signal amplitude as folding progressed. The decrease of the central EPR feature from folding protein is shown as a function of the inverse of flow velocity (lower horizontal axis) and of the calibrated delivering time (upper horizontal axis) between the mixer and the center of the observation EPR zone. The fit (dashed lines) at 20 °C indicates the presence of two components of the folding processes, one with a 0.12 ± 0.02 ms exponential decay and the other with a 6.2 ± 0.8 ms exponential decay. The former had an amplitude which was ~37 % of the signal change, and the latter had an amplitude which was ~63 % of the signal change. (**B**) Shows the decreasing fractional amplitude taken at 7 °C, where there was only one component in the 0.1 to 2 ms range whose decay time was 0.5 ± 0.05 ms; this phase accounted for ~25 % of the overall signal change. Figure reprinted from Grigoryants et al. (2000) with permission from the Biophysical Society.

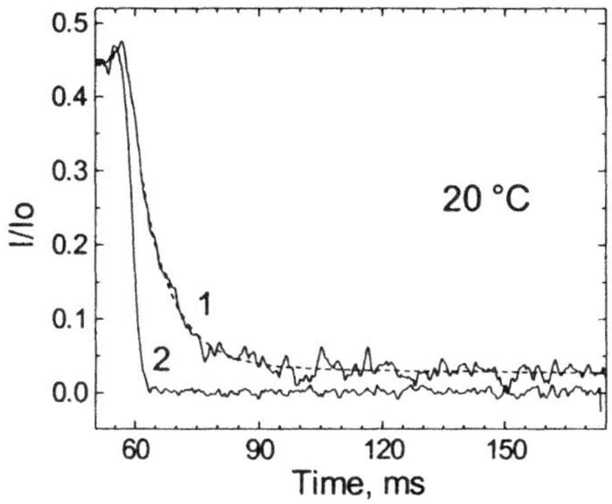

Figure 17. This figure shows the stopped-flow kinetics (curve 1) measured at 20 °C represented by the fractional signal intensity I/I_o under the same solution mixing conditions for C102-SL as in Figure 16A. In this case there was a rapid transient showing a time decay of 7 ± 2 ms, consistent with the slow 6 ms process measured by flow EPR in Figure 16A. Because there is a finite fast braking time for the flow, we also provide the profile of a very fast reaction (curve 2) of TEMPO with dithionite (1 mM TEMPO + 1 M dithionite) which represents the apparatus slowing function. The experimental apparatus slowing function is shown because we rely on RAM braking and on fluid friction in our narrow tubes rather than a stopping syringe to stop flow. Figure reprinted from Grigoryants et al. (2000) with permission from the Biophysical Society.

3.3.2 EPR-detected Folding Kinetics of Externally Located Cysteine-Directed Spin-Labeled Mutants of Iso-1-Cytochrome *c*

We next systematically probed protein folding in yeast iso-1-cyt *c* at cysteine-directed spin-labeled locations. The locations studied were not previously directly probed by other techniques, and we observed them on a time scale stretching from 50 microseconds to seconds. The following mutation-tolerant, externally located cysteine labeling sites shown in Fig. 9 above and Table 1 below were chosen (in helices: T8C, E66C, N92C; in loops: E21C, V28C, H39C, D50C, K79C). These externally located spin labels generally retained high local mobility upon the protein to which they were attached. In contrast to labeling at naturally occurring C102, these did not destabilize the fold or perturb the thermal melting temperature (DeWeerd et al., 2001). These mutant yeast cytochromes were created in our lab by our application of recombinant DNA technology. Using a high yield heterologous *E. coli* plasmid system (Pollock et al., 1998; Morar et al., 1999), we obtained ~100 mg quantities of these mutant cytochromes for kinetic work.

Dilution of denaturant (i.e., GdnHCl) induced folding, and the folding caused a kinetic change in the spin label EPR signal as folding altered the motion of the spin label. Under folding conditions, including the presence of imidazole to eliminate kinetic trapping due to heme-misligation, a phase of folding on the 20-30 ms time scale at room temperature was everywhere found, as shown in Figure 18. The time constants and amplitudes of folding phases for all spin labeled derivatives were given in Table 2 of DeWeerd et al. (2001). The N-C helical regions, which include T8C-SL and N92C-SL, have been identified by hydrogen exchange measurements as the location of early helix formation occurring on this time scale (Roder et al., 1988; Elöve et al., 1994; Englander et al., 1998). However, *all* regions of the protein, not just the N & C locale, experienced the folding phase in the 20-30 ms time window as reported by spin labels.

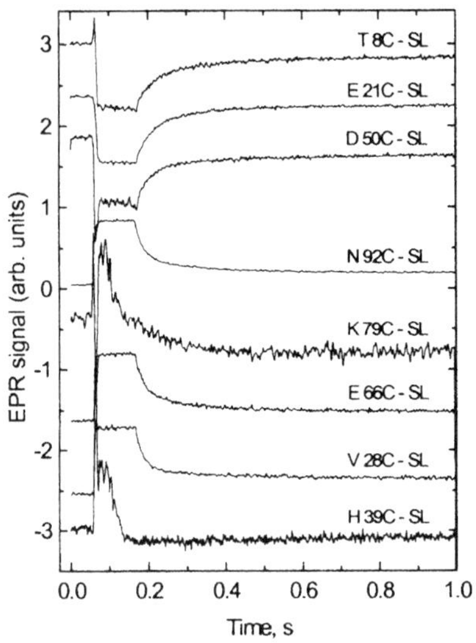

Figure 18. This figure presents refolding behavior as followed by ball mixer stopped-flow EPR over a 1 s range showing fast folding in the 20-30 ms range for the following samples: T8C-SL, E21C-SL, D50C-SL, N92C-SL, K79C-SL, E66C-SL, V28C-SL, and H39C-SL. Initially unfolded protein of approximate 400 μM concentration in pH 5.0, 1.7 M GdnHCl, 200 mM imidazole was mixed in a 1:1 fashion with 0.05 M acetate buffer, 200 mM imidazole at pH 5.0 to initiate folding. K79C-SL and H39C-SL required 100 shots at 10 μL/shot so that the time for flow during a shot was half as long and the flow stopped sooner for these two samples. The other samples required approximately 50 shots at 20 μL/shot. The traces are shown normalized to approximately the same amplitude to make a convenient presentation. The overall amplitudes of the intrinsic recovery signal will vary from one site to the next because the signal difference between folded and unfolded protein will vary from one labeling site to the next. The traces which show the most noise are actually the ones from sites where the difference between folded and unfolded EPR signals was the smallest. Figure reprinted with permission from DeWeerd et al. (2001); Copyright 2001 American Chemical Society.

The global nature of the 20-30 ms phase, which our site-directed spin labels reported, is a simple yet important finding for the folding of cytochrome *c*. A common criticism of spin labels is that they perturb the phenomena that they are supposed to measure; the general evidence of our study is that our spin labeling strategy reported common themes and not highly perturbed, disconnected kinetic events.

Although it was expected and generally found that the derivative signal would become more intense for the unfolded protein because the spin label is more mobile in unfolded protein, there were several instances (T8C-SL, E21C-SL, D50C-SL) where the signal became larger for the folded protein than for the unfolded protein in the time range beyond a millisecond (Figure 18). The paramagnetic ferric heme is itself attached at amino acids 14, 17, and 18. Investigation is underway to determine if for T8C-SL, E21C-SL, D50C-SL it is paramagnetic relaxation in the unfolded state or greater impediment to probe motion in the unfolded state which causes their signals in the time range beyond a millisecond to increase as the protein folds.

The ultrafast submillisecond component reported by flow EPR from D50C-SL showed probe immobilization on the 50 μs time scale, as indicated in Figure 19, even at 5 °C. This immobilization was an order of magnitude faster immobilization than C102-SL showed (~500 μs at 7 °C) at a comparable temperature (Grigoryants et al., 2000). It is possible that D50 lies in or near an initiation site where a loop rapidly forms on the 50 μs time scale (Hagen et al., 1996); the region where D50 is located is not one where early helical structure occurs (Roder et al., 1988), nor is it in a hydrophobic region. At V28C-SL, H39C-SL, E66C-SL, and K79C-SL as well as D50C-SL stopped-flow EPR indicated a substantial percentage of burst phase. Therefore a comprehensive investigation by rapid-mix, submillisecond flow EPR is underway on V28C-SL, H39C-SL, E66C-SL, K79C-SL, and D50C-SL to characterize the time scale(s), spatial extent, and activation energy of the submillisecond folding/pre-folding kinetics in the 50 μs-5 ms time window.

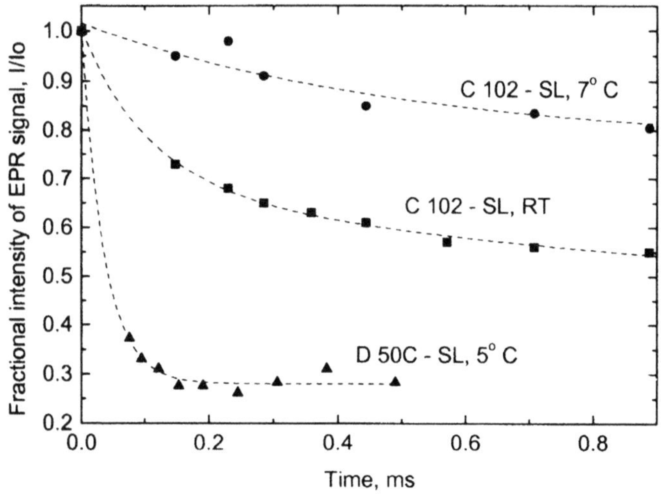

Figure 19. This trace compares the ultrarapid submillisecond folding/compacting as measured by rapid flow EPR of D50C-SL at 5 °C with that of C102-SL at room temperature and at 7 °C. For the D50C-SL sample 400 μM spin labeled protein in 1.7 M guanidinium hydrochloride (GdnHCl) was mixed 1:1 with dilute 0.1 M sodium acetate to obtain a 0.85 M GdnHCl concentration. Data for C102-SL is the first millisecond of data shown in Figure 16. The decay time for folding of D50C-SL was estimated at 40 μs. The dead time of the ball mixer (2001 model) used with D50C-SL was approximately 65 μs. Figure reprinted with permission from DeWeerd et al. (2001); Copyright 2001 American Chemical Society.

4. APPLICATIONS OF STOPPED-FLOW AND FLOW EPR TO NATURALLY OCCURRING TRANSIENT RADICALS

Klimes et al. (1980) presented a stopped-flow system based on laboratory development and construction at the Academy of Sciences of the GDR, Berlin. For the purpose of rapidly driving a pulse of fluid, this system used two stroke-magnets which were powered by a condenser discharge through a magnet solenoid. The initial version (Klimes et al., 1980) drove the fluid from a two-jet tangential mixer into a thick-walled, small volume flat cell that was designed to fit into a standard metallized EPR cavity. During the resultant pulsed flow, whose velocity varied during the pulse, the fastest dead time from mixer to cavity center was estimated at 0.8 ms. An overall dead time after stopping was estimated at 4 ms, the geometrically determined dead volume between mixer and cell center was 20 μL, and the usage of each of two reacting substances was 70 μL per shot. A recent version of this device (marketed as a stopped-flow accessory by Galenus GmbH, Berlin-Adlershof, Germany) now uses a pressure resistant cylindrical tube insert (1.3 mm i.d.; 7.3 mm o.d.) connected to the mixer. This device has the distinct advantage that it integrates with a standard multipurpose rectangular X-band cavity. This device was applied to resolving naturally occurring radical species from ribonucleotide reductase (Lassmann et al., 1992; Sahlin et al., 1995), and to resolving kinetics of incorporation (Marx et al., 1997) and flipping (Marx et al., 2000) of spin-labeled phospholipids in membranes. Although the flow dead time is in the millisecond range, it was noted that a stopping valve to stop flow caused mechanical disturbance which limited the detection of an ESR signal to times longer than 10 ms (Marx et al., 1997; Marx et al., 2000).

A separate flow accessory, explicitly dedicated to continuous-flow, *in situ* EPR but not stopped-flow was co-developed by G. Lassmann, who was formerly an associate of the Academy of Sciences, GDR. This device is now marketed as a dielectric mixing resonator by Bruker Instruments (Part No. ER 4117 D-MV). The dielectric for this system is a sapphire ring which focuses the B_1 field on the sample. The sample flows in a thick walled, narrow bore quartz tube (initially 0.3 mm i.d., more recently 0.4 mm i.d.). The exact filling factor has not been estimated but the combination of B_1 focusing by the sapphire and by the thick-walled quartz EPR tube led to a sensitivity increase by a factor of 10 compared to aqueous samples in the standard multipurpose rectangular TE_{102} cavity. The EPR-active length of the tube was 10 mm with a geometrically calculated dead volume from mixer to EPR active zone of 0.35 μL and a geometrically calculated 0.7 ms dead time at the fastest 0.5 mL/s flow rate. A simple Y-type mixer was integrated to the EPR sample tube just outside the resonator shield, and fluid was pumped by a Harvard syringe pump and delivered to the mixer by HPLC-type PEEK tubing. The mixing was purposefully made incomplete (in our terminology, inefficient) so that mixing would take place within the EPR cell rather than in the mixer. The advantage of incomplete mixing is that a transient species whose lifetime is less than the dead time may be detected. Thus this device is highly appropriate for learning of early transient species but less appropriate for learning of their explicit kinetic behavior because the time-zero for a chemical reaction is not definite.

A major application of the rapid flow ER 4117 D-MV was in determining the structure of transient radicals formed by rapid oxidation processes such as Fenton chemistry in aqueous solution. Recent work focused on a transient histidine radical created by attack from hydroxyl radicals ($HO^{\bullet}$) created with a Ti^{3+}/H_2O_2 Fenton system at both low pH (Lassmann et al., 1999) and neutral pH (Lassmann et al., 2000). The higher sensitivity dielectric mixing resonator enabled economical detection of isotopically labeled species (viz., selectively deuterated histidine) whose spectra were required for assignment of hyperfine features. Excellent signal to noise

(Figure 20) was obtained at a flow rate of approximately 0.5 mL/s with two accumulated 20 s scans which consumed about 150 mg of ββ-deuterated histidine. The actual concentration of histidine radical during flow was estimated at 50 μM (Lassmann et al., 2000).

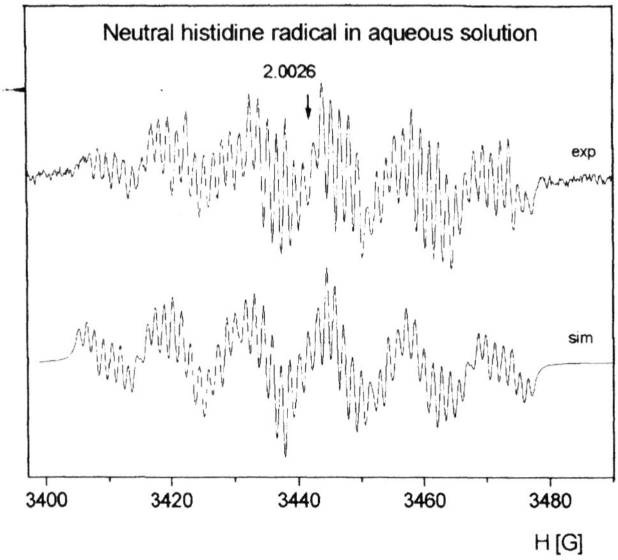

Figure 20. EPR spectrum of the transient neutral histidine radical in solution as rapidly formed upon oxidation of histidine in a Ti^{3+}/EDTA/H_2O_2 - Fenton system at pH 7, recorded under fast continuous flow (flow rate 24 mL/min). These spectra were from the 5-oxohistidine radical in aqueous solution. The simulated spectrum is below the experimental one. Figure reprinted with permission from Lassmann et al. (2000); Copyright 2000 American Chemical Society.

Neutral Radical (A) **Addition of HO$^\bullet$ to C5 (B)**

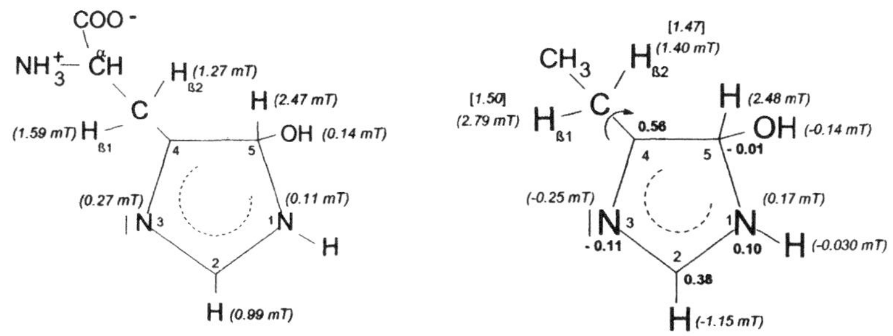

Figure 21. (**A**) shows the experimental isotropic hyperfine coupling constants of the neutral histidine OH-addition radical. (**B**) shows the corresponding DFT data of hyperfine coupling constants and spin densities of the neutral HO$^\bullet$-addition radical of the histidine model 4-ethyl imidazole. Hyperfine coupling constants are given in italics and spin densities are given in bold next to the corresponding nuclei. For the β proton hyperfine coupling constants: ( ), static values; [ ], averaged values of freely rotating side chain. Figure reprinted with permission from Lassmann et al. (2000); Copyright 2000 American Chemical Society.

The reason for interest in the histidine radical is that histidine radicals have been postulated as intermediates in enzymatic reactions, but they have not been characterized under ambient conditions because of their transient nature and because of their frequent proximity to paramagnetic metal ion relaxers. In addition, explicit oxohistidine radicals may well be formed physiologically under conditions of oxidative stress by HO$^\bullet$ radicals emanating from Fe^{2+} Fenton chemistry or by oxidizing products of peroxide, peroxynitrite, or superoxide. Figure 20 shows the extremely detailed EPR spectrum of the neutral histidine radical. The isotropic hyperfine constants of two β protons, three ring protons, and two nitrogen nuclei of the 5-oxohistidine radical (in both cation and neutral forms) were elucidated by EPR. Figure 21A shows the experimental isotropic hyperfine coupling constants of the neutral 5-oxohistidine radical, and 21B shows the DFT (density functional theory) correlation of hyperfine coupling constants and spin densities. DFT was used to discriminate between possible carbon 2, 4, or 5 positions of attack of the hydroxyl on the histidine ring. Good agreement of theory and experiment was found between the experimental hyperfine couplings and DFT hyperfine coupling predictions for the histidine radical as caused by hydroxyl attack at the C5 position.

Peroxynitrite is a highly reactive molecule formed from nitric oxide and superoxide under conditions of oxidative stress; it has both biodamaging and bioregulatory actions. In the laboratory of O. Augusto the ER4117 D-MTV was used to directly detect carbonate radicals in unambiguous fashion within milliseconds of their formation from carbon dioxide and peroxyinitrite, all without the complication of a spin trap (Bonini et al., 1999). In further probing the reactivity of peroxynitrite, direct detection of peroxynitrite-induced sulfinyl and disulfide radicals from glutathione and cysteine was obtained with the ER4117 D-MTV (Bonini and Augusto, 2001).

5. FUTURE DEVELOPMENTS AND APPLICATIONS OF FLOW AND STOPPED-FLOW EPR

At present the limit on fastest flow and shortest dead times with our flow device is the stress tolerance of our thick-walled glass driver syringes. With stainless steel syringes we would expect to increase the fluid velocity to decrease the dead time for fast flow below 50 μs. If the mixer can be made on a smaller scale, for example by nanofabrication, high frequency, low volume Q or W band resonators may in the future provide shorter dead times.

For site-directed spin labeled systems we foresee the continued application of flow and stopped-flow EPR to determine the time scale and early location of folding. Using bi-labeled systems, we expect overall kinetically evolving spectra to be particularly helpful with resolving distances between folding subunits and kinetic variation in those distances. To obtain overall spectra on samples with submillisecond ages, an accurate rapid scan device is being incorporated with rapid flow. With spin labeled substrate, flow and stopped-flow EPR should provide evidence of rapid immobilization of the enzyme-substrate complex extending into the submillisecond regime and evidence for subsequent conformational change after substrate binding. It is conceivable that a rapid mixing device such as that of Grigoryants et al. (2000) may be integrated with a high power pulsed EPR probe (Borbat and Freed, 2000; Borbat et al., 2002) to observe overall rapid spectral development, T_1 and T_2 behavior, and kinetic variation of distances and distance distributions between bi-labels. In the field of radicals of oxidative stress, we would look for rapid flow and stopped-flow to resolve the more early radicals which have to date only been inferred.

6. ACKNOWLEDGEMENTS

This work has been supported by grants from the National Institutes of Health (GM 35103), the National Science Foundation (MCB-9817598) and the American Heart Association. Acknowledgment is made to the Donors of the Petroleum Research Fund, administered by the American Chemical Society, for partial support of this research (ACS-PRF Grant No. 34132-

AC4). The following scientists provided considerable help in different facets of our development of rapid mix EPR: Dr. Donald Borg, Prof. Wayne L. Hubbell, Dr. Andrzej Sienkiewicz, Dr. Vladimir Grigoryants, Dr. Jacquelyn S. Fetrow, and Dr. Kim DeWeerd. We are extremely grateful to Dr. Günter Lassmann for providing a very considerable body of useful information on the flow and stopped-flow systems which he developed and which are commercially available.

7. REFERENCES

Bonini, M. G., R. Radi, G. Ferrer-Sueta, A. M. Ferreira, and O. Augusto. 1999. Direct EPR detection of the carbonate radical anion produced from peroxynitrite and carbon dioxide. *J Biol Chem* **274**:10802-10806.

Bonini, M. G., and O. Augusto. 2001. Carbon Dioxide Stimulates The Production Of Thiyl, Sulfinyl, And Disulfide Radical Anion From Thiol Oxidation By Peroxynitrite. *J Biol Chem* **276**:9749-9754.

Borbat, P., and J. H. Freed. 2000. Double Quantum ESR and Distance Measurements. *In*: Biological Magnetic Resonance: Distance Measurements in Biological Systems by EPR. L. J. Berliner, G. R. Eaton, and S. S. Eaton, editors. Plenum, New York.383-459.

Borbat, P. P., H. S. Mchaourab, and J. H. Freed. 2002. Protein Structure Determination Using Long-Distance Constraints From Double-Quantum Coherence ESR: Study Of T4 Lysozyme. *J Am Chem Soc* **124**:5304-5314.

Borg, D. C. 1964a. Continuous Flow Methods Adapated For Epr Apparatus. *In*: Rapid Mixing and Sampling Techniques in Biochemistry. B. Chance, R. H. Eisenhardt, Q. H. Gibson, and K. K. Lonberg-Holm, editors. Academic Press, New York.135-149.

Borg, D. C. 1964b. An Improved Flow System For Electron Paramagnetic Resonance Spectrometry Of Aqueous Solutions. *Nature* **201**:1087-1090.

Borg, D. C., and J. J. Elmore. 1967. Continuous Flow Apparatus For EPR Spectroscopy At 35 GHz. *In*: Magnetic Resonance in Biological Systems. A. Ehrenberg, B. G. Malmstrom, and T. Vanngard, editors. Pergamon Press, Oxford.383-387.

Brems, D. N., and E. Stellwagen. 1983. Manipulation Of The Observed Kinetic Phases In The Refolding Of Denatured Ferricytochromes C. *J Biol Chem* **258**:3655-3660.

Bromberg, S. E., and I. Y. Chan. 1992. Enhanced Sensitivity For High-Pressure EPR Using Dielectric Resonators. *Rev. Sci. Instrum.* **63**:3670-3676.

Buettner, G. R. 1987. Spin Trapping: ESR Parameters Of Spin Adducts. *Free Rad. Biol. Med.* 3:259-303.

Chance, B. 1940a. The Accelerated Flow Method For Rapid Reactions. Part I. Analysis. *Journal of the Franklin Institute* **229**:455-476.

Chance, B. 1940b. The Accelerated Flow Method For Rapid Reactions. Part II. Design, constructions, and tests. *Journal of the Franklin Institute* **229**:737-766.

Cohn, S. B. 1968. Microwave Bandpass Filters Containing High-Q Dielectric Resonators. *IEEE Trans. Microwave Theory Tech* **16**:218-227.

Colón, W., G. A. Elöve, L. P. Wakem, F. Sherman, and H. Roder. 1996. Side Chain Packing Of The N- And C-Terminal Helices Plays A Critical Role In The Kinetics Of Cytochrome C Folding. *Biochemistry* **35**:5538-5549.

Colón, W., L. P. Wakem, F. Sherman, and H. Roder. 1997. Identification Of The Predominant Non-Native Histidine Ligand In Unfolded Cytochrome C. *Biochemistry* **36**:12535-12541.

Creighton, T. E. 1994. The Energetic Ups And Downs Of Protein Folding [News]. *Nat Struct Biol* **1**:135-138.

Davies, P. W. 1962. *In*: Physical Techniques in Biological Research. Academic Press, New York.137-179.

DeWeerd, K., V. Grigoryants, Y. Sun, J. S. Fetrow, and C. P. Scholes. 2001. EPR-Detected Folding Kinetics Of Externally Located Cysteine-Directed Spin-Labeled Mutants Of Iso-1-Cytochrome C. *Biochemistry* **40**:15846-15855.

Dykstra, R. W., and G. D. Markham. 1986. A Dielectric Sample Resonator For Enhanced Sensitivity Of EPR Spectroscopy. *J. Mag. Reson.* **69**:350-355.

Ebert, B., G. V. Semisotnov, and N. A. Rodionova. 1990. Studies On Globular Protein Refolding Kinetics By ESR Stopped Flow Spectroscopy. *Stud. Biophys.* **137**:125-132.

Elöve, G. A., A. K. Bhuyan, and H. Roder. 1994. Kinetic Mechanism Of Cytochrome C Folding: Involvement Of The Heme And Its Ligands. *Biochemistry* **33**:6925-6935.

Elöve, G. A., A. F. Chaffotte, H. Roder, and M. E. Goldberg. 1992. Early Steps In Cytochrome C Folding Probed By Time-Resolved Circular Dichroism And Fluorescence Spectroscopy. *Biochemistry* **31**:6876-6883.

Englander, S. W., T. R. Sosnick, L. C. Mayne, M. Shtilerman, P. X. Qi, and Y. Bai. 1998. Fast And Slow Folding In Cytochrome C. *Acc. Chem. Res.* **31**:737-744.

Fan, C., J. F. Bank, R. G. Dorr, and C. P. Scholes. 1988. An Electron Nuclear Double Resonance Investigation Of Redox-Induced Electronic Structural Change At $Copper_A^{2+}$ In Cytochrome C Oxidase. *J. Biol. Chem.* **263**:3588-3591.

Fenton, H. J. H. 1894. *J. Chem. Soc. Trans.* **65**:899-910.

Fetrow, J. S., T. S. Cardillo, and F. Sherman. 1989. Deletions And Replacements Of Omega Loops In Yeast Iso-1-Cytochrome C. *Proteins* **6**:372-381.

Feynman, R. P., R. B. Leighton, and M. Sands. 1964. The Feynman Lectures on Physics. Addison-Wesley, Reading, MA.

Froncisz, W., and J. S. Hyde. 1982. The Loop-Gap Resonator: A New Microwave Lumped Circuit ESR Sample Structure. *J. Magn. Reson.* **47**:515-521.

Froncisz, W., C. S. Lai, and J. S. Hyde. 1985. Spin-Label Oximetry: Kinetic Study Of Cell Respiration Using A Rapid-Passage T_1-Sensitive Electron Spin Resonance Display. *Proc. Natl. Acad. Sci. U. S. A.* **82**:411-415.

Grigoryants, V. M., A. V. Veselov, and C. P. Scholes. 2000. Variable Velocity Liquid Flow EPR Applied To Submillisecond Protein Folding. *Biophys J* **78**:2702-2708.

Hagen, S. J., J. Hofrichter, A. Szabo, and W. A. Eaton. 1996. Diffusion-Limited Contact Formation In Unfolded Cytochrome C: Estimating The Maximum Rate Of Protein Folding. *Proc Natl Acad Sci U S A* **93**:11615-11617.

Halliwell, B., and J. M. C. Gutteridge. 1989. Free Radicals in Biology and Medicine. Second Edition. Clarendon Press, Oxford.

Hubbell, W. L., and C. Altenbach. 1994. Investigation Of Structure And Dynamics In Membrane Proteins Using Site-Directed Spin Labeling. *Curr. Opin. Struct. Biol.* **4**:566-573.

Hubbell, W. L., D. S. Cafiso, and C. Altenbach. 2000. Identifying Conformational Changes With Site-Directed Spin Labeling. *Nat Struct Biol* **7**:735-739.

Hubbell, W. L., W. Froncisz, and J. S. Hyde. 1987. Continuous And Stopped-Flow EPR Spectrometer Based On A Loop Gap Resonator. *Rev. Sci. Instrum.* **58**:1879-1886.

Hubbell, W. L., A. Gross, R. Langen, and M. A. Lietzow. 1998. Recent Advances In Site-Directed Spin Labeling Of Proteins. *Curr. Opin. Struct. Biol.* **8**:649-656.

Jaworski, M., A. Sienkiewicz, and C. P. Scholes. 1997. Double-Stacked Dielectric Resonator For Sensitive EPR Measurements. *J Magn Reson* **124**:87-96.

Jiang, J. J., J. F. Bank, W. W. Zhao, and C. P. Scholes. 1992. The Method Of Time-Resolved Spin-Probe Oximetry: Its Application To Oxygen Consumption By Cytochrome c Oxidase. *Biochemistry* **31**:1331-1339.

Klimes, N., G. Lassmann, and B. Ebert. 1980. Time-Resolved EPR Spectroscopy. Stopped-Flow EPR Apparatus For Biological Application. *J. Magn. Reson.* **37**:53-59.

Lai, C. S., L. E. Hopwood, J. S. Hyde, and S. Lukiewicz. 1982. ESR Studies Of Oxygen Uptake By Chinese Hamster Ovary Cells During The Cell Cycle. *Proc. Natl. Acad. Sci. U. S. A.* **79**:1166-1170.

Lassmann, G., L. Thelander, and A. Graslund. 1992. EPR Stopped-Flow Studies Of The Reaction Of The Tyrosyl Radical Of Protein R2 From Ribonucleotide Reductase With Hydroxyurea. *Biochem Biophys Res Commun* **188**:879-887.

Lassmann, G., L. A. Eriksson, F. Himo, F. Lendzian, and W. Lubitz. 1999. Electronic Structure Of A Transient Histidine Radical In Liquid Aqueous Solution: EPR Continuous-Flow Studies And Density Functional Calculations. *J. Phys. Chem. A* **103**:1283-1290.

Lassmann, G., L. A. Eriksson, F. Lendzian, and W. Lubitz. 2000. Structure Of A Transient Neutral Histidine Radical In Solution: EPR Continuous-Flow Studies In A Ti^{3+}/EDTA-Fenton System And Density Functional Calculations. *J. Phys. Chem. A* **104**:9144-9152.

Leszczynski, J. F., and G. D. Rose. 1986. Loops In Globular Proteins: A Novel Category Of Secondary Structure. *Science* **234**:849-855.

Louie, G. V., and G. D. Brayer. 1990. High-Resolution Refinement Of Yeast Iso-1-Cytochrome c And Comparisons With Other Eukaryotic Cytochromes c. *J Mol Biol* **214**:527-555.

Marx, U., G. Lassmann, H. G. Holzhutter, D. Wustner, P. Muller, A. Hohlig, J. Kubelt, and A. Herrmann. 2000. Rapid Flip-Flop Of Phospholipids In Endoplasmic Reticulum Membranes Studied By A Stopped-Flow Approach. *Biophys J*:2628-2640.

Marx, U., G. Lassmann, K. Wimalasena, P. Muller, and A. Herrmann. 1997. Rapid Kinetics Of Insertion And Accessibility Of Spin-Labeled Phospholipid Analogs In Lipid Membranes: A Stopped-Flow Electron Paramagnetic Resonance Approach. *Biophys J* **73**:1645-1654.

Mascarenhas, R., Y. H. Wei, C. P. Scholes, and T. E. King. 1983. Interaction In Cytochrome c Oxidase Between Cytochrome a3 Ligated With Nitric Oxide And Cytochrome a. *J. Biol. Chem.* **258**:5348-5351.

Morar, A. S., D. Kakouras, G. B. Young, J. Boyd, and G. J. Pielak. 1999. Expression Of ^{15}N-Labeled Eukaryotic Cytochrome c In Escherichia Coli. *J Biol Inorg Chem* 4:220-222.

Nall, B. T., and T. A. Landers. 1981. Guanidine Hydrochloride Induced Unfolding Of Yeast Iso-2 Cytochrome c. *Biochemistry* 20:5403-5411.

Osterhout, J. J., Jr., and B. T. Nall. 1985. Slow refolding kinetics in yeast iso-2 cytochrome c. *Biochemistry* **24**:7999-8005.

Plurde, K. J., D. F. Linn, H. M. O'Bryan, Jr., and J. Thomson, Jr. 1975. $Ba_2Ti_9O_{20}$ as a Microwave Dielectric Resonator. *J. Am. Ceram. Soc.* **58**:418-427.

Plurde, K. J., and C.-L. Ren. 1981. Application Of Dielectric Resonators In Microwave Components. *IEEE Trans. Microwave Theory Tech* **29**:754-770.

Pollock, W. B., F. I. Rosell, M. B. Twitchett, M. E. Dumont, and A. G. Mauk. 1998. Bacterial expression of a mitochondrial cytochrome c. Trimethylation Of Lys72 In Yeast Iso-1-Cytochrome C And The Alkaline Conformational Transition. *Biochemistry* **37**:6124-6131.

Qu, K., J. L. Vaughn, A. Sienkiewicz, C. P. Scholes, and J. S. Fetrow. 1997. Kinetics And Motional Dynamics Of Spin-Labeled Yeast Iso-1-Cytochrome c: 1. Stopped-Flow Electron Paramagnetic Resonance As A Probe For Protein Folding/Unfolding Of The C-Terminal Helix Spin-Labeled At Cysteine 102. *Biochemistry* **36**:2884-2897.

Regenfuss, P., R. M. Clegg, M. J. Fulwyler, F. J. Barrentes, and T. M. Jovin. 1985. Mixing Liquids In Microseconds. *Rev. Sci. Instrum.* **56**:283-290.

Richtmyer, D. 1939. Dielectric Resonators. *J. Appl. Physics* **10**:391-398.

Roder, H., G. A. Elöve, and S. W. Englander. 1988. Structural Characterization Of Folding Intermediates In Cytochrome C By H-Exchange Labelling And Proton NMR. *Nature* **335**:700-704.

Sahlin, M., G. Lassmann, S. Potsch, B. M. Sjoberg, and A. Graslund. 1995. Transient Free Radicals In Iron/Oxygen Reconstitution Of Mutant Protein R2 Y122F. Possible Participants In Electron Transfer Chains In Ribonucleotide Reductase. *J Biol Chem* **270**:12361-12372.

Shastry, M. C., and H. Roder. 1998. Evidence For Barrier-Limited Protein Folding Kinetics On The Microsecond Time Scale. *Nat Struct Biol* **5**:385-392.

Shastry, M. C. R., J. M. Sauder, and H. Roder. 1998. Kinetic And Structural Analysis Of Submillisecond Folding Events In Cytochrome c. *Acc. Chem. Res.* **31**:717-725.

Sienkiewicz, A., A. M. da Costa Ferreira, B. Danner, and C. P. Scholes. 1999. Dielectric Resonator-Based Flow And Stopped-Flow EPR With Rapid Field Scanning: A Methodology For Increasing Kinetic Information. *J Magn Reson* **136**:137-142.

Sienkiewicz, A., K. Qu, and C. P. Scholes. 1994. Dielectric Resonator-Based Stopped-Flow EPR. *Rev. Sci. Instrum.* **65**:68-74.

Sosnick, T. R., L. Mayne, R. Hiller, and S. W. Englander. 1994. The Barriers In Protein Folding. *Nat Struct Biol* **1**:149-156.

Sosnick, T. R., M. D. Shtilerman, L. Mayne, and S. W. Englander. 1997. Ultrafast Signals In Protein Folding And The Polypeptide Contracted State. *Proc Natl Acad Sci U S A* **94**:8545-8550.

Stevens, T. H., C. T. Martin, H. Wang, G. W. Brudvig, C. P. Scholes, and S. I. Chan. 1982. The Nature Of CuA In Cytochrome c Oxidase. *J Biol Chem* **257**:12106-12113.

Tsong, T. Y. 1976. Ferricytochrome c Chain Folding Measured By The Energy Transfer Of Tryptophan 59 To The Heme Group. *Biochemistry* **15**:5467-5473.

Tullius, T. D. 1988. DNA Footprinting With Hydroxyl Radical. *Nature* **332**:663-664.

van Camp, H. L., Y. H. Wei, C. P. Scholes, and T. E. King. 1978. Electron Nuclear Double Resonance Of Cytochrome Oxidase: Nitrogen And Proton ENDOR From The 'Copper' EPR Signal. *Biochim Biophys Acta* **537**:238-246.

Vanderkooi, J. M., G. Maniara, T. J. Green, and D. F. Wilson. 1987. An Optical Method For Measurement Of Dioxygen Concentration Based Upon Quenching Of Phosphorescence. *J Biol Chem* **262**:5476-5482.

Walsh, W. M., and L. W. Rupp. 1986. Enhanced ESR Sensitivity Using A Dielectric Resonator. *Rev. Sci. Instrum.* **57**:2278-2279.

Wilson, D. F., W. L. Rumsey, T. J. Green, and J. M. Vanderkooi. 1988. The Oxygen Dependence Of Mitochondrial Oxidative Phosphorylation Measured By A New Optical Method For Measuring Oxygen Concentration. *J Biol Chem* **263**:2712-2718.

Windrem, D. A., And W. Z. Plachy. 1980. The diffusion-solubility of oxygen in lipid bilayers. *Biochim Biophys Acta* **600**:655-665.

Winkler, J. R., and H. B. Gray. 1998. Protein Folding. *Acc. Chem. Res.* **31**:698.

Yamazaki, I., H. S. Mason, and L. Piette. 1960. Identification by Electron Paramagnetic Resonance Spectroscopy, Of Free Radicals Generated From Substrates By Peroxidase. *J. Biol. Chem.* **235**:2444-2449.

Yamazaki, I., and L. H. Piette. 1990. ESR Spin-Trapping Studies On The Reaction Of Iron(2+) Ions With Hydrogen Peroxide-Reactive Species In Oxygen Toxicity In Biology. *J. Biol. Chem.* **265**:13589-13594.

Chapter 4

Application of Angle-Selected Electron Nuclear Double Resonance to Characterize Structured Solvent in Small Molecules and Macromolecules

Devkumar Mustafi and Marvin W. Makinen
Department of Biochemistry and Molecular Biology, The University of Chicago, Cummings Life Science Center, 920 East 58th Street, Chicago, IL 60637 USA

Abstract: Applications of angle-selected electron nuclear double resonance (ENDOR) spectroscopy using the vanadyl (VO^{2+}) cation or nitroxyl spin-labels as paramagnetic probes are reviewed in which structured solvent in small molecule and macromolecular complexes have been identified and characterized. By determination of the principal hyperfine (hf) coupling components (A) of magnetic nuclei in the vicinity of the paramagnetic probe, electron-nucleus distances are estimated according to the dipolar equation, and the relative coordinates of the nuclei are assigned on the basis of the orientation dependence of magnetic interactions. The precision in determining electron-nucleus distances over an approximate 3-8 Å range for VO^{2+} and 4-11 Å range for spin-labels is generally ≤ 5% based on ENDOR line widths and is exceeded only by that associated with X-ray diffraction of single crystals. The detailed structure of metal-bound or hydrogen-bonded solvent in small molecules, nucleotide complexes, and proteins in frozen glassy solutions is described. In particular, comparison of the structural details of solvent hydrogen-bonded to spin-labeled β-lactam antibiotics free in solution and sequestered in the active site of a spin-labeled penicilloyl acylenzyme reaction intermediate of TEM-1 β-lactamase is shown to bring improved understanding of the origin of the reactivity of β-lactam antibiotics and the mechanism of action of β-lactamases. Future directions that may allow a richer level of structural detail for assessment of macromolecular structure are also briefly discussed.

1. INTRODUCTION

The structure and function of biological macromolecules are intimately associated with solvent water. The dynamical mobility of water interacting with macromolecules ranges from ordered and essentially static, detectable by X-ray and neutron diffraction methods at ambient temperature over periods of days, to that of bulk solvent with rapid interconversion (~ 1000 s^{-1}) of structural arrangements and reorientation of hydrogen-bonds. Special interest has long been directed towards sites in macromolecules with structured water stabilized through hydrogen-bonding. Such sites may underlie the catalytic function of enzymes whereby water molecules are strategically positioned to react as nucleophiles or to facilitate rapid translocation of protons in clusters through hydrogen-bonding and electrostatic interactions.

To assign the chemical role of water molecules strategically positioned in such sites requires measurement of their atomic coordinates and description of their structural interactions with protein residues. This objective, for enzyme reaction intermediates, requires preparation of the enzyme in a catalytically functional state and application of a physical method to determine atomic positions. Because the lifetimes of catalytically competent enzyme reaction intermediates are much shorter than the time of data collection, X-ray methods have generally taken advantage of the stability of enzyme-inhibitor complexes to mimic enzyme-substrate interactions. However, non-productive structural relationships are responsible for the long-term stability of enzyme-inhibitor complexes in contrast to the labile, short-lived, productive enzyme-substrate interactions responsible for the catalytic conversion of substrate to product. For this reason, enzyme-inhibitor complexes can offer only a partial description of the catalytically competent structure of the active site. Furthermore, introduction of inhibitor ligands into enzyme active sites is often associated with displacement of strategically positioned water molecules. This chapter describes how these drawbacks can be avoided by combined application of electron nuclear double resonance (ENDOR) spectroscopy with cryoenzymology to isolate catalytically competent enzyme reaction intermediates for structural analysis.

The underlying physical process of interest in the application of ENDOR spectroscopy is hyperfine (hf) coupling, the interaction between an unpaired electron with a nearby magnetic nucleus. ENDOR provides the most precise method to measure the influence of the (oscillating) magnetic field produced by the unpaired electron on the nucleus, resulting in a shift in its resonance frequency from that in the absence of the electron. This measurement can lead to highly precise estimates of the separation between the unpaired electron and the nucleus and provides thereby a sensitive method to determine electronic and molecular structure. In this review we describe ENDOR studies from our laboratory that have led to assignments of the coordination geometry of metal-bound or hydrogen-bonded solvent in small molecules, nucleotide complexes, and proteins. We emphasize instances in which comparison of the details of small molecule systems with their counterparts in proteins and enzymes has brought improved understanding of macromolecular function. We also discuss briefly how extensions of this spectroscopic approach may enhance structural analysis and thereby bring new insights into the study of structure-function relationships of biological macromolecules.

For further discussions of ENDOR and related double resonance methods, the reader should consult Atherton (1993), whose textbook on electron paramagnetic resonance (EPR) places a heavy emphasis on ENDOR in comparison to others, or specialized monographs on other forms of double resonance spectroscopy (Kevan and Kispert, 1976; Dorio and Freed, 1979; Kurreck et al., 1988). Ch. 15 of volume 23 describes electron spin-echo envelope modulation (ESEEM) spectroscopy, a related electron magnetic resonance method complementary to ENDOR. Since ENDOR is usually applied to determine dipolar electron-nucleus distances between paramagnetic sites and nearby magnetic nuclei, Ch. 8, describing applications of pulsed EPR for electron-electron distance measurements, should be also consulted. Both continuous wave (cw) and pulsed ENDOR spectrometers have been commercially available for several decades. Studies described from our laboratory have been carried out with a cw X-band Bruker ESP 300E spectrometer equipped with a TM_{110} cylindrical cavity, Bruker ENDOR accessory, and Oxford Instruments ESR 910 liquid helium cryostat.

Since the first demonstration of ENDOR to measure hf coupling of ^{29}Si nuclei with impurity sites in silicon (Feher, 1956, 1957), it has been widely applied to free radicals in solution and in the solid state. Moreover, its application has resulted in precise determination of the coordinates of protons and deuterons of metal-bound and hydrogen-bonded water molecules in single crystals of diamagnetic metal ion complexes into which Mn^{2+} (van Ormondt and Visser, 1968), Gd^{3+} (Hutchison and McKay, 1977; Fields and Hutchison, 1985), Nd^{3+} (Hutchison and Orlowski, 1980), and VO^{2+} (Atherton and Shackleton, 1980, 1984) have been incorporated as the paramagnetic probe. In these studies the uncertainty in determining coordinates of protons and deuterons was 0.002-0.099 Å for metal-nucleus distances over a 2-5 Å range. The major source of error was the uncertainty in determining the direction of the laboratory magnetic field in the crystal. This precision compares favorably with that achieved in high resolution X-ray diffraction studies of small molecules in which the positions of hydrogen atoms are directly determined on the basis of scattering data after extensive crystallographic least-squares refinement. ENDOR of single crystals containing substitutionally dilute paramagnetic species has been also applied to determine positions of ^{19}F (Baker et al., 1968), ^{13}C (Rudin et al., 1982), and ^{14}N and ^{15}N (Feher et al., 1973; Scholes et al., 1982) with comparable precision.

While the precision achieved through single crystal ENDOR analysis is impressive, its application to biological problems is limited. The usual dimensions of crystals of biological macromolecules and concentration of paramagnetic sites place severe limitations on signal-to-noise and the total number of spins in the spectrometer cavity. Also, the systems of interest, *e.g.*, macromolecular complexes or intermediates of enzyme-catalyzed reactions, may not be obtainable in single crystal form. For these reasons, a more conventional approach has been applied with use of polycrystalline or frozen glassy samples in which EPR spectra reflect a "powder" average over all molecular orientations with respect to the laboratory magnetic field.

The orientation dependence of magnetic hyperfine interactions was first described for EPR spectra of crystals and powder samples of free radicals in the absence of g anisotropy (Blinder, 1960). Subsequently Rist and Hyde (1968, 1970) demonstrated in the case of g anisotropy that, at some settings of the magnetic field, the so-called "turning points" in the EPR spectrum, "single-crystal-type" ENDOR spectra can be obtained from molecules essentially in a single orientation. Hoffman and coworkers (1984) presented a general formulation of the orientation dependence of ENDOR spectra of polycrystalline samples for which g anisotropy dominates. Furthermore, it has been shown that ENDOR spectra can be obtained throughout the EPR spectrum when frequency modulation of the swept radiofrequency (rf) field is employed (Hurst et al., 1985). Thus, one is not confined only to the limited number of specific orientations found at the turning points of the EPR absorption spectrum.

The first applications of "angle-selected ENDOR" to determine structure and to assign relative coordinates of protons with respect to the paramagnetic center were published for polycrystalline dithiophosphate and dimethyldithiocarbamate complexes of nickel(II) doped with Cu^{2+} (Yordanov and Zdravkova, 1986; Yordanov et al., 1986), and for VO^{2+} adducts of Schiff bases (Attanasio, 1986, 1989) and Gd^{3+}-complexes (Yim and Makinen, 1986) in frozen glassy solutions. At about the same time, insightful descriptions of the mathematical theory underlying the interpretation of ENDOR spectra of randomly oriented molecules appeared from the laboratories of Kreilick (Henderson et al., 1985; Hurst et al., 1985) and Hoffman (Hoffman et al., 1984,1985; Hoffman and Gurbiel, 1989). The methodology has been subsequently extended in our laboratory to a variety of VO^{2+}-coordination complexes (Mustafi and Makinen, 1988; Mustafi et al., 1992; Mustafi and Nakagawa, 1994, 1996; Makinen and Mustafi, 1995; Jiang and Makinen, 1995; Mustafi et al., 2000; Makinen and Brady, 2002) and nitroxyl spin-labels (Wells and Makinen, 1988; Mustafi et al., 1990b,

1990c, 1993, 1995, 1997, 2001, 2002; Wells et al., 1990; Mustafi and Makinen, 1995; Jiang et al., 1998; Makinen, 1998; Makinen et al., 1998) to determine small molecule and macromolecular structure. The uncertainty in electron-nucleus distances estimated from ENDOR line widths in these studies is ≤ 5% over an approximate 3-8 Å range for VO^{2+}-complexes (Makinen and Mustafi, 1995) and 4-11 Å range for spin-labels (Makinen et al., 1998).

2. ENDOR ASSIGNMENT OF MOLECULAR STRUCTURE AND CONFORMATION WITH VO^{2+} AND NITROXYL SPIN-LABELS

2.1 EPR Absorption Dependent upon Molecular Orientation

The physical process underlying ENDOR spectroscopy is the hf interaction between an unpaired electron and a magnetic nucleus. This is due to the interaction of the unpaired electron spin of the paramagnetic site with nearby magnetic nuclei, producing shifts in their resonance frequencies. This phenomenon gives rise to broadening of EPR spectra, as illustrated in Figs. 1 and 2, by comparison of the spectra of VO^{2+} in protiated and perdeuterated methanol and of the nitroxyl spin-label 2,2,5, 5-tetramethyl-1-oxypyrroline-3-carboxylic acid in its protiated and perdeuterated forms, respectively. In Fig. 1 the EPR absorption features of $[VO(H_2O)_5]^{2+}$ are broader in a solvent of natural abundance isotope composition than in the perdeuterated solvent, reflecting the stronger but unresolved electron-nuclear hf couplings of nearby protons of solvent molecules in the inner and outer coordination shells. On the other hand, the line broadening of nitroxyl spin-labels, as seen in Fig. 2, is entirely dominated by the covalent hydrogens in the pyrrolinyl ring and the methyl groups adjacent to the N–O group. There is no perceptible change in EPR line width of the spin-label with change of solvent. These examples illustrate the major limitation of EPR in probing structural details of a paramagnetic site, namely, the failure to resolve ligand hf or superhyperfine (shf) couplings of nearby magnetic nuclei that interact with the unpaired electron spin. Nonetheless, quantitative evaluation of shf couplings is essential for structure determination. In order to assign nuclear coordinates and to estimate electron-nucleus separations, it is necessary to measure the strength of shf interactions and to determine their relationships to magnetic axes in the molecule and to the external, static laboratory magnetic field $\mathbf{H}_0$. This objective is best achieved by application of ENDOR spectroscopy.

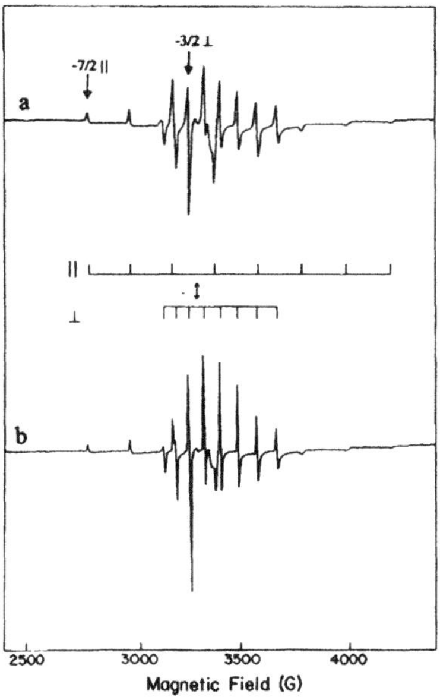

Figure 1. First-derivative EPR absorption spectra of VO^{2+} in 50:50 (v/v) (a) H_2O:methanol and (b) D_2O:d_4-methanol solvent mixtures. The parallel (||) and perpendicular (⊥) EPR absorption components (−7/2, −5/2, · · · , +5/2, +7/2) are identified in sequence in the low- to high-field direction. The double-headed arrow indicates the magnetic field position corresponding to the forbidden $\Delta m_I = \pm 1$ transition. The spectra were obtained with a Bruker ER200D spectrometer operating at 9.5 GHz with the sample temperature maintained at 10 K with an Oxford Instruments 910 ESR liquid helium cryostat. Reprinted from Mustafi and Makinen (1988) with permission by the American Chemical Society.

In Table 1 we compare the spectroscopic parameters describing the principal components of the $\mathbf{g}_e$ tensor and nuclear hyperfine tensor of the VO^{2+} ion and of the N–O group of nitroxyl spin-labels. For both paramagnetic species, each with one unpaired electron, EPR and ENDOR spectra can be described by the spin Hamiltonian in Eq. (1), which includes the electronic Zeeman interaction (H_1) and the nuclear Zeeman interaction (H_2) while hf interactions of the unpaired electron with different classes of nuclei are contained within H_3. In Eq. (1), β_e and β_n represent the Bohr electron and nuclear magneton, $\mathbf{g}_e$ the electronic Zeeman interaction tensor, g_n the nuclear g-factor, $\mathbf{H}_0$ the external (laboratory) magnetic field, $\mathbf{S}$ and $\mathbf{I}$ the electron and nuclear spin operators, respectively, and $\mathbf{A}$ the electron-nucleus hf tensor.

$$\begin{array}{cccccc} H_s & = |\beta_e|\mathbf{H}_0 \cdot \mathbf{g}_e \cdot \mathbf{S} & + & g_n\beta_n\mathbf{H}_0 \cdot \mathbf{I} & + & \mathbf{S} \cdot \mathbf{A} \cdot \mathbf{I} \\ & H_1 & + & H_2 & + & H_3 \end{array} \qquad (1)$$

Table 1. Parameters[a] Characterizing EPR Spectra of VO^{2+} and Nitroxyl Spin-Labels

Paramagnetic specie	g_x	g_y	g_z	A_x	A_y	A_z
$[VO(H_2O)_5]^{2+}$	1.9781	1.9781	1.9332	70.68	70.68	182.59
Spin-label[b]	2.0089	2.0064	2.0027	4.58	4.58	32.84

[a]hfc components are given in units of cm^{-1} x 10^4. [b]2,2,5,5-tetramethyl-1-oxypyrrolinyl-3-carboxylic acid.

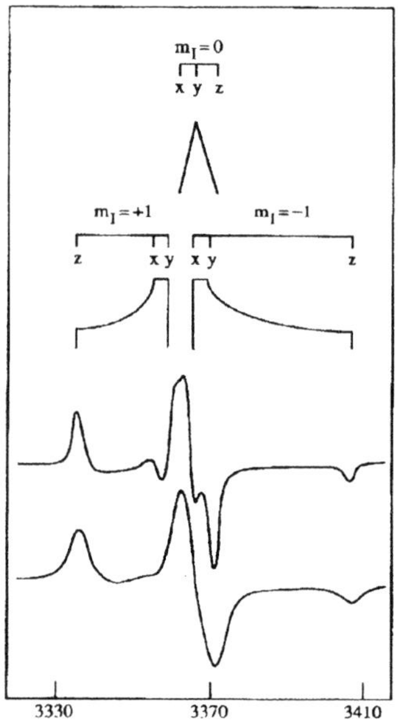

Figure 2. Schematic diagram of the rigid limit EPR powder pattern for the first-derivative EPR absorption spectra of methyl *N*-(2,2,5,5-tetramethyl-1-oxypyrrolinyl-3-carboxyl)-L-alaninate and of methyl *N*-(2,2,5,5-[$^2H_{12}$]tetramethyl-1-oxy-[3-^{2}H]pyrrolinyl-3-carboxyl)-L-alaninate in frozen d_4-methanol. In the upper half of the figure, stick diagrams identify different components of the electronic spin transitions according to g_x, g_y, and g_z and values of m_I. The upper spectrum is of the d_{13}-spin-labeled compound while the lower spectrum is of the spin-labeled compound of natural abundance isotope composition. EPR parameters used to calculate the stick diagrams at a microwave frequency of 9.45 GHz are given in Table 1. Reprinted from Makinen et al. (1998) with permission.

As a rapidly tumbling species in fluid solution at ambient temperature, the VO^{2+} ion gives rise to eight sharp EPR transitions because of the hf interaction of the unpaired electron with the 100% naturally abundant (I = 7/2) ^{51}V nucleus. Correspondingly, the nitroxyl radical gives rise to three EPR transitions due to the interaction of the unpaired spin with the (I = 1) ^{14}N nucleus. In the limit of a poly-crystalline sample or frozen glassy solution near the temperature of liquid helium, as illustrated in Figs. 1 and 2, the EPR spectra are accordingly composites from different sets of molecules designated according to the projections of the ^{51}V or ^{14}N nuclear moments onto the laboratory magnetic field $\mathbf{H}_0$. For both unpaired spin systems, the separation of the principal components of the $\mathbf{g}_e$ tensor as a result of anisotropic hf interactions provides the basis to apply angle-selected ENDOR (Rist and Hyde, 1970).

For VO^{2+} with an axially symmetric $\mathbf{g}_e$ tensor and an axially symmetric hf tensor due to the ^{51}V nucleus, there are eight EPR absorption lines parallel to the symmetry axis of the $\mathbf{g}_e$ tensor and eight EPR absorption lines perpendicular to this axis (Gersmann and Swalen, 1962; Kivelson and Lee, 1964; Albanese and Chasteen, 1978). As seen in Fig. 1, the low-field and high-field absorptions in the spectrum are resolved while in the central region parallel and perpendicular components heavily overlap. The –7/2 parallel and –3/2 perpendicular components designated by arrows correspond to sets of molecules for which the V = O bond lies parallel or perpendicular to $\mathbf{H}_0$, respectively.

For the nitroxyl spin-label in Fig. 2, the low-field feature in the EPR spectrum arises from the interaction of the g_z component of the $\mathbf{g}_e$ tensor with the $m_I = +1$ projection of the ^{14}N nucleus. Since the g_z component is perpendicular to the x,y-plane of the pyrrolinyl ring, molecules for which the spin-label ring lies perpendicularly to $\mathbf{H}_0$ give rise to this feature of the EPR spectrum. Similar geometrical relationships apply also to the high-field absorption feature with $m_I = -1$. On the other hand, for an X-band spectrometer, the prominent central feature of the frozen solution spectrum of a spin-label arises from heavily overlapping absorption components parallel and perpendicular to the molecular plane. Since the g_x, g_y, and g_z components within the central feature of the EPR absorption spectrum of the nitroxyl group are overlapping at X-band frequencies, this feature is best described, therefore, as arising from a collection of randomly oriented molecules (Wells and Makinen, 1988; Makinen et al., 1998).

The principal values of $\mathbf{A}$ for the vanadium nucleus and $\mathbf{g}_e$ that characterize vanadyl complexes can be extracted from EPR spectra of frozen solutions or polycrystalline samples, as summarized for $[VO(H_2O)_5]^{2+}$ in Table 1. For vanadyl complexes it has been shown that the values of the principal spectroscopic parameters $A_{||}$, $A_{\perp}$ and $g_{||}$, $g_{\perp}$, in particular $A_{||}$ and $g_{||}$, reflect the number and types of equatorial donor-ligand atoms according to chemical element (Chasteen, 1981). Further analysis of this relationship has shown that structural distortion of the complex from square pyramidal to trigonal pyramidal geometry does not alter the contribution of an equatorial donor-atom to $A_{||}$ (Cornman et al., 1995, 1997; Smith et al., 2002).

In the case of the nitroxyl group, in which the unpaired spin density is shared primarily between the N and O atoms (Hayat and Silver, 1973; Davis et al., 1975), the isotropic hf coupling represented by A_0 denotes the fraction of spin density associated with the ^{14}N atom. This provides a measure of the effective dipolar position of the unpaired spin density associated with the N–O group (Wells and Makinen, 1988; Mustafi et al., 1990a; Makinen et al., 1998). The fraction of spin density associated with the N atom and the effective dipolar position of the unpaired spin may vary slightly according to the dielectric constant of the surrounding medium (Jost and Griffith, 1978). The maximum possible shift amounts to no more than 0.041 Å along the N–O bond between solvents of high and low dielectric constant (Makinen et al., 1998).

2.2 Mathematical Formulation of the Physical Basis of ENDOR

ENDOR spectroscopy is performed by "pumping" the electronic transitions of the paramagnetic system under high microwave power and irradiating the system simultaneously with a strong rf field. When the frequency of the rf field is scanned under these conditions and resonance of a nucleus interacting with the unpaired electron is reached, a forbidden transition equivalent to simultaneous electron and nuclear "spin flips" is stimulated, giving rise to increased EPR signal amplitude. Thus, the ENDOR method has its basis in detection of nuclear resonance absorption by observing changes in the intensity of an EPR line.

ENDOR spectroscopy under continuous wave (cw) conditions, as in the studies described here, is optimally carried out with paramagnetic probes with relatively long electronic spin-lattice relaxation times (T_{1e}), allowing microwave power saturation of the paramagnetic center to be achieved with temperatures ranging from that of liquid helium to that of liquid nitrogen. Since the EPR absorption of VO^{2+} and of nitroxyl spin-labels can be detected at room temperature, subsequent lowering of the temperature readily results in achieving an electronic spin-lattice relaxation time favorable for ENDOR spectroscopy. Also, in our experience, cw ENDOR is optimally detected in systems characterized by narrow EPR absorption features with high peak-to-peak amplitudes. With respect to this latter characteristic, we have found VO^{2+} and nitroxyl spin-labels to be ideal paramagnetic probes for ENDOR spectroscopy (Makinen and Mustafi, 1995; Makinen et al., 1998).

ENDOR spectroscopy provides the most precise method of measuring the strength of electron-nucleus hf interactions. Consequently, its application results in very high resolving power for study of molecular structure. Since an ENDOR spectrum is a nuclear resonance spectrum, one must determine the total field at the nucleus to calculate the transition frequency. Beginning with the spin Hamiltonian in Eq. (1), it can be shown (Hurst et al., 1985) that the first-order transition frequencies $\nu_\pm$ for a nucleus are given by Eq. (2)

$$\nu_\pm = [\sum_{i=1}^{3}[(m_s/g_e)(\sum_{j=1}^{3} g_j l_j A_{ji}) - l_i \nu_n]^2]^{1/2} \tag{2}$$

where the l_i are the direction cosines of $\mathbf{H}_0$ in the molecular axis system, A_{ji} is the orientation-dependent value of the hf coupling, ν_n is the nuclear Larmor frequency, m_s is the electron spin quantum number, and g_e represents the effective g value defined by the relation $g_e = [l_1^2 g_{xx}^2 + l_2^2 g_{yy}^2 + l_3^2 g_{zz}^2]^{1/2}$. In systems of low g-anisotropy with $S = 1/2$, the axis of quantization of the unpaired electron spin can be assumed as the direction of the applied magnetic field $\mathbf{H}_0$. If the applied field is oriented parallel to the principal axis j of the hf tensor $\mathbf{A}$, Eq. (2) simplifies to Eq. (3) where the separation of $\nu_\pm$ about ν_n is called the ENDOR shift.

$$\nu_\pm = \nu_n \pm |A_{jj}|/2 \tag{3}$$

For symmetric separations, the hf coupling is, thus, twice the value of this frequency spacing. Eq. (3) applies to the condition $\nu_n > |A|/2$, which is characteristic of ^1H and ^{19}F. For some nuclei, *e. g.*, ^{31}P, the condition $\nu_n < |A|/2$ may apply, in which case Eq. (3) becomes $\nu_\pm = |A|/2 \pm \nu_n$.

Within the strong-field approximation, the observed hf coupling A is given by Eq. (4) as a function of r and α where h is the Planck constant, r is

$$A = \frac{g_N |\beta_N| g_e |\beta_e|}{hr^3}(3\cos^2\alpha - 1) + A_{iso} \tag{4}$$

the modulus of the electron-nucleus position vector $\mathbf{r}$, and α is the angle between $\mathbf{H}_0$ and $\mathbf{r}$. For VO^{2+} and nitroxyl spin-label systems, the observed principal hfc components $A_{||}$ and $A_\perp$ correspond, respectively, to the maximum and minimum ENDOR shifts in the spectrum. The principal hfc components due to dipole-dipole interaction $A_{||}^D$ and $A_\perp^D$ correspond to the first term of Eq. (4) for values of $\alpha = 0°$ and $90°$, respectively. Under the conditions $|A_{iso}| << |A_{||}|$ and $|A_\perp|$, and $A_{||}^D > 0 > A_\perp^D$, the traceless dipolar hfc components $A_{||}^D$ and $A_\perp^D$ can be calculated under the constraint $(A_{||} + 2A_\perp) = 3A_{iso}$. The signature of the isotropic hfc constants of methyl and the ring hydrogens with respect to that of the nitroxyl nitrogen has been assigned on the basis of TRIPLE spectroscopy (Mustafi and Joela, 1995). These results validate application of the above expressions to assign the magnitude and signs of $A_{||}$ and $A_\perp$.

The vanadyl ion and nitroxyl spin-labels exhibit low g-anisotropy, as seen in Table 1. Since the pseudo-contact contribution to the isotropic hf coupling, represented as the second right-hand term in Eq. (4), is negligible in cases of low *g*-anisotropy (McConnell and Chestnut, 1958), we have made the approximation that A_{iso} arises entirely from the Fermi contact term. Because the unpaired electron is localized to the metal $3d_{xy}$ orbital in the VO^{2+} ion (Ballhausen and Gray, 1962) or primarily to the N–O group in spin-labels (Hayat and Silver, 1973; Mustafi et al., 1991), the transfer of unpaired spin density to other atoms in the system is small. This aspect is of particular importance, for it ensures valid application of the point-dipole approximation, allowing structure determination with high precision.

Snetsinger *et al.* (1992) have shown that attributing the observed hf coupling of an unpaired electron in a metal $3d_{xy}$ orbital entirely to the dipole-dipole interaction with a nuclear spin of $I = 1/2$ over a distance of 2.09 Å results in a calculated separation of 2.15 Å. This result applies directly to the VO^{2+} ion, and the difference of 0.06 Å can be considered essentially inconsequential. Since the contribution of the isotropic hfc is less for larger electron-nucleus distances, the error will be correspondingly smaller. The shortest vanadium-proton distance that we have measured is of the order 2.6 Å, corresponding to the protons of solvent molecules in the inner coordination sphere of the $[VO(H_2O)_5]^{2+}$ ion (Mustafi and Makinen, 1988).

In the case of nitroxyl spin-labels, the important electron-nucleus distances for structure analysis correspond invariably to nuclei not immediately attached to the pyrrolinyl ring. However, even in the case of hydrogens attached to ring carbon atoms, the error remains negligible. The electron–nucleus separation measured by ENDOR for the vinyl proton in 2,2,5,5-tetramethyl-1oxypyrroline-3-carboxamide was 3.78 ± 0.01 Å (Mustafi et al., 1990c, 1991). The distance based on X-ray coordinates (Turley and Boer, 1972) yields 3.79 Å. In general, we have shown that errors associated with the point-dipole and strong-field approximations are less than 5% for $r \geq 5$ Å in both types of paramagnetic systems. Furthermore, assignment of the effective dipolar position of the unpaired spin of the N–O group on the basis of ENDOR-determined electron-nucleus distances has been made to within ± 0.04 Å (Mustafi et al., 1991). We, thus, conclude that errors originating from the point-dipole and strong-field approximations are negligible (Makinen and Mustafi, 1995; Makinen et al., 1998).

2.3 Characteristics of ENDOR Spectra at Selected Molecular Orientations

To determine structural information by ENDOR, one must (*i*) identify the surrounding nuclei contributing to hf interactions and (*ii*) determine the dependence of ENDOR spectra on the settings of the external magnetic field with respect to magnetic axes in the molecule. For VO^{2+}-complexes and nitroxyl spin-labels, the g_{zz} or $g_{||}$ component of the $\mathbf{g}_e$ tensor is critical for selection of molecular orientation. In the case of VO^{2+}, setting $\mathbf{H}_0$ to the –7/2 parallel component of the EPR spectrum (*cf.*, Fig. 1) selects those molecules for which the V=O bond or $g_{||}$ component of the $\mathbf{g}_e$ tensor is parallel to the laboratory magnetic field, *i.e.*, the equatorial *x,y*-plane is perpendicular to $\mathbf{H}_0$. Similarly for spin-labels, setting $\mathbf{H}_0$ to the low-field component of the EPR spectrum (*cf.*, Fig. 2) selects those molecules for which the g_{zz} component of the $\mathbf{g}_e$ tensor is parallel to $\mathbf{H}_0$, *i.e.*, the molecular *x,y*-plane is similarly perpendicular to $\mathbf{H}_0$. Correspondingly, setting $\mathbf{H}_0$ to the –3/2 perpendicular component of the EPR spectrum of VO^{2+} (*cf.*, Fig. 1) selects molecules for which the *x,y*-molecular plane is parallel to the laboratory field. In the case of nitroxyl spin-labels, the central prominent absorption feature arises from molecules of all orientations, including those for which the molecular plane is perpendicular to $\mathbf{H}_0$.

Figs. 3 and 4 compare the expected pattern of observed ENDOR splittings as a function of the positions of protons and the orientation of $\mathbf{H}_0$ with respect to $g_{||}$ of VO^{2+} or with respect to g_{zz} of spin-labels, respectively. When $\mathbf{H}_0$ is perpendicular to the molecular plane, $A_{||}$ of a proton located along the symmetry axis or $A_{\perp}$ of a proton located near or in the *x,y*-plane will be observed. On the other hand, if $\mathbf{H}_0$ is parallel to the *x,y*-plane, $A_{\perp}$ of a proton located along the symmetry axis and both $A_{||}$ and $A_{\perp}$ of a proton located near or in the *x,y*-plane are observed. On this basis, the ENDOR splittings for both types of paramagnetic probes can be classified into three categories:

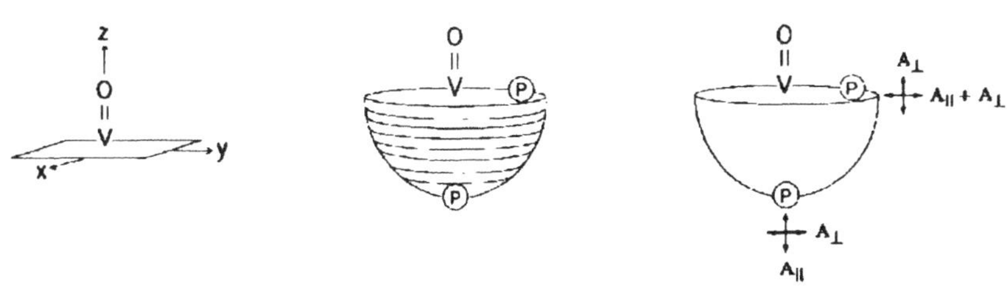

Figure 3. Schematic illustration of the relationships of the symmetry axis of the VO^{2+} ion to the principal axes of the **g** tensor and of the hf tensors of nearby protons. *Left diagram*: Direction of the molecular axes with respect to the V=O bond. *Central diagram*: Relative positions of protons near the molecular *x,y*-plane and near or along the *z*-axis. Each circle represents an orientation of $\mathbf{H}_0$ within the *g*-axis system. *Right diagram*: Principal hfc components that are detected for equatorially or axially positioned protons according to whether $\mathbf{H}_0$ is aligned parallel or perpendicular to the *z*-axis.

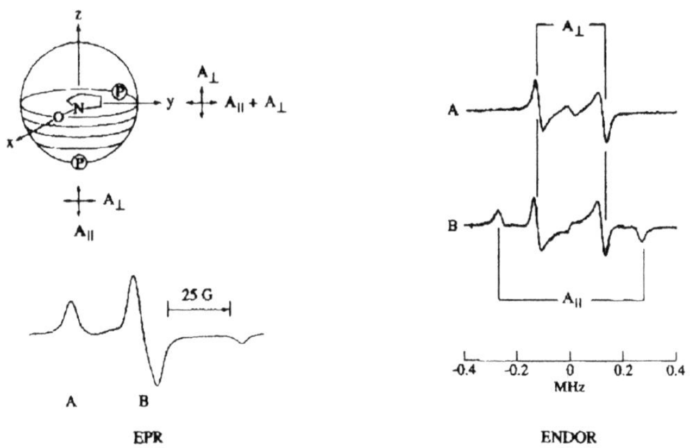

Figure 4. Magnetic interactions governing angle-selected ENDOR in a nitroxyl spin-label. *Upper left-hand portion*: Relationships of molecular axes to the principal axes of the $\mathbf{g}_e$ tensor of the nitroxyl group and of the A tensors of nearby protons located in the *x,y*-plane or on the *z* axis. *Lower left-hand portion*: Frozen solution EPR spectrum of a nitroxyl spin-label indicating $|\mathbf{H}_0|$ settings (**A**, **B**) for microwave saturation. (*cf.*, Fig. 2). *Right-hand portion*: ENDOR spectra of *N*-(2,2,5,5-tetramethyl-1-oxypyrrolinyl-3-carbonyl)-L-alaninate enriched with deuterium except at H^{α} recorded for settings A and B of the static magnetic field $\mathbf{H}_0$. The abscissa measures the observed ENDOR shift (observed frequency of resonance minus the proton Larmor frequency). Reprinted from Mustafi et al. (2001) with permission.

i. splittings that are observed only for the parallel orientation ($\mathbf{H}_0$ parallel to the g_{zz} or $g_{||}$ axis);
ii. splittings that are observed only for the perpendicular orientation ($\mathbf{H}_0$ perpendicular to $g_{||}$); and
iii. splittings that are observed as common to both parallel and perpendicular orientations.

When *g*-anisotropy is small compared to the average *g* value, the maximum hf interaction energy occurs for a field oriented along the electron-nucleus vector (A_z axis), and the minimum occurs when the field is in the A_x,A_y plane. For a proton in the *x,y*-plane, well resolved features corresponding to $A_{\perp}$ are observed which persist with an essentially constant ENDOR shift at all $\mathbf{H}_0$ values (Hurst et al., 1985). This observation is an important diagnostic feature since the observed $A_{||}$ couplings reach a maximum splitting dependent on $\mathbf{H}_0$ (Yim and Makinen, 1986; Mustafi et al., 1990b). Both $A_{||}$ and $A_{\perp}$ are observed when the magnetic field orientation $\mathbf{H}_0$ is near or in the molecular *x,y*-plane. This relationship provides the basis to analyze hf interactions in terms of nuclear coordinates for both types of paramagnetic species. For VO^{2+} and nitroxyl spin-labels, we have observed hitherto only axially symmetric hf interactions such that each class of protons gives rise to only $A_{||}$ and $A_{\perp}$ couplings. Furthermore, the principal axes of the hf and $\mathbf{g}_e$ tensors are observed to be coincident. These relationships may not always obtain. Surprisingly, in the case of $Gd(CH_3COO)_3$ in frozen glassy solutions, the covalent protons of the acetate ligand exhibit axially symmetric hf couplings while the protons of inner-shell coordinated water molecules within the same complex do not (Yim and Makinen, 1986). Similar observations have been made by others (Gochev and Yordanov, 1993; Zdravkova and Yordanov, 1994).

3. ENDOR CHARACTERIZATION OF STRUCTURED SOLVENT IN SMALL MOLECULE COMPLEXES AND IN PROTEINS

3.1 VO^{2+} as a Structural Probe of Ca^{2+} and Mg^{2+} Sites in Proteins and Nucleotides

The molecular structure of $[VO(H_2O)_5]^{2+}$ was first elucidated through single crystal ENDOR studies (Atherton and Shackleton, 1980, 1984). However, results of early magnetic resonance studies supported the intuitive expectation that the *penta-aquo* vanadyl cation $[VO(H_2O)_5]^{2+}$ in solution would exhibit square pyramidal geometry with tetragonal symmetry, as had already been shown by X-ray for *bis*(acetylacetonato)oxovanadium(IV) (Dodge et al., 1961). The results of early nuclear magnetic resonance (NMR) studies using ^{17}O-enriched water were consistent with a square pyramidal complex; however, the presence of the fifth axial ligand could not be unambiguously established (Wuthrich and Connick, 1968). Albanese and Chasteen (1978) analyzed the EPR spectrum of VO^{2+} in frozen aqueous medium and were the first to show quantitatively that the dipolar broadening produced by protons of inner sphere coordinated water molecules were consistent with $[VO(H_2O)_5]^{2+}$ as a complex of square pyramidal geometry, using vanadium-oxygen bond distances and valence angles determined crystallographically (Ballhausen et al., 1968). From proton ENDOR spectra of VO^{2+} incorporated into host single crystals of $Mg(NH_4)_2(SO_4)_2 \cdot 6\,H_2O$, in which the VO^{2+} ion replaced Mg^{2+} plus one water molecule, Atherton and Shackleton (1980, 1984) determined the principal hfc components for all ten protons of the $[VO(H_2O)_5]^{2+}$ species. They showed that the traceless components of the principal hf tensors are nearly axially symmetric, expected for point-dipole interactions. Because the anisotropy of the various magnetic interactions was explored by rotation about crystal axes with respect to the applied magnetic field $\mathbf{H}_0$, coordinates of each proton in the crystal could be assigned.

A more extensive study of the solvation structure of the VO^{2+} ion was made by ENDOR spectroscopy of frozen solutions of VO^{2+} in methanol and water-methanol mixtures on the basis of both ^{1}H and ^{13}C ENDOR (Mustafi and Makinen, 1988). In this study the structure of solvated VO^{2+} was assigned through molecular modeling constrained by ENDOR-determined electron-nucleus distances. Although resonance features arising separately from methyl and hydroxyl protons overlap under these conditions, a total of seven pairs of resonance features due to hydroxyl protons with five pairs due to methyl protons were identified by selective deuteration. The principal hfc components for each class of protons were identified from ENDOR spectra with $\mathbf{H}_0$ settings at the −7/2 parallel and −3/2 perpendicular EPR absorption features (*cf.*, Fig. 1), corresponding to magnetic field orientations parallel and perpendicular to the V=O bond, respectively.

For each class of protons, the principal hfc components for the $[VO(CH_3OH)_5]^{2+}$ complex were axially symmetric. In Table 2 are listed the dipolar and isotropic hfc components, the metal-nucleus distances estimated according to Eq. (4), and brief comments for each class of nuclei to indicate the structural relationships of the ligand to the VO^{2+} ion. Because the unpaired electron is localized to the metal $3d_{xy}$ orbital (Ballhausen and Gray, 1962), the isotropic contributions of ligand nuclei in the equatorial plane are significantly higher than for the axially coordinated ligand. Nevertheless, the values of A_{iso} from this study correspond very closely to values determined on the basis of single crystal ENDOR studies (Atherton and Shackleton, 1980).

Fig. 5 illustrates the solvation structure of $[VO(CH_3OH)_5]^{2+}$ with ENDOR assigned inner- and outer-sphere coordinated methanol molecules (Mustafi and Makinen, 1988). In this structure, the outer-sphere coordinated molecules were assigned orientations that could account for plausible hydrogen-bonding interactions with inner-sphere coordinated methanol molecules but which were compatible with the ENDOR-determined vanadium-nucleus distances in axial or equatorial positions. It was shown by analysis of ENDOR spectra that $[VO(CH_3OH)_5]^{2+}$ was a unique complex formed only in neat methanol. In water-methanol mixtures, two types of species were identified: one with axially coordinated water *trans* to the vanadyl oxygen and the other with axially coordinated methanol. Both types of complexes were shown to have only equatorially coordinated water molecules. The coordination geometry of $[VO(CH_3OH)_5]^{2+}$ in neat methanol and of $[VO(H_2O)_5]^{2+}$ and of $[VO(H_2O)_4(CH_3OH)]^{2+}$ in water-methanol cosolvent mixtures was best accounted for as square-pyramidal with tetragonal symmetry. The structural detail obtained in this ENDOR study approached the precision associated with small molecule X-ray crystallographic studies.

Table 2 Summary of the Principal hfc Components (in MHz) and Estimated Metal-Nucleus Distances (in Ångstroms) for $[VO(CH_3OH)_5]^{2+}$

$A_{\parallel}$	$A_{\perp}$	A_{iso}	$A_{\parallel}^{D}$	$A_{\perp}^{D}$	r (Å)	Comment[a]
^{1}H-ENDOR						
6.12	3.21	-0.10	6.22	-3.11	2.90	*ax*. OH
4.34	1.40	0.50	3.83	-1.91	3.42	V=O···H
15.67	0.70	4.76	10.91	-5.46	2.52	*eq*. OH (in plane)
13.38	1.03	3.77	9.61	-4.78	2.60	*eq*. OH (out-of-plane)
1.64	0.31	0.34	1.30	-0.65	4.80	outer-sphere, *ax*. OH
7.61	0.12	2.45	5.16	-2.57	3.20	outer-sphere, *eq*. OH
2.36	1.40	-0.15	2.51	-1.25	3.92	*ax*. CH_3
0.80	0.45	0.03	0.83	-0.42	5.60	V=O···$HOCH_3$
3.21	1.96	-0.23	3.44	-1.73	3.54	*eq*. CH_3
1.83	0.16	0.50	1.33	-0.66	4.85	outer-sphere, *ax*. CH_3
^{13}C-ENDOR						
0.99	0.44	0.04	0.95	-0.48	3.43	*ax*. CH_3
2.04	0.67	0.23	1.81	-0.90	2.78	*eq*. CH_3
0.23	0.13	-0.01	0.24	-0.12	5.45	outer-sphere *eq*. CH_3

[a]Axial and equatorial (abbreviated as *ax*. and *eq*., respectively) protons of OH and CH_3 and ^{13}C-nucleus of the methyl group were assigned on the basis of ENDOR spectra according to chemical origin and magnetic field ($\mathbf{H}_0$) settings. From Mustafi and Makinen (1988).

As shown in Fig. 1, inhomogeneous broadening observed in the spectrum of $[VO(H_2O)_5]^{2+}$ provides a particularly pertinent example of the difficulty to assign coordination environment on the basis of EPR alone. Axial and equatorial ligands do not make equivalent contributions to the shf broadening. In the case of $[VO(H_2O)_5]^{2+}$, Albanese and Chasteen (1978) were the first to point out that shf broadening by protons of axial ligands is weak compared to that of equatorial ligands. We have, furthermore, observed that the shf contributions of axial ligands can be masked by equatorial ligands. For instance, the line widths of EPR spectra of nucleotide complexes of VO^{2+} formed with ADP (Mustafi et al., 1992) or 5'-GMP (Jiang and Makinen, 1995) and having $[VO(NXP)_2H_2O)]$ composition are insensitive to exchange of perdeuterated solvent. Nonetheless, for such complexes, in which the nucleotide phosphate groups are equatorially coordinated to VO^{2+}, the presence of an axially coordinated solvent molecule could be demonstrated by ENDOR. Similarly, Schweiger and coworkers were able to demonstrate on the basis of ^{13}C ESEEM that benzaldehyde was axially coordinated through its carbonyl oxygen to VO^{2+} in *bis*(1*R*-3-heptafluorobutyrylcamphorate)oxovanadium(IV), a Diels-Alder catalyst (Togni et al., 1993). By applying $\mathbf{H}_0$ to the −7/2 parallel EPR absorption feature in pulsed ESEEM studies, they were able to show that the ^{13}C hfc components of an axial aldehyde group were very similar to those of $^{13}CH_3OH$ axially coordinated to VO^{2+}, as illustrated in Fig. 5.

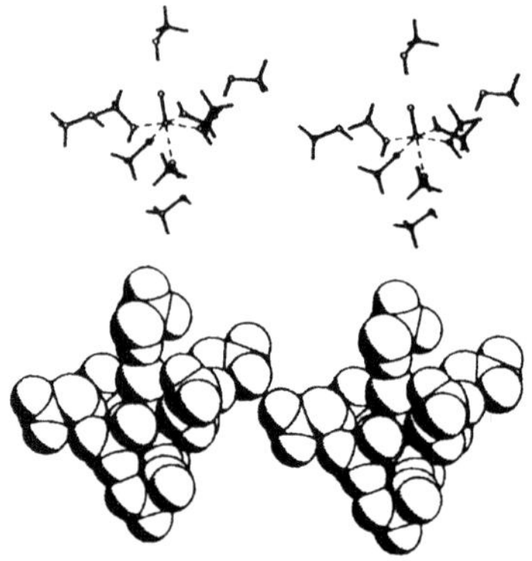

Figure 5. Stereo diagram of the coordination structure of VO^{2+} in methanol determined on the basis of ENDOR and molecular modeling. The upper diagram illustrates the complex in stick skeletal form. Broken lines connect the inner-sphere methanol molecules coordinated to the vanadium. The lower diagram illustrates the complex in space-filling form (in the same projection as in the upper diagram), and was drawn to scale for van der Waals radii of 1.53 Å (C), 1.4 Å (O), 1.2 Å (H), and 1.35 Å (V). From Mustafi and Makinen (1988) with permission of the American Chemical Society.

The VO^{2+} ion occupies a position between Ca^{2+} and Mg^{2+} according to ionic charge density defined on the basis of Z/R, where Z is the number of electrons and R the ionic radius (Williams, 1985). This characteristic undoubtedly accounts for its successful application as a paramagnetic probe of Ca^{2+}- and Mg^{2+}-binding sites in proteins and nucleotides although VO^{2+} has been used as a substitute also for Zn^{2+} and Fe^{3+} sites in proteins (Chasteen, 1981, 1983,1990). We have found VO^{2+} to be a particularly useful probe of Ca^{2+}-binding sites in proteins, revealing the chemical origins of both equatorial and axial ligands (Mustafi and Nakagawa, 1994, 1996; Mustafi et al., 2000). In these studies the resonance features of $[VO(H_2O)_5]^{2+}$ in solution have provided an important basis for analyzing coordination environment in proteins. We have shown that the X-ray structure of Ca(α-ethylmalonate)$_2$, generally considered a model compound of γ-carboxy-L-glutamic acid (Gla) residues in proteins (Zell et al., 1985), does not account for the Ca^{2+}-binding properties of the A and B isoforms of mammalian nephrocalcin. This protein containing four distinct Ca^{2+}-binding sites is the important factor secreted into renal tubules retarding stone formation in the mammalian kidney. Isoforms A and B, exhibiting the tightest Ca^{2+}-binding affinity ($K_d \sim 10^{-8}$ M), contain 3-4 equivalents of Gla residues each while isoforms C and D have none and are associated with lower Ca^{2+}-binding affinity by two orders of magnitude (Nakagawa et al., 1981, 1983, 1985).

Fig. 6 compares the proton ENDOR spectra of VO^{2+}-complexes of nephrocalcin isoforms B and D in natural and perdeuterated aqueous buffer. In each panel, the ENDOR spectrum of $[VO(H_2O)_5]^{2+}$ is also compared with that of the nephrocalcin complex. The results show that in isoform B, as in isoform A (Mustafi et al., 2000), only protein residues with non-exchangeable hydrogens are detected as ligands with complete exclusion of solvent water from the inner coordination sphere of the metal ion. The coordination geometry suggested by the Ca(α-ethylmalonate)$_2$ complex places the Gla residues as bidentate ligands in the equatorial plane with an axial water (Zell et al., 1985). Clearly, the structure of the Ca(α-ethylmalonate)$_2$ complex cannot account for the coordination environment of the Ca^{2+} sites in nephrocalcin isoforms A and B. For isoforms C and D, on the other hand, two water molecules are detected by ENDOR in the inner coordination sphere of the metal ion (Mustafi et al., 2000), but these isoforms do not contain Gla residues.

Nephrocalcin isoforms A and B undergo a more prominent conformational change upon binding Ca^{2+} or VO^{2+} than do isoforms C and D, as measured through circular dichroism (Mustafi and Nakagawa, 1994, 1996; Mustafi et al., 2000). Under the assumption that the interior regions of isoforms A and B acquire a lower dielectric constant through the conformational change, becoming less polar, it is unusual that water is excluded from the inner coordination sphere of the bound Ca^{2+} ions. It is of interest that parvalbumin contains two Ca^{2+}-binding sites, of which the high affinity site ($K_d \leq 10^{-7}$ M) similarly has no inner-sphere coordinated water (Kretsinger and Nockolds, 1973). Since nephrocalcin hinders the aggregation and growth of microcrystals of calcium oxalate into renal stones, elucidating the molecular basis by which stone formation is inhibited presents a challenging problem.

Since Mg^{2+} binds only to the phosphate groups of nucleotides (Happe and Morales, 1966) and is generally required for enzyme-catalyzed phosphoryl transfer reactions in cells, it is important to determine structures of Mg^{2+}-nucleotide complexes or to employ spectroscopic probes that closely simulate Mg^{2+} interactions with phosphate groups. In our laboratory we have observed through EPR and ENDOR spectroscopy that VO^{2+} is coordinated only to the phosphate groups of nucleotides, similar to that observed for Mg^{2+}-nucleotide complexes (Mustafi et al., 1992; Jiang and Makinen, 1995).

Since it was possible to demonstrate in these studies that VO^{2+}-phosphate binding was inhibited by Mg^{2+} with no evidence of non-specific binding of the VO^{2+} ion to the ribose hydroxyl groups or to the base hetero-atoms, it is likely that VO^{2+} may serve as the most suitable paramagnetic probe to simulate Mg^{2+}-nucleotide interactions in spectroscopic studies.

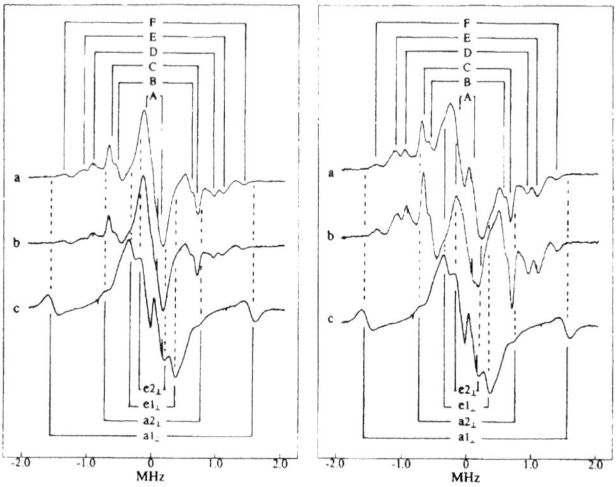

Figure 6. Proton ENDOR spectra of VO^{2+} complexes in frozen aqueous solutions. In the left-hand panel, spectra correspond to the following: **a**, VO^{2+} : nephrocalcin isoform B (4:1 molar ratio) in protiated aqueous buffer; **b**, VO^{2+} : isoform B (4:1 molar ratio) in deuterated aqueous buffer; and **c**, $[VO(H_2O)_5]^{2+}$ complex. In the right-hand panel, the spectra correspond to the same conditions but for nephrocalcin isoform D. In both sets of ENDOR spectra, the magnetic field was set to −3/2 perpendicular EPR absorption feature, and proton ENDOR absorptions from inner-sphere coordinated water molecules in axial and equatorial positions, labeled $\mathbf{a1}_\perp$, $\mathbf{a2}_\perp$, and $\mathbf{e1}_\perp$, $\mathbf{e2}_\perp$, respectively, are identified by stick diagrams. For the complex of isoform B, no resonance features for axial or equatorial coordinated water molecules are detected, as indicated by the broken vertical lines. For the VO^{2+}-complex of isoform D, proton ENDOR features for inner shell equatorial water molecules are seen, as indicated by the solid vertical lines between spectra **a** and **b**. In each panel ENDOR line pairs deriving from amino acid residues of the protein, labeled A-F, are indicated by stick diagrams with solid vertical lines for spectra **a**. The ENDOR line pairs are equally spaced about the free proton Larmor frequency of 13.9 MHz. The abscissa indicates the ENDOR shift. Reprinted from Mustafi et al. (2000) with permission of the CMB Association.

Of the five types of nucleic acid bases in DNA and RNA, guanine is unique because its nucleosides and nucleotides are capable of forming self-structured assemblies in solution through hydrogen-bonding to yield G – G base pairs and G-quartets. The latter are square planar arrays of the guanine bases with Hoogsteen hydrogen bonding interactions with each other. These assemblies are important because the 3'-overhang regions of DNA strands in the cell are rich in guanine and serve as the point of attachment of the DNA strand to protein subunits comprising the mitotic spindle apparatus in cell division (Blackburn, 2000). It is thought that G-quartet formation in these regions of DNA strands may be important in binding to the protein.

It has long been known that G-quartet assemblies can be formed from 5'-GMP in solution as a square-planar array of hydrogen-bonded guanine bases (Gellert et al., 1962). Equilibria controlling the formation of G-quartets and of stacked quartets to form octets and higher order assemblies are sensitive to pH and the presence of sodium and potassium ions (Pinavaia et al., 1978). Although helical fibers of stacked quartets of 5'-GMP were characterized crystallographically (Zimmerman, 1976) and by infrared spectroscopy (Audet et al., 1991), the pucker of the ribose ring and conformation of the base moiety in these helical arrays were not established.

Jiang and Makinen (1995) demonstrated on the basis of NMR studies that the $[VO(GMP)_2(H_2O)]$ complex in the presence of excess 5'-GMP enters into self-structured quartet and octet assemblies through hydrogen-bonding like the free nucleotide. They were able to show that the ENDOR shifts of the protons assigned to the ribose moiety and to the 8-H position in the nucleic acid base were identical for the monomeric form of $[VO(GMP)_2(H_2O)]$ as well as for the metal-nucleotide complex incorporated into quartet and octet assemblies. This observation indicated that the conformation of the metal-nucleotide complex was unchanged upon its incorporation into G-quartet assemblies.

Table 3 summarizes the ENDOR transition frequencies and their assignments for the $[VO(GMP)_2(H_2O)]$ complex in solution. Fig. 7 illustrates the results of torsion angle search calculations. Only a small family of conformations accommodate the ENDOR-determined vanadium-proton distances within van der Waals hard-sphere constraints. Different conformations of the ribose ring were tested. The ENDOR determined electron-proton distances restricted the ribose conformation to a *C3'- endo* pucker. Modeling studies using the ENDOR distances as constraints were able to rule out the *C2'- endo* conformation, which is the other prevalent conformation found for monomeric ribonucleotides (Saenger, 1984). On the other hand, for G-quartets formed with $[VO(GMP)_2(H_2O)]$ in solution, the guanine base was restricted to the *anti* conformation. The X-ray structure of double-stranded $d(G_4T_4G_4)$, which forms hydrogen-bonded arrays of G-quartets in crystals, shows the guanine base to occupy both *syn* and *anti* conformations within each planar quartet array (Kang et al., 1992). However, the quality of the electron density map did not allow an unambiguous assignment of ribose pucker. In deoxyribonucleotides the absence of the 2'-hydroxyl group allows more conformational flexibility with respect to ring pucker than in ribonucleotides. The structure of the $[VO(GMP)_2(H_2O)]$ complex incorporated into a G-quartet, as derived through EPR, ENDOR, and NMR investigations (Jiang and Makinen, 1995) is illustrated in Fig. 8.

Table 3. Summary of Principal Hyperfine Coupling Components (in MHz) and Estimated Metal-Proton Distances (in Ångstroms) in the $[VO(GMP)_2(H_2O)]$ Complex

$A_{\parallel}$	$A_{\perp}$	A_{iso}	$A_{\parallel}^{D}$	$A_{\perp}^{D}$	r (Å)[a]	proton assignment[b]
0.94	0.33	0.09	0.85	-0.42	5.65	Gua H(8)
1.58	0.35	0.29	1.29	-0.65	4.92	Rib H(1')
2.01	0.78	0.15	1.86	-0.93	4.35	Rib H(5')
2.12	0.17	0.59	1.53	-0.76	4.64	Rib H(O2')
3.51	2.82	-0.71	4.22	-2.11	3.31	Rib H(4')
5.08	2.30	0.16	4.92	-2.45	3.15	Axial H_2O

[a]Estimated error in r is 0.1-0.2 Å. [b]Proton assignments are based on ENDOR determinations and molecular modeling studies. From Jiang and Makinen (1995).

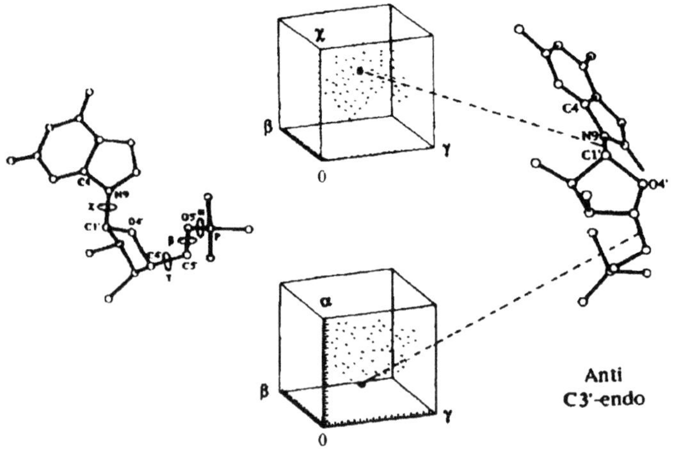

Figure 7. Angle maps showing conformational space accessible to the 5'-GMP moiety in $[VO(GMP)_2(H_2O)]$ under van der Waals hard-sphere constraints only (low density dots) and upon application of the distance constraints in Table 3 (high density dots connected by broken lines to the conformer accommodating vanadium-proton distance constraints within their line width based uncertainties). Because of the closed ring structures of the nucleic acid base and of the ribose, only four dihedral angles define the conformation of 5'-GMP (standard designations shown in the left-hand structure). The right-hand structure depicts the ENDOR assigned conformation as *C3'- endo*. In the upper angle map, the axes correspond to 0–360° of rotation for the dihedral angles β[C(4')–C(5')–O(5')–P(1")], γ[C(3')–C(4')–C(5')–O(5')], and χ[C(2')–C(1')–N(9)–C(8)] over which the search calculations were carried out in 1° increments. The lower angle map shares axes for dihedral angles β and γ while the third axis represents 0–360° of rotation for the dihedral angle α[C(5')–O(5')–P(1")–O(1")].

Hyperfine couplings of the vanadium center with ^{31}P nuclei estimated on the basis of our EPR studies of nucleotide structure with VO^{2+} as the paramagnetic probe have been amply confirmed through pulsed ESEEM and HYSCORE experiments (Dikanov et al., 1999, 2002). Fig. 9 shows the ligand hf structure underlying the $-3/2 \perp$ EPR absorption feature for a frozen solution containing excess ADP with respect to VO^{2+}. The resonance features in perdeuterated solvent (D_2O) exhibit a quintet pattern with 6.62 G interval spacings and approximate intensity ratio of 1:4: 6:4:1. Since ligand hfc produces a total of $(2nI + 1)$ absorption lines, this hf pattern can be attributed only to $n = 4$ structurally equivalent ($I = 1/2$) ^{31}P atoms, in view of the stoichiometry of VO^{2+} and ADP forming a complex with $[VO(ADP)_2(H_2O)]$ composition according to EPR titrations (Mustafi et al., 1992). This result demonstrates that VO^{2+} is coordinated to the 5'-diphosphate groups of two nucleotides with the phosphate oxygens in the equatorial plane. Similar conclusions were drawn for VO^{2+} complexed to adenosine 5'-(α,β-methylene)-diphosphate and to ATP.

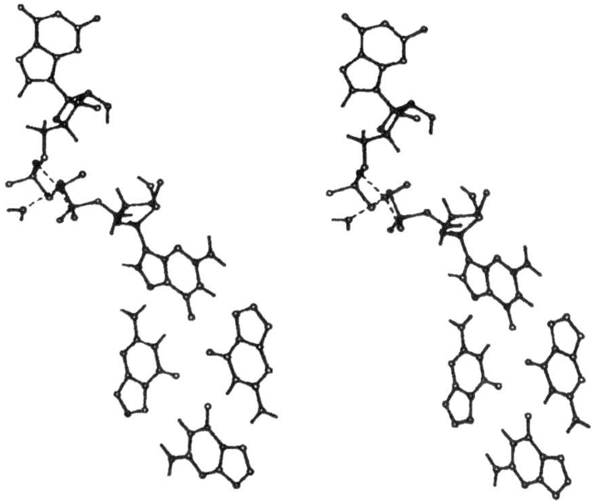

Figure 8. Stereo diagram showing how the ENDOR assigned structure of $[VO(GMP)_2(H_2O)]$ is accommodated into a G-quartet. The structure of the vanadyl complex itself in monomers, quartets, and higher order aggregates was determined by ENDOR. The ENDOR based $[VO(GMP)_2(H_2O)]$ complex was then modeled into a G-quartet with use of the X-ray coordinates of $[d(G_4T_4G_4)]_2$ (Kang et al., 1992). Reprinted from Jiang and Makinen (1995) with permission of the American Chemical Society.

As pointed out above, the ^{31}P shf coupling of only equatorial ligands is observed as a broadening contribution to the EPR line width of VO^{2+} complexes of nucleotides, and the weaker dipolar contributions of the axial solvent molecule are entirely masked. Assignments of tridentate triphosphate coordination to VO^{2+} by ATP have been made recently on the basis of ESEEM investigations (Dikanov et al., 2002). While confirming our estimates of ^{31}P shf couplings for equatorially coordinated phosphate groups, obtained, for instance, through spectra as in Fig. 9, the authors state that we failed to consider the possibility of tridentate triphosphate coordination with ATP because $^1H/^2H$ solvent exchange did not result in a change in the EPR line width. As has been pointed out earlier, Albanese and Chasteen (1978) were the first to show that shf broadening by equatorial ligands dominates the lineshape, masking the dipolar contributions of axial solvent molecules. We know of no instance in which ^{31}P shf coupling from axial ligands has been observed in VO^{2+} complexes structurally defined as tridentate triphosphate complexes by diffraction methods.

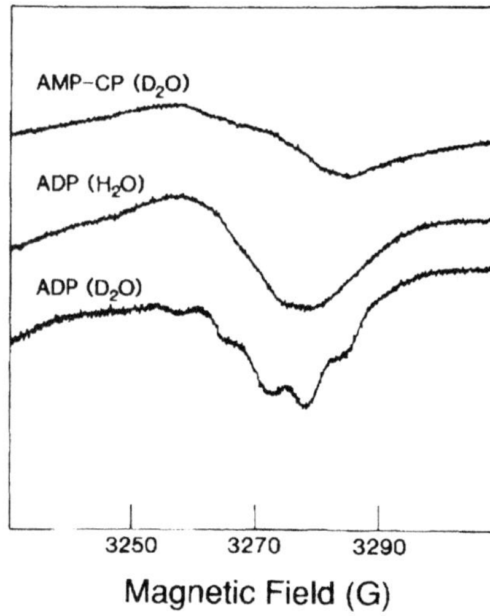

Figure 9. ^{31}P shf structure associated with the $-3/2 \perp$ EPR absorption feature of the frozen solution EPR spectrum of VO^{2+} complexed to ADP and adenosine 5'-(α,β-methylene)-diphosphate (AMP-CP) in aqueous methanol (50:50 v/v) cosolvent mixtures. The nucleotide:vanadyl ratio was in excess of 2:1 to ensure saturation of VO^{2+} with nucleotide ligand. Reprinted from Mustafi et al. (1992) with permission of the American Chemical Society.

3.2 Nitroxyl Spin-Labels as Probes of Structured Solvent

3.2.1 Assignment of conformation and structure on the basis of ENDOR constraints

The chemical bonding structure of nitroxyl spin-labels. Nitroxyl spin-labels are widely used in biophysical studies of macromolecules to characterize conformational changes and secondary structure (Hubbell et al., 1998; Langen et al., 2001), to measure inter-probe distances (Pfannebecker et al., 1996; Rakowsky et al., 1998; see Ch. 8 in this volume.), and to assess pH (Khramtsov and Weiner, 1988; Khramtsov and Volodarsky, 1998). Spin-labels are also used to determine the local concentration of molecular dioxygen in tissues by *in vivo* EPR imaging (Swartz and Halpern, 1998; see Ch. 11 and 12 in volume 23.). Structural formulas of nitroxyl spin-labels employed in biophysical studies are illustrated in Fig. 10. The diagram compares the variation in their cyclic structure as four-, five-, and six-membered ring compounds, with heteroatoms in the ring, with totally saturated carbons, or with unsaturated olefinic bonds.

In general, spin-labels with saturated ring structures have interconverting conformers in solution, and their proton ENDOR spectra show only broadened, overlapping features (Mustafi et al., 1990c). Also, the five-membered pyrrolidinyl spin-labels (**III**) consist of racemic mixtures because the C(3)-position, to which functional groups are covalently attached to the spin-label, constitutes a chiral center. Hitherto there has been only one investigation in which pyrrolidinyl spin-labels were resolved into their respective chiral enantiomers (Flohr and Kaiser, 1972; Ament et al., 1973). For *p*-nitrophenyl esters of 2,2,5,5-tetramethyl-1-oxy-pyrrolidine-3-carboxylic acid, that becomes the acyl moiety upon reaction with α-chymotrypsin to form an acylenzyme, rate-limiting deacylation was shown to be measurably different for the separated enantiomers of the spin-label probe. Also, through single crystal EPR studies, it was shown that the *R*(+)-acyl group was more favorably oriented for nucleophilic attack by the hydrolytic water molecule in the active site (Bauer and Berliner, 1979).

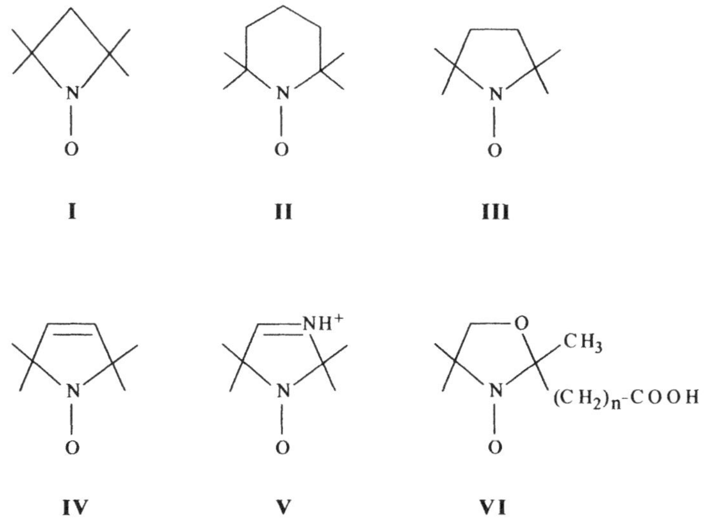

Figure 10. Schematic drawing of chemical bonding structures of four-, five-, and six-membered nitroxyl spin-label compounds.

It is important in ENDOR studies that the spin-label is not associated with conformational equilibria and that the nitroxyl N–O group is structurally invariant with respect to the molecular x,y-plane of the spin-label within the thermal motion of the nonhydrogen atoms. For distance measurements, therefore, we have employed as ENDOR probes only derivatives of spin-label **IV** illustrated in Fig. 10. The atoms constituting the side chain attached at the olefinic C(3') position are essentially coplanar with the ring atoms according to X-ray data (Turley and Boer, 1972; Makinen et al., 1998) and exhibit no ENDOR detectable conformational variation dependent on solvent environment (Mustafi et al., 1990c, 1991).

To assign the molecular conformation of the spin-labeled molecule on the basis of ENDOR-determined electron-nucleus distances, the coordinates of the constituent nonhydrogen atoms are constructed from X-ray defined molecular fragments, illustrated, for example, through Fig. 11 by construction of 7-*N*-(2,2,5,5-tetramethyl-1-oxypyrrolinyl-3-carboxyl)-cephalosporanic acid (SLCEP), a kinetically specific, spin-labeled cephalosporin substrate of class C β-lactamases (Mustafi et al., 1997).

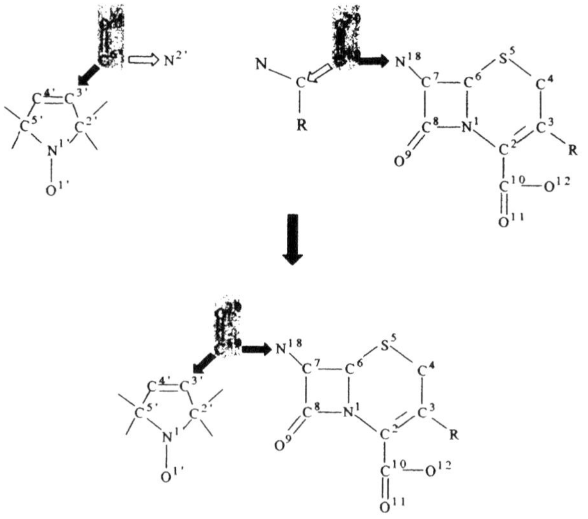

Figure 11. Illustration of the method of generating atomic coordinates for the molecular model of 7-*N*-(2,2,5,5-tetramethyl-1-oxypyrrolinyl-3-carboxyl)-cephalosporanic acid by joining molecular fragments of X-ray-defined molecules, in this case, 2,2,5,5-tetramethyl-1-oxy-pyrrolinyl-3-carboxamide (Turley and Boer, 1972) and cephaloglycin (Sweet and Dahl, 1970). Hydrogen atoms are added according to idealized parameters. See text for discussion.

Of importance in joining molecular fragments to generate coordinates of the spin-labeled molecule is to retain bond distances and valence angles for the atoms constituting the point of union that are reflective of their electronic structure in the resultant compound of interest. In this instance, since the bond distances and valence angles of the carboxamide group of the spin-label are nearly identical to those of the acylamido linkage of cephalosporin derivatives at the 7-*N*-position (Mustafi et al., 1997), to generate coordinates for SLCEP requires simply to preserve the C(3')–C(6') and C(6')–O(2') bond distances and the valence angles around C(3') and C(6') in spin-label **IV** and to assign the same bond distance between C(6') and the N(18) of the cephalosporin moiety as in cephaloglycine (Sweet and Dahl, 1970). The procedure has been described previously (Wells and Makinen, 1988; Wells et al., 1990, 1994; Mustafi et al., 1993, 1997; Mustafi and Makinen, 1994, 1995; Makinen et al., 1998).

Analysis of molecular structure and conformation. A series of torsion angle search calculations are then carried out around all rotatable bonds within van der Waals hard-sphere space constrained by the ENDOR-determined electron-nucleus distances and their line width based uncertainties. ENDOR spectra of SLCEP are presented in Fig. 12 to illustrate how resonance assignments are made to extract electron-nucleus distances as computational constraints. For this molecule, H(18) (*cf.*, atomic numbering scheme in Fig. 11) represents a classic example of a solvent exchangeable proton. In the top spectrum, the ENDOR features of H(18) labeled $a_{\|}$, $a_{\perp}$ were identified by their disappearance upon addition of an excess of D_2O or d_4-methanol. On the other hand, the features for H(7) labeled $b_{\|}$, $b_{\perp}$ were assigned on the basis of their similarity to the H(6) proton in SLPEN (6-*N*-(2,2,5,5-tetra-1-oxypyrrolinyl-3-carboxyl)-penicillanic acid) (Mustafi and Makinen, 1995) and H^{α} in spin-labeled amino acids, which has been assigned by deuterium substitution (Mustafi et al., 1990b; Wells et al., 1990; Joela et al., 1991). The lower spectrum is of SLCEP to which *one* equivalent of d_3-methanol has been added. Two line pairs not observed in the upper spectrum are identified and are assigned, therefore, to the parallel and perpendicular hfc components of the hydroxyl proton of the methanol molecule. These resonance features will be discussed later.

Table 4 summarizes the hf couplings extracted from ENDOR spectra of SLCEP, their corresponding electron-nucleus distances, and their line width determined uncertainties. Torsion angle analysis requires first determining the sterically accessible conformational space of the molecule through 0°-360° of rotation around each rotatable bond. Appropriate radii of atoms for these computations are, for example, the parameters published by Marshall and co-workers, generally applicable to polypeptides and nucleic acid residues (Iijima et al., 1987).

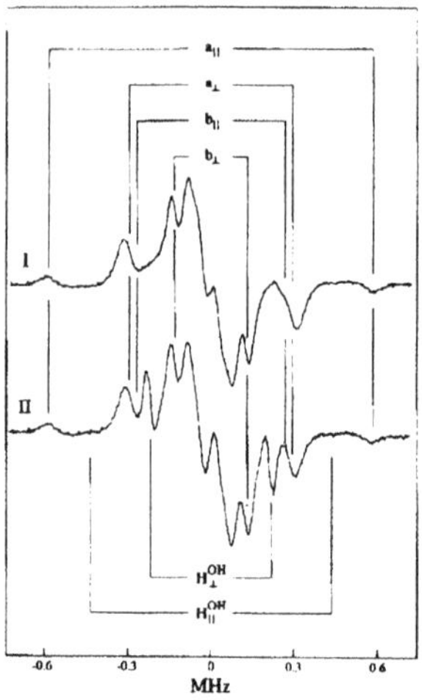

Figure 12. Proton ENDOR spectra of SLCEP in different solvents to illustrate determination of electron-proton distances for assignment of molecular conformation. Solvent conditions are (**I**), 50:25:25 (v/v) d_6-DMSO:d_6-toluene:d-chloroform to which d_4-methanol was added in 1:1 stoichiometry with respect to SLCEP and (**II**), 50:25:25 (v/v) d_6-DMSO:d_6-toluene:d-chloroform to which d_3-methanol was added in 1:1 stoichiometry. These spectra were taken with $\mathbf{H}_0$ at setting B of the EPR spectrum. The line pairs labeled a and b are assigned to H(18) and H(7), respectively (*cf.*, atomic numbering in Fig. 11). Because the $a_{||}$, $a_{\perp}$ and $b_{||}$, $b_{\perp}$ resonance features appear in the spectra for setting B, H(18) and H(7) must lie in or close to the molecular x,y-plane of the spin-label (*cf.*, Fig. 4). The spectra also identify a tightly bound methanol molecule forming a 1:1 adduct with spin-labeled cephalosporin through hydrogen-bonding. Both parallel and perpendicular hfc components of the methanol hydroxyl proton are identified in spectrum **II**. Therefore, the hydroxyl proton must similarly lie in or close to the plane of the spin-label. To detect the proton resonances assigned to the hydrogen-bonded methanol molecule, as shown in spectrum **II**, solvents were dried before use. Reprinted from Mustafi et al. (1997) with permission of the American Chemical Society.

For SLCEP the low density dots in the angle map in Fig. 13 represent van der Waals hard-sphere allowed conformational space for 0°-360° of rotation around bonds (**1**) C(19)–N(18), (**2**) N(18)–C(7), and (**3**) C(3')–C(19) where **1-3** represent the coordinate axes of the angle map. Within hard sphere constraints, the torsion angle search calculations are then carried out to identify which conformers accommodate the ENDOR distance constraints and their respective uncertainties summarized in Table 4. In Fig. 13 an arrow points to a small family of conformers represented by high density dots that satisfy the ENDOR distance constraints. Only one small family of conformers is accommodated by the distance constraints within the line width determined uncertainties. From the electron-nucleus distances corresponding to these torsion angles, a stereo diagram of the ENDOR-defined conformation of SLCEP is given in Fig. 14.

Table 4 Principal hfc Components (A, MHz) and Estimated Electron-Proton Distances (r, Å)[a] in SLCEP.

Proton	$A_{\parallel}$	$A_{\perp}$	A_{iso}	$A_{\parallel}^{D}$	$A_{\perp}^{D}$	r_{ENDOR}	r_{model}
Cephalosporin moiety of SLCEP							
H(18)	1.167	0.581	0.002	1.165	-0.583	5.13	5.12
H(7)	0.537	0.268	0.000	0.537	-0.268	6.66	6.66
H(6)	–	0.150	–	–	-0.150	8.08	8.08
H(4)	0.16	0.085	-0.001	0.167	-0.084	9.77	9.58
Solvent hydrogen-bonded to N(18)							
H^{OH}	0.875	0.439	-0.001	0.876	-0.438	5.66	–

[a]The uncertainties for each value of r based on the line width of the corresponding resonance feature are H(18), ±0.02 Å; H(7), ±0.03 Å; H(6), ±0.14 Å; H(4), ±0.22 Å, and H^{OH}, ±0.03 Å. From Mustafi et al. (1997).

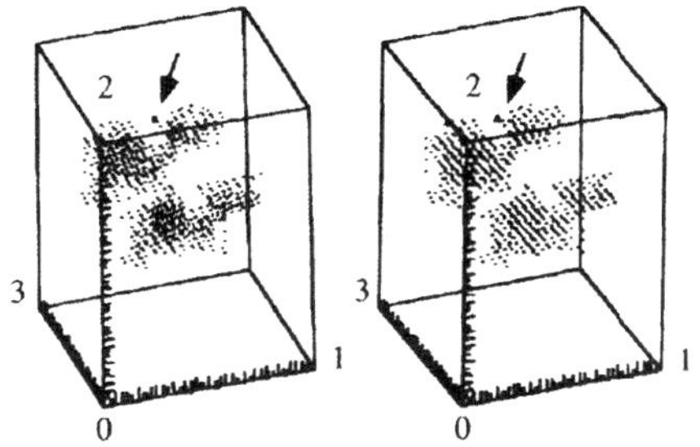

Figure 13. Stereo diagram of angle map showing conformational space accessible for SLCEP under van der Waals hard-sphere constraints alone (low density dots) and upon application of the distance constraints in Table 4 for H(18) and H(7) within van der Waals space (arrow pointing to a very small family of conformers). The axes **1**-**3** of the angle map correspond to 0°-360° of rotation around the three bonds (**1**) C(19)–N(18), (**2**) N(18)–C(7), and (**3**) C(3')–C(19) over which the search calculations were carried out, generally in 2° increments. See the numbering scheme of SLCEP in Fig. 11 for atom designations.

A single family of conformers accommodated by experimentally determined ENDOR constraints is not always the result. For instance, in the case of the analogous spin-labeled penicillin, two different families of conformers were identified compatible with ENDOR constraints (Mustafi and Makinen, 1995), of which one was exactly overlapping with the X-ray structure of amoxycillin (Boles et al., 1978) and benzylpenicillin (Fattah et al., 1993). Correspondingly more than two conformers of (2,2,5,5-

tetramethyl-1-oxypyrrolinyl-3-methyl)methanethiolsulfonate (SL-MTS), a spin-label compound employed in protein structure analysis through cysteine mutagenesis (Bolin et al., 1998; Hubbell et al., 1998; Bolin and Millhauser, 1999), were identified (Mustafi et al., 2002) and two conformers of methyl *N*-(2,2,5,5-tetramethyl-1-oxypyrrolinyl-3-carboxyl)-L-tryptophanate were identified according to ENDOR spectra (Wells et al., 1990).

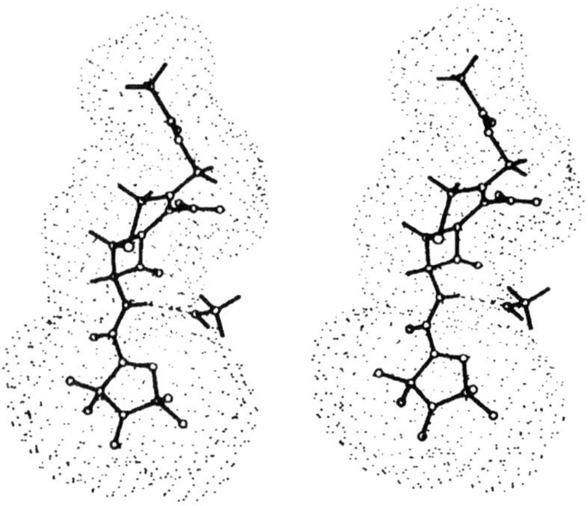

Figure 14. Stereo diagram of the ENDOR-determined conformation of SLCEP and the hydrogen-bonded methanol molecule representative of the small family of conformers identified in Fig. 13 that accommodate all ENDOR constraints. The dotted surface represents the solvent accessible surface of spin-labeled cephalosporin, calculated according to the Lee-Richards algorithm (Lee and Richards, 1971). The hydrogen-bond between the pseudopeptide NH group and O_{MeOH} is indicated by a broken line. Reprinted from Mustafi et al. (1997) with permission of the American Chemical Society.

If more than one conformer satisfies the ENDOR constraints, the relative occupancies of the conformers must be evaluated according to other physical criteria. For analysis of small molecules, we have found that a change in solvent resulting in alteration of dielectric constant may noticeably alter the relative amplitudes of ENDOR features, indicating shifts in conformational equilibria. Pertinent examples are SLMTS and spin-labeled methyl L-tryptophanate. In these cases, the change from methanol, a solvent of high dielectric constant, to chloroform/toluene, a solvent of low dielectric constant, not only removed the potential for hydrogen-bonding of polar groups with solvent but also favored almost total occupancy by the conformer of lower potential energy (Wells et al., 1990; Mustafi et al., 2002). A further criterion that can be invoked to rule out multiple conformers satisfying steric and ENDOR distance constraints are eclipsed torsion angle relationships observed through molecular graphics analysis. Conformers with eclipsed bonds are of high potential energy and are unlikely to be significantly populated. This criterion was invoked, for instance, in analysis of the conformers contributing to the ENDOR spectra of *N*-(2,2,5,5-tetramethyl-1-oxypyrrolinyl-3-carboxyl)-L-phenylalaninal, an aldehyde transition-state substrate analog of α-chymotrypsin (Jiang et al., 1998). Lastly, a check is made to ensure that the conformers satisfying the distance constraints are spectroscopically compatible with the two different patterns of ENDOR shifts and their dependence on $\mathbf{H}_0$.

Ordinarily it would be preferable to resolve instances of more than one conformer accommodating distance constraints through computational methods involving potential energy minimization or trajectory calculations in which the dynamical motion is simulated. While detailed force-field parameters have been derived for nitroxyl spin-labels (di Matteo and Barone, 1999; Improta et al., 2000), they are largely applicable to only saturated spin-labels associated with interconverting conformers. However, recently force-field parameters have been derived for the 2,2,5,5-tetramethyl-1-oxypyrroline-3-carboxyl molecular fragment (Structure **IV** in Fig. 10) suitable for energy minimization calculations and molecular dynamics (MD) simulations (Van Zele et al., 2001). Their accuracy was tested by simulating MD trajectories of methyl *N*-(2,2,5,5-tetramethyl-1-oxypyrrolinyl-3-carboxyl)-L-tryptophanate in methanol and chloroform, solvents of high and low dielectric constant, respectively, in which different conformers had been observed and structurally defined by ENDOR. Importantly, the force-field parameters are compatible with a variety of computational program suites such as CHARMM (Brooks et al., 1983; MacKerell et al., 1998), AMBER (Weiner and Kollman, 1981; Pearlman et al., 1995), and GROMOS (Scott et al., 1999).

Table 5 compares the MD averaged and ENDOR-determined electron-proton distances of the spin-labeled methyl L-trytophanate solute molecule in two different solvents. Not only was there close agreement of the electron-nucleus distances evaluated by both ENDOR and MD methods, as evident in Table 5, but also there was complete agreement between the number and types of conformers observed by MD and by ENDOR in each solvent. That is, two conformers were detected in methanol by ENDOR, termed conformers A and C, corresponding to the classical perpendicular and anti-perpendicular orientations of the indole side chain of the trytophanyl moiety, respectively (Wells et al., 1990). Only these two conformers were found as stable structures in MD simulations (Van Zele et al., 2001). The two MD averaged structures calculated from trajectories for each conformer are compared in Fig. 15. It can be seen in Table 5 that conformer C in chloroform is equivalent to conformer C in methanol within the root-mean-square fluctuations of the electron-nucleus distances. Not only was there good agreement between the ENDOR-determined and MD-averaged electron-nucleus distances, as summarized in Table 5, but there was also good agreement of the MD-averaged and ENDOR-based values of the dihedral angles within each solvent system.

Table 5. Comparison of MD Averaged and ENDOR-Determined Electron-Proton Distances in Spin-Labeled Methyl L-Tryptophanate

	Distance (Ångstroms)				
	Conformation A[a]		Conformation C[b]		
	Methanol		Methanol	$CHCl_3$	
Proton[c]	MD	ENDOR	MD	MD	ENDOR
HN	5.13 ± 0.17	5.11 ± 0.06	5.18 ± 0.17	5.05 ± 0.17	5.08 ± 0.04
HA	6.77 ± 0.18	6.70 ± 0.07	6.83 ± 0.14	6.77 ± 0.15	6.68 ± 0.08
HB1	7.34 ± 0.32	7.54 ± 0.08	7.27 ± 0.31	7.31 ± 0.25	7.01 ± 0.10
HB2	8.23 ± 0.28	8.29 ± 0.30	8.26 ± 0.24	8.11 ± 0.19	8.20 ± 0.18
HD1	6.35 ± 0.78	–[d]	7.79 ± 0.70	7.41 ± 0.52	–[d]
HE1	6.64 ± 1.06	–[d]	7.48 ± 0.97	6.73 ± 0.71	–[d]
HE3	8.29 ± 0.83	–[d]	6.36 ± 0.95	6.24 ± 0.52	–[d]
HZ2	7.79 ± 1.32	–[d]	7.02 ± 0.99	6.07 ± 0.75	–[d]
HZ3	9.15 ± 1.26	9.77 ± 1.02[e]	6.34 ± 1.39	5.99 ± 0.76	5.45 ± 0.10[e]
HH2	8.98 ± 1.44	10.28 ± 1.02[e]	6.70 ± 1.25	5.91 ± 0.79	7.63 ± 0.33[e]
HCM	9.70 ± 0.53	9.05 ± 0.22	9.68 ± 0.56	9.65 ± 0.49	9.67 ± 0.35

[a]In methanol the spin-labeled methyl L-tryptophanate is observed by ENDOR spectroscopy to partition into conformers A and C. Conformer A corresponds to the perpendicular conformer of the indole side-chain (χ_2[CA–CB–CG–CD_1] ~ 90°) and is of higher population in a solvent of high dielectric constant. Conformer C corresponds to the anti-perpendicular orientation (χ_2[CA–CB–CG–CD_1] ~ –90°) of the indole side chain (Wells et al., 1990).
[b]In chloroform/toluene only conformer C is observed with χ_2[CA–CB–CG–CD_1] ~ –90°.
[c]Atomic numbering for the tryptophanyl moiety corresponds to IUPAC-IUB conventions [*Biochemistry* **9**, 3471-3479 (1970)].
[d]The resonances of covalent hydrogens belonging to the aromatic indole group are not resolved in ENDOR experiments (Wells et al., 1990), and, therefore, no estimates of these electron-proton distances are given.
[e]In ENDOR experiments fluorine-substituted tryptophan derivatives were employed to assign orientation of the indole side chain since the resonance features of hydrogens attached to the aromatic ring system are not resolved. Proton ENDOR spectra are otherwise identical for fluorine substituted and parent tryptophan compounds (Wells, et al. 1990). On this basis, the corresponding electron-proton distances given here are derived from the ENDOR assigned conformation of the side chain using a C–H bond vector of length 1.00 Å in place of a C–F bond vector of length 1.30 Å. From Van Zele et al. (2001).

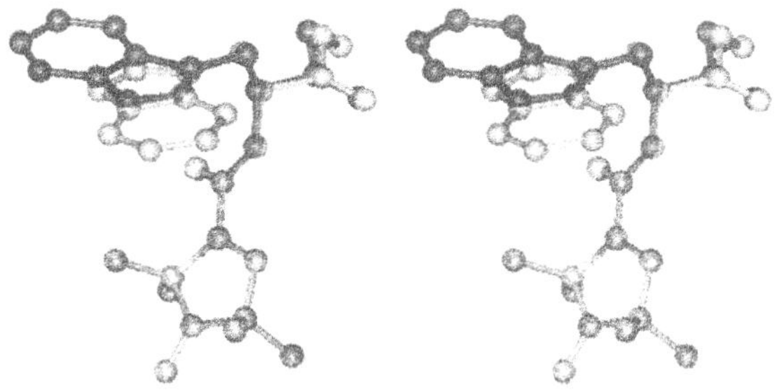

Figure 15. Stereo diagram of MD-averaged spin-labeled methyl L-tryptophanate in high and low dielectric solvents. The spin-label moieties for each conformation, including the carbonyl group adjacent to the spin-label ring, were superimposed to highlight the change in the orientation of the indole side chain. The molecule rendered black represents the conformer of higher population found in high dielectric solvents ($\chi_2 \sim +90°$) while that rendered gray corresponds to the conformer of higher population in low dielectric solvents ($\chi_2 \sim -90°$). See Table 5 for definition of χ_2. Reprinted from Van Zele et al. (2001) with permission.

Since bond lengths of nonhydrogen atoms and valence angles are often not fixed in MD simulations of small molecules, simulation of the dynamical motion of spin-labeled molecules constrained by ENDOR data with application of the force- field parameters for the spin-label moiety derived by Van Zele *et al.* (2001) represents a significantly improved means for structural analysis compared to the rigid body approach with fixed molecular fragments that has been hitherto necessary (Makinen, 1998; Makinen et al., 1998). On this basis, structure analysis would be comparable to present day applications of simulated annealing calculations of protein and polypeptide structure in which nuclear Overhauser distances are incorporated as restraints in NMR studies (Herrmann et al., 2002). ENDOR distance constraints, while fewer in number, are, however, of significantly higher precision than the inter-nuclear distances determined by NMR (Zhao and Jardetzky, 1994). Further exploration of this approach in ENDOR structural analysis should be pursued and applied to macromolecular systems since it offers the most accurate and precise method of defining local structural details, for instance, those in active sites of enzyme reaction intermediates that are likely to be of catalytic significance.

3.2.2 Structured solvent molecules hydrogen-bonded to spin-labeled β-lactam antibiotics

The penam and cepham fused ring structures of first generation β-lactam antibiotics belonging, respectively, to 6-aminopenicillanic acid and 7-aminocephalosporanic acid are illustrated in Fig. 16. The free amine forms of the β-lactam antibiotics exhibit no inhibitory action against pathogenic bacteria (Bush et al., 1995; Massova and Mobashery, 1998). However, derivatization of the amine group confers antimicrobial activity. The spin-labels **IV** in Fig. 10, used to derivatize the amine group of penicillin and cephalosporin in our ENDOR studies, are sterically and structurally analogous to a variety of acylamido groups found in clinically useful antibiotics.

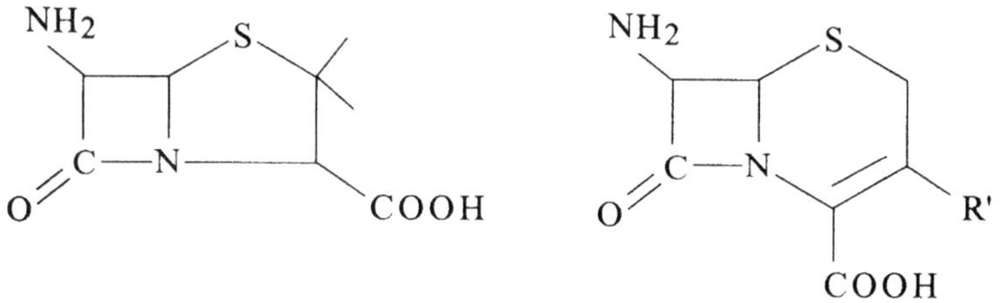

Figure 16. Chemical bonding structures of 6-aminopenicillanic acid (left) and 7-aminocephalosporanic acid (right), illustrating penam and cepham fused ring structures, respectively, of first generation β-lactam antibiotics.

We have observed through ENDOR studies that the –NH– group of the acyl-amido linkage of free, spin-labeled antibiotics in solution exhibits a pronounced tendency for hydrogen-bonding to solvent molecules (Mustafi and Makinen, 1995; Mustafi et al., 1997). In Fig. 12 the two narrow line pairs belonging to the OH group of the hydrogen-bonded methanol molecule appearing upon addition of only one equivalent of methanol in an anhydrous aprotic solvent indicate that methanol forms a tightly bound adduct with the SLCEP molecule. Since the parallel and perpendicular resonances of H^{OH} appear in the spectrum for setting B, the hydroxyl proton must lie in the plane of the spin-label according to the requirements of angle selection, as summarized in Fig. 4. A search for sites on the SLCEP molecule capable of forming a tightly bound adduct with the OH group of the methanol molecule such that the electron-proton distance of 5.66 ± 0.03 Å is satisfied and that the proton lies in the plane of the spin-label leaves only the acylamido –NH– group as a possible candidate for hydrogen-bonding. Moreover, by searching sterically accessible conformational space, we determined that the dipolar electron-proton distance to H^{OH} is compatible with the OH group hydrogen-bonded to the acylamido NH group *only* on the *endo* or concave surface of the *β*-lactam group (Mustafi et al., 1997). In Fig. 14 the dotted surface representing the solvent accessible surface (Lee and Richards, 1971) shows that the OH group of the methanol molecule positioned according to ENDOR structural constraints is sterically accommodated. Water molecules are found hydrogen-bonded to the acylamido –NH– group on the *endo* surface in the X-ray structures of cephaloglycin (Sweet and Dahl, 1970) and amoxicillin (Boles et al., 1978). By ENDOR we have similarly assigned a hydrogen-bonded solvent molecule on the *endo* surface of SLPEN (Mustafi and Makinen, 1995).

The β-lactam C–N bond of antibiotics free in solution is cleaved through solvolytic reactions and is the specific point of attack by β-lactamase enzymes. It is of particular interest to compare the known stereochemistry of solvolytic and enzyme-catalyzed reactions. The steric approach of the hydrolytic solvent molecule has not been resolved for free β-lactam antibiotics in solution and remains conjectural. Possible pathways for nucleophilic attack of the β-lactam carbonyl carbon are illustrated in Fig. 17. Attack from the *exo* or concave surface is stereoelectronically forbidden while nucleophilic attack from the *endo* or concave surface is stereoelectronically allowed (Deslongchamps, 1983; Benner, 1988). The hydrogen-bonded solvent molecules located on the *endo* or concave surface of β-lactam antibiotics defined by X-ray crystallographic data have been ignored in proposals of the mechanistic pathway for solvolysis (Page, 1987). Consequently it has been thought that nucleophilic attack can occur only on the *exo* or convex surface of the β-lactam ring because it was assumed that the *endo* surface could not sterically accommodate a solvent molecule. However, ENDOR identification of hydrogen-bonded solvent only on the *endo* surface for both SLCEP and SLPEN indicates that the solvolytic reaction of free β-lactam antibiotics in solution could proceed via *endo* nucleophilic attack, consistent with stereoelectronic rules (Mustafi and Makinen, 1995; Mustafi et al., 1997).

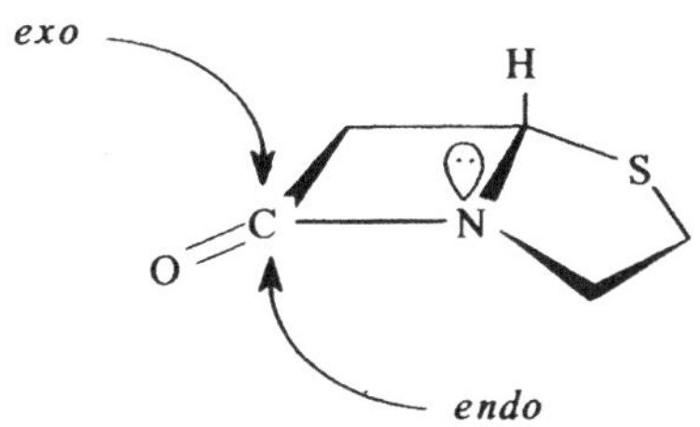

Figure 17. Schematic drawing of the molecular structure of β-lactam antibiotics to illustrate the *exo* or convex surface (from top side, sterically-preferred path, but stereoelectronically forbidden path for nucleophilic attack), and *endo* or concave surface (from bottom side, sterically more restricted, but stereoelectronically allowed path for nucleophilic attack). Note that in the case of *endo* attack, the path of the incoming nucleophile is antiperiplanar to the lone pair on the β-lactam nitrogen atom, as required for stereoelectronically allowed hydrolysis. On the other hand, the *exo* approach of an electron-rich nucleophile undergoes repulsive interactions with the lone pair orbital. Protonation of the β-lactam nitrogen prior to nucleophilic attack renders *exo* and *endo* surfaces stereoelectronically equivalent.

Interestingly, in TEM-1 β-lactamase, and, therefore, presumably in all other homologous class A β-lactamases, the active site Ser70 side chain and the deacylating water molecule approach the carbonyl carbon atom via the *exo* surface of the β-lactam substrate (Strynadka et al., 1992). Protonation of the β-lactam nitrogen prior to nucleophilic attack renders *exo* and *endo* surfaces stereoelectronically equivalent (Mustafi and Makinen, 1995) and protonation of the β-lactam nitrogen prior to *exo* nucleophilic attack is energetically favored (Atanasov et al., 2000). In class C *β*-lactamases, however, the approach of the hydrolytic water in the acylenzyme reaction intermediate is thought to proceed from the *endo* surface (Massova and Mobashery, 1998). Therefore, the hydrolytic mechanism of class C β-lactamases is likely to differ from that associated with class A enzymes. Further details of the mechanistic pathway of *β*-lactam hydrolysis in class C enzymes have yet to be defined.

3.3 ENDOR Detection of the Hydrolytic Water in a Spin-labeled Acylenzyme Reaction Intermediate of TEM-1 *β*-lactamase

3.3.1 Cryokinetic isolation of enzyme reaction intermediates

Much of our understanding of the structural basis of enzyme catalytic action is derived from X-ray analysis of enzyme-inhibitor complexes. Such complexes mimic in part enzyme reaction intermediates, but they are inherently chemically stable because of non-productive spatial relationships. On the other hand, true enzyme reaction intermediates are chemically labile because of catalytically productive spatial relationships. While the active site interactions induced by an inhibitor that is structurally similar to a substrate may overlap in part with those required for catalysis, the two different sets of enzyme-ligand interactions cannot be identical. Thus, in any enzyme-inhibitor complex at least one critical structural interaction will be absent that is required for catalysis. Nonetheless, mechanisms of action of enzymes have been generally derived through modeling based on active site spatial relationships observed in enzyme-inhibitor complexes.

There are salient instances in which mechanistic conclusions on the basis of only X-ray structure analysis have proven to be misleading. The recent identification of a covalent reaction intermediate of hen egg-white lysozyme (Vocadlo et al., 2001), in contrast to the general acid catalyzed mechanism involving no covalent interactions between the substrate and enzyme proposed on the basis of X-ray defined inhibitor complexes (Phillips, 1967; Stryer, 1988), serves as a prominent reminder. In addition, EPR and ENDOR characterization of a covalent (mixed anhydride) acylenzyme intermediate of carboxypeptidase A (Makinen et al., 1979; Kuo et al., 1983; Mustafi and Makinen, 1994) stands in contradistinction to the general base catalyzed mechanism favored through X-ray studies (Christianson and Lipscomb, 1986, 1989) since an enzyme-product complex was shown not to account for the spectroscopic results. Furthermore, nucleophile trapping experiments have demonstrated a mixed anhydride linkage for both esterolytic and proteolytic substrates of carboxypeptidase A (Sander and Witzel, 1985, 1986). These differences highlight the importance of identifying true reaction intermediates for structural studies. For these reasons, characterization of true, catalytically competent enzyme reaction intermediates should always be the preferred approach to define the structural basis of enzyme action.

We have found application of cryoenzymologic methods in combination with angle-selected ENDOR to be highly suitable for defining three-dimensional active site structure in catalytically competent intermediates of enzyme-catalyzed reactions. Cryoenzymology entails establishing, by experiment, conditions in the 0° C to –90° C range under which transient intermediates of enzyme-catalyzed reactions can be more readily detected and temporally resolved than at room temperature because of their intrinsically longer half-lives at low temperatures (Douzou, 1977; Makinen and Fink, 1977; Fink and Geeves, 1979; Fink and Cartwright, 1981). Cryokinetic isolation of an enzyme reaction intermediate is carried out in fluid, cosolvent mixtures at low temperatures which form glasses upon freezing. In contrast to frozen water or ice, which is crystalline, glasses are by definition isotropic in structure, and, therefore, are not expected to be associated with perturbation of protein conformation. In our studies ENDOR must be carried out on solid state systems (to detect the dipolar hf interactions through which structural relationships are defined) and at low temperatures (to saturate relaxation processes with microwave power that govern detection of resonance absorption). Consequently, preparation of enzyme reaction intermediates in cryosolvent mixtures followed by freeze-quenching of the reaction mixture is ideal for ENDOR spectroscopy.

The basis of cryoenzymology is very similar in principle to cryo-crystallography. In cryo-crystallography, cryoprotectant cosolvents and flash-cooling methods are employed, particularly when synchrotron radiation is used to collect diffraction data. The protein crystal is equilibrated with an aqueous, cryoprotectant, cosolvent mixture followed by flash-freezing in liquid nitrogen (Garman and Schneider, 1997; Rodgers, 1997). The cryoprotectant cosolvents employed are generally polyol substances such as glycerol, ethylene glycol, or 2-methyl-2,4-pentanediol which can be also used in cryoenzymology (Douzou, 1977). Because of the high aqueous solvent content of protein crystals (Matthews, 1968), it is believed that freezing of the cryoprotectant cosolvent mixture results in glass formation within the solvent channels of the crystal (Rodgers, 1997). This is essentially not different from freezing of enzyme complexes and enzyme reaction intermediates in cryosolvent mixtures for ENDOR. Since 1995, no less than

40% of X-ray determined protein structures have been carried out by application of cryo-crystallographic methods with data collection at 100 K or lower (Garman and Schneider, 1997). Teng and Moffat (1998) have pointed out that factors that can potentially perturb protein structure using cryo-cooled crystals, such as rate of crystal cooling, choice of cryosolvent, change in crystal mosaic spread, and binding of cryosolvent to the protein in the crystal, to name a few, have not been evaluated in detail. On the other hand, we have not observed deleterious effects of cryosolvent mixtures on protein structure upon freezing of cryosolvent mixtures, provided appropriate precautions are taken during introduction of the organic cosolvent to avoid protein denaturation. Since denaturation of proteins is subject to high activation barriers, this problem can be avoided by coordinating lowering of the temperature with introduction of the organic cosolvent in small aliquots (Douzou, 1977; Fink and Geeves, 1979).

There are important advantages to structurally characterize intermediates of enzyme-catalyzed reactions through a combined approach of cryoenzymology applied in conjunction with ENDOR spectroscopy:

(*i*) It is often possible, depending on the forward and backward rate constants of the enzyme-catalyzed reaction, to achieve near-stoichiometric conversion of free enzyme to reaction intermediate. This cannot be as easily achieved through use of rapid-flow, freeze-quench methods (Makinen and Fink, 1977).

(*ii*) The half-life of a species entering a unimolecular reaction (generally typical of each sequential step in a one-substrate, enzyme-catalyzed reaction after formation of the Michaelis complex) is given by the relationship $t_{1/2} = \ln 2 / k$ where k is the rate constant governing the step. Thus, the half-life can be made significantly longer than the mixing time by achieving a small enough value of k at low temperatures. This condition ensures that the concentrations of breakdown products are negligible. Furthermore, for unimolecular reactions with enzyme in excess, it is straightforward to identify conditions under which the substrate is virtually totally bound, *i.e.*, saturated by the enzyme, in forming a reaction intermediate. This is especially important with use of substrates that are synthetically designed to serve as the paramagnetic structural probe. Thus, overlapping spectroscopic effects due to free and bound substrate or product can be avoided.

(*iii*) Cryokinetic characterization of the enzyme-catalyzed reaction under conditions of *enzyme in excess* allows direct application of the reaction conditions for preparing reaction intermediates for ENDOR studies. Thus, there is no question whether the conditions employed for kinetic characterization are relevant to those employed for isolation of reaction intermediates for structure determination.

Furthermore, cryostabilized reaction intermediates do not represent "trapped, low-energy structures." For instance, we have demonstrated by ENDOR spectroscopy that the eclipsed conformation of the substrate moiety in the spin-labeled tryptophanyl acylenzyme intermediate of α-chymotrypsin stabilized in cryosolvents (Wells et al., 1994) is identical to that detected in a complex of the enzyme formed at room temperature with a spin-labeled transition-state analog (Jiang et al., 1998; Makinen, 1998). Such eclipsed dihedral angle relationships are indicative of conformations of high potential energy and are not characteristic of the ground state species. When cryo-stabilized enzyme reaction intermediates are formed according to kinetic criteria, they represent catalytically competent species even though they have been generated in cosolvent mixtures at low temperatures.

3.3.2 The catalytic role of sequestered water in the active site of TEM-1 β-lactamase

Resistance to β-lactam antibiotics has become a serious public health threat because of the widespread distribution of β-lactamases in pathogenic bacteria and because of their continual ability to produce mutant enzymes to avoid drug action (Neu, 1992). Therefore, it is important to understand not only structural determinants of antibiotic substrate recognition among the various classes of *β*-lactamases but also structural differences that underlie their mechanisms of action. The antimicrobial action of penicillin and cephalosporin antibiotics is due to the reactivity of the four-membered β-lactam ring (*cf.*, Fig. 16), on which basis inhibition of cell wall synthesis occurs in bacteria. The DD-transpeptidase enzyme that generates the peptidoglycan polymer of the bacterial cell wall forms an acylenzyme with β-lactam antibiotics with a half-life ≥ 12 hours. This results in cell death. Resistance to β-lactam antibiotics by pathogenic bacteria has developed through evolving highly efficient β-lactamase enzymes that are evolutionarily derived from the DD-transpeptidase and are secreted into the periplasmic space of the bacterium. Thus, they are exposed to the antibiotic before it can penetrate to reach the membrane-bound DD-transpeptidase enzyme. Thus, while β-lactam antibiotics are mechanism-based inactivators of the target DD-transpeptidase enzyme, they are catalytically specific substrates of the β-lactamases.

The interaction of the β-lactamases of the serine hydrolase variety with *β*-lactam antibiotics is schematically illustrated in Fig, 18. Although the structures of a number of β-lactamases have been solved to high resolution (Jelsch et al., 1993; Knox, 1995; Paetzel et al., 2000), there is relatively little agreement about the chemical roles of active site residues except for that of the active site nucleophilic serine residue. The side chain attacks the carbonyl carbon of the *β*-lactam group. In class A *β*-lactamases, of which the TEM-1 enzyme is a prominent example, it is anticipated, in analogy to serine proteases, that a general-base participates in catalysis in the active site by abstracting a proton from the side chain of Ser70 during formation of the transient tetrahedral adduct. However, the identity of the residue serving this function and governing catalytic cleavage with a $pK_a \sim 4.4$ has remained conjectural.

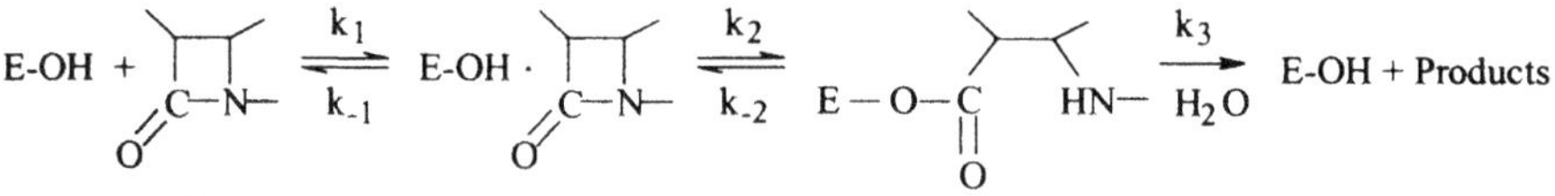

E **ES** **EY**

Figure 18. Schematic illustration of the hydrolytic reaction underlying the interaction of penicillin-recognizing enzymes of the serine hydrolase type with *β*-lactam antibiotics. Formation of the Michaelis complex (ES) is followed by acylation of the nuclephilic serine side chain in the active site. The acylenzyme (EY) undergoes rate-limiting hydrolysis leading to destruction of the antibiotic potency of the *β*-lactam compound. For the DD-transpeptidase representing the target enzyme of the antibiotic, k_3 is very low, $\sim 10^{-3}$ s^{-1}, while for *β*-lactamases, k_3 can be of the order of 2000 s^{-1}.

It has been suggested through X-ray studies of a deacylation defective mutant of TEM-1 β-lactamase that Lys73, while hydrogen-bonded to Ser70 in the free enzyme, acts as a neutral base in the Michaelis complex and functions as a proton acceptor (Strynadka et al., 1992). However, NMR titration studies of TEM-1 β-lactamase biosynthetically enriched with ε-^{13}C lysine show that the pK_a of Lys73 is >10, whereby Damblon and coworkers (1996) suggest by elimination that the only residue left in the active site to function as a general-base and to account for the pK_a is Glu166. By cryokinetic isolation of a true catalytically competent acylenzyme reaction intermediate, we were able to identify by ENDOR the hydrolytic water molecule hydrogen-bonded to the side chain of Glu166, confirming its role as the general base catalyst in the active site (Mustafi et al., 2001). We were also able to demonstrate that the proposed functional role of Lys73 as a proton acceptor (Strynadka et al., 1992) was not structurally compatible with the active site geometry of the acylenzyme intermediate.

Fig. 19 illustrates ENDOR spectra of chemically modified amino acid side chains that served as probes of active site structure in the deuterated enzyme. The resonance feature in the upper set of spectra is assigned to the CH_3– group of acetyl-Tyr105 through its disappearance upon reaction of the enzyme with d_3-acetylimidazole (the next nearest tyrosinyl residue is > 17 Å, rendering it undetectable). Similarly, the lower set of spectra in Fig. 19 belongs to the Glu240Cys mutant, in which the mutant cysteinyl side chain has an attached CH_3S– group. The decrease in intensity for the deuterated analog identifies the resonance feature belonging to the CH_3S– group.

Similar results were also obtained for the acyl-enzyme of the Met272-Cys mutant.

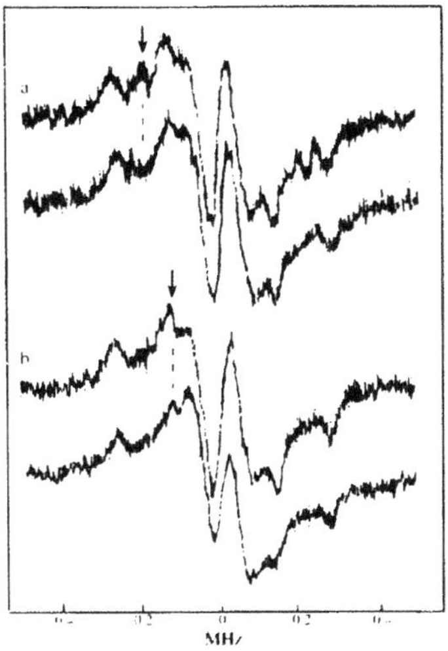

Figure 19. Comparison of proton ENDOR spectra of spin-labeled acylenzyme reaction intermediates formed with acetylated (wild type) TEM-1 β-lactamase (upper set of spectra, labeled **a**) and with the Glu240Cys mutant in which the Cys240 side chain has been modified with methylmethanethiolsulfonate (MMTS) (lower set of spectra, labeled **b**). In each case, enzyme biosynthetically enriched with deuterium (88-90% ^{2}H) was used. The spectrum for each acylenzyme intermediate reacted with acetylimidazole, correspondingly, MMTS (top spectrum in each set) is compared to the spectrum obtained for the enzyme reacted with d_3-acetylimidazole, or d_3-MMTS (bottom spectrum in each set). The resonance feature specific for each chemically modified side chain, highlighted by an arrow, is absent in the spectrum of the deuterated analog. Reprinted from Mustafi et al. (2001) with permission.

Fig. 20 illustrates the spin-labeled penicilloyl moiety in the active site of TEM-1 β-lactamase and its structural relationships to the three side chains of active site residues that have been modified for use as ENDOR probes. Their respective electron- nucleus distances are indicated in the figure. In separate experiments the resonance of the H(6) proton on the penicilloyl moiety of the acylenzyme yielded an electron-nucleus distance of 6.54 ± 0.10 Å (Mustafi et al., 2001). This constraint, together with the distance constraints to chemically modified amino acid side chains in the active site, allowed virtually no additional degrees of freedom to accommodate the substrate in its catalytically competent conformation due to the hard-sphere, van der Waals radii of other active site residues and the fixed geometry of the ester bond between the acyl moiety of the substrate and the side chain of Ser70. While the resonances described above required the use of heavily deuterated enzyme (Sosa-Peinado et al., 2000) in perdeuterated solvent, identification of sequestered solvent molecules in the active site required protiated solvent. Fig. 21 compares ENDOR spectra of deuterium enriched spin-labeled reaction intermediates of the wild type and Glu166Asn enzymes in protiated solvent. The resonance features, labeled H' and H", cannot be attributed to bulk solvent and, therefore, must arise from solvent exchangeable protons of amino acid residues or solvent molecules sequestered in the protein. Of critical importance here is the comparison of the wild type spectra to those of the Glu166Asn mutant enzyme. This mutant lacks the Glu-166 side chain and, therefore, is unable to catalyze deacylation. The H" resonance in Fig. 21 is common to both the wild type and Glu166Asn mutant enzymes, yielding an electron-nucleus distance of 5.61 ± 0.10 Å radius. This resonance was assigned to the $-NH_2$ group of Asn132, the only residue with solvent exchangeable hydrogens in the active site satisfying both the distance constraint and the angle selection requirement of lying in the plane of the spin-label (Mustafi et al., 2001).

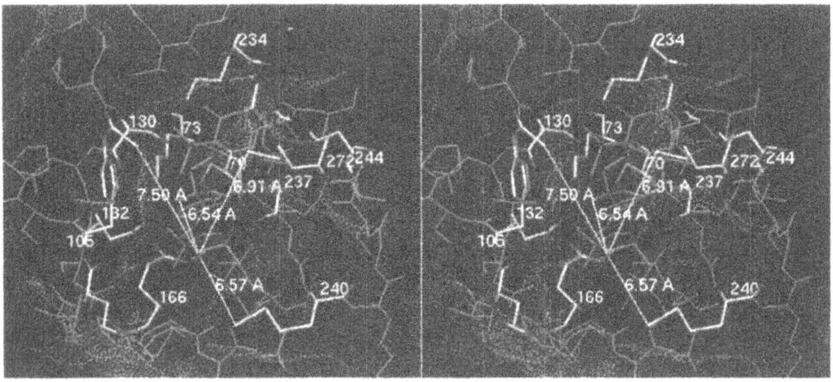

Figure 20. Stereo view of the active site of the acylenzyme of TEM-1 β-lactamase formed with spin-labeled penicillin (SLPEN). The conformation of the substrate is constrained by the ENDOR-determined electron-H(6) distance in the spin-labeled penicilloyl moiety. Electron-proton distances from the unpaired electron of the nitroxyl group to the methyl group of acetyl-Tyr105 (7.59 Å) and to the thiomethoxy groups attached to the mutant cysteinyl side chains in the Glu240Cys (6.57 Å) and Met272Cys (6.91 Å) enzymes are also indicated. The dotted surface represents the calculated Lee-Richards solvent accessible surface (Lee and Richards, 1971) of the active site. Reprinted from Mustafi et al. (2001) with permission.

On the other hand, the H' feature yielding a 6.65 ± 0.10 Å electron-nucleus distance is not observed for the acylenzyme of the Glu166Asn mutant. It must arise, therefore, in a region of the wild type enzyme that differs from the mutant, *i.e.*, the immediate environment of the Glu166 side chain. Since we found no other residues with exchangeable protons satisfying the ENDOR spectroscopic constraints, we ascribed the H' resonance to sequestered water (Mustafi et al., 2001). To identify the location of the sequestered water, we searched for possible hydrogen-bonding contacts that could stabilize a sequestered water molecule within van der Waals hard-sphere constraints satisfying the electron-proton distance of 6.65 ± 0.10 Å and lying close to the molecular plane of the spin-label, as required by the dependence of the resonance features on $\mathbf{H}_0$. Accordingly, the H' resonance was best ascribed to a water molecule hydrogen-bonded to the $O^{\varepsilon 1,2}$ atoms of the Glu166 side chain, as illustrated in Fig. 22. The ENDOR-defined water molecule sits just on the extended van der Waals surface (Lee and Richards, 1971) of the substrate. The water oxygen lies at an approximate 2.9 Å distance from the carbonyl carbon of the scissile ester bond between the substrate and O^{γ} of Ser70 and forms an angle of ~105° with the C = O bond, precisely that expected for O · · · C = O nucleophilic attack (Burgi et al., 1973, 1974).

Deacylation, as the rate-limiting step for TEM-1 catalyzed hydrolysis of penicillin substrates, is controlled by an ionizing group with pK_{a1} ~ 4.4, precisely that expected for a glutamyl side chain. The results in Figs. 21 and 22 provide direct structural confirmation that the catalytic role of Glu166 is that of a general-base, activating the water molecule for breakdown of the acylenzyme. Moreover, molecular graphics inspection of the active site shows that the ϵ-amino group of Lys73, proposed as a general base catalyst for acylation through the action of a hydrogen bonded water molecule (Strynadka et al., 1992), cannot form a hydrogen-bond to the ENDOR defined water molecule in Fig. 22 (the distance of closest approach being ~4 Å). This observation, therefore, rules against the proposed role of Lys73.

On the basis of electrostatic calculations, it has been shown that Glu166 can serve as the general base catalyst in both acylation and deacylation steps of the reaction, catalyzing extraction of the hydroxyl proton of Ser70 to form the acylenzyme and activating a water molecule as the nucleophile for its breakdown (Atanasov et al., 2000). Thus, ENDOR results provide direct confirmation of the role of Glu166 in catalysis and serve to define the structural basis of β-lactamase action. Only in a catalytically competent form of the TEM-1 enzyme has it been possible to identify the hydrolytic water molecule. The results again emphasize the need to structurally characterize true intermediates of enzyme-catalyzed reactions instead of enzyme-inhibitor complexes or enzymes rendered catalytically incompetent through mutagenesis.

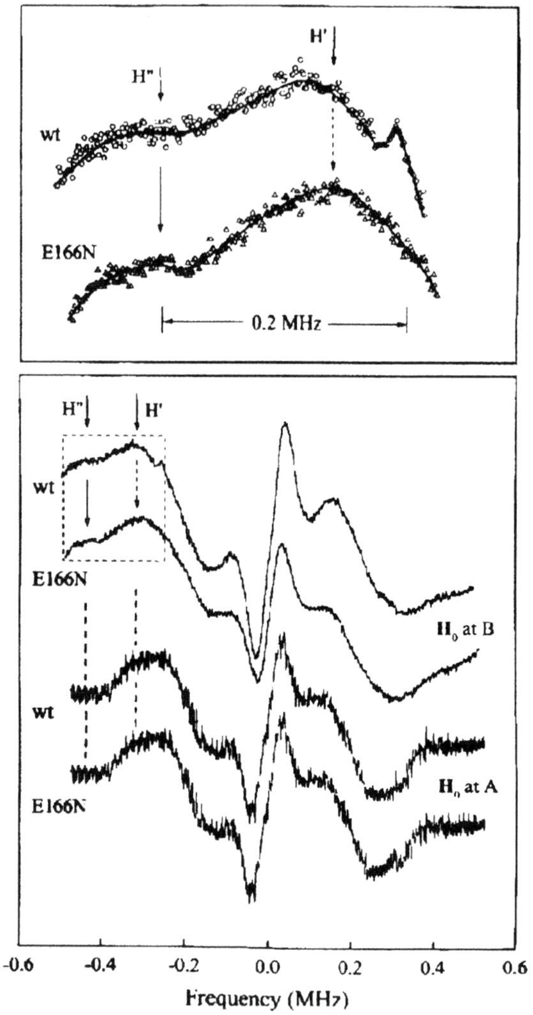

Figure 21. Comparison of proton ENDOR spectra of the acylenzyme intermediate formed with SLPEN and deuterium enriched wild type (wt) TEM-1 β-lactamase and deuterium enriched Glu166Asn mutant enzyme in protiated cryosolvent buffer. In the lower panel, spectra of the mutant (E166N) and wt enzymes are shown for A and B settings of $\mathbf{H}_0$ as lower and upper sets, respectively. Two line pairs observed in the B setting spectra of the wt and E166N mutant species are indicated by stick diagrams, labeled H' and H", respectively (also illustrated at higher gain in the upper panel). Comparison of their respective line shapes and positions shows that the feature with the larger splitting, labeled H", is present in spectra of both the wt and Glu166Asn mutant enzyme while the feature with the smaller splitting, labeled H', is seen only for the wt enzyme. With permission from Mustafi et al. (2001).

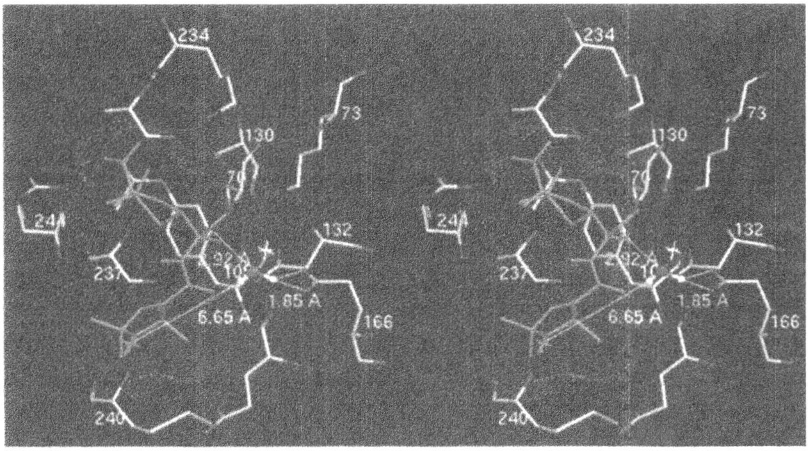

Figure 22. Stereo view of active site of TEM-1 β-lactamase illustrating the ENDOR-defined location of the H' proton assigned to a sequestered water molecule in the active site. The ENDOR-determined distance from the unpaired electron to one of the water protons and structural relationships to the carboxylate oxygens of Glu166 and to the carbonyl carbon of the ester bond of SLPEN formed with the side chain of Ser70 are indicated. The dotted surface represents the extended van der Waals surface of the spin-labeled acyl group showing that the water molecule is accommodated sterically by the substrate. The star just above the ENDOR-defined water molecule indicates the position of an X-ray-defined water molecule in the free enzyme (Jelsch et al., 1993). Reprinted from Mustafi et al. (2001) with permission.

4. FUTURE PERSPECTIVES AND CONCLUDING REMARKS

In this review we have emphasized characterization of metal-bound and hydrogen-bonded solvent in small molecule and macromolecule environments to highlight the precision and level of structural detail that can be achieved through application of angle-selected ENDOR. The studies reviewed also show that comparable precision can be achieved in defining structure and conformation of active site residues and of the substrate in cryokinetically isolated reaction intermediates of enzymes. In this respect, ENDOR spectroscopy applied in conjunction with cryosolvent methods offers many advantages that cannot be achieved through application of other spectroscopic methods capable of three-dimensional structure determination, for instance, multi-dimensional NMR. While the latter is well suited for defining the relative spatial distribution of atoms in a protein in solution (at present ≤ 30 kDa) through nuclear Overhauser measurements, given the covalent bonding structure of the constituent amino acid residues, the uncertainties are significantly larger than in ENDOR, up to 50%, for internuclear separations ≤ 5.0 Å (Zhao and Jardetzky, 1994; Gradwell and

Feeney, 1996; Zabell and Post, 2002). Furthermore, the time required for NMR data collection of macromolecules in solution and the viscosity of cryosolvent mixtures at low temperatures are incompatible with structural analysis of true intermediates of enzyme-catalyzed reactions or of other chemically labile systems.

Application of cw ENDOR is associated, nonetheless, with inherent limitations despite the high precision afforded for structure analysis. Three-dimensional structure determination by cw ENDOR is most straightforwardly carried out with $I = 1/2$ nuclei. Assignments of hydrogen resonances are generally dependent on carrying out synthetic chemical procedures for site-specific incorporation of deuterium to be used in parallel experiments. Such added chemical complexity is often time-consuming and arduous. In addition, signal-to-noise in data collection becomes an important consideration since the intensity of resonance features is dependent not only on the electron-nucleus distance, anisotropy of relaxation processes, and the number of nuclei contributing to the resonance, but also on the nuclear moment. Because of this latter factor, use of ^{13}C, an important nuclide in multi-dimensional NMR experiments, has been limited in cw ENDOR, and most cw ENDOR studies have been restricted to ^{1}H and ^{19}F for structural analysis of the type described here. While improved resolution of overlapping resonance features may be achieved by cw ENDOR instrumentation applied at higher microwave frequencies through greater separation of g-values, structure analysis is not likely to be significantly improved unless the number of magnetic nuclei in the molecule of interest used for structure analysis can be increased.

Because the number of electron-nucleus distances determined in cw ENDOR experiments is relatively small, structure analysis at present rests heavily on bond distance and valence angle information determined independently for large fragments of the molecular complex when ENDOR-determined electron-nucleus distances are applied as constraints in torsion angle search calculations. These fragments provide the molecular scaffolding to which the "ENDOR-active" nuclei are covalently attached. Improvement in structure analysis beyond that achieved hitherto with cw ENDOR, therefore, requires a means to increase significantly the number of independently determined electron-nucleus distances. Of most importance, therefore, are pulsed EPR and ENDOR methods that can provide a means to increase the number of electron-nucleus distances through detection of ^{2}H, ^{13}C, ^{14}N , ^{15}N, and ^{31}P since these nuclides are not easily employed in cw ENDOR. Two recent reviews describe pulsed EPR and ENDOR spectroscopy and their applications to biological systems (Cammack et al., 1999; Prisner et al., 2001). Also, a series of investigations have been reported in which pulsed EPR and ENDOR methods have been applied to characterize structures of enzyme active sites and to determine protein-ligand interactions (Gurbiel, et al., 1996; Gromov et al., 1999; Walsby et al., 2001). Pulsed methods have also been used to characterize structured solvent in macromolecular systems (Goldfarb, et al., 1996; DeRose et al., 1996).

With respect to the objective of increasing the number of dipolar electron-nucleus distances through a combination of cw and pulsed EPR and ENDOR methods, the goal in structural analysis should be to decrease as much as possible the dependence of the analysis on stereochemical data that define the molecular scaffold to which the "ENDOR-active" nuclei are covalently attached. For instance, while the conformation of SLCEP in Fig. 14 was assigned successfully through use of ENDOR-determined distance constraints, the analysis relies heavily on input from X-ray diffraction studies providing bond distances and valence angles of atoms that are not ENDOR-active. In this respect, it should be noted that direct detection of ^{13}C NMR in the vicinity of paramagnetic centers is favored through paramagnetic relaxation processes in contrast to ^{1}H (Banci et al., 1991). On this basis it may be possible to take advantage of the differential relaxation characteristics of ^{13}C and ^{1}H with stochastic ENDOR (Brueggeman and Niklas, 1994). This is essentially a pulsed ENDOR technique. While approaches to extract geometrical information based on this differential property have yet to be developed, it has been shown that NMR detection of ^{1}H-decoupled ^{13}C near paramagnetic centers in proteins can be used to identify residue connectivities (Machonkin et al., 2002).

The three-dimensional structure and conformation of a (diamagnetic) chemotactic tripeptide uniformly enriched with ^{13}C and ^{15}N has been determined on the basis of simulated annealing calculations restrained by inter-nuclear distance and torsion angle measurements obtained through solid-state, magic-angle spinning NMR experiments (Rienstra et al., 2002). Extension of a similar approach should be feasible for paramagnetic solid-state systems through a combination of cw and pulsed EPR and ENDOR methods. This combined approach need not be restricted only to macromolecules with naturally occurring paramagnetic sites. Applications of ENDOR structural probes such as nitroxyl spin-labels and the VO^{2+} cation would be of clear advantage in view of the success with which they have been employed in both small molecule and macromolecular systems hitherto. On this basis, combined application of cw and pulsed ENDOR methods is likely to yield a significantly enhanced basis for structure assignments.

5. ACKNOWLEDGMENTS

This work has been supported by grants of the National Science Foundation (MCB-0092524) and of the National Institutes of Health (DK57599).

6. REFERENCES

Albanese, N. F. and Chasteen, N. D. (1978). Origin of Electron Paramagnetic Resonance Line Widths in Frozen Solutions of Oxovanadium(IV) Ion. *J. Phys. Chem.* **82**, 910-914.

Ament, S. S., Wetherin, J. B., Moncrief, J. W., Flohr, K., Mochizuk, M., and Kaiser, E.T. (1973). Determination of Absolute-Configuration of (+)3-Carboxy-2,2,5,5-Tetramethyl-1-Pyrrolidinyloxy. *J. Am. Chem. Soc.* **95**, 7896-7897.

Atanasov, B. P., Mustafi, D., and Makinen, M. W. (2000). Protonation of the β-Lactam Nitrogen is the Trigger Event in the Catalytic Action of Class A β-Lactamases. *Proc. Natl. Acad. Sci. USA* **97**, 3160-3165.

Atherton, N. M. (1993). *Principles of Electron Spin Resonance*, Royal Society of Chemistry, London, U. K. Atherton, N. M. and Shackleton, J. F. (1980). Proton ENDOR of $[VO(H_2O)_5]^{2+}$ in $Mg(NH_4)_2\ (SO_4) \cdot 6H_2O$. *Mol. Phys.* **39**, 1471-1485.

Atherton, N. M. and Shackleton, J. F. (1984). Proton Hyperfine Couplings in $[VO(H_2O)_5]^{2+}$.The Validity of the Point-Dipole Approximation. *Chem. Phys. Lett.* **103**, 302-304.

Attanasio, D. (1986). Structural Information from ENDOR Spectroscopy: The Frozen Solution Proton Spectra of Some VO(IV) Complexes. *J. Phys. Chem.* **90**, 4952-4957.

Attanasio, D. (1989). Structural Information from Powder ENDOR Spectroscopy. *J. Chem. Soc., Faraday Trans. 1* **85**, 3927-3937.

Audet, P., Simard, C., and Savoie, R. (1991). A Vibrational Spectroscopic Study of the Self-Association of 5'-GMP in Aqueous Solution. *Biopolymers* **31**, 243-251.

Baker, J. M., Davies, E. R., and Hurrell, J. P. (1968). Electron Nuclear Double Resonance in Calcium Fluoride Containing Yb^{3+} and Ce^{3+} in Tetragonal Sites. *Proc. Royal Soc.(London)* **A308**, 403-431.

Ballhausen, C. J. and Gray, H. B. (1962). The Electronic Structure of the Vanadyl Ion. *Inorg. Chem.* **1**, 111-122.

Ballhausen, C. J., Djurinskij, B. F., and Watson, K. J. (1968). Polarized Absorption Spectra of 3 Crystalline Polymorphs of $VOSO_4 \cdot 5H_2O$. *J. Am. Chem. Soc.* **90**, 3305-3309.

Banci, L., Bertini, I., and Luchinat, C. (1991). *Nuclear and Electronic Relaxation*, VCH Publishers, Weinheim, Germany.

Bauer, R. S. and Berliner, L. J. (1979). Spin Label Investigations of α-Chymotrypsin Active-Site Structure in Single-Crystals. *J. Mol. Biol.* **128**, 1-19.

Benner, S. A. (1988). Stereoelectronic Analysis of Enzymatic Reactions in *Mechanistic Principles of Enzyme Activity*, J. Liebman and A. Grunberg, eds., VCH Publishers, Inc., New York, pp. 27-74.

Blackburn, E. H. (2000). Telomere States and Cell Fates. *Nature* **408**, 53-56.

Blinder, S. M. (1960). Orientation Dependence of Magnetic Hyperfine Structure in Free Radicals. *J. Chem. Phys.* **33**, 748-752.

Boles, M. O., Girven, R. J., and Gane, P. A. C. (1978). Structure of Amoxycillin Trihydrate and a Comparison with Structures of Ampicillin. *Acta Crystallogr.* **B34**, 461-466.

Bolin, K. A., Hanson, P., Wright, S. J., and Millhauser, G. L. (1998). An NMR Investigation of the Conformational Effect of Nitroxide Spin Labels on Ala-rich Helical Peptides. *J. Magn. Reson.* **131**, 248-253.

Bolin, K. A. and Millhauser, G. L. (1999). Alpha and 3(10): The Split Personality of Polypeptide Helices. *Accts. Chem. Res.* **32**, 1027-1033.

Brooks, B. R., Bruccoleri, R. E., Olafson, B. D., States, D. J., Swaminathan, S., and Karplus, M., (1983). CHARMM: A Program for Macromolecular Energy, Minimization, and Dynamics Calculations. *J. Comput. Chem.* **4**, 187-217.

Brueggeman, W. and Niklas, J. R. (1994). Stochastic ENDOR. *J. Magn. Reson.* Series **A** **108**, 25-29.

Burgi, H. B., Dunitz, J. D., and Shefter, E. (1973). Geometrical Reaction Coordinates. 2. Nucleophilic Addition to a Carbonyl Group. *J. Am. Chem. Soc.* **95**, 5065-5067.

Burgi, H. B., Dunitz, J. D., and Shefter, E. (1974). Chemical-Reaction Paths. 4. Aspects of O C=O Interactions in Crystals. *Acta Crystallogr.* **B30**, 1517-1527.

Bush, K., Jacoby, G. A., and Medeiros, A. A. (1995). A Functional Classification Scheme for β-Lactamases and Its Correlation with Molecular Structure. *Antimicrob. Agents. Chemother.* **39**, 1211-1233.

Cammack, R., Gay, E., and Shergill, J. K. (1999). Studies of Hyperfine Interactions in [2Fe-2S] Proteins by EPR and Double Resonance Spectroscopy. *Coord. Chem. Rev.* **192**, 1003-1022.

Chasteen, N. D. (1981). Vanadyl(IV) EPR Probes. Inorganic and Biochemical Aspects. in *Biological Magnetic Resonance*, Vol. 3, L. J. Berliner and J. Reuben, eds., Plenum Press, New York, pp. 53-119.

Chasteen, N. D. (1983). The Biochemistry of Vanadium. *Structure & Bonding (Berlin)* **53**, 105-138.

Chasteen, N. D., Ed. (1990). *Vanadium in Biological Systems*, Kluwer Academic, Boston, Massachusetts.

Christianson, D. W. and Lipscomb, W. N. (1986). X-Ray Crystallographic Investigation of Substrate Binding to Carboxypeptidase A at Subzero Temperature. *Proc. Natl. Acad. Sci. USA* **83**, 7568-7572.

Christianson, D. W. and Lipscomb, W. N. (1989). Carboxypeptidase A. *Accts. Chem. Res.* **22**, 62-69.

Cornman, C. R., Zovinka, E. P., Boyajian, Y. D., Geiserbush, R. M., Boyle, P. D., and Singh, P. (1995). Structural and EPR Studies of Vanadium Complexes of Deprotonated Amide Ligands. Effects on the V-51 Hyperfine Coupling-Constant. *Inorg. Chem.* **34**, 4213-4219.

Cornman, C. R., Geiser-Bush, K. M., Rowley, S. P., and Boyle, P. D. (1997). Structural and Electron Paramagnetic Resonance Studies of the Square Pyramidal to Trigonal Bipyramidal Distortion of Vanadyl Complexes Containing Sterically Crowded Schiff Base Ligands. *Inorg.Chem.* **36**, 6401-6408.

Damblon, C., Raquet, X., Lian, L. Y., Lamotte-Brasseur, J., Fonze, E., Charlier, P., Roberts, G. C. K., and Frère, J. M. (1996). The Catalytic Mechanism of β-Lactamases: NMR Titration of an Active Site Lysine Residue of the TEM-1 Enzyme. *Proc. Natl. Acad. Sci. USA* **93**, 1747-1752.

Davis, T. D., Christofferson, R. E., and Maggiora, G. N. (1975). *Ab Initio* Calculations on Large Molecules Using Molecular Fragments. Nitroxide Spin Label Characterizations. *J. Am. Chem. Soc.* **97**, 1347-1354.

DeRose, V. J., Liu, K. E., Lippard, S. J., and Hoffman, B. M. (1996). Investigation of the Dinuclear Fe Center of Methane Monooxygenase by Advanced Paramagnetic Resonance Techniques: On the geometry of DMSO binding. *J. Am. Chem. Soc.* **118**, 121-134.

Deslongchamps, P. (1983). *Stereoelectronic Effects in Organic Chemistry*, Pergamon Press, Oxford, U.K.

di Matteo, A. and Barone, V. (1999). Development and Validation of Effective Computational Strategies for the Study of Metal Nitroxide Complexes. *J. Phys. Chem.* **A103**, 7676-7685.

Dikanov, S. A., Liboiron, B. D., Thompson, K. H., Vera, E., Yuen, V. G., McNeill, J. H., and Orvig, C. (1999). *In vivo* Electron Spin-Echo Envelope Modulation (ESEEM) Spectroscopy: First Observation of Vanadyl Coordination to Phosphate in Bone. *J. Am. Chem. Soc.* **121**, 11004-11005.

Dikanov, S. A., Liboiron, B. D., and Orvig, C. (2002). Two-Dimensional (2D) Pulsed Electron Paramagnetic Resonance Study of VO^{2+}-Triphosphate Interactions: Evidence for Tridentate Triphosphate Coordination, and Relevance to Bone Uptake and Insulin Enhancement by Vanadium Pharmaceuticals. *J. Am. Chem. Soc.* **124**, 2969-2978.

Dodge, R. P., Tempelton, D.H., and Zalkin, A. (1961). Crystal Structure of Vanadyl Bisacetylacetonate. Geometry of Vanadium in Fivefold Coordination. *J. Chem. Phys.* **35**, 55-67.

Dorio, M. M. and Freed, J. H., Eds. (1979). *Multiple Electron Resonance Spectroscopy*, Plenum Press, New York.

Douzou, P. (1977). *Cryobiochemistry*, Academic Press, New York.

Fattah, J., Twyman, J. M., Heyes, S. J., Watkin, D. J., Edwards, A. J., Prout, K., and Dobson, C. M. (1993). Combination of CP MAS NMR and X-Ray Crystallography. Structure and Dynamics in a Low-Symmetry Molecular-Crystal, Potassium Penicillin-V. *J. Am. Chem. Soc.* **115**, 5636-5650.

Feher, G. (1956). Observation of Nuclear Magnetic Resonances via the Electron Spin Resonance Line. *Phys. Rev.* **103**, 834-835.

Feher, G. (1957). Electronic Structure of F-Centers in KCl by the Electron Spin Double Resonance Technique. *Phys. Rev.* **105**, 1122-1123.

Feher, G., Isaacson, R. A., Scholes, C. P., and Nagel, R. (1973). Electron Nuclear Double Resonance (ENDOR) Investigation on Myoglobin and Hemoglobin. *Ann. N. Y. Acad. Sci.* **222**, 86-101.

Fields, R. A. and Hutchison, Jr., C. A. (1985). The Determination of Hydrogen Coordinates in Lanthanum Nicotinate Dihydrate Crystals by Gd^{3+}-Proton Double Resonance. *J. Chem. Phys.* **82**, 1711-1722.

Fink, A. L. and Geeves, M. A. (1979). Cryoenzymology: The Study of Enzyme Catalysis at Subzero Temperatures. *Methods Enzymol.* **63**, 336-370.

Fink, A. L. and Cartwright, S. J. (1981). Cryoenzymology. *CRC Crit. Rev. Biochem.* **11**, 145-207.

Flohr, K. and Kaiser, E. T. (1972). Enantiomeric Specificity in Chymotrypsin-Catalyzed Hydrolysis of 3-Carboxy-2,2,5,5-Tetramethylpyrrolidin-1-Oxy *para*-Nitrophenyl Ester. *J. Am. Chem. Soc.* **94**, 3675.

Garman, E. F. and Schneider, T. R. (1997). Macromolecular Cryocrystallography. *J. Appl. Crystallogr.* **30**, 211-237.

Gellert, M., Lipsett, M. N., and Davies, D. R. (1962). Helix Formation by Guanylic Acid. *Proc. Natl. Acad. Sci. USA* **48**, 2013-2018.

Gersmann, H. R. and Swalen, J. D. (1962). Electron Paramagnetic Resonance Spectra of Copper Complexes. *J. Chem. Phys.* **36**, 3221-3233.

Gochev, G. P. and Yordanov, N. D. (1993). Polycrystalline ENDOR Crystallography, a New Methodological Approach. *J. Magn. Reson.* **A102**, 180-182.

Goldfarb, D., Bernardo, M., Thomann, H., Kroneck, P. M. H., and Ullrich, V. (1996). Study of Water Binding to Low-Spin Fe(III) in Cytochrome P450 by Pulsed ENDOR and Four-pulse ESEEM Spectroscopies. *J. Am. Chem. Soc.* **118**, 2686-2693.

Gradwell, M. J. and Feeney, J. (1996). Validation of the Use of Intermolecular NOE Constraints for Obtaining Docked Structures of Protein-Ligand Complexes. *J. Biomol. NMR* **7**, 48-58.

Gromov, I., Marchesini, A., Farver, O., Pecht, I., and Goldfarb, D. (1999). Azide Binding to the Trinuclear Copper Center in Laccase and Ascorbate Oxidase. *Eur. J. Biochem.* **266**, 820-830.

Gurbiel, R. J., Doan, P. E., Gassner, G. T., Macke, T. J., Case, D. A., Ohnishi, T., Fee, J. A., Ballou, D. P., and Hoffman, B. M. (1996). Active Site Structure of Rieske-Type Proteins: Electron Nuclear Double Resonance Studies of Isotopically Labeled Phthalate Dioxygenase from *Pseudomonas cepacia* and Rieske Protein from *Rhodobacter capsulatus* and Molecular Modeling Studies of a Rieske Center. *Biochemistry* **35**, 7834-7845.

Happe, J. A. and Morales, M. (1966). Nitrogen-15 Nuclear Magnetic Resonance Evidence that Mg^{2+} Does Not Complex with the Nitrogen Atoms of Adenosine Triphosphate. *J. Am. Chem. Soc.* **88**, 2077-2078.

Hayat, H. and Silver, B. L. (1973). Oxygen-17 and Nitrogen-14 σ–π Polarization Parameters and Spin Density Distribution in Nitroxyl Group. *J. Phys. Chem.* **77**, 72-78.

Henderson, T. A., Hurst, G. C., and Kreilick, R. W. (1985). Angle-Selected ENDOR Spectroscopy. 2. Determination of Proton Coordinates from a Polycrystalline Sample of *Bis*(2,4-pentanedionato)copper(II). *J. Am. Chem. Soc.* **107**, 7299-7303.

Herrmann, T., Guntert, P., and Wuthrich, K. (2002). Protein NMR Structure Determination with Automated NOE Assignment Using the New Software CANDID and the Torsion Angle Dynamics Algorithm DYANA. *J. Mol. Biol.* **319**, 209-227.

Hoffman, B. M., Martinsen, J., and Venters, R. A. (1984). General Theory of Polycrystallline ENDOR Patterns. *g* and Hyperfine Tensors of Arbitrary Symmetry and Relative Orientation. *J. Magn. Reson.* **59**, 110-123.

Hoffman, B. M., Venters, R. A., and Martinsen, J. (1985). General Theory of Polycrystalline ENDOR Patterns. Effects of Finite EPR and ENDOR Component Linewidths. *J. Magn. Reson.* **62**, 537-542.

Hoffman, B. M. and Gurbiel, R. J. (1989). Polycrystalline ENDOR Patterns from Centers of Axial EPR Spectra. General Formulas and Simple Analytic Expressions for Deriving Geometric Information from Dipolar Couplings. *J. Magn. Reson.* **82**, 309-317.

Hubbell, W. L., Gross, A., Langen, R., and Lietzow, M. A. (1998). Recent Advances in Site-Directed Spin-Labeling of Proteins. *Curr. Opin. Struct. Biol.* **8**, 649-656.

Hurst, G. C., Henderson, T. A., and Kreilick, R. W. (1985). Angle Selected ENDOR Spectroscopy. 1. Theoretical Interpretation of ENDOR Shifts from Randomly Oriented Transition Metal Complexes. *J. Am. Chem. Soc.* **107**, 7294-7299.

Hutchison, Jr., C. A. and McKay, D. B. (1977). Determination of Hydrogen Coordinates in Lanthanum Nicotinate Dihydrate Crystals by Nd^{3+}-Proton Double-Resonance. *J. Chem. Phys.* **66**, 3311-3330.

Hutchison, Jr., C. A. and Orlowski, T. E. (1980). The Determination of Deuterium Atom Coordinates and Nuclear-Quadrupole Interactions in Lanthanum Nicotinate Dihydrate Crystals by Nd^{3+}-Deuterium Double-Resonance. *J. Chem. Phys.* **73**, 1-14.

Iijima, H., Dunbar, J. B., and Marshall, G. R. (1987). Calibration of Effective van der Waals Atomic Contact Radii for Proteins and Peptides. *Proteins: Struct., Func., Genet.* **2**, 330-339.

Improta, R., di Matteo, A., and Barone, V. (2000). Effective Modeling of Intrinsic and Environmental Effects on the Structure and Electron Paramagnetic Resonance Parameters of Nitroxides by an Integrated Quantum Mechanical/Molecular Mechanics/Polarizable Continuum Model Approach. *Theor. Chem. Accts.* **104**, 273-279.

Jelsch, C., Maury, L., Masson, J. M., and Samama, J. P. (1993). Crystal Structure of *Escherichia coli* TEM1 β-Lactamase at 1.8 Å Resolution. *Proteins: Struct., Func., Genet.* **16**, 364-383.

Jiang, F. S. and Makinen, M. W. (1995). NMR and ENDOR Conformational Studies of the Vanadyl Guanosine 5'-Monophosphate Complex in Hydrogen-Bonded Quartet Assemblies. *Inorg. Chem.* **34**, 1736-1744.

Jiang, F. S., Tsai, S. W., Chen, S., and Makinen, M. W. (1998). ENDOR-Determined Structure of a Complex of α-Chymotrypsin with a Spin-Labeled Transition-State Inhibitor Analogue. *J. Phys. Chem.* **B102**, 4619-4627.

Joela, H., Mustafi, D., Fair, C. C., and Makinen, M. W. (1991). Structure and Confirmation of Spin-Labeled Methyl L-Phenylalanate and L-Phenylalanine Determined by Electron Nuclear Double resonance spectroscopy. *J. Phys. Chem.* **95**, 9135-9144.

Jost, P. C. and Griffith, O. H. (1978). The Spin-Labeling Technique. *Methods Enzymol.* **49**, 369-418.

Kang, C., Zhang, X. H., Ratliff, R., Moyzis, R., and Rich, A. (1992). Crystal-Structure of 4-Stranded Oxytricha Telomeric DNA. *Nature* **356**, 126-131.

Kevan, L. and Kispert, L. D. (1976). *Electron Spin Double Resonance Spectroscopy*, John Wiley and Sons, New York.

Khramtsov, V. V. and Weiner, L. (1988). Proton Exchange in Stable Nitroxyl Radicals: pH Sensitive Spin Probes. Chap. 2 in *Imidazoline Nitroxides*, Vol. II, L. B. Volodarsky, ed., CRC Press, Boca Raton, Florida, pp. 37-80.

Khramtsov, V. V. and Volodarsky, L. B. (1998). Use of Imidazoline Nitroxides in Studies of Chemical Reactions in *Biological Magnetic Resonance*, Vol. 14, L. J. Berliner, ed., Plenum Press, New York, pp. 109-180.

Kivelson, D. and Lee, S.-K. (1964). ESR Studies and the Electronic Structure of Vanadyl Ion Complexes. *J. Chem. Phys.* **41**, 1896-1903.

Knox, J. R. (1995). Extended-Spectrum and Inhibitor-Resistant TEM-Type β-Lactamases: Mutations, Specificity, and Three-Dimensional Structure. *Antimicrob. Agents Chemother.* **39**, 2593-2601.

Kretsinger, R. H. and Nockolds, C. E. (1973). Carp Muscle Calcium-Binding Protein. II. Structure Determination and General Description. *J. Biol. Chem.* **248**, 3313-3326.

Kurreck, H., Kirste, B., and Lubitz, W. (1988). Electron Nuclear Double Resonance Spectroscopy of Radicals in Solution: Application to organic and Biological Chemistry, VCH Publishers, New York.

Langen, R., Oh, K. J., Cascio, D., and Hubbell, W. L. (2001). Crystal Structures of Spin Labeled T4 Lysozyme Mutants: Implications for the Interpretation of EPR Spectra in Terms of Structure. *Biochemistry* **39**, 8396-8405.

Lee, B. and Richards, F. M. (1971). The Interpretation of Protein Structures: Estimation of Static Accessibility. *J. Mol. Biol.* **55**, 379-400.

Machonkin, T. E., Westler, W. M., and Markley, J. L. (2002). C-13{C-13} 2D NMR: A Novel Strategy for the Study of Paramagnetic Proteins with Slow Electronic Relaxation Rates. *J. Am. Chem. Soc.* **124**, 3204-3205.

MacKerell, A. D., Bashford, D., Bellott, M., Dunbrack, R. L., Evanseck, J. D., Field, M. J., Fischer, S., Gao, J., Guo, H., Ha, S., Joseph-McCarthy, D., Kuchnir, L., Kuczera, K., Lau, F. T. K., Mattos, C., Michnick, S., Ngo, T., Nguyen, D. T., Prodhom, B., Reiher, W. E., Roux, B., Schlenkrich, M., Smith, J. C., Stote, R., Straub, J., Watanabe, M., Wiorkiewicz-

Kuczera, J., Yin, D., and Karplus, M. (1998). All-Atom Empirical Potential for Molecular Modeling and Dynamics Studies of Proteins. *J. Phys. Chem.* **B102**, 3586-3616.

Makinen, M. W. and Fink, A. L. (1977). Reactivity and Cryoenzymology of Enzymes in the Crystalline State. *Annu. Rev. Biophys. Bioeng.* **6**, 301-343.

Makinen, M. W., Kuo, L. C., Dymowski, J. J., and Jaffer, S. (1979). Catalytic Role of the Metal Ion of Carboxypeptidase A in Ester Hydrolysis. *J. Biol. Chem.* **254**, 356-366.

Makinen, M. W. and Mustafi, D. (1995). The Vanadyl Ion: Molecular Structure of Coordinating Ligands by Electron Paramagnetic Resonance and Electron Nuclear Double Resonance Spectroscopy in *Metal Ions in Biological Systems*, Vol. 31, H. Sigel and A. Sigel, eds., Marcel Dekker, Inc., New York, pp. 89-127.

Makinen, M. W. (1998). Electron Nuclear Double Resonance Determined Structures of Enzyme Reaction Intermediates: Structural Evidence for Substrate Destabilization. *Spectrochim. Acta. Part A, Mol. Biomol. Spectrosc.* **54**, 2269-2281.

Makinen, M. W., Mustafi, D., and Kasa, S. (1998). ENDOR of Spin Labels for Structure Determination: From Small Molecules to Enzyme Reaction Intermediates in *Biological Magnetic Resonance*, Vol. 14, L. J. Berliner, ed., Plenum Press, New York, pp. 181-249.

Makinen, M. W. and Brady, M. J. (2002). Structural Origins of the Insulin-Mimetic Activity of Bis(acetylacetonato)oxovanadium(IV). *J. Biol. Chem.* **277**, 12215-12220.

Massova, I. and Mobashery, S. (1998). Kinship and Diversification of Bacterial Penicillin-Binding Proteins and β-Lactamases. *Antimicrob. Agents Chemother.* **42**, 1-17.

Matthews, B. W. (1968). Solvent Content of Protein Crystals. *J. Mol. Biol.* **33**, 491-497.

McConnell, H. M. and Chestnut, D. B. (1958). Theory of Isotropic Hyperfine Interactions inπ-Electron Radicals. *J. Chem. Phys.* **28**, 107-117.

Mustafi, D. and Makinen, M. W. (1988). ENDOR-Determined Solvation Structure of VO^{2+} in Frozen-Solutions. *Inorg. Chem.* **27**, 3360-3368.

Mustafi, D., Boisvert, W. E., and Makinen, M. W. (1990a). Structure and Conformation of the Nitroxyl Spin-Label Ethyl 3-(2,2,5,5-tetramethylpyrrolinyl-1-oxyl)-propen-2-oate Determined by Electron Nuclear Double Resonance: Comparison with the Structure of a Spin-Label Substrate of Carboxypeptidase A. *Biopolymers* **29**, 45-55.

Mustafi, D., Sachleben, J. R., Wells, G. B., and Makinen, M. W. (1990b). Structure and Conformation of Spin-Labeled Amino Acids in Frozen Solutions Determined by Electron Nuclear Double Resonance. 1. Methyl *N*-(2,2,5,5-Tetramethyl-1-Oxypyrrolinyl-3-Carbonyl)-L-Alanate, a Molecule with a Single Preferred Conformation. *J. Am. Chem.Soc.* **112**, 2558-2566.

Mustafi, D., Wells, G. B., Joela, H., and Makinen, M. W. (1990c). Assignment of Proton ENDOR Resonances of Nitroxyl Spin-Labels in Frozen Solution. *Free Radical Res. Commun.* **10**, 95-101.

Mustafi, D., Joela, H., and Makinen, M. W. (1991). The Effective Position of the Electronic Point Dipole of the Nitroxyl Group of Spin Labels Determined by ENDOR Spectroscopy. *J.Magn. Reson.* **91**, 497-504.

Mustafi, D., Telser, J., and Makinen, M. W. (1992). Molecular Geometry of Vanadyl Adenine Nucleotide Complexes Determined by EPR, ENDOR, and Molecular Modeling. *J. Am. Chem. Soc.* **114**, 6219-6226.

Mustafi, D., Boisvert, W. E., and Makinen, M. W. (1993). Synthesis of Conjugated Polyene Carbonyl Derivatives of Nitroxyl Spin-Labels and Determination of Their Molecular Structure and Conformation by Electron Nuclear Double Resonance. *J. Am. Chem. Soc.* **115**, 3674-3682.

Mustafi, D. and Makinen, M. W. (1994). Catalytic Conformation of Carboxypeptidase A.Structure of a True Enzyme Reaction Intermediate Determined by Electron Nuclear Double Resonance. *J. Biol. Chem.* **269**, 4587-4595.

Mustafi, D. and Nakagawa, Y. (1994). Characterization of Calcium-Binding Sites in the Kidney Stone Inhibitor Glycoprotein Nephrocalcin with Vanadyl Ions: Electron Paramagnetic Resonance and Electron Nuclear Double Resonance Spectroscopy. *Proc. Natl. Acad. Sci. USA* **91**, 11323-11327.

Mustafi, D. and Joela, H. (1995). Origin of the Temperature Dependent Isotropic Hyperfine Coupling of the Vinylic Proton of Oxypyrrolinyl Nitroxyl Spin-Labels. *J. Phys. Chem.* **99**, 11370-11375.

Mustafi, D. and Makinen, M. W. (1995). Structure, Conformation, and Probable Mechanism of Hydrolysis of a Spin-Labeled Penicillin Revealed by Electron Nuclear DoubleResonance Spectroscopy. *J. Am. Chem. Soc.* **117**, 6739-6746.

Mustafi, D. and Nakagawa, Y. (1996). Characterization of Ca^{2+}-Binding Sites in the Kidney Stone Inhibitor Glycoprotein Nephrocalcin Using Vanadyl Ions: Different Metal Binding Properties in Strong and Weak Inhibitor Proteins Revealed by EPR and ENDOR. *Biochemistry* **35**, 14703-14709.

Mustafi, D., Knock, M. M., Shaw, R. W., and Makinen, M. W. (1997). Conformational Changes in Spin-Labeled Cephalosporin and Penicillin upon Hydrolysis Revealed by Electron Nuclear Double Resonance Spectroscopy. *J. Am. Chem. Soc.* **119**, 12619-12628.

Mustafi, D., Nakagawa, Y., and Makinen, M. W. (2000). ENDOR Studies of VO^{2+}: Probing Protein-Metal Ion Interactions in Nephrocalcin. *Cell. Mol. Biol.* **46**, 1345-1360.

Mustafi, D., Sosa-Peinado, A., and Makinen, M. W. (2001). ENDOR Structural Characterization of a Catalytically Competent Acylenzyme Reaction Intermediate of Wild-Type TEM-1 β-Lactamase Confirms Glutamate-166 as the Base Catalyst. *Biochemistry* **40**, 2397-2409.

Mustafi, D., Sosa-Peinado, A., Gupta, V., Gordon, D. J., and Makinen, M. W. (2002). Structure of Spin-Labeled Methylmethanethiolsulfonate in Solution and Bound to TEM-1 β-Lactamase Determined by Electron Nuclear Double Resonance Spectroscopy. *Biochemistry* **41**, 797-808.

Nakagawa, Y., Margolis, H. C., Yokoyama, S., Kezdy, F. J., Kaiser, E. T., and Coe, F. L. (1981). Purification and Characterization of a Calcium Oxalate Monohydrate Crystal Growth Inhibitor from Human Kidney Tissue Culture Medium. *J. Biol. Chem.* **256**, 3936-3944.

Nakagawa, Y., Abram, V., Kezdy, F. J., Kaiser, E. T., and Coe, F. L. (1983). Purification of the Principal Inhibitor of Calcium Oxalate Monohydrate Crystal Growth in Human Urine. *J.Biol. Chem.* **258**, 12594-12600.

Nakagawa, Y., Otsuki, T., and Coe, F. L. (1985). Elucidation of the Multiple Forms of Nephrocalcin by P-31 NMR. *FEBS Lett.* **250**, 187-190.

Neu, H. C. (1992). The Crisis in Antibiotic Resistance. *Science* **257**, 1064-1073.

Paetzel, M., Danel, F., de Castro, L., Mosimann, S. C., Page, M. G. P., and Strynadka, N. C. J. (2000). Crystal Structure of the Class D β-Lactamase OXA-10. *Nature Struct. Biol.* **7**, 918-925.

Page, M. I. (1987). The Mechanisms of Reactions of β-Lactam Antibiotics in *Adv. Phys. Org. Chem.*, Vol. 23, D. Bethell, ed., Harcourt Brace Jovanovich Publishers, London, pp. 165-270.

Pearlman, D. A., Case, D. A., Caldwell, J. W., Ross, W. S., Cheatham, T. E., Debolt, S., Ferguson, D., Seibel, G., and Kollman, P. (1995). AMBER, a Package of Computer-Programs for Applying Molecular Mechanics, Normal-Mode Analysis, Molecular-Dynamics and Free-Energy Calculations to Simulate the Structural and Energetic Properties of Molecules. *Comput. Phys. Commun.* **91**, 1-41.

Pfannebecker, V., Klos, H., Hubrich, M., Volkmer, T., Heuer, A., Wiesner, U., and Spiess, H. W. (1996). Determination of End-to-End Distances in Oligomers by Pulsed EPR. *J. Phys. Chem.* **100**, 13428-13432.

Phillips, D. C. (1967). The Hen Egg-White Lysozyme Molecule. *Proc. Natl.Acad. Sci. USA* **57**, 484-495.

Pinavaia, T. J., Marshall, C. L., Mettler, C. L., Fisk, C. L., Miles, H. T., and Becker, E. D. (1978). Alkali Metal Ion Specificity in the Solution Ordering of a Nucleotide, 5'-Guanosine Monophosphate. *J. Am. Chem. Soc.* **100**, 3625-3627.

Prisner, T., Rohrer, M., and MacMillan, F. (2001). Pulsed EPR Spectroscopy: Biological Applications. *Ann. Rev. Phys. Chem.* **52**, 279-313.

Rakowsky, M. H., Zecevic, A., Eaton, G. R., and Eaton, S. S. (1998). Determination of High-Spin Iron(III)-Nitroxyl Distances in Spin-Labeled Porphyrins by Time-Domain EPR. *J. Magn. Reson.* **131**, 97-110.

Rienstra, C. M., Tucker-Kellogg, L., Jaroniec, C. P., Hohwy, M., Reif, B., McMahon, M. T., Tidor, B., Lozano-Perez, T., and Griffin, R. G. (2002). *De novo* Determination of Peptide Structure with Solid-State Magic-Angle Spinning NMR Spectroscopy. *Proc. Natl. Acad. Sci.USA* **99**, 10260-10265.

Rist, G. H. and Hyde, J. S. (1968). Ligand ENDOR of Cu-8-Hydroxyquinolate Substituted into a Single Crystal and a Powder of Phthalimide. *J. Chem. Phys.* **49**, 2449-2451.

Rist, G. H. and Hyde, J. S. (1970). Ligand ENDOR of Metal Complexes in Powders. *J. Chem. Phys.* **52**, 4633-4643.

Rodgers, D. W. (1997). Practical Cryocrystallography. *Methods Enzymol.* **276**, 183-203.

Rudin, M., Schweiger, A., and Gunthard, H. H. (1982). On the Electronic-Structure of *N,N'*-Ethylene-*bis*(acetylacetonatiminato)Co(II), Co(II)Acacen. 2. ENDOR and Double ENDOR of Ligand Nuclei. *Mol. Phys.* **46**, 1027-1044.

Saenger, W. (1984). *Principles of Nucleic Acid Structure*, Springer Verlag, New York.

Sander, M. E. and Witzel, H. (1985). Direct Chemical Evidence for the Mixed Anhydride Intermediate of Carboxypeptidase A in Ester and Peptide Hydrolysis. *Biochem. Biophys. Res. Commun.* **132**, 681-687.

Sander, M. E. and Witzel, H. (1986). Direct Chemical Evidence for an Anhydride Intermediate of Carboxypeptidase A in Ester and Peptide Hydrolysis in *Zinc Enzymes*, I. Bertini, C. Luchinat, W. Maret, and M. Zeppezauer, eds., Birkhaeuser, Boston, Massachusetts, pp. 207-214.

Scholes, C. P., Lapidot, A., Mascarenhas, R., Inubushi, T., Isaacson, R. A., and Feher, G. (1982). Electron Nuclear Double Resonance (ENDOR) from Heme and Histidine Nitrogens in Single-Crystals of Aquometmyoglobin. *J. Am. Chem. Soc.* **104**, 2724-2735.

Scott, W. R. P., Hunenberger, P. H., Tironi, I. G., Mark, A. E., Billeter, S. R., Fennen, J., Torda, A. E., Huber, T., Kruger, P., and van Gunsteren, W. F. (1999). The GROMOS Biomolecular Simulation Program Package. *J. Phys.Chem.* **A103**, 3596-3607.

Smith, T. S., LoBrutto, R., and Pecoraro, V. L. (2002). Paramagnetic Spectroscopy of Vanadyl Complexes and Its Applications to Biological Systems. *Coord. Chem. Rev.* **228**, 1-18.

Snetsinger, P. A., Chasteen, N. D., Cornelius, J. B., and Singel, D. J. (1992). Probing the Iron Center of the Low-Spin Cyanide Adduct of Transferrin by ESEEM Spectroscopy. *J. Phys. Chem.* **96**, 7917-7922.

Sosa-Peinado, A., Mustafi, D., and Makinen, M. W. (2000). Overexpression and Biosynthetic Deuterium Enrichment of TEM-1 β-Lactamase for Structural Characterization by Magnetic Resonance Methods. *Protein Expres. Purif.* **19**, 235-245.

Stryer, L. (1988). *Biochemistry*, 3rd edit., W. H. Freeman and Company, New York.

Strynadka, N. C. J., Adachi, H., Jensen, S. E., Johns, K., Sielecki, A., Betzel, C., Sutoh, K., and James, M. N. J. (1992). Molecular Structure of the Acylenzyme Intermediate in β-Lactam Hydrolysis at 1.7 Å Resolution. *Nature* **359**, 700-705.

Swartz, H. M. and Halpern, H. (1998). EPR Studies of Living Animals and Related Model Systems (*In Vivo* EPR) in *Biological Magnetic Resonance*, Vol. 14, L. J. Berliner, ed., Plenum Press, NewYork, pp. 367-404.

Sweet, R. M. and Dahl, L. F. (1970). Molecular Architecture of the Cephalosporins. Insights into Biological Activity Based on Structural Investigations. *J. Am. Chem. Soc.* **92**, 5489-5507.

Teng, T. Y. and Moffat, K. (1998). Cooling Rates during Flash Cooling. *J. Appl. Crystallogr.* **1**, 252-257.

Togni, A., Rist, G., Rihs, G., and Schweiger, A. (1993). EPR, H-1 and C-13 ENDOR, N-14 ESEEM, and X-Ray Crystallographic Studies of Oxovanadium(IV) *Bis*((1*R*)-3-(Hepta-fluorobutyryl)Camphorate) - a Catalyst for Asymmetric Hetero-Diels-Alder Reactions. *J. Am. Chem. Soc.* **115**, 1908-1915.

Turley, J. W. and Boer, F. P. B. (1972). The Crystal Structure of the Nitroxide Free Radical 2,2,5,5-Tetramethyl-3-carbamidopyrroline-1-oxyl. *Acta Crystallogr.* **B28**, 1641-1644.

van Ormondt, D. and Visser, H. (1968). Ligand ENDOR of Mn^{++} in $La_2Mg_3(NO_3)_{12} \cdot 24H_2O$. *Phys. Lett.* **A26**, 343-344.

Van Zele, C. J., Cunningham, M. A., and Makinen, M. W. (2001). Validation of Nitroxyl Spin-Label Force-Field Parameters through Molecular Dynamics Simulations. *J. Comput. Chem.* **22**, 1113-1123.

Vocadlo, D. J., Davies, G. J., Laine, R., and Withers, S. G. (2001). Catalysis by Hen Egg-White Lysozyme Proceeds *via* a Covalent Intermediate. *Nature* **412**, 835-838.

Walsby, C. J., Hong, W.,Broderick, W. E., Cheek, J., Ortillo, D., Broderick, J. B., and Hoffman, B. M. (2002). Electron Nuclear Double Resonance Spectroscopic Evidence that S-Adenosylmethionine Binds in Contact with the Catalytically Active [4Fe-4S](+) Cluster of Pyruvate Formate-Lyase Activating Enzyme. *J. Am. Chem. Soc.* **124**, 3143-3151.

Weiner, P. K. and Kollman, P. A. (1981). AMBER - Assisted Model-Building with Energy Refinement - a General Program for Modeling Molecules and Their Interactions. *J. Comput. Chem.* **2**, 287-303.

Wells, G. B. and Makinen, M. W. (1988). ENDOR Determined Molecular Geometries of Spin-Labeled Fluoroanilides in Frozen Solution. *J. Am. Chem. Soc.* **110**, 6343-6352.

Wells, G. B., Mustafi, D., and Makinen, M. W. (1990). Structure and Conformation of Spin-Labeled Amino Acids in Frozen Solutions Determined by Electron Nuclear Double Resonance. 2. Methyl *N*-(2,2,5,5-Tetramethyl-1-Oxypyrrolinyl-3-Carbonyl)-L-Tryptophanate, a Molecule with Multiple Conformations. *J. Am. Chem. Soc.* **112**, 2566-2574.

Wells, G. B., Mustafi, D., and Makinen, M. W. (1994). Structure at the Active-Site of an Acylenzyme of α-Chymotrypsin and Implications for the Catalytic Mechanism. An Electron Nuclear Double Resonance Study. *J. Biol. Chem.* **269**, 4577-4586.

Williams, R. J. P. (1985). The Symbiosis of Metal and Protein Functions. *Eur. J. Biochem.* **150**, 231-248.

Wuthrich, K. and Connick, R. F. (1968). Nuclear Magnetic Resonance Studies of the Coordination of Vanadyl Complexes in Solution and the Rate of Elimination of Coordinated Water Molecules. *Inorg. Chem.* **7**, 1377-1388.

Yim, M. B. and Makinen, M. W. (1986). ENDOR Study of Gd^{3+} Complexes in Frozen Solutions. *J. Magn. Reson.* **70**, 89-105.

Yordanov, N. D. and Zdravkova, M. (1986). H-1 and P-31 ENDOR Studies on Powdered Samples of Magnetically Dilute Copper(II) *O,O'*-Disubstituted Dithiophosphate Complexes. *Chem. Phys. Lett.* **127**, 487-491.

Yordanov, N. D., Zdravkova, M., and Shopov, D. (1986). An Improved Method for ENDOR Study of Powdered Samples. *Chem. Phys. Lett.* **124**, 191-195.

Zabell, A. P. R. and Post, C. B. (2002). Docking Multiple Conformations of a Flexible Ligand into a Protein Binding Site Using NMR Restraints. *Proteins: Struct., Func., Genet.* **46**, 295-307.

Zdravkova, M. and Yordanov, N. D. (1994). ENDOR Crystallography – Current Practical Applications. *Appl. Magn. Reson.* **6**, 83-105.

Zell, A., Einspahr, H., and Bugg, C. E. (1985). Model for Calcium-Binding to γ-Carboxyglutamic Acid Residues of Proteins – Crystal Structure of Calcium α-Ethylmalonate. *Biochemistry* **24**, 533-537.

Zhao, D. Q. and Jardetzky, O. (1994). An Assessment of the Precision and Accuracy of Protein Structures Determined by NMR. Dependence on Distance Errors. *J. Mol. Biol.* **239**, 601-607.

Zimmerman, S. B. (1976). X-Ray Study by Fiber Diffraction Methods of a Self-Aggregate of Guanosine-5'-Phosphate with Same Helical Parameters as Poly(Rg). *J. Mol. Biol.* **106**, 663-672.

Chapter 5

Solution-ENDOR of Some Biologically Interesting Radical Ions

Fabian Gerson and Georg Gescheidt
Department of Chemistry, University of Basel, Klingelbergstrasse 80, CH-4056 Basel, Switzerland

Abstract: A simple phenomenological treatment of the solution-ENDOR spectroscopy is presented. It is followed by a brief report on such studies carried out on some radical ions belonging to two classes of biologically interesting compounds, quinones and porphyrinoids.

1. SOLUTION-ENDOR SPECTROSCOPY

1.1 Introduction

For studies of organic radicals, the by far most important multiresonance technique is electron-nuclear double resonance (ENDOR) discovered by Feher in 1956 on a phosphorus-doped silicon system (Feher,1956; Feher, 1998). Several years later, it was applied to radicals in solution by Hyde and Maki (Hyde and Maki, 1964; Hyde, 1965; Hyde, 1974), as well as by Möbius and his colleagues (Biehl et al., 1971; Möbius and Dinse, 1972; Möbius, 1998) who also introduced TRIPLE-resonance techniques (Biehl et al., 1975; Möbius and Biehl, 1979). The reason for the application of ENDOR spectroscopy to radicals in solution lagging behind that to paramagnetic species in solids was partly due to the lack of interest in the liquid phase by physicists who first used this technique. Even more important were problems of instrumentation. In addition to the conventional EPR apparatus and a special cavity with radiofrequency (RF) coils, the ENDOR technique requires a RF source to saturate the NMR transitions. For liquids, the RF power must be much higher than for solids, and so must be the efficiency of the cooling system (Atherton, 1979). Although ENDOR has not attained a popularity comparable to EPR, it is now used by an increased number of research groups, especially since ENDOR accessories have become commercially available from the Varian Associates (Hyde, 1998) in the seventies and from the Bruker GmbH (Schmalbein, 1998) in the eighties. The ENDOR technique has been briefly dealt with in several early monographs on EPR spectroscopy (Ayscough, 1967; Carrington and McLachlan, 1967; Scheffler and Stegmann, 1970; Wertz and Bolton, 1972; Atherton, 1973) and, in some length, in a few books specialized in multiresonance (Kevan and Kispert, 1976; Dorio and Freed, 1979). An excellent introduction into the ENDOR technique, as used for organic radicals in solution, is to be found in a review article (Kurreck et al., 1984) and, in more detail, in a book by same authors (Kurreck et al., 1988). The latter also contains a comprehensive account of the pertinent ENDOR studies up to 1988. The physical fundamentals underlying this double-resonance technique can be grasped by considering the so called transient-ENDOR effect in the way presented by Kurreck et al. and adopted in a recent monograph on EPR spectroscopy (Gerson and Huber, 2003). The following treatment is a condensed version of a section in this monograph.

1.2 Physical Fundamentals

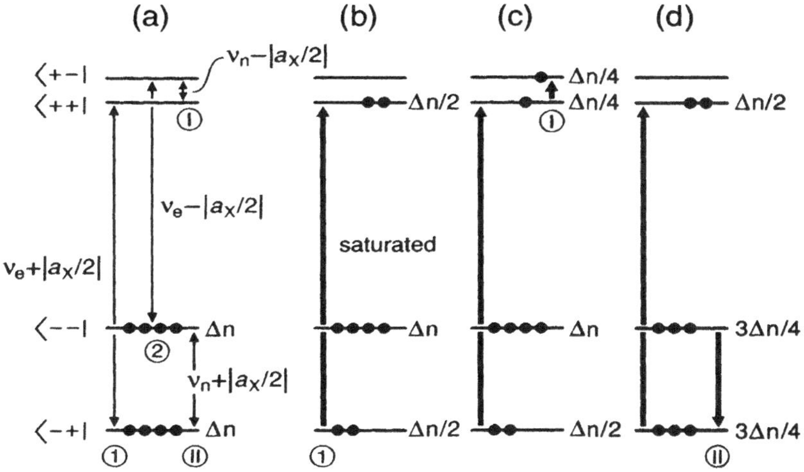

Figure 1. Schemes relevant to the transient-ENDOR effect for a paramagnetic system consisting of one unpaired electron and one magnetic nucleus with $I = 1/2$ and $g_n > 0$. (a) Energy levels in absence of the saturation. (b) Effect of saturation of the ESR transition ① on the populations. (c) and (d) Effect of the saturation of the NMR transitions (I) and (II), respectively, on the populations. Reproduced partly from (Kurreck et al., 1988) and (Gerson and Huber, 2003) by permission of VCH Publishers and Wiley-VCH, respectively.

The relevant schemes are shown in Figure 1. They depict four Zeeman-energy levels which, at a given field strength, B, of the magnetic field are characteristic of a paramagnetic system consisting of one unpaired electron and one magnetic nucleus X, such as proton, with the spin-quantum number $I = 1/2$ and a positive nuclear factor g_n. The four levels, $|+\,+\rangle$, $|+\,-\rangle$, $|-\,+\rangle$, and $|-\,-\rangle$, are specified by the signs of the magnetic spin-quantum numbers, $+1/2$ (spin up; α) or $-1/2$ (spin down; β), whereby the first sign applies to the number, M_S, of electron and the second to that, M_I, of nucleus. The excess, Δn, of the electron-spin population in the levels $|-\,+\rangle$ and $|-\,-\rangle$ relative to $|+\,+\rangle$ and $|+\,-\rangle$, required for the EPR absorption, is symbolized by four dots, each dot standing for Δn/4. The Zeeman splittings are given as frequencies, ν_e for the electron and ν_n for the nucleus, and the hyperfine-coupling constant, a_X, of the nucleus X also has the dimension of ν. Division by $\gamma = 28.04$ MHz/mT, the gyromagnetic ratio of the electron, converts this a_X value in MHz into the coupling constant in mT, the unit of B. According to the selection rules, $\Delta M_S = \pm 1$ and $\Delta M_I = 0$ for the electron and $\Delta M_S = 0$ and $\Delta M_I = \pm 1$ for the nucleus, two EPR (① and ②) and two NMR transitions ((I) and (II)) are allowed.

The schemes (a) – (d) in Figure 1 hold for $\nu_n > |a_X/2|$, which is usually the case with protons in π-radicals. In this case, the level $|+\,+\rangle$ lies below $|+\,-\rangle$, and the NMR transition (I) has the frequency $\nu_n - |a_X/2|$. On the other hand, for $\nu_n < |a_X/2|$, the level $|+\,-\rangle$ is shifted below $|+\,+\rangle$, and the frequency of (I) becomes $|a_X/2| - \nu_n$. In either case, the level $|-\,+\rangle$ is situated below $|-\,-\rangle$, so that the NMR transition (II) has the frequency $\nu_n + |a_X/2|$. The frequencies of the EPR transitions ① and ② are throughout $\nu_e + |a_X/2|$ and $\nu_e - |a_X/2|$, respectively, thus differing by $|a_X|$ as expected.

In an ENDOR experiment, one EPR transition is selected for further procedure; it is the transition ① in scheme (a). After having been locked at its frequency, $\nu_e + |a_X/2|$, this transition is saturated by an intense microwave (MW) irradiation. Consequently, as indicated in scheme (b), the populations in the two levels relevant to ①, $|-\,+\rangle$ and $|+\,+\rangle$, become equal. Both levels then exhibit an excess Δn/2, and the intensity of the pertinent EPR signal is strongly reduced. In the next step, the system is subjected to an intense irradiation with radiofrequency (RF), which is scanned from 0 to higher values. At two frequencies, the NMR transitions become saturated, first (I) at $\nu_n - |a_X/2|$ and, subsequently, (II) at $\nu_n + |a_X/2|$. As a result, the populations in the pairs of the affected levels are equalized, as shown in schemes (c) and (d) for the transitions (I) and (II), respectively. Saturation of the transition (I) leads to an excess Δn/4 in each of the levels $|+\,-\rangle$ and $|+\,+\rangle$, while such a process in (II) yields $3\Delta n/4$ in $|- +\rangle$ and $|- -\rangle$. Thus, either of the two saturation processes makes the population in the level $|- +\rangle$ by $\Delta n/4$ higher than in $|+ +\rangle$, so that in either case the EPR transition (1) is desaturated, and the intensity of the EPR signal exhibits an increase, the so called ENDOR enhancement. Such an enhancement is, however, not directly verified, but its occurrence is confirmed by the NMR absorptions observed for the transitions (I) and (II).

1.3 Spectra

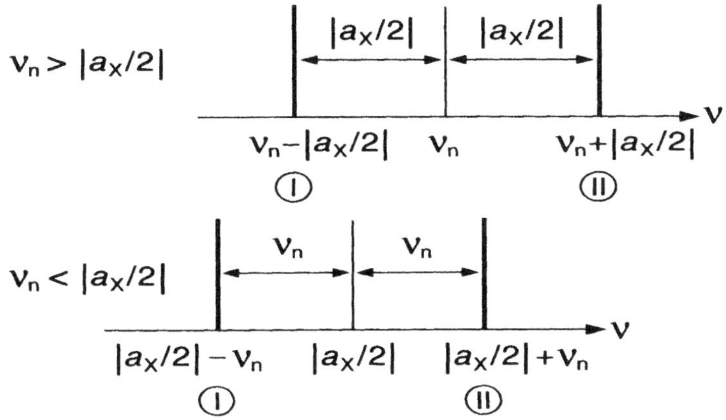

Figure 2. Schematic presentation of an ENDOR spectrum arising from one nucleus X or a set of equivalent nuclei X with the coupling constant a_X. Reproduced from (Gerson and Huber, 2003) by permission of Wiley-VCH.

The ENDOR signals which arise from the NMR transitions (I) and (II), while scanning the RF ν, are schematically shown in Figure 2. Contrary to the NMR experiment, their intensity is due to a population excess, Δn, of electron spins, which is by several orders of magnitude larger than the analogous excess number of nuclear spins. Thus, the sensitivity of ENDOR is much higher than that of NMR, although it is lower than that of EPR. It can readily be verified that any magnetic nucleus X or a set of such equivalent nuclei with the coupling constant a_X gives rise to a single pair of ENDOR signals, irrespective of the nuclear spin-quantum number I and the nuclear factor g_n. This pair of signals generally appears in a separate NMR frequency range characteristic of X and ν_n. For $\nu_n > |a_X/2|$, which holds for the ENDOR spectra reproduced in this chapter, the two signals appear at $\nu_n \pm |a_X/2|$; they are centered on the frequency, ν_n, of the "free" nucleus X and separated by the coupling constant a_X (Figure 2, top). On the other hand, for $\nu_n < |a_X/2|$, the two signals occur at $|a_X/2| \pm \nu_n$; they are centered on $|a_X/2|$ and separated by $2\nu_n$ (Figure 2, bottom). The ENDOR signals can be recorded as absorption A or as the first derivative dA/dν as function of ν, depending on whether modulation is applied to the magnetic field or to the frequency. The latter procedure was used to record the ENDOR spectra in Figures 3 – 7 shown in the following sections.

Although ENDOR is less sensitive than EPR, this deficiency is amply made good by the enormous increase in spectral resolution. As the width, $\Delta\nu$, of ENDOR signals (line-width) of ca. 0.3 MHz is comparable to ΔB (= $\Delta\nu/\gamma$) of ca. 0.01 mT, generally achieved for EPR lines in a well-resolved spectrum of an organic radical in fluid solution, the increase in resolution by ENDOR relative to EPR spectroscopy is due to a drastic decrease in the number of lines. With each further set of equivalent nuclei X giving rise to pairs of ENDOR signals, the number of lines grows *additively*, and not *multiplicatively* as in EPR spectra. Irrespective of the number (n_1, n_2, ... n_k) of nuclei in each set (1, 2, ... k) with I_1, $I_2 \ldots I_k$, the total number of ENDOR lines for k sets is thus 2k and not $(2n_1 I_1 +1)(2n_2 I_2 +1) \ldots (2n_k I_k +1)$.

A disadvantage of ENDOR spectroscopy is, that, unlike NMR, the intensity of a signal is not a reliable measure for the number of interacting nuclei giving rise to it. This is because the ENDOR enhancement and, therewith, the intensity of the ENDOR signals depends on whether the nuclear-spin relaxation responsible for the saturation of the NMR transitions Ⓘ and Ⓘⓘ can compete with the electron-spin relaxation effective in saturating the selected EPR transition ①. Usually, the electron-spin relaxation, which takes care of inversions of electron spins ($\Delta M_S = \pm 1$), is much more efficient than the nuclear-spin relaxation which causes inversions of nuclear spins ($\Delta M_I = \pm 1$), so that the former must be slowed down by appropriate experimental conditions. When cross-relaxation processes with $\Delta(M_S + M_I) = 0$ can be neglected, as is often the case with protons in organic and bioorganic radicals, this slowing down is achieved by using viscous solvents and/or low temperatures. The ENDOR experiment is impeded by an enhanced electron-spin relaxation, e.g. in the presence of heavy nuclei (which are often contained in transition metals of bioorganic molecules) or by the dynamic Jahn-Teller effect relevant to radical with an axial symmetry (rotational axis C_n with n equal or larger than 3) in a degenerate ground state. An enhanced electron-spin relaxation is mostly conspicuous in the corresponding EPR spectrum, as hyperfine lines are broadened and difficult to saturate.

The ENDOR technique proved to be particularly useful for radicals of low symmetry with a large number of overlapping and/or incompletely resolved EPR lines (Gerson et al., 1975). Because of its lower sensitivity, it requires somewhat larger radical concentration than EPR spectroscopy, and its application to transient radicals is, therefore, more problematic. In order to increase the signal-to-noise ratio, the ENDOR spectra are usually accumulated by repeated recording and addition. Radical ions, which are electrolytically generated inside the cavity, cannot be studied by the ENDOR technique, as the electrodes interfere with the RF coils.

The number of the nuclei giving rise to an ENDOR signal must be verified by close examination, preferably by simulation, of the corresponding EPR spectrum. Other procedures can likewise be used to this aim, e.g. isotopic substitution, which also serves for the assignment of coupling constants to sets of equivalent nuclei. ENDOR spectroscopy is particularly suited for such an assignment, because the signals of different isotopes appear in separate frequency regions. Relative numbers of nuclei responsible for the signals can be determined by techniques such as special-TRIPLE resonance, while the relative signs of the coupling constants a_X are obtainable by general-TRIPLE resonance.

1.4 Triple Resonance

This technique requires a second powerful RF source. In the special-TRIPLE-resonance or double-ENDOR experiment (Kurreck et al., 1984; Kurreck, et al., 1988), the sample is irradiated simultaneously with two RF fields, in addition to the saturating MW irradiation, so that both NMR transitions Ⓘ and Ⓘⓘ (Figure 1) are excited at the same time. According to schemes (c) and (d) of this Figure, such procedure should double the ENDOR enhancement, because it leads simultaneously to $\Delta n/4$ in $|+\,+\rangle$ and $3\Delta n/4$ in $|-\,+\rangle$, thus yielding for the levels involved in the relevant EPR transition ① a difference $\Delta n/2$ instead of $\Delta n/4$ achieved with a single RF source. The main advantage of special-TRIPLE resonance is that the signal intensities reproduce better the relative number of nuclei giving rise to them than it does ENDOR spectroscopy. The special-TRIPLE-resonance signal, associated with the coupling constant a_X appears separated from the origin (NMR frequency $\nu = 0$) by $|a_X/2|$ in unit of ν. As shown in Figure 3 for the phenalenyl radical (Kurreck et al., 1984), the intensities of the special-TRIPLE-resonance signals from the six protons in the 1,3,4,6,7,9-positions relative to those from the three protons in the 2,5,8-positions exhibit more exactly the expected ratio 2.

Whereas in the special-TRIPLE-resonance experiment the NMR transitions of the same set of protons are irradiated (homonuclear-TRIPLE resonance), in its general-TRIPLE counterpart, transitions of different sets of nuclei are saturated simultaneously (heteronuclear-TRIPLE resonance). One NMR transition is "pumped" with the first (unmodulated) RF frequency, while the second (modulated) RF field is scanned over the whole range of NMR resonances. The "pumping" causes characteristic intensity changes of the high- and low-frequency signals relative to those observed by ENDOR. When the high- (low-) frequency signal is pumped, its intensity strongly reduced, while that of the low- (high-) frequency partner, associated with the same coupling constant, is enhanced, because, for the latter signal, the pumping corresponds to a special-TRIPLE-resonance experiment.

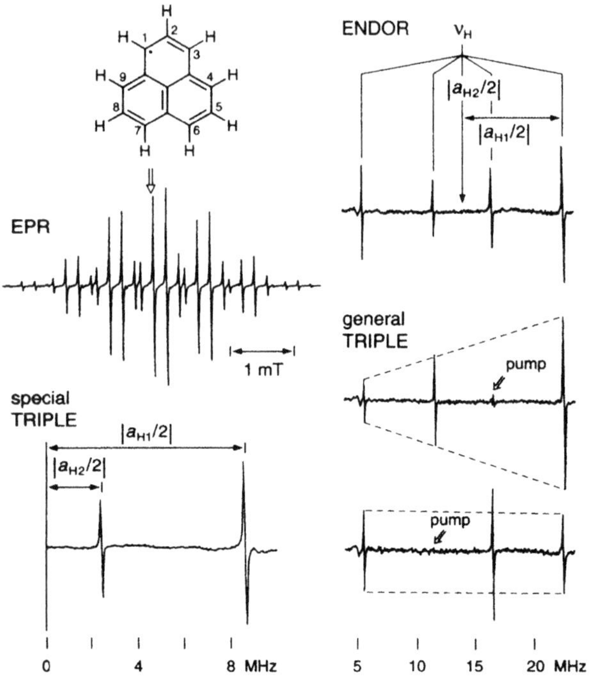

Figure 3. EPR (left, center), ^{1}H-ENDOR (right, top), and the corresponding special-TRIPLE-resonance (left, bottom) and general-TRIPLE-resonance (right, center and bottom) spectra of the phenalenyl radical; solvent mineral oil, temperature 300 K. The arrow above the EPR spectrum indicates the line selected for saturation, while those in the general-TRIPLE-resonance spectra mark the ENDOR signal chosen for pumping. ν_H = 14.2 MHz is the frequency ν_n of the free proton. Taken with a Bruker-ER-200-D spectrometer and a Bruker ENB-ENDOR cavity. Reproduced from (Kurreck, et al., 1988) by permission of VCH Publishers.

Figure 3 demonstrates the effect of pumping the signals separated by the smaller coupling constant, $|a_{H2,5,8}|$, of the three protons in the 2,5,8-positions of phenalenyl radical (Kurreck, et al., 1984). When the high-frequency signal at $\nu_n + |a_{H2,5,8}/2|$ is pumped, it nearly disappears, while its low-frequency partner at $\nu_n - |a_{H2,5,8}/2|$ is substantially strengthened. An opposite effect on the intensities is observed on pumping this low-frequency signal. Simultaneously, striking intensity changes are observed for the pair of signals separated by the larger coupling constant, $|a_{H1,3,4,6,7,9}|$, of the six protons in the 1,3,4,6,7,9-positions, although neither of these signals is subjected to pumping. As is evident from Figure 3, such changes follow patterns which are diametrically opposed to those induced on the signals separated by $|a_{H2,5,8}|$. The intensity ratio of the high-frequency signal at ν_n + $|a_{H1,3,4,6,7,9}/2|$ to its low-frequency partner at $\nu_n - |a_{H1,3,4,6,7,9}/2|$ increases relative to that in the ENDOR spectrum when the high-frequency signal at $\nu_n + |a_{H2,5,8}/2|$ is pumped. In contrast, this ratio decreases when the pumping is carried out on the low-frequency signal at $\nu_n - |a_{H2,5,8}/2|$. This behavior points to opposite signs of the two coupling constants, i.e. $a_{H1,3,4,6,7,9}$ = −17.64 and $a_{H2,5,8}$ = +5.08 MHz, corresponding to −0.629 and +0.181 mT, respectively (Gerson, 1966).

2. QUINONES

2.1 Introduction

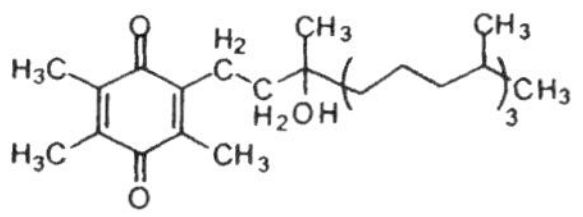

1

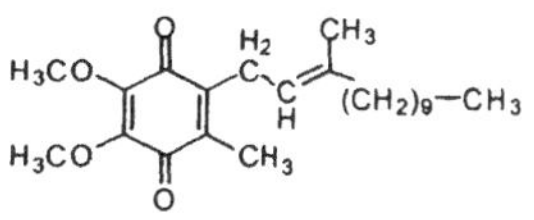

2

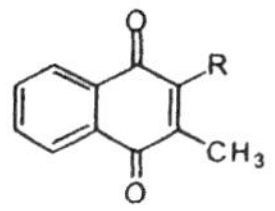

3 R = $CH_2CH{=}C(CH_3)[(CH_2)_3CH(CH_3)]_3CH_3$
4 R = H

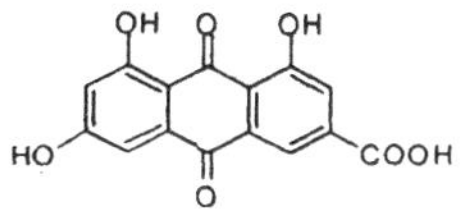

5

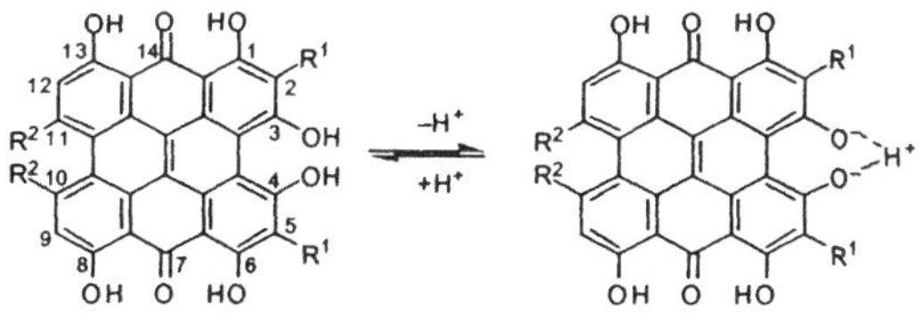

6H R¹ = H R² = CH_3 6⁻
7H R¹ = Br R² = CH_3 7⁻
8H R¹ = R² = H 8⁻
9 is 6H with all OH replaced by OAc

Quinones are the most common electron acceptors and electron-transfer mediators in biological processes, in which their radical anions (semiquinone anions) play an important role (Morton, 1965). An early ENDOR study of biologically interesting semiquinone anions in fluid solution was published (Das et al., 1970) a few years after ENDOR spectra of organic radicals in solution had been reported for the first time (Hyde and Maki, 1964). The pertinent semiquinones were the radical anions of α-tocopherol (vitamin E quinone; **1**) and ubiquinone (**2**), which are both derivatives of *p*-benzoquinone, as well as those of 2-methyl-3-phytyl-1,4-naphthoquinone (vitamin K_1 quinone; **3**) and menadione (vitamin K_3 quinone; **4**) which is another derivative of 1,4-naphthoquinone. Recently, a solution-ENDOR study of the semiquinone anion from emodic acid (**5**), an oxidation product of emodin, was reported (Rahimipour et al., 2001b); emodin, a derivative of 9,10-anthraquinone, is used as a laxative and has other pharmacological applications (Hartmann and Goldstein, 1989). Here, we describe solution-ENDOR spectra of the semiquinone anion and dianion from hypericin (**6**H) (Gerson et al., 1995), a derivative of biphenoquinone, which has been isolated from St. John's wort and displays antiviral, antidepressive, and photodynamic activity (Muldner and Zoller, 1984; Suzuki et al., 1984). Solution-ENDOR spectra of a radical anion and two radical dianions from hypericin derivatives (Rahimipour et al., 2001a) are also presented.

2.2 Hypericin

Hypericin is claimed to exist as 7,14- and 1,6-dihydroxy tautomers, of which the former (**6**H) is more stable and prevails in non-concentrated solutions (Dax et al., 1999; Etzlstorfer and Falk, 2000; Freeman et al., 2001). In neutral and alkaline media, it readily deprotonates to yield its conjugate base $\mathbf{6}^-$. Both **6**H and $\mathbf{6}^-$ readily accept an additional electron, and ENDOR spectroscopy serves as a straightforward tool to find out whether deprotonation also occurs at the stage of the one-electron-reduced species.

Figure 4 shows the ^{1}H-ENDOR spectrum of the radical dianion $\mathbf{6}^{\cdot 2-}$ generated from the sodium salt of $\mathbf{6}^-$ with potassium in tetrahydrofuran (THF); identical spectra were observed under the same conditions from other salts of $\mathbf{6}^-$, and very similar ones were obtained upon reduction of the sodium salt with zinc in *N,N*-dimethylformamide (DMF) and DMF/H_2O (10:1) (Gerson et al., 1995). Use of DMF-d_8/D_2O (10:1) led to the radical dianion $\mathbf{6}$-$d_5^{\cdot 2-}$, in which all five OH protons were replaced by deuterons; the pertinent ^{1}H- and ^{2}H-ENDOR spectra are also reproduced in Figure 4. Contrary to the EPR and ENDOR spectra of the persistent radical dianion $\mathbf{6}^{\cdot 2-}$, those of the radical anion $\mathbf{6}\text{H}^{\cdot -}$, its conjugated acid, were observable only under strictly anhydrous conditions, because $\mathbf{6}\text{H}^{\cdot -}$ rapidly deprotonated to $\mathbf{6}^{\cdot 2-}$ in the presence of traces of water. A ^{1}H-ENDOR spectrum of $\mathbf{6}\text{H}^{\cdot -}$, recorded immediately upon reaction of **6**H with potassium in a carefully dried THF, is likewise displayed in Figure 4.

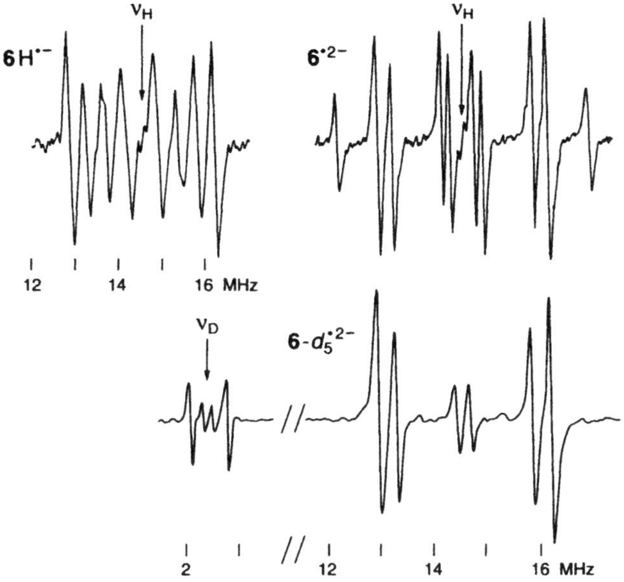

Figure 4. Radical ions from hypericin. ^{1}H-ENDOR spectrum of the anion **6H**$^{\cdot-}$ (top, left); solvent THF, counterion K^+. ^{1}H-ENDOR spectrum of the dianion **6**$^{\cdot 2-}$ (top, right); solvent THF, counterions Na^+ and K^+. ^{1}H- and ^{2}H-ENDOR spectra of the dianion **6**-$d_5^{\cdot 2-}$ (bottom); solvent DMF-d_8/D_2O, counterions Na^+ and Zn^{2+}. Temperature 298 K throughout. $\nu_H = 14.56$ and $\nu_D = 2.23$ MHz are the frequencies ν_n of the free proton and deuteron, respectively. Taken with a Bruker-ESP-300 spectrometer. Reproduced from (Gerson et al., 1995) by permission of *J. Am. Chem. Soc.*

Table 1. Hyperfine Data, $a_{H\mu}$ and $a_{D\mu}$ in mT, for the Radical Dianion and Radical Anion from Hypericin[a]

	6$^{\cdot 2-}$			**6**-$d_5^{\cdot 2-}$	**6H**$^{\cdot-}$	
μ =	K/THF	Zn/DMF	Zn/DMF/H_2O	Zn/DMF-d_8/D_2O	K/THF	Zn/DMF
1,6-OH	−0.027[b]	−0.028[b]	−0.029[b]	−0.005[c]	−0.026[b]	−0.026[b]
2,5	−0.005	−0.006	−0.010	−0.009	−0.005	−0.005
3,4-OH	+0.167[d]	+0.160[d]	+0.157[d]	+0.024[e]	−0.059[b]	−0.056[b]
8,13-OH	−0.017[b]	−0.016[b]	−0.015[b]	−0.002[c]	−0.026[b]	−0.026[b]
9,12	−0.094	−0.093	−0.090	−0.091	−0.091	−0.087
10,11	+0.114[f]	+0.115[f]	+0.116[f]	+0.116[f]	+0.119[f]	+0.121[f]

[a]The values refer to α-protons unless otherwise indicated; $g = 2.00275 \pm 0.00005$ throughout. [b]Two OH protons. [c]Two OD deuterons. [d]One OH proton. [e]One OD deuteron. [f]Six methyl β-protons.

The hyperfine data for **6**$^{\cdot 2-}$, **6**-$d_5^{\cdot 2-}$, and **6H**$^{\cdot-}$, obtained under various conditions, are listed in Table 1. Relying on simulation of the EPR spectra of

$\mathbf{6}^{\cdot 2-}$, its coupling constants, $a_{H\mu}$, of one and six protons were straightforwardly assigned to the single proton bridging the 3,4-O atoms and to the β-protons of the two 10,11-methyl substituents, respectively. Moreover, deuteration allowed to distinguish the proton pairs in the 1,6- and 8,13-OH groups from the sets of two α-protons at the π-centers μ = 2,5 and 9,12. (According to the conventional nomenclature of EPR spectroscopy, protons directly linked to the π-centers μ are called α, while those separated from such centers by one sp^3-hybridized C atom are denoted β.) The remaining ambiguities were resolved by UB3LYP/6-31G* calculations (Rahimipour et al., 2001a). The coupling constants for $\mathbf{6}H^{\cdot-}$ are similar to those for $\mathbf{6}^{\cdot 2-}$ with the exception of the value due to the two protons in the non-dissociated 3,4-OH groups of $\mathbf{6}H^{\cdot-}$ which strongly differs from that of the single proton bridging the pertinent two O atoms in $\mathbf{6}^{\cdot 2-}$. The signs of the coupling constants were determined by the general-TRIPLE-resonance experiments with the reasonable assumption that the values of the α-protons are negative. All signs were confirmed by theoretical calculations. The hyperfine data for $\mathbf{6}^{\cdot 2-}$ and $\mathbf{6}H^{\cdot-}$ are compatible with a twisted helical geometry and an effective C_2 symmetry.

2.3 Hypericin Derivatives

These derivatives are 2,5-dibromohypericin (**7**H), 10,11-desmethylhypericin (**8**H), and 1,3,4,6,8,13-hexaacetylhypericin (**9**). In contrast to the parent hypericin (**6**H), the π-system of **8**H can adopt a planar geometry, while the 3,4-OAc groups of **9** are not amenable to deprotonation. The three compounds were reduced under various experimental conditions (Rahimipour et al., 2001a), and the ^{1}H-ENDOR spectra of the paramagnetic species thus obtained with zinc in DMF are shown in Figure 5. These spectra allowed to identify the species in question as the radical dianions $\mathbf{7}^{\cdot 2-}$ and $\mathbf{8}^{\cdot 2-}$ and the radical anion $\mathbf{9}^{\cdot-}$, respectively. Their hyperfine data are given in Table 2.

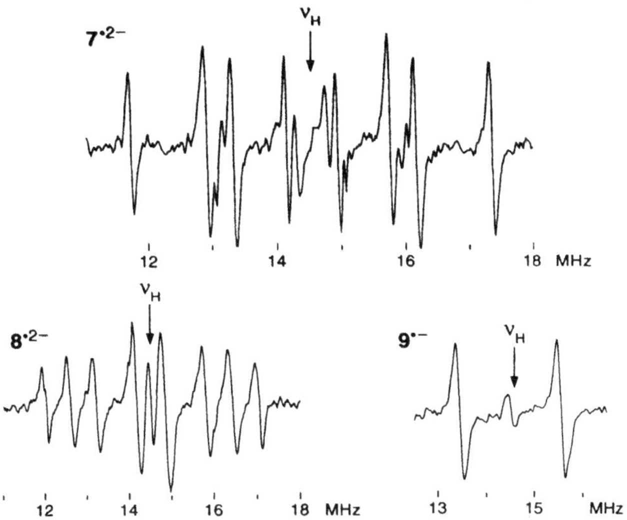

Figure 5. Radical ions from hypericin derivatives. ^{1}H-ENDOR spectra of the dianions $\mathbf{7}^{\bullet 2-}$ (top) and $\mathbf{8}^{\bullet 2-}$ (left, bottom) and of the anion $\mathbf{9}^{\bullet -}$ (right, bottom); solvent DMF, counterion Zn^{2+}, temperature 273 K. ν_H = 14.56 MHz is the frequency ν_n of the free proton. The unsplit signals at ν_H in the spectra of $\mathbf{8}^{\bullet 2-}$ and $\mathbf{9}^{\bullet -}$ arise from protons with a very small coupling constant; the pertinent (absolute) values are 0.003 mT for 1,6-OH protons in $\mathbf{8}^{\bullet 2-}$ (Table 2) and <0.005 mT for the acetyl protons in $\mathbf{9}^{\bullet -}$ (not given in Table 2). Taken with a Bruker-ESP-300 spectrometer. Reproduced from (Rahimipour et al., 2001a) by permission of *Photochem. Photobiol.*

Table 2. Hyperfine Data, $a_{H\mu}$ in mT, for the Radical Dianions and Radical Anion from Hypericin Derivatives[a]

μ =	$\mathbf{7}^{\bullet 2-}$	$\mathbf{8}^{\bullet 2-}$	$\mathbf{9}^{\bullet -}$
1,6-OH	−0.029[b]	−0.003[b]	–
2,5	–	−0.028[c]	−0.078[c]
3,4-OH	+0.204[d]	+0.180[d]	–
8,13-OH	−0.017[b]	−0.022[b,c]	–
9,12	−0.087	−0.137	−0.083[c]
10,11	+0.116[c]	−0.094	+0.075[c,e]

[a]The values refer to α-protons unless otherwise indicated; *g* factor not reported. [b]Two OH protons. [c]Values derived by simulation of the EPR spectrum and based on the observed single pair of ENDOR signals (Figure 5). [d]One OH proton. [e]Six methyl β-protons.

Assignments and signs of the coupling constants $a_{H\mu}$, which compare favorably with the corresponding values for $\mathbf{6}^{\bullet 2-}$ and $\mathbf{6}H^{\bullet -}$, were corroborated by UB3LYP/6-31G* calculations. The spin distribution in these radical dianions and radical anion is not markedly altered by the structural modifications of hypericin, as the bulk of spin population remains accommodated by the central biphenoquinone moiety. With respect to biological processes it is important to note that the structure of $7^{\cdot 2-}$, $8^{\cdot 2-}$, like that of $6^{\cdot 2-}$, remains unaltered when organic solvents are replaced by an aqueous buffer solution.

3. PORPHYRINOIDS

3.1 Introduction

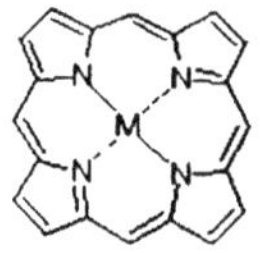

10H₂ M = 2 H
10M M = Metal

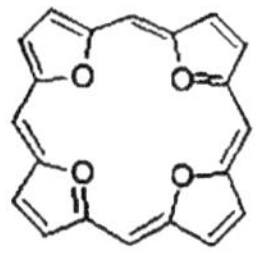

11

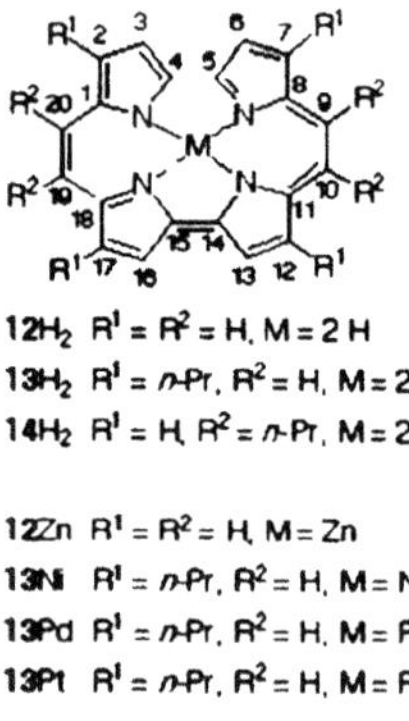

12H₂ $R^1 = R^2 = H$, M = 2 H
13H₂ R^1 = *n*-Pr, R^2 = H, M = 2 H
14H₂ R^1 = H, R^2 = *n*-Pr, M = 2 H

12Zn $R^1 = R^2 = H$, M = Zn
13Ni R^1 = *n*-Pr, R^2 = H, M = Ni
13Pd R^1 = *n*-Pr, R^2 = H, M = Pd
13Pt R^1 = *n*-Pr, R^2 = H, M = Pt

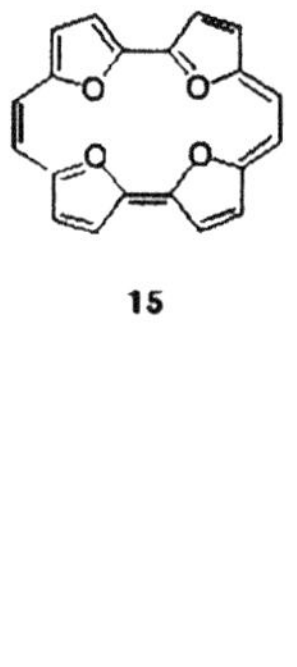

15

Porphyrins in general, and metalloporphyrins in particular, are involved in many biological processes and, owing to their deep colors, have been called the pigments of life (Battersby and McDonald, 1979; Battersby and Frobel, 1982). The prominent property of these macrocyclic π-systems is the easy acceptance and donation of electrons, which leads to many redox stages. Because of their effective D_{4h} symmetry, the radical ions of the unsubstituted porphyrin (**10**H_2) and metalloporphyrins (**10M**) (Seth and Bocian, 1994), as well as those of tetraoxaporphyrin (**11**) (Bachmann et al., 1992; Bachmann, 1996), have a degenerate ground state and are subject to the dynamic Jahn-Teller effect which enhances the electron-spin relaxation. As a consequence, the EPR lines of these ions are excessively broadened and difficult to saturate, so that their solution-ENDOR spectra could not be observed. On the other hand, the effective symmetry is lowered to D_{2h} in the isomeric porphycene (**12**H_2) (Vogel et al., 1986), metalloporphycenes (**12M**), and tetraoxaporphycene (**15**) (Vogel et al., 1988) which also have applications to photobiology (Toporowicz et al., 1989). The corresponding radical ions have thus a nondegenerate ground state, and they gave rise to highly resolved EPR spectra along with readily observable solution-ENDOR signals (Schlüpmann et al., 1990; Bachmann et al., 1993; Bachmann, 1996), as is described below.

3.2 Porphycenes and Metalloporphycenes

The porphycenes, of which the radical anions were studied, include the free base (**12**H_2) and its 2,7,12,17- and 9,10,19,20-tetra-*n*-propyl derivatives (**13**H_2 and **14**H_2, respectively), as well the metalloporphycenes **12**Zn, **13**Ni, **13**Pd, and **13**Pt (Schlüpmann et al., 1990).

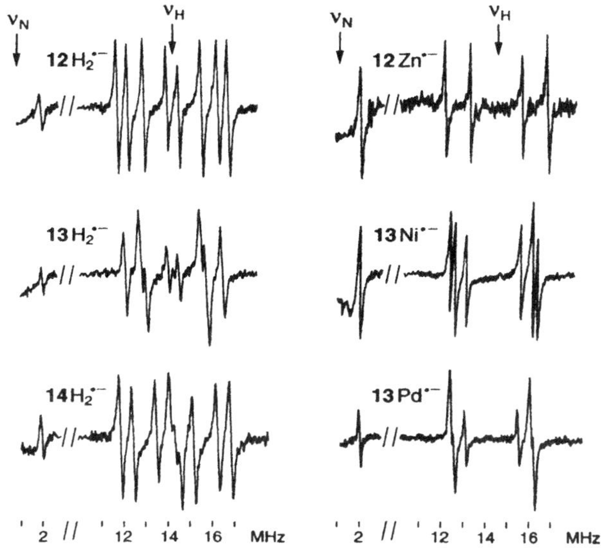

Figure 6. Radical ions of porphycenes and metalloporphycenes. ^{1}H- and ^{14}N-ENDOR spectra of the anions **12**$H_2^{\bullet-}$, **13**$H_2^{\bullet-}$, **14**$H_2^{\bullet-}$ (left, from top to bottom) and **12**$Zn^{\bullet-}$, **13**$Ni^{\bullet-}$, and **13**$Pd^{\bullet-}$ (right, from top to bottom); solvent THF, counterion Na^+, temperature 220–240 K. $\nu_H = 14.38$ and $\nu_N = 1.00$ MHz are the frequencies ν_n of the free proton and ^{14}N nucleus, respectively. For the ^{14}N-coupling constant, a_N, the low-frequency signal at 0.04 MHz was not observed, because it is below the range of the spectrometer. Taken with a home-built instrument. Reproduced from (Schlüpmann et al., 1990) by permission of *J. Am. Chem. Soc.*

Figure 6 presents the ^{1}H- and ^{14}N-ENDOR spectra of the corresponding radical anions (except those of **13**$Pt^{\bullet-}$), which were generated from their neutral precursors with sodium in THF. The MW power required for saturation of the EPR line, a prerequisite for a successful ENDOR experiment, had to be raised on going from **12**$Zn^{\bullet-}$ to **13**$Ni^{\bullet-}$ and to **13**$Pd^{\bullet-}$. ENDOR signals of **13**$Pt^{\bullet-}$ were not detected even with the MW power as high as 200 mW. The increased reluctance to saturation is a consequence of the more efficient spin-orbit coupling and the enhanced electron-spin relaxation with the growing atomic order of the metal. This phenomenon was also manifested by the broadening of the EPR lines (that of **13Pt**$^{\bullet-}$ is a single unresolved signal) and the higher g factor, as indicated in Table 3 which gives the hyperfine data for the six radical anions with observable ^{1}H- and ^{14}N-coupling constants, $a_{H\mu}$ and a_N.

Table 3. Hyperfine Data, $a_{H\mu}$ and a_N in mT, for the Radical Anions of Phorphycenes and Metalloporphycenes[a]

μ =	$\mathbf{12H_2^{\bullet-}}$	$\mathbf{13H_2^{\bullet-}}$	$\mathbf{14H_2^{\bullet-}}$	$\mathbf{12Zn^{\bullet-}}$	$\mathbf{13Ni^{\bullet-}}$	$\mathbf{13Pd^{\bullet-}}$
2,7,12,17	−0.146	+0.105[b]	−0.136	−0.165	+0.126[b]	+0.127[b]
3,6,13,16	−0.182	−0.157	−0.178	−0.171	−0.144	−0.136
9,10,19,20	−0.096	−0.090	+0.060[c]	−0.083	−0.087	−0.086
H(N)	+0.018	+0.017	+0.015	–	–	–
^{14}N	−0.071	−0.071	−0.070	−0.068	−0.071	−0.075
g[d]	2.0025	2.0024	2.0025	2.0026	2.0036	2.0046

[a]The values refer to α-protons unless otherwise indicated. [b]Eight methylene β-protons of the four *n*-propyl substituents. [c] Four methylene β-protons of the four *n*-propyl substituents (the eight protons of these groups are nonequivalent and split into two sets of four; the value of the four protons in the second set is much smaller and it has not yet been definitively established). [d]g = 2.0112 for **13Pt**$^{\bullet-}$.

The $a_{H\mu}$ and a_N values are compatible with the symmetry D_{2h} being effective on the hyperfine time-scale, especially with the equivalency of the four ^{14}N nuclei and the fast tautomerization of the two N-protons in $\mathbf{12H_2^{\bullet-}}$, $\mathbf{13H_2^{\bullet-}}$, and $\mathbf{14H_2^{\bullet-}}$. Assignments of the coupling constants $a_{H\mu}$ relied (i) on special-TRIPLE-resonance experiments and simulation of the EPR spectra, from which the numbers of protons giving rise to the ENDOR signals were derived, (ii) on deuteration in the 9,10,19,20-positions of **13Ni**$^{\bullet-}$ (Renner et al., 1989), and (iii) on the internal consistency of the hyperfine data in the series, by which sets with equal numbers of protons were distinguished. The relative signs of all coupling constants were derived from the general-TRIPLE-resonance spectra with the reasonable assumption that the values of the α-protons are negative. Both assignments and signs of the coupling constants were confirmed by all-valence-electrons-self-consistent-field MO calculations (RHF/INDO-/SP) for $\mathbf{12H_2^{\bullet-}}$, $\mathbf{13H_2^{\bullet-}}$, and $\mathbf{14H_2^{\bullet-}}$. The changes in the π-spin distribution along the series are moderate, and metallation has only a minor effect. On going from $\mathbf{12H_2^{\bullet-}}$ to **12Zn**$^{\bullet-}$ and from $\mathbf{13H_2^{\bullet-}}$ to **13Ni**$^{\bullet-}$ or **13Pd**$^{\bullet-}$, there is an increase in the π-spin populations ρ_μ at the centers μ = 2,7,12,17 and a corresponding decrease at μ = 3,6,13,16 and 9,10,19,20.

3.3 Tetraoxaporphycene

The radical cation of tetraoxaporphycene (**15**) is isoelectronic with the radical anion of porphycene (**12H$_2$**) considered in the previous section. This cation is one of the five distinct redox stages, namely the dication **15^{2+}** (isolated as BF_4^- salt), the radical cation **15$^{\cdot+}$**, the neutral compound **15**, the radical anion **15$^{\cdot-}$**, and the dianion **15^{2-}** (Bachmann, et al., 1993; Bachmann 1996). They are interconvertible by reduction or oxidation with appropriate reagents, as indicated in the Reaction Scheme.

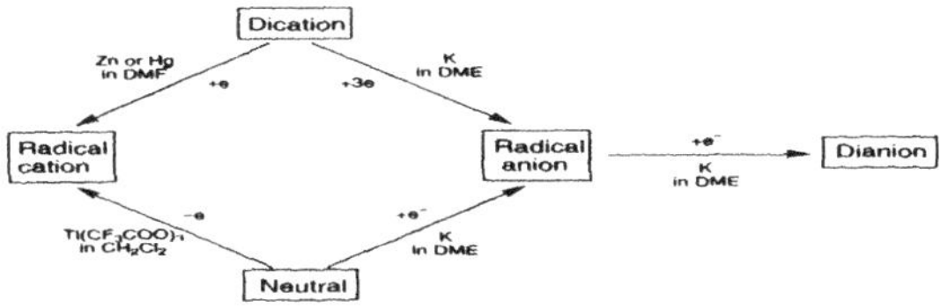

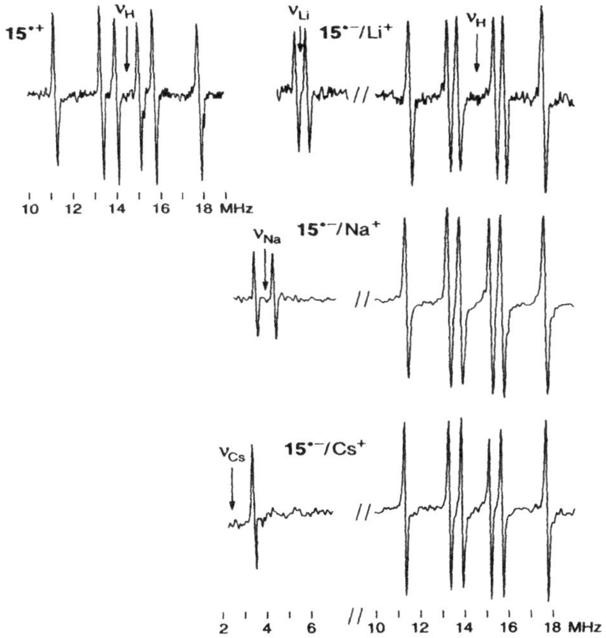

Figure 7. Radical ions of tetraoxaporphycene. ^{1}H-ENDOR spectrum of the cation **15$^{\cdot+}$** (top, left); solvent CH_2Cl_2, counterion ClO_4^-, temperature 243 K. ^{1}H-, ^{7}Li-, ^{23}Na- and ^{133}Cs-ENDOR spectra of the anion **15$^{\cdot-}$** (right, from top to bottom); solvent MTHF, counterion Li^+, Na^+, or Cs^+, respectively, temperature 198 K. $\nu_H = 14.56$, $\nu_{Li} = 5.66$, $\nu_{Na} = 3.85$, and $\nu_{Cs} = 2.39$ MHz are the frequencies ν_n of the free proton and ^{7}Li, ^{23}Na, and ^{133}Cs nuclei, respectively. (For the ^{133}Cs-coupling constant a_{Cs}, the low-frequency signal at 1.33 MHz was not observed, because of the low sensitivity of the apparatus in this region). Taken with a Bruker-ESP-300 spectrometer. Reproduced from (Bachmann, 1996) by permission of the author and from (Bachmann et al., 1993) by permission of *J. Am. Chem. Soc.*

Figure 7 (top, left) shows the ^{1}H-ENDOR spectrum of the radical cation **15**$^{\cdot+}$, generated from the dication with zinc in DMF. The ^{1}H-, ^{7}Li-, ^{23}Na-, and ^{133}Cs-ENDOR spectra of the radical anion **15**$^{\cdot-}$, also reproduced in Figure 7 (right), were observed upon reaction of the neutral compound with the respective alkali metal in 2-methyltetrahydrofuran (MTHF), in which solvent, the radical anion is tightly ion-paired with its counterion Li^+, Na^+, K^+, or Cs^+. (^{39}K-ENDOR signals were not detected, because of the low sensitivity of the apparatus in the frequency range 0–2 MHz.) The hyperfine data $a_{H\mu}$ for **15**$^{\cdot+}$ and a loosely ion-paired **15**$^{\cdot-}$, as well as $a_{H\mu}$, a_{Li}, a_{Na}, a_K, and a_{Cs} for the tightly paired **15**$^{\cdot-}$, are listed in Table 4.

Table 4. Hyperfine Data, $a_{H\mu}$, a_{Li}, a_{Na}, a_K, and a_{Cs} in mT, for the Radical Cation and Radical Anion of Tetraoxaphorphycene[a]

μ =	**15**$^{\cdot+}$	**15**$^{\cdot-}$, tight ion pair with				**15**$^{\cdot-}$, loose ion pair
		Li^+	Na^+	K^+	Cs^+	
2,7,12,17	–0.145	–0.091	–0.087	–0.085	–0.085	–0.085
3,6,13,16	–0.169	–0.060	–0.050	–0.047	–0.043	–0.037
9,10,19,20	–0.123	–0.215	–0.221	–0.223	–0.228	–0.234
Counterion	–	–0.019[b]	–0.030[c]	–0.009[d]	–0.076[e]	–
g	2.0024	2.0032	2.0032	2.0032	2.0032	2.0032

[a]The values refer to α-protons unless otherwise indicated. [b] ^{7}Li. [c] ^{23}Na. [d] ^{39}K. [e] ^{133}Cs.

Assignments of the coupling constants $a_{H\mu}$ to the three sets of four α-protons were guided by the results of Hückel-McLachlan calculations and, in the case of **15**$^{\cdot+}$, also by analogy to the corresponding values for the isoelectronic **12**$H_2^{\cdot-}$. General-TRIPLE-resonance experiments indicate that all coupling constants $a_{H\mu}$ have the same sign, which must be negative as for the α-protons. They point to a likewise negative sign for a_{Li}, a_{Na}, and a_{Cs}, which should also be shared by a_K.

The singly occupied π-orbitals (SOMOs) of the radical cation **15**$^{\cdot+}$ and the radical anion **12**$H_2^{\cdot-}$ have vertical nodal planes through the heteroatoms (O and N, respectively) and can thus be considered as a MO of a 20-membered π-perimeter with the π-spin population ρ_μ of 1/20 at each center μ. In fact, the average value of the coupling constants $a_{H\mu}$ observed for the three sets of four α-protons in **15**$^{\cdot+}$ is –0.146 mT, which corresponds to such spin population. The deviations from an average are +0.001, –0.023, and +0.023 mT (Table 4), as compared with –0.005, –0.041, and +0.045 mT for the isoelectronic **12**$H_2^{\cdot-}$ (average value –0.141; Table 3). The π-spin distribution in the perimeter is thus more strongly perturbed by the N- or NH-bridging in **12**$H_2^{\cdot-}$ than by the corresponding O atoms in **15**$^{\cdot+}$.

Loose ion pairs of the radical anion **15**$^{\cdot-}$ with its alkali-metal counterions occurred in 1,2-dimethoxyethane (DME) at low temperature. Upon warming, they became tighter, due to the decreasing polarity of the solvent.

In MTHF, a solvent of poor cation-solvating power, tight ion pairs were observed in the whole range of investigation (183–298 K), as indicated by the appearance of a hyperfine splitting from the alkali-metal nucleus. The counterion is situated on a twofold axis, above or below the molecular plane, where it contacts the lone-electron pairs of all O atoms. The small absolute values and the negative sign of the coupling constants of the alkali-metal nuclei, as well as the missing effect of the ion-pairing on the *g* factor (Table 4), are accounted for by the position of the counterion in a vertical nodal plane of the SOMO. (For the SOMO of $\mathbf{15}^{\bullet-}$, in contrast to that of $\mathbf{15}^{\bullet+}$, such a plane does not pass through the O atoms but crosses the C9–C10 and C19–C20 bonds.) The ratio $|a_{Li}| : |a_{Na}| : |a_{K}| : |a_{Cs}|$ roughly corresponds to that of the atomic parameters calculated for a ns-spin population of +1 at the Li, Na, K, and Cs atoms (Morton and Preston, 1978), which confirms the similar structure of the ion pairs. The effect of the ion-pairing on the spin distribution in $\mathbf{15}^{\bullet-}$ is a slight increase in the π-spin populations ρ_μ at the centers μ = 2,7,12,17 and 3,6,13,16 and some decrease at μ = 9,10,19,20; this effect diminishes with the growing size of the counterion (Table 4).

4. REFERENCES

Atherton, N. M. (1973). *Electron Spin Resonance*, Halsted Press, London.

Atherton, N. M. (1979). Solution ENDOR, in *Multiple Electron Resonance Spectroscopy*, Dorio, M. M. and Freed, J. H., eds., Plenum Press, New York, chapt. 4., 143–168.

Ayscough, P. B. (1967). *Electron Spin Resonance in Chemistry*, Methuen & Co Ltd, London.

Bachmann, R., Gerson, F., Gescheidt, G., and Vogel, E. (1992). Five Redox Stages of Tetraoxaporphyrin: A UV/Visible/Near-IR, ESR, and MO-Theoretical Study, *J. Am. Chem. Soc.* **114**, 10855–10860.

Bachmann, R., Gerson, F., Gescheidt, G., and Vogel, E. (1993). Tetraoxaporphycene: ESR/ENDOR, UV/Visible/Near-IR, and MO-Theoretical Study of Its Five Redox Stages, *J. Am. Chem. Soc.* **115**, 10286–10292.

Bachmann, R. (1996). *ESR/ENDOR- und UV/VIS/NIR-Untersuchungen an porphyrinoiden Verbindungen*, Ph.D. Thesis, Universität Basel.

Battersby, A. R. and McDonald, E. (1979). Origin of the Pigments of Life: the Type-III Problem in Porphyrin Biosynthesis, *Acc. Chem. Res.* **12**, 14–22.

Battersby, A. R. and Frobel, K. (1982). Porphyrin Biosynthesis - Vitamin B_{12}, Heme, and Related Porphyrins, *Chem. unserer Zeit* **16**, 124–134.

Biehl, R., Dinse, K.-P., and Möbius, K. (1971). ENDOR Investigation of Biphenyl and Terphenyl Anion Radicals in Solution, *Chem. Phys. Lett.* **5**, 605–609.

Biehl, R., Plato, M., and Möbius, K. (1975). General TRIPLE Resonance on Free Radicals in Solution – Determination of Relative Signs of Isotropic Hyperfine Coupling Constants, *J. Chem. Phys.* **63**, 3515–3522.

Carrington, A. and McLachlan, A. D. (1967). *Introduction to Magnetic Resonance*, Harper und Row, New York, and John Weatherhill Inc., Tokyo.

Das, M. R., Connor, H. D., Leniart, D. S., and Freed, J. H. (1970). An Electron Nuclear Double Resonance and Electron Spin Resonance Study of Semiquinones Related to Vitamins K and E, *J. Am. Chem. Soc.* **92**, 2258–2268.

Dax, T. G., Falk, H., and Kapinus, E. I. (1999). A Structural Proof for the Hypericin 1,6-Dioxo Tautomer, *Monatsh. Chem.* **130**, 827–831.

Dorio, M. M. and Freed, J. H., eds. (1979). *Multiple Electron Resonance Spectroscopy*, Plenum Press, New York.

Etzlstorfer, C. and Falk, H. (2000). Concerning the Association of Hypericin Tautomers and Their Hypericinate Ions, *Monatsh. Chem.* **131**, 333–340.

Feher, G. (1956). Observation of Nuclear Magnetic Resonances via the Electron Spin Resonance Line, *Phys. Rev.* **103**, 834–835.

Feher, G. (1998). The Development of ENDOR and Other Reminescences of the 1950's, in *Foundation of Modern EPR*, G. R. Eaton, S. S. Eaton, and K. M. Salikhov, eds., World Scientific, Singapore and New Jersey, chapt. H.8., 548–556.

Freeman, D., Konstantinovskii, L., and Mazur, Y. (2001). The Structure of Hypericin in Solution. Searching for Hypericin's 1,6-Tautomer, *Photochem. Photobiol.* **74**, 206–210.

Gerson, F. (1966). Notiz über das ESR-Spektrum des Phenalenyl-Radikals, *Helv. Chim. Acta* **49**, 1463–1467.

Gerson, F., Jachimowicz, J., Möbius, K., Biehl, R., Hyde, J. S., Leniart, D. S. (1975). Application of ENDOR Spectroscopy to Radicals of Low Symmetry: Radical Anion of 2-Phenylcycl[3.2.2]azine, *J. Magn. Reson.* **18**, 471–484.

Gerson, F., Gescheidt, G., Häring, P., Mazur, Y., Freeman, D., Spreitzer, H., and Daub, J. (1995). Electron-Acceptor Properties of Hypericin and Its Salts: An ESR/ENDOR and Electrochemical Study, *J. Am. Chem. Soc.* **117**, 11861–11866.

Gerson, F. and Huber, W. (2003). *Electron Spin Resonance Spectroscopy of Organic Radicals*, Wiley-VCH, Weinheim (Germany).

Hartmann, P. E. and Goldstein, M. A. (1989). Superoxide Generation by Photomediated Redox Cycling of Anthraquinones, *Environ. Mol. Mutagen.* **14**, 42–47.

Hyde, J. S. and Maki, A. H. (1964). ENDOR of Free Radicals in Solution, *J. Chem. Phys.* **40**, 3117–3118.

Hyde, J. S. (1965). ENDOR of Free Radicals in Solution, *J. Chem. Phys.* **43**, 1806–1818.

Hyde, J. S. (1974). Paramagnetic Relaxation, *Ann. Rev. Phys. Chem.* **25**, 407–435.

Hyde, J. S. (1998). EPR at VARIAN: 1954–1974, in *Foundation of Modern EPR*, G. R. Eaton, S. S. Eaton, and K. M. Salikhov, eds., World Scientific, Singapore and New Jersey, chapt. K.1, 704, 707, 709.

Kevan, L. and Kispert, L. D. (1976). *Electron Spin Double Resonance*, John Wiley & Sons, New York.

Kurreck, H., Kirste, B., and Lubitz, W. (1984). ENDOR Spectroscopy – A Promising Technique for Investigating the Structure of Organic Radicals, *Angew. Chem. Int. Ed. Engl.* **23**, 173–194.

Kurreck, H., Kirste, B. and Lubitz, W. (1988). *Electron Nuclear Double Resonance Spectroscopy of Radicals in Solution*, VCH Publishers, Weinheim (Germany).

Möbius, K. and Dinse, K.-P. (1972). ENDOR of Organic Radicals in Solution, *Chimia* **26**, 461–470.

Möbius, K. and Biehl, R. (1979). Electron–Nuclear–Nuclear TRIPLE Resonance of Radicals in Solution, in *Multiple Electron Resonance Spectroscopy*, M. M. Dorio and J. H. Freed, eds., Plenum Press, New York, chapt. 14., 475–507.

Möbius, K. (1998). ENDOR in Liquids, in *Foundation of Modern EPR*, G. R. Eaton, S. S. Eaton, and K. M. Salikhov, eds., World Scientific, Singapore and New Jersey, chapt. H.9., 557–576.

Morton, J. A., ed. (1965). *Biochemistry of Quinones*, Academic Press, New York.

Morton, J. R. and Preston, K. F. (1978). Atomic Parameters for Paramagnetic Resonance Data, *J. Magn. Reson.* **30**, 577–582.

Muldner, H. and Zoller, M. (1984). Antidepressive Effect of a Hypericum Extract Standardized to an Active Hypericine Complex. Biochemical and Clinical Studies, *Arzneim. Forsch.* **34**, 918–920.

Rahimipour, S., Palivan, C., Freeman, D., Barbosa, F., Fridkin, M., Weiner, L., Mazur, Y., and Gescheidt, G. (2001a). Hypericin Derivatives: Substituent Effects on Radical-anion Formation, *Photochem. Photobiol.* **74**, 149–156.

Rahimipour, S., Bilkis, I., Peron, V., Gescheidt, G., Barbosa, F., Mazur, Y., Koch, Y., Weiner, L., and Fridkin, M. (2001b). Generation of Free Radicals by Emodic Acid and its [D-Lys^6]GnRH-conjugate, *Photochem. Photobiol.* **74**, 226–236.

Renner, M. W., Forman, A., Wu, W., Chang, C. K., and Fajer, J. (1989). Electrochemical, Theoretical, and ESR Characterizations of Porphycenes. The π Anion Radical of Nickel(II) Porphycene, *J. Am. Chem. Soc.* **111**, 8618–8621.

Scheffler, K. and Stegmann, H. B. (1970). *Elektronenspinresonanz*, Springer-Verlag, Berlin.

Schlüpmann, J., Huber, M., Toporowicz, M., Plato, M., Köcher, M., Vogel, E., Levanon, H., and Möbius, K. (1990). Liquid-Phase ESR, ENDOR, and TRIPLE Resonance of Porphycene Anion Radicals, *J. Am. Chem. Soc.* **112**, 6463–6471.

Schmalbein, D. (1998). A Bruker EPR History, in *Foundation of Modern EPR*, G. R. Eaton, S. S. Eaton, and K. M. Salikhov, eds., World Scientific, Singapore and New Jersey, chapt. K.2., 721.

Seth, J. and Bocian, D. F. (1994). Electron Paramagnetic Resonance Studies of Metalloporphyrin Anion Radicals. Effects of Solvent, Counterion, Temperature, and Isotopic Substitution on the Jahn-Teller Active ^2E_g Ground State, *J. Am. Chem. Soc.* **116**, 143–153.

Suzuki, O., Katsumata, Y., Oya, M., Bladt, S., and Wagner, H. (1984). Inhibition of Monoamine Oxidase by Hypericin, *Planta Med.* **50**, 272–274.

Toporowicz, M., Ofir, H., Levanon, H., Vogel, E., Köcher, M., Pramod, K., and Fessenden, R. W. (1989). Triplet State of Metalloporphycenes: Zinc-PCl, Palladium-PC2, Platinum-PC2, and Nickel-PC2, *Photochem. Photobiol.* **50**, 37–43.

Vogel, E., Köcher, M., Schmickler, H., and Lex, J. (1986). Porphycene, a New Type of Porphyrin Isomer, *Angew. Chem. Int. Ed. Engl.* **25**, 257–259.

Vogel, E., Sicken, M., Röhrig, P., Schmickler, H., Lex, J., and Ermer, O. (1988). Tetraoxaporphycene Dication, *Angew. Chem. Int. Ed. Engl.* **27**, 411–414.

Wertz, J. E. and Bolton J. R. (1972). *Electron Spin Resonance*, McGraw-Hill Inc., New York.

Chapter 6

Electron-Electron Double Resonance

Lowell D. Kispert
Chemistry Department, The University of Alabama, Box 870336, Tuscaloosa, Al 35487

Abstract: Electron Electron Double Resonance (ELDOR) consisting of a strongly saturating continuous wave (CW) pump microwave source and a nonsaturating observing source, can be used in a field swept display to monitor saturation-transfer mechanisms such as Heisenberg Exchange, nitrogen nuclear relaxation, and rotational diffusion. 2D pulse ELDOR techniques known as DEER or PELDOR using two separate microwave frequencies or a similar "2 + 1" technique using a single frequency have been configured for probing dipolar interactions up to 8 nm. Spin-echo ELDOR techniques have been developed to study slow motions in a wide range of biological problems. Two-dimensional Fourier transform techniques permit all combinations of pump and probe frequencies with nanosecond time resolution making it possible to study the microscopic orientations of a system. Multiquantum ELDOR techniques have been developed to measure the transfer of magnetization and not a reduction factor as measured in a field swept ELDOR.

1. INTRODUCTION

ELDOR is an acronym that stands for electron-electron double resonance, and requires two microwave frequencies within one resonator; one is called a "pump" microwave source and another is the observing microwave source. In the CW mode, the observing microwave frequency that monitors the change in EPR intensity of a line is fixed at a nonsaturating power level. The "pump" microwave source set at a strongly saturating power level irradiates a portion of the EPR spectrum either the same line or a matching line related by a hyperfine coupling, and the effect on the spectrum is monitored by the observing source. The effect observed is due to a transfer of saturation between spins irradiated by the pump source and those spins detected by the observing source. At short times, this is directly related to the transition probability between the spectral positions. "Saturation transfer" means that the z component of the population difference at the observing frequency is no longer at the Boltzmann population difference. In the CW mode the ELDOR effect is measured by the ELDOR reduction factor R given by

$$R = \frac{\text{(signal with pump off)} - \text{(signal with pump on)}}{\text{(signal with pump off)}} \quad (1)$$

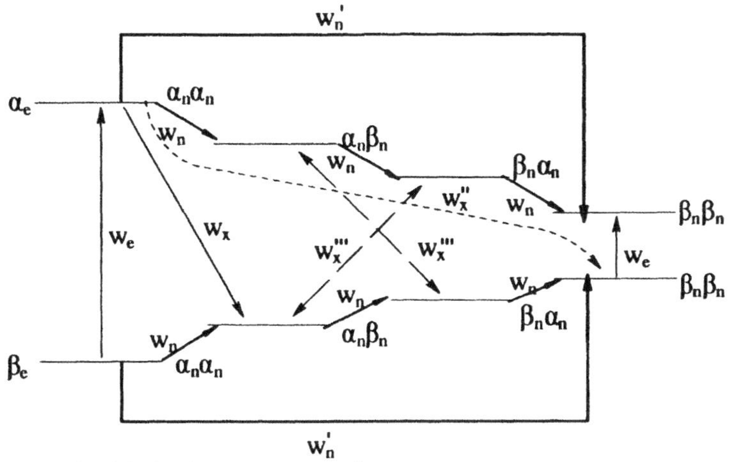

Figure 1. The eight level energy diagram for $\underset{\sim}{S} = 1/2$, $\underset{\sim}{I}_1 = 1/2$, $\underset{\sim}{I}_2 = 1/2$, $A > 0$, $g_n > 0$, and $|A|/2 > \nu_n$

In Figure 1, is given an eight level energy diagram showing the cross-relaxation pathways W_x, W_x'', and W_x''' that occur for a radical when there are two nonequivalent protons interacting with an unpaired electron. W_n and W_e are the lattice-induced nuclear spin-flip and electron-flip transitions probabilities, respectively. These transition probabilities are related to relaxation times by the following relations: $W_n = T_n^{-1}$, $W_e = T_{1e}^{-1}$, $W_x = T_x^{-1}$, $W_x'' = (T_x'')^{-1}$. Applying a strong saturating pump source to the left most transition can be detected with a non-saturating observing source as a change in the right most line intensity. The change in intensity depends on the relative values of the cross-relaxation pathways W_x, W_x'', W_x''', W_e, W_n and W_n'' whose magnitudes are dependent on a given relaxation or combination of relaxations. There are many mechanisms by which saturation of one line can be transferred to another line, resulting in both complexities in the ELDOR spectra and new tools to study mechanisms and their rates.

The theory of ELDOR was developed by Hyde et al. (1968) and Freed (1979, 1979a) who showed that a plot of R^{-1} versus (pump power)$^{-1}$ is linear and the intercept yields the (saturation - transfer rate)/electron spin-lattice relaxation rate. Thus, the R factor is a measure of how competitive saturation transfer is with respect to the spin-lattice relaxation time (T_{1e}). Values of T_{1e} are preferably measured by time domain methods. In general the ELDOR measurements provide a means to study a variety of dynamic processes that govern relaxation between the pumped and observed spins. For example, extensive ELDOR studies have been reported for nitroxide radical spin labels, which typically possess T_{1e} values between 0.3 and 10 μs. Over this range of times there are three saturation-transfer mechanisms that are effective. They are the Heisenberg Exchange (HE) - an intermolecular relaxation depending on concentration and two intermolecular relaxations; nitrogen nuclear relaxation and very slow rotational diffusion.

Irradiated organic crystals were also especially suited for CW ELDOR studies (Hyde et al., 1968a, Kispert, 1979). The CW reduction factors (R) recorded for radicals in organic crystals were found to be a measure (Kevan and Kispert, 1976) of the rate of intramolecule motion, hyperfine coupling anisotropy, intramolecular proton spin exchange, intramolecular admixture of nuclear spin states, quadrupole interaction, deuterium hyperfine couplings, resolution of small hyperfine couplings, spin-flip transitions, intermolecular spin diffusion; fluorine coupling anisotropy, hydrogen-deuterium exchange, tunneling methyl groups, triplet states and excition motion.

Using pulsed ELDOR techniques to generate echo modulation, it is possible (Schweiger and Jeschke, 2001) to determine distances between two electron spins, measure polarization transfer to another region of the spectrum, measure weak couplings between electron spins, measure conformation statistics, measure broad distribution of distances, and measure end to end distances of spin-labeled polypeptides.

1.1 Historical

The first application of dual microwave frequencies was published by Bowers and Mims, in 1959 on paramagnetic relaxation in nickel fluorosilicate followed soon after by Sorokin, Lahser and Gelles in 1960 who showed that individual cross-relaxation times for nitrogen centers in diamond could be determined. Subsequently Unruh and Culvahouse (1963) carried out pulsed ELDOR measurements on Co^{2+} in lanthanide crystals from 1.18 - 4.2 $^{\circ}$K and determined the temperature dependence of the relaxation rates. Moran (1964) introduced continuous pumping into ELDOR technology and showed that static spin packets were distributed within an inhomogeneous envelope. When the pump power was increased, forbidden transitions were observed.

Roughly eight years after the first CW ELDOR experiments were begun in solids, the first solution application of CW ELDOR was reported both in James Hyde's laboratory at Varian (Hyde, Chien and Freed, 1968) and in Russia (Benderskii et al., 1968). Nechtschein and Hyde (1970) also showed that pulsed ELDOR was possible. Typically, ELDOR was applied in the CW mode (Bruno and Freed, 1974; Dorio and Freed, 1979), to study slow motions of nitroxide radicals, dynamics, and irradiation damage in organic crystals (Kispert, 1979), disordered matrices (Kevan, 1979), polymers (Dorio, 1979) and iron-sulfur proteins (Sands, 1979). The CW method proved useful in the study of similar systems until pulsed ELDOR (PELDOR) configured for probing dipolar interactions (Milov et al., 1981) and a spin-echo ELDOR technique for studying slow motions was developed independently by Hornak and Freed (1983) and by Dzuba et al. (1982, 1984), enable the technique to be especially useful in the study of a wide range of biological problems. Pulsed ELDOR techniques have considerable advantages over CW ELDOR methods. These advantages are.

1. The absence of the radiation fields including the absence of a dc field modulation during the rotational diffusion of the molecules and during the evolution time of the spins,
2. the direct measurement of relaxation rates rather than just their ratios as obtained by CW-ELDOR and
3. the cancellation of inhomogeneous broadening effects.

These features result in greater accuracy in the data analysis and eliminates the need to include the radiation fields in the analysis of the data. Later improvements (Gorcester and Freed, 1988a; Gorcester and Freed, 1986, 1988) with 2D FT ELDOR permitted two-dimensional displays and the possibility of all combinations of pump and probe frequencies displayed in a single 2D spectrum. This meant greater efficiencies in the data acquisition and removed the need to apply additional techniques to analyze the problem. The 2D ELDOR technique has required the development of new microwave and digital electronics to make it possible to obtain the broad band irradiation and detection at microwave frequencies with nanosecond time resolution.

Analysis of 2D-ELDOR spectra allows the coupling between different molecular orientations of a system to be determined. This gives rise to considerable insight into the microscopic details of the rotational process. In the rigid limit, 2D-ELDOR leads to the appearance of forbidden auto peaks (Gorcester et al., 1990) as well as the appearance of cross peaks, due to coherence transfer by the hyperfine interaction. If there are two nuclei, the 2D-ELDOR spectrum of polycrystalline samples can give information on their relative orientation. Further double electron-electron resonance (DEER) techniques were used (Pannier et al., 1998), 200; Pfannebecker et al., 1996; Larsen and Singel, 1993) or PELDOR (Milov et al., 1981, 1998) based on a solid-state concept where use is made of selective pulses at two well-separated microwave frequencies. This approach was applicable to systems such as bilabeled biomolecules with broad spectra (≥50 G). The "2 + 1" technique (Kurshev et al., 1988, 1989) is similar to the 3-pulse DEER technique except for its use of only a single frequency.

Further improvements were made (Borbat and Freed, 1997) in pulsed two-dimensional FT-ESR spectrometers with the design of multi-frequency spectrometers at 9.2 and 17.3 GHz, where the higher frequency gave 4 times the signal-to-noise ratio observed at 9.2 GHz.

A multiquantum ELDOR technique based on coherent effects was introduced (Mchaourab et al., 1991) which provided a signal that is a measure of the transfer of magnetization. Double quantum coherence (DQC) based upon allowed pathways (Borbat and Freed, 1999) has been suggested to have some advantages over DEER measurements: strong pulses (used to observe the allowed pathway) should yield signals at least an order of magnitude greater, an advantage when working with small amounts of bilabeled biomolecules or when attempting to measure distances up to 80 Å. This advantage has not been proven experimentally where DQC and DEER measurements were carried out on the same sample with the same spectrometer. It has however, been possible to measure EPR distances in a bilabeled biomolecule with DEER techniques as shown by Persson et al. (2000) as well as between two copper centers in azurin (Huber et al., 2002). Unpublished reports at conferences have reported preliminary results of DEER measurements on proteins with 450 residues and on DNA.

The most precise measurement of a dipolar coupling to date was a 51 Å shape-persistent biradical with a well-defined distance using a deadtime free single-frequency technique for refocussing (SIFTER) dipolar couplings (Jeschke et al., 2000a). These experiments were based on the solid-echo and Jeener-Brockaert sequences which are well established sequences in dipolar NMR spectroscopy of solids. For distances larger than 3 nm, SIFTER appears to provide better resolution than with DEER techniques. Similar precision has been possible with DEER for shape-persistent biradicals of phenylene-ethylnylene based biradicals (Jeschke et al., 2002a). Advanced techniques for extraction of distances from DEER is also discussed.

It is premature at this time to say much about the relative benefits of the various methods. To date only one paper has made any such comparisons (Persson et al., 2001) and even then for only a few methods.

1.2 Reviews

The use of the ELDOR (also known as DEER or PELDOR) technique has been detailed in books, book chapters and in various reviews. They are summarized here.

The early instrumentation, principles, and applications of CW ELDOR measurements to radical systems in crystals, powders and solution were detailed in a book by Kevan and Kispert (1976) and this is a good start for the beginner in double resonance methods. Early reviews were given of the published literature by Atherton (1972, 1974, and 1976), by Hyde (1974), and Moebius (1977 and 1979). A book edited by Dorio and Freed (1979) covered the use of ELDOR measurements to study organic radicals in iron-sulfur proteins (Sands, 1979), crystals (Kispert, 1979), disordered matrices (Kevan, 1979), polymers (Dorio, 1979), instrumentation (Leniart, 1979) and examples of biological interest (Sarna, 1979). In crystalline systems, Kispert, (1979), showed that it is possible to use ELDOR measurements to study the reaction mechanism of radical formation in irradiated organic crystals. Additional reviews were published by Moebius (1981), Evans and Rowlands (1984), Hyde and Feix (1989) and Sarna (1979).

Following Freed's (1979) suggestion to use spin echoes in the study of slow motions coupled with the availability of fast microwave switches, amplifiers and loop gap resonators, Hornak and Freed (1983) developed the spin-echo ELDOR technique and applied it to slow motions. The literature that followed was summarized and the essential features detailed in Chapter 3 by Gorcester et al. (1990) in a book entitled "Modern Pulsed and Continuous-Wave Electron Spin Resonance" edited by Kevan and Bowman (1990). For those needing to learn more about 2D-ELDOR techniques and in general 2D EPR techniques and spectral analysis – this chapter is highly recommended reading. More recently the DEER experiments have been summarized and the essential features explained by Jeschke et al. (2000) in Chapter 11 (Vol. 19) of the Book on Biological Magnetic Resonance. In the same book in Chapter 10, the "2 + 1" pulse sequence is detailed and examples provided by Raitsimring (2000), while in Chapter 9, Multiple Quantum ELDOR has been clearly explained by Borbat and Freed (2000). Pulsed ELDOR has been discussed in some detail in Chapter 13 of a book by Schweiger and Jeschke (2001). These chapters are highly recommended for reading before attempting to read the literature in these areas. They provide the best up-to-date analysis of these techniques.

2. INSTRUMENTAL TECHNIQUES

A brief summary of the instrumentation used to carry out ELDOR measurements and the applications of ELDOR measurements to the solution of various problems is given below. In general the CW measurements require low-power microwave devices while high-power TWT amplifiers are required for pulse measurements.

2.1 Field Swept CW ELDOR

Because field modulation is used to detect EPR, it has been shown (van der Drift, 1975) that during the field sweep, the magnetic field (H_0) and the pumping frequency (ν_p) must be coupled together by the resonance condition of the pumped line so that a correct value for the ELDOR reduction factor, extrapolated to infinite pumping power, is obtained. Other recording schemes (a) H_0 swept and ν_o -ν_p equal to the hyperfine distance in the spectrum and held constant during the sweep as well as (b) ν_p is swept and ν_o is a constant and H_0 is set either at the center of the observed line or at the top of the first derivative line, yield incorrect relaxation parameters. Recording schemes (a) and (b) cause the R(ELDOR reduction) factor to contain the effect of the pumping upon the modulated observing signal and secondly the effect of modulating the pumping line upon the observing absorption signal. However, by fixing the two microwave frequencies at the desired separation and sweeping the field an R value nearly independent of modulation at infinite power can be achieved. The value of R determined this way is technically simpler: two spectra are obtained, one at zero pump power and one at the desired pump power, permitting immediate determination of R. This method is normally referred to as field-swept ELDOR.

2.2 Modulation Scheme for CW ELDOR

To achieve an R factor that is independent of the modulation field, one method is to program a computer to control both the magnetic field sweep and the pump frequency sweep using the g factor of the pumped line. In another method the resonance condition is held by locking the pump frequency to the resonance signal of the pumped EPR line which requires the introduction of extra pump power. This can introduce false ELDOR signals. A scheme has been developed out (Mehlkopf et al., 1983) where the pump power and the magnetic field modulation amplitude can be independently optimized for the lock signal and the ELDOR spectrum and the frequencies of the false ELDOR signals are limited to even multiples of the frequency of the wanted ELDOR signal. These are then easily eliminated.

2.3 Loop-Gap Resonator for ELDOR

A significant improvement in the signal-to-noise ratio by a factor of 20 on a molar basis was achieved when the bimodal cavity used in CW experiments up to 1985 at the X-band frequency was replaced by a loop-gap resonator at the X-band in a configuration designed by Hyde et al. (1985). On a number-of-spins basis the improvement was a factor of 70 due to the smaller volume of the resonator. When the separation between the pump and observing frequencies exceeds 100 MHz other resonators designed by Froncisz and Hyde; a bimodal loop-gap resonator with a total tuning range of 500 MHz (Tsapin et al., 1992) and doubly-tuned resonators (Mehdizadeh et al., 1983) are preferable. The loop-gap resonators with a low Q and high field factor are more practical than the earlier used bimodal cavity resonators (Hyde et al., 1968) because they had a high Q which made it more difficult to tune and decouple. More recently, a bimodal loop-gap resonator consisting of two identical one-loop-one gap resonators in coaxial juxtaposition was designed (Piasecki et al., 1996) for use at S band (1-3 GHz). This bimodal resonator could be used to carry out saturation recovery, modulation of saturation, CW ELDOR, CW EPR of a spin-label line sample and probing field distribution with DPPH, features that other loop-gap resonators could not attain. The resonator is characterized by a very high filling factor, a low quality factor and a high microwave field. It is a two-port resonant structure supporting two modes. The currents in two loops are parallel and in the other antiparallel. Adding capacitors between the loops, the frequencies of the two loops can be made to coincide. The bimodal loop-gap resonator has a very high filling factor relative to a bimodal cavity resonator. Details have been worked out concerning variable coupling to each mode, tuning the resonant frequency of one mode to that of the other, and how to adjust the isolation between modes. A crossed-loop resonator for ELDOR at S-band (Rinard et al., 1996) and L-band (Rinard et al., 2000) has been described and provider exceptional isolation between the microwave modes.

2.4 Magnetic Field-Pulsed ELDOR Spectroscopy

An early attempt to carry out ELDOR measurements utilizing a pulsed EPR spectrometer was made by Rengan et al., 1979 to operate in the saturation-recovery mode. The spectrometer was used for measurements of electron spin-lattice relaxation times. The procedure was to saturate one of the hyperfine lines by a pulse of high microwave power. At the end of the microwave pulse, the magnetic field was shifted within a time shorter than the electronic T_1 to the resonant value of another hyperfine component. The saturation transfer that had taken place was measured by monitoring the EPR signal intensity as a function of time as is usually done in a saturation - recovery experiment - measured as a function of microwave pulse width and peak saturating pulse. The response time of the spectrometer was 3 μsec and it was possible to study transients on the order of 10 μs was possible to study.

2.5 2-D Pulsed ELDOR

The advent of the spin-echo technique made it possible to perform ELDOR measurements with a spin-echo spectrometer. Rapid stepping of the dc magnetic field between pumping and observing resonant positions generated a two-dimensional spectrum. Use of Fourier transform (FT) ESR techniques made it possible to record 2D ELDOR spectra analogous to 2D NMR displays. Special instrumenta-tion and techniques are needed to carry out such experiments. Gorcester and Freed (1986, 1988) designed a spectrometer with a flexible design making it possible to display both 2D FT and 2D field swept systems.

The 2D FT experiment permits the display of motional magnetization transfer. The 2D FT experiment allows for the recording of ELDOR spectra (Hornak and Freed, 1983) with only a single frequency source since the spectral band produced by a finite pulse is coherently related.

2D ELDOR technique (Gorcester et al., 1990) makes use of the $\pi/2$ pulses applied in the sequence $\pi/2$ - τ_1 - $\pi/2$ - T - $\pi/2$ - τ_2 with the mixing time T held constant. The initial or preparation $\pi/2$ pulse generates the initial transverse magnetization. The phase of this pulse determines the phase of the amplitude modulation during the subsequent evolution period, τ_1. The second $\pi/2$ pulse is the beginning of the mixing period where the longitudinal magnetization components associated with each EPR line can be exchanged. This provides a mechanism whereby different components are mixed carrying different processional - frequency information such as motional magnetization - transfer. Applying the third $\pi/2$ pulse rotates the magnetization into the xy plane. Components that were initially processing with angular frequency $\omega_1 = \omega_\alpha$, now process at a new frequency equal to ω_b. The pulse sequence is repeated for a series of equally spaced values of τ_1. The FID is collected, the phase of the preparation pulse is advanced by 90° followed by collecting a second FID.

Examples of 2D ELDOR have been reported (Gorcester and Freed, 1986, 1988; Patyal et al., 1990). Magnetization transfer induced by

Heisenberg spin exchange during the mixing period gives rise to cross-correlations that appear as cross-peaks. A direct measure of the Heisenberg exchange (HE) rates can be deduced from the relative intensities of the peaks in the 2 D ELDOR spectrum. The quantitative determination of the exchange rates was detailed. It is also possible to distinguish between HE and electron-electron dipolar (EED) relaxation between probe molecules (Gorcester et al., 1990). Considerable improvement in signal-to-noise was accomplished by using time-domain data-processing techniques based upon linear prediction with singular value decomposition (LPSVD) (Gorcester and Freed, 1988). Details and examples have been reported (Gorcester and Freed). The 2D ELDOR approach yields the relaxation rates directly while the CW ELDOR measurements yield only ratios such as ω_{HE}/W_e where W_e is the electron spin-flip rate and ω_{HE} is the exchange rate. The advantage of 2D-ELDOR experiments over spin echo (SE) experiments is that the SE experiments require the use of substantially larger volumes to obtain adequate sensitivity (the signal voltage) for the SE is an order of magnitude less intense than the FID's in the 2D ELDOR experiments. Also a determination of W_n and ω_{HE} is obtained directly from the 2D ELDOR experiments while extensive least square fitting is required to obtain this information from SE measurements.

2.6 2-D Spectrometer X, Ku

A four fold improvement in signal-to-noise at 17.35 GHz occurred with the construction of a two-dimensional Fourier transform ESR (2D FT ESR), operating at 9.25 (X) and 17.35 GHz (Ku), (Borbat, Crepeau and Freed, 1997). A discussion is given of the technical problems associated with multifrequency 2D FT spectroscopy. The Ku bridge consists of an efficient heterodyne system wherein 9.25 GHz is the intermediate frequency. The sensitivity at the Ku-band is increased by almost an order of magnitude. A full 2D ELDOR spectrum could be collected in less than 20 min. for samples containing 0.5 to 5 mmol of nitroxide spin-probe in the slow-motional regime. Use is made of a bridged loop-gap resonator (BLGR) and a dielectric ring resonator (DR) to obtain broad spectral coverage at the Ku band. By using shorter microwave pulses of about 3 ns duration, a more uniform spectra excitation is observed. Successful simulation of the 2D ELDOR spectra at both 9.25 and 17.35 GHz for spin-labeled phospholipid probes in DMPG membrane vesicles using the same set of model parametry at both frequencies was achieved even though the spectra were very different in appearances at the two frequencies. Improved sensitivity and shorter dead-time at the K_u band made it possible to obtain orientation dependent 2D

ELDOR of the cholestane spin probe. Details of the technical problems associated with the 2D FT spectrometer were given.

2.7 DEER Technique

The three-pulse DEER experiment consists of a two-pulse echo sequence $\pi/2 - \tau - \pi$ (Hahn echo) at a fixed observer microwave frequency and an additional microwave π pulse at a fixed pump frequency at time τ after the $\pi/2$ pulse whose position is varied between the positions of the two observer pulses (Milov et al., 1981; Milov et al., 1984). Bimodal loop-gap resonators as well as the overcoupled monomodal resonator available with the commercial Bruker pulsed spectrometer (Jeschke et al., 2000) are used. A complete review of the principle behind this pulse sequence and its application to the measurements of distances in the range between 1.5 and 8 nm has been given by Jeschke et al., 2000. Conformational distributions limit the precision using the three-pulse sequence in non-crystalline systems. To remedy this situation, a four-pulse DEER method has been devised to determine spin pairs with a broad distribution of small (< 1.5 nm) and intermediate (1.5 - 2.5 nm) distances. In this case, the signal contributions to the spin echo decay almost completely within a few tens of nanoseconds. In practice a dead time of at least three times the microwave pulse length can hardly be avoided so the three pulse sequence is extended by one more observer pulse (Pannier et al., 2000). The four-pulse sequence consists of a refocussed echo sequence ($\pi/2 - \tau_1 - \pi - \tau_1$ - echo - $\tau_2 - \pi - \tau_2$ - echo) at the observer frequency and a π pulse at the pump frequency at time τ after the first π pulse at the observer sequence. Time τ is varied, τ_1, and τ_2 are fixed. Limitation and comparisons to other methods have been detailed and reviewed by Jeschke et al., 2000. A review has also been given of the use of DEER in the study of nitroxides and cluster sizes and intercluster distances in monomers.

Both the three-pulse and four-pulse DEER experiment require either an ELDOR extension of a commercial pulse EPR spectrometer like the Bruker E-580 FT-EPR machine or a home built pulse ELDOR spectrometer.

2.8 "2 + 1 Pulse Sequence"

Constructed similar to the DEER technique is the "2 + 1" pulse sequence (Kurshev et al., 1988, 1989) also configured for probing dipolar interactions. In the DEER technique the first and third pulse produce a primary spin echo signal at ω_1, which is detected while the position of the second pulse applied at $\omega_2 \neq \omega_1$ placed between the first and third is varied. In contrast, the "2 + 1" technique operates with all three pulses at the same carrier frequency.

However, DEER techniques are better for measurements than the "2+1" approach because unwanted echoes remain coherent when the same microwave source using the same frequency is used for all three pulses.

This problem never occurs in DEER. Raitsimring et al. (2000) suggested using a different source for the "+1" pulse to simplify phase cycling. In other words use the same spectrometer as for DEER. If this change is used for the "2+1" method it is an ELDOR experiment with coinciding frequencies.

The "2 + 1" technique has been used to measure dipolar interactions between similar paramagnetic centers where the EPR spectral width (< 60 G) is similar to the pulse amplitude so spectrum may be excited by a single pulse (~30 G). The pulse DEER technique is more useful for determining distances of paramagnetic centers with well separated EPR spectra. One of the centers can be designated as the observed spin and the other as the pumped spin. Details of how the various phases of the echo signal are handled in both techniques have been reported by Raitsimring (2000). The "2 + 1" technique allows for rigorous manipulation of the dipolar interactions.

2.9 Multiquantum ESR (MQ-ESR)

In multiquantum ESR (MQ-ESR) spectroscopy (Mchaourab et al., 1991; Hyde et al., 1995), two microwave frequencies generated from a common oscillator irradiate a sample. The two frequencies are usually separated by 10 kHz, selected to be much less than the homogeneous linewidth. Adsorption and emission of photons at both frequencies leads to oscillation of the spin population at $k(\omega_1 - \omega_2)$ and to the production of intermodulation sidebands at $\omega_0 \pm (k + ½) (\omega_1 - \omega_2)$ where ω_0 is the average frequency and k is a positive integer. These new microwave frequencies, generated by the spin system, are recorded. This requires spectrally-pure irradiation sources with spurious signals at 70 db. It has been shown (Mchaourab et al., 1991; Christides et al., 1996; Mchaourab, 1994; Hyde et al., 1995) that using the MQ-ESR approach has several advantages. (1) Detection of pure absorption lines, not derivative-like shapes, (2) spectral intensities proportional to T_1 and (3) reduced linewidths. For the ELDOR measurement, the effect of pumping one transition by a pair of frequencies is observed by detecting sidebands generated on another transition using a weak observing microwave field. These MQ-ELDOR experiments have become technically feasible because of the development of loop-gap resonators. The measurement provides a signal that is a measure of the transfer of magnetization and not a reduction factor as measured in a field-swept

ELDOR. The magnetization is coded by sinusoidal modulation and its steady-state spectral diffusion is monitored.

2.10 Spin Label Oximetry

The term spin label oximetry refers to the use of nitroxide radical spin labels to monitor oxygen transport. Both the longitudinal (T_1) and transverse (T_2) relaxation times of the spin label are altered by bimolecular collisions with oxygen. Various experiments can be used to observe these effects. Particular examples described by Hyde et al. (1990) include (a) saturation-recovery time-domain EPR using high observing power from a S band saturation-recovery apparatus and (b) ELDOR as an oxygen sensitive display. The ELDOR method is a simpler approach since the saturation-recovery time-domain requires a pulse instrument in the laboratory.

Introduction of molecular oxygen shortens the effective T_1 and diminishes the ELDOR effect. To observe oxygen transport, conditions must exist to give a strong ELDOR effect in the absence of oxygen. For example a 10^{-3} M concentration of the spin label CTPO (3-carbamoyl-2,2,5,5-tetramethyl-3-pyrrolin-1-yloxy) in water is optimum to observe strong ELDOR signals. Oxygen transport is studied by controlled introduction of various oxygen mixtures such as 30% and 40% molecular oxygen and studying ELDOR reductions. A plot of (R_{∞}^{-1}-1) vs. [O_2] concentration will yield a straight line where the extrapolated intercept at 0% [O_2] equals $2W_e/\omega_X$ and the slope equals K_o/ω_X Here R_{∞} = reduction factor at infinite microwave power, K_o = rate constant for Heisenberg exchange between oxygen and the spin label, ω_X is the bimolecular collision rate of oxygen with spin labels, and $(2\ W_e)^{-1} = T_1$. Studying oximetry using ELDOR has an advantage in that only changes in W_e affect the ELDOR signal when bimolecular collisions take place. Changes in T_2 do not enter the problem nor does one need to know about the T_2 of the spin label. The effect depends only on transfer of saturation and analysis of the m_s component of magnetization using the rate equation

$$(R_{\infty}^{-1} - 1) = \frac{2W_e}{\omega_X} + \frac{K_o[O_2]}{\omega_X} \qquad (2)$$

2.11 Time locked subsampling

Time locked subsampling (TLSS) has been used in three forms of pulsed EPR measurement at X-band frequency; saturation recovery, pulsed ELDOR and free induction decay. It was developed very recently (Froncisz et al., 2001) to monitor bimolecular collisions of oxygen with spin labels. This method involves (1) the translation of the signal from a microwave carrier to an intermediate frequency (IF) carrier where the IF offset between the signal oscillator and local oscillator was synthesized (2) sampling the IF carrier four times in an odd number of cycles, (3) signal averaging for best signal-to-noise ratio, (4) separating the even and odd digitized words into two separate signal channels; signals in-phase and in-quadrature (Q) with respect to the IF carrier, and (5) detecting the envelope of I and Q by changing the signs of alternate words in each of the two channels.

A central advantage of TLSS detection of pulsed ELDOR is the ability to acquire both the dispersion and the absorption simultaneously with essentially perfect relative phasing. It is now possible with computer control of the spectrometer and acquisition and storage of transient signals to record spectra continuously as a function of magnetic field across the entire spectrum. This will enable one to study the superposition of two spin label spectra with different motions and different accessibilities to oxygen for site directed spin labeling. Such three - dimensional data could permit the sorting out of the oxygen accessibility of each spin label environment. These measurements would require specialized modification to existing pulsed ELDOR equipment and is available at the National Biomedical EPR Center, Medical College of Wisconsin, Milwaukee, Wisconsin.

2.12 Power Saturation Method-SR-ELDOR

Power saturation methods by Haas, (1993) have been described and used along with SR-ELDOR (saturation recovery ELDOR) measurements to extract multiple rates from recovery curves. It is not possible to separate T_{1e}^{-1}, T_{1n}^{-1}, and the characteristic rotational correlation rate from one another with SR-EPR measurements alone when multicomponent exponential decays occur. However use of SR-ELDOR techniques, these rates are measured directly.

2.13 Pulsed ESR/Pulsed ELDOR

A pulsed ELDOR method was devised by Schosseler, Wacker and Schweiger, (1994) to excite allowed and forbidden transitions, thereby burning spectral holes into the EPR line. The holes caused by excitation of the forbidden transitions correspond to nuclear transition frequencies of the spin system. This makes it possible to determine small hyperfine couplings in disordered systems.

2.14 2D-Correlation Spectroscopy

A pulsed 2-dimensional (2D) Fourier transform ESR spectrometer has been described by Gorcester and Freed (2001) and its application to study motionally narrowed nitroxides using a two-pulse experiment (COSY) or a three-pulse experiment (2D ELDOR). Cross correlations between hyperfine lines were observed. Heisenberg exchange rates deduced by this method for PD-tempone agreed with those deduced by spin echo T_2 measurements as a functions of nitroxide concentration. This broadband irradiation of the entire spectrum offers substantial savings in data acquisition time and potentially yields more information than the analogous field-swept experiment.

2.15 2D-Techniques Theory

A comprehensive theory as sophisticated as that used for analyzing CW-EPR spectra has been published by Lee, Budil and Freed (1994) for interpreting two-dimensional Fourier Transform (2-D-FT). The theory includes motional rates from fast to slow motions, and microscopic as well as macroscopic molecular ordering. The theory is appropriate for COSY, SECSY and 2-dimensional-ELDOR where either the free-induction decay (FID) or echo decay is sampled. This theory has been applied to simulated experiments of nitroxide spin labels in membrane vesicles where microscopic molecular ordering but macroscopic disorder (MOMD) applies. Recovery of homogenous linewidths from FID-based COSY experiments on complex fluids was also demonstrated. Application to nitroxide probes illustrated that rotational correlation times as slow as milliseconds can be measured.

Employing 2-D-ELDOR measurements, Maresch et al. (1992) showed it was possible to measure magnetization transfer (electron spin diffusion, nuclear relaxation and slow rotational diffusion) throughout the EPR spectrum by using a narrow-band microwave excitation pulse followed by a rapid magnetic field step. Both the pumping and detecting fields were swept. It was observed that the magnetization transfer between states with close molecular orientations but different nitrogen nuclear spin projections dominated. A formalism developed by Gamliel and Freed (1990) for computing 2D ESR lineshapes with nuclear modulation (ESR-COSY, ESR-SECSY and 2D ELDOR) has been reported for polycrystalline and single-crystal samples. The method gives more detailed structural information than the corresponding ESEEM spectra.

3. DYNAMICS OF BIOMOLECULES IN LIQUID CRYSTALS, GLASSY SOLIDS, POLYMERS AND CRYSTALS

Analysis of ELDOR spectra can yield significant information regarding molecular reorientation of radicals, more often referred to as slow motion dynamics or molecular dynamics. Numerous examples exist in the literature and examples reported since 1980 are given here.

Studies have been reported that the CW ELDOR pattern provides information about molecular reorientation measured in real time in liquids (Nordio and Segre, 1980; Closs et al., 1982; Eastman et al., 1970; Hyde et al., 1969), polymers (Piven and Benderskii, 1984; Chien, 1979, Yang and Chien, 1978; 1978a; Kerillov et al., 1976; Dorio and Chien, 1975a, 1975b), liquid crystals (Xu et al., 1996), glassy solids (Dubinskii et al., 1994; Saalmueller et al., 1995; Lin, and Kevan, 1977; Lin et al., 1976; Yoshida et al., 1973; 1973a, 1972), crystals (Lee et al., 1993; Hwang et al., 1981); Pace, 1979; Hwang et al., 1979; Mottley et al., 1979; Mukai et al., 1979; Geoffroy et al., 1979; Kispert et al., 1979; Kispert, et al., 1978; Perkins, 1977; Perkins et al., 1977; Kispert, et al., 1976; Mottley et al., 1976; Mottley et al., 1975; 1975a; 1975b; Kispert et al., 1975; Mottley et al., 1975b; Percival et al., 1975; Lund et al., 1975; Dorio and Chien; 1975; Dalton et al., 1974; Robinson et al., 1974; Kispert and Wang, 1974; Iwasaki et al., 1974; Kispert et al., 1974; Kispert and Chang, 1973; Kispert et al., 1973, 1973b; Kispert et al., 1972), and complex fluids (Doi and Kuwata, 1979; Stitter et al., 1976; Saxena and Freed, 1997. Saxena and Freed, 1997; Antsiferova et al., 1987; Hornak and Freed, 1983; Dulcic and Poric, 1982; Dammers et al., 1982; Van der Drift and Smidt, 1982; Van der Drift et al., 1981; Van der Drift et al., 1980). Intermolecular nuclear-spin exchange was detected by CW ELDOR in ion-pair systems in solutions (Doi and Kuwata, 1979a). Matrix ELDOR signals were observed for nitroxide radicals in an amorphous polystyrene matrix (Dorio and Chien, 1976).

A 20 year old problem in understanding the longitudinal relaxation mechanisms (T_{1e}^{-1}) for nitroxide spin labels was solved by pulsed ELDOR methods (Hass et al., 1993; Robinson et al., 1994). The spin-lattice relaxation rates of the electron and the nitrogen nucleus were measured and the rotational correlation time range covered was from picoseconds to milliseconds. These rates were explained by the isotropic rotational Brownian dynamics modulating the interactions between the electron spin and the molecular angular momentum, the nitrogen and electron spins and the solvent protons with both the electron and the nitrogen spins.

Detailed geometry of the molecular reorientation can be deduced by combining the measured electron spin-lattice relaxation and nuclear spin lattice relaxation with the anisotropies of the electron phase. 2D-ELDOR studies have been reported (Dubinskii et al., 1994) for nitroxide spin labels in the solid state where motional behavior occurs over a wide range of correlation times from 10^{-4} to 10^{-12} s. The 2D-ELDOR spectra can be simulated based on analytical solutions of the spin-relaxation behavior for small-angle fluctuations so that the experimental data can be quantitatively analyzed. 2D-ELDOR exhibits exchange (Lee et al., 1993) cross peaks as well as coherence peaks from nuclear modulation. By varying the mixing time, the two effects can be separated.

By applying 2D-ELDOR sequences, (Van der Struijf and Levine, 1998) the rotational motion of proteins in lipid bilayer systems have been deduced. Saturation-Recovery ELDOR (SR-ELDOR) has been used (Veksli and Rakvin, 1997; Saxena and Freed, 1997) to detect slow-motional dynamics in the millisecond region. Molecular dynamics of the end label of a liquid crystalline polymer was measured by 2D-ELDOR studies. Microscopic order but macroscopic disorder (MOMD) was examined (Xu et al., 1996). The motion of spin probes and spin labels in amorphous polymers has been studied below the glass transition temperature with a two-dimensional field-step ELDOR. (Saalmueller et al., 1996). Internal motions of the chelating ring structure of alkali metal-o-dimesitoylbenzene radical complexes have been deduced by ELDOR studies (Van der Drift and Smidt, 1982). Structures of the radicals were deduced in single crystals by ELDOR spectroscopy (Hwang et al., 1981). The relative orientations of weakly coupled paramagnetic centers have been deduced (Maryasov et al., 1998).

3.1 Distance Measurements

Pulsed ELDOR methods have been used to determine dipole-dipole interaction in polypeptide-biradicals, (Milov et al., 1999b) and the structure of radical pairs and the dipole distance (Kawamore et al., 1998) between paramagnetic species in photosystem II. Maryasov et al. (1998) also showed based on simulations that it might be possible to deduce the relative orientations of weakly coupled paramagnetic centers. Further details of using ESR techniques to determine distance is discussed in Ch 8 of this volume.

For distances between 1.5 and 8 nm, procedures to use ELDOR or DEER measurements have been described (Jeschke et al., 2000b). Distances can be deduced by ELDOR using the procedure outlined recently (Jeschke et al., 2000a; Froncisz et al., 2001). Cluster sizes and cluster-to-cluster distances (Pannier et al., 2000, 2001) were deduced (Pannier et al., 2000a) by four-pulse ELDOR. Experimental details were given to reduce the deadtime.

The end-end distance in a series of tempo diradicals has been measured (Martin et al., 1998; Larsen and Singel, 1993).

3.2 Spin-Labeled Studies

Saturation-recovery electron-electron double resonance, the technique of pumping one spin orientation and observing it at another, was used to measure the rotational rate of a molecule connecting two portions of an ESR spectrum of perdeutered [^{15}N] TEMPOL (Haas et al., 1993; Haas et al., 1992; Mailer et al., 1992 and CTPO in glycerol-H_2O mixtures and spin-labeled hemoglobin. Saturation transfer between hyperfine components of nitroxide spin labels in liquids has been examined (Hyde et al., 1984 and Van der Drift et al., 1984). Inversion-recovery studies have been carried out for nitroxide spin labels in solution (Koptyug et al., 1996). The electron and nitrogen relaxation times T_1 agree poorly with electron-nuclear dipole mechanism. Inversion-recovery studies have been carried out for nitroxide spin labels in solution (Koptyug et al., 1996). In addition, PELDOR studies were used to study the kinetics of phase relaxation due to dipole-dipole spin coupling of Fremy's radical ions and neutral nitroxyl radicals TEMPON in glassy frozen solutions at 77 K. Dipole-dipole spin phase relaxation for charged radicals was deduced (Milov and Tsvetkov, 2000). Saturation recovery has also been measured for slowly tunneling spin labels (Smigel et al., 1974, 1974a; Hyde et al., 1975). Saturation-recovery EPR and ELDOR methods have been shown (Marsh, 1992) to be useful in the analysis of CW saturation studies and it can be used to determine exchange frequencies in the presence of nuclear relaxation.

3.3 Spin Probes in Liquid Crystals

2D-ELDOR and electron spin-echo (ESE) measurements were reported for the spin-probe PD-tempone in smectic a liquid crystals, as a function of director orientation and temperature (Gorcester et al., 1989). A measure of the solute dynamics was established. For instance, at 288-323 K, intense 2D-ELDOR cross peaks were observed only for $\Delta M_I = \pm 1$ which indicated ^{14}N spin-relaxation and negligible Heisenberg exchange. Dipolar spectral densities were obtained from angular dependent ^{14}N spin-relaxation rates at the hyperfine frequency. By combining ESE and 2D-ELDOR measurements, the dipolar and Zeeman-dipolar spectral densities could be obtained at zero frequency. Order director fluctuations in the smectic phase are suppressed at frequencies on the order of 10 MHz. The behavior of the observed spectral densities at zero frequency suggests an additional contribution to solute reorientation due to cooperative hydrocarbon chain fluctuations.

A 2-D ELDOR study (Sastry et al., 1996) of a rigid rodlike cholestane spin label in a liquid crystal solvent N-(*p*-butoxybenzylidene)-*p*-octylaniline) over a wide temperature range 96° to 25 °C showed a greatly enhanced sensitivity to rotational dynamics. Over this temperature range the liquid crystal solvent exhibits isotropic (±), nematic, smectic A, smectic B and crystal phases. A three dimensional experiment can be presented by recording 2D-ELDOR spectra as a function of mixing time (T_m). A slowly relaxing local structure (SRLS) gave a better fit to the data than a model of Brownian reorientation in a macroscopic aligning potential. In the SRLS model, a dynamic cage of solvent molecules relaxes on a slower time scale than the cholestane spin-label. This provides a local orienting potential in addition to that of the macroscopic aligning potential in the liquid crystal phase. An estimate of the cage potential in the different phases was deduced as well as the rotational diffusion tensor of the cholesane spin label. A similar study of the small globular spin probe perdeuterated tempone in the same liquid crystal (Sastry et al., 1996a) was also reported. A model of a SRLS behavior fit significantly better than a standard Brownian reorientation. It was found that as the temperature is reduced, the spin probe molecules are partially expelled from the hard core (dipolar) region of the liquid crystal molecules and move toward the more flexible aliphatic chain region due to the increased core packing from smectic layer formation, thus experiencing a more fluid local cage structure.

3.4 Spin Labels in Membranes

The ELDOR technique was used by Popp and Hyde (1982), to study lipid lateral diffusion (Lai et al., 1986) in model membranes using the [^{14}N]-16-doxyl stearate spin label. Diffusion rates were measured at elevated temperatures but difficulties were encountered at physiological temperature due to the effects of electron-nuclear dipolar (END) relaxation. Problems due to the intermolecular END mechanism were overcome (Feix et al. 1984) by introducing ^{14}N:^{15}N spin-label pairs and measuring the interactions between diffusion-mediated Heisenberg spin exchange. This approach made it possible to examine interacting probes on different types of probes, between lipid-bound and protein-bound spin labels and between different isomeric lipid labels. The replacement of the bimodal cavities with a loop-gap resonator (LGR) (Hyde et al. 1985) made it possible to increase the signal-to-noise ratio (S/N) in ELDOR measurements by a factor of almost 20. A rate equations approach for analyzing saturation recovery data from ^{14}N:^{15}N pairs was introduced by Yin and Hyde (1987) so that the limitations imposed by spectral overlap (Yin, Feix and Hyde, 1988) in ELDOR experiments could be solved. The use of ^{14}N:^{15}N spin-label pairs remains a powerful technique for examining molecular interactions in biological systems [Sorokin et al., 1996), Feix et al. (1987); Yin et al. (1987); Hyde and Feix, (1988).] Lai, et al. 1986; Renk et al., 1988.

For example, ELDOR techniques (Feix et al., 1987) were also used to measure the interaction of ^{14}N:^{15}N stearic acid spin-labeled pairs in fluid-phase model membrane bilayers composed of a variety of phospholipids. Lateral diffusion and vertical fluctuations toward a membrane diffusion were measured as a function of alkyl chain length of the host lipid. Only a slight effect was observed for lateral diffusion but vertical fluctuation is quite sensitive to host lipid unsaturation. Bimolecular collision (Yin and Hyde, 1989) rates of ^{14}N containing nitroxide radical labeled stearic acid with similar ^{15}N species in DMPC liposomes were examined and found to be 20% greater for labels at the C-16 position than for the C-12 position. This suggested a difference at these two positions. Measurements of ^{15}N spin-labels at C-5 and ^{14}N label at C-16 showed that the bimolecular collisions between C-5 and C-16 occurred with half the frequency of C-16-C-16 collisions (Feix et al., 1984). Vertical fluctuations were very pronounced and dependent on pH and temperature.

The effects of oxygen on the EPR spectra of nitroxide spin-label probes (Popp and Hyde, 1981) in DMPC was examined by ELDOR measurements. The ELDOR reduction was much greater when a deoxygenated sample of 16 SASL in DMPC was used. The secondary structure of a double spin-labeled peptide incorporated inside a tetrameric supramolecular assembly of unlabeled peptide molecules has been studied by ELDOR measurements.

3.5 Spin Labeled Peptides

Heisenberg spin exchange (HSE) has been measured in frozen glassy samples of spin-labeled peptides (Miick and Millhauser, 1994). HSE directly measures the rate of collision between nitroxides and nitroxide spin-labeled bimolecules. Distance measurements were made between two spin-labels in 2-substituted aminoisobutyl residues. A distribution of distances (Milov et al., 2000) in spin labeled peptides (Milov et al., 1999) was found. Structural and distance information was also deduced for the secondary structure of a peptide (Milov et al., 2001). For instance 25% of the peptide biradicals were found to be separated by a distance of 20 Å but most of the spin-labeled peptides (65 - 85%) had conformations with distances between spin labels of greater than 20 Å. The self-aggregation (Milov et al., 2000) of a spin-labeled peptide frozen at 77 K in mixed solvents of varying polarity has been studied by ELDOR measurements and shown to be sensitive to distances between 2.3 and 3.3 nm. Other studies involving the structural examination of peptides have been reported (Milov et al., 2000a, 2000b, 1999a, 2000).

3.6 Photosystem II

In 1996, a pulsed ELDOR measurement was applied to measure the dipole interactions between paramagnetic species on the donor side of photosystem II (Hara et al., 1996). The distance deduced between the Mn cluster and the redox-active tyrosiene residue YD was estimated to be 27 ± 0.2 Å in the S2 state of the oxygen-evolving photosystem II.

A later study used pulsed ELDOR methods (Kawamori et al., 1998) to measure the distance between the electron acceptor quinone QA^- and the YD^+ radical pairs in cyanide-treated PSII. It was found to equal 38.5 ± 0.7 Å. The distance between QA^- and Yz^+ was found to be 38 ± 1 Å. Further pulsed ELDOR studies (Mino et al., 2000) of oriented Ca^{2+}-depleted PS II membranes indicate that the vector connecting the doublet-signal center with the YD butyl radical and the plane of the thylakoid membrane are at an angle of 8°.

The ELDOR method was used (Kuroiwa et al., 2000) to measure the dipole distance between cytochrome b559 and the primary acceptor quinone observed at g = 2.0045 (dH_{pp}= 9 G) in photosystem II where the non-heme Fe^{2+} was substituted by Zn^{2+}. The dipole distance of 40 ± 1 Å was deduced by fitting to the observed ELDOR time profile. The orientation of QA to cyt b559 and the membrane normal was determined. Studies of doublet signals and singlet-like signal was observed for oriented Ca^{2+}-depleted PSII membranes [Mino et al., 2000].

3.7 Biological Samples

Pulsed ELDOR studies [Enemark et al., 2000] have also been reported for the Mo(V)/Fe(III) state of sulfite oxidase in chicken liver.

It has been shown (Shin and Hubbell, 1992) that the electrostatic potentials at biological surfaces can be deduced from ELDOR measurements of the collision frequency of a charged nitroxide in solution with a nitroxide fixed to the biological surface of a phospholipid bilayer. The values deduced are consistent with those predicted by the Gary-Chapman Theory. ELDOR measurements were also deduced near the surface of DNA and were consistent with those calculated using the nonlinear Poisson-Boltzmann equation.

ELDOR measurements were used (Rank et al., 1988) to examine the specific interactions of the 5 and 16 ^{15}N-spin-labeled stearic acid with a ^{14}N

spin-labeled retinal chromophore of a rhodopsin analog. No interaction between the ^{15}N and ^{14}N spin-labels was detected indicating that the ring portion of the chromophore must be highly sequestered from the phospholipid bilayer in both rhodopsin and metarhodopsin II forms. ELDOR studies (Yin et al., 1987) were also used to study the effects of cholesterol on lateral diffusion and vertical fluctuations in lipid bilayers. ELDOR studies were also carried out (Lai et al., 1986) to measure the lateral diffusion constants(D) of lipids in the surface membrane of intact human blood platelets. A marked increase in D was observed upon storing blood platelets suggesting a loss of cholesterol during storage from which a correlation between lipid lateral diffusion and cholesterol levels in cell membrane is inferred. Electrostatic potentials near the surface of DNA were calculated and found to be in good agreement with the potentials measured by ELDOR (Hecht et al., 1995). The phospholipid asymmetry and flip-flop rate in rod outer segment disk membrane has been studied by both a spin-label method and an ELDOR measurement. Rapid trans membrane diffusion suggests that the process is mediated by proteins in the disk membrane (Wu and Hubbell, 1993).

Milov et al. (2001) have shown that if the PELDOR method is combined with the CW-ESR technique, it is possible to study the frozen glassy solutions of the double TOAC spin-labeled Trichogin GA IV diluted by the unlabeled peptides. The double spin-labeled peptides aggregate in nonpolar environment. The intermolecular distance between the spin labels of the peptide has been found to equal 15.7 Å. The results were consistent with four amphilic helical peptide molecules forming a vesicular system with the polar amino acid chains pointing to the interior and the apolar side chains to the exterior of the cluster.

4. PRACTICAL ASPECTS OF MEASUREMENTS

Many of the pulsed techniques reviewed in this chapter can be carried out using the commercially available Bruker ESP 380 E Pulse EPR spectrometer operating at X-band frequencies. Recently a second generation Bruker X-band (ELEXSYS E580 FT/CW) spectrometers and the E 680 FT/CW pulsed W-band spectrometers have become available. These come with pulsed ELDOR optional features, saturation-recovery ELDOR, and DEER. The caution in carrying out pulsed measurements is that a well trained operator must be present to successful make use of these instruments. Realistically, for the average or beginning user, one should contact a National ESR Center for these measurements.

For SR-ELDOR, TLSS, MQ-ESR or spin label oximetry measurements contact: The National Biomedical ESR Center at the Medical College of Wisconsin, Milwaukee, WI, (www:http://www.biophysics.mcw.edu/bri-epr), phone (414) 456-4008.

For 2D pulse measurements contact the National Biomedical Center for Advanced ESR Technology (ACERT) at Cornell University, Ithaca, NY (www:http://www.ccmr.cornell.edu/~acert, phone (607) 255-3647.

Pulsed Bruker X-band or the homemade S band pulse measurements are carried out at the University of Denver (contact geaton@DU.edu for more information).

Use of the Bruker ESP 380 E pulse ESR spectrometer can be arranged at the WR Wiley Environmental Molecular Sciences Laboratory, Pacific Northwest National Laboratory, Richland, Washington, Contact Mike Bowman at Michael Bowman @PNL.Gov, phone (509) 376-3299.

Purchase of a Bruker pulse X and/or W-band spectrometers with ELDOR, saturation-recovery ELDOR and DEER techniques can be arranged. Contact Bruker Instruments, EPR Division; 19 Fortune Drive, Billerica, MA 01821-3991 or phone (978) 663-7406, e-mail: epr@Bruker.com or see world wide web:http://www.Bruker-biospin.com.

The facility at the Max-Planck Institute for Polymer Research in Mainz, Germany (Jeschke, Spiess) are examples of facilities where pulsed ELDOR (DEER, 2+1 pulse sequence) can be carried out.

With the recent demonstration of 2D ELDOR, DEER, 2+1 pulse sequence, MQ-ELDOR, 2D-correlation spectroscopy, and SIFTER methods; the use of ELDOR techniques in the measurement of distances, oximetry and diffusion in biological materials will greatly expand the use of EPR methods in the solution of significant structural biology problems and will become a major instrument used in the structural determinations. The limitation will be in the availability of trained personnel to maintain, operate and interpretation of the spectra. The availability of commercially instruments will follow.

5. REFERENCES

Antsiferova, L. I., Ivanova, A. N., and Kulagina, T. P. (1987) On the Feasibility of Molecular Motion Studies by the Electron-Electron Double Resonance Method. *Mol. Phys.* **61**, 1299-307.

Atherton, N. M. (1972) ENDOR and ELDOR. *Electron Spin Resonance*, **1**, 32-46.

Atherton, N. M. (1974) ENDOR [Electron Nuclear Double Resonance] and ELDOR [Electron-Electron Double Resonance]. *Electron Spin Reson.*, **2**, 36-51.

Atherton, N. M. (1976) ENDOR and ELDOR. *Electron Spin Reson.* **3**, 23-34.

Benderskii, V. A., Blumenfeld, L. A., Stunzhas, P. A., and Sokolov, E. A. (1968). Double Electron-Electron Resonance of Triplet Excitions in Ion-Radical Salts. *Nature* **220**, 365-367.

Borbat, P. P., and Freed, J. H. (1999). Multiple-Quantum ESR and Distance Measurement. *Chem. Phys. Lett.*, **313**, 145-154.

Borbat, P. P., and Freed, J. H. (2000). Double-Quantum ESR and Distance Measurements in *Biological Magnetic Resonance: Distance Measurements in Biological Systems by EPR.* Berliner, J. L., Eaton, S. S. and Eaton, G. R. eds., Kluwer Academic, N.Y., Ch. 10.

Borbat, P. P., Crepeau, R. H., and Freed, J. H. (1997) Multifrequency Two-Dimensional Fourier Transform ESR: an X/Ku-band Spectrometer. *J. Magn. Reson.* **127**, 155-167.

Bruno, G. V., and Freed J. H. (1974). ESR Lineshapes and Saturation in the Slow Motional Region: ELDOR. *Chem. Phys. Lett.* **25**, 328-332.

Bowers, K. D., and Mims, W. B. (1959) Paramagnetic Relaxation in Nickel Fluorosilicate *Phys. Rev.* **115**, 285-295.

Chien, J. C. W. (1979) Electron-Electron Double Resonance Studies of Chain Motion in Polymers. *Polym. Prepr., Am. Chem. Soc., Div. Polym. Chem.* **20**, 187-8.

Christidis, T. C., Mchaourab, H. S., and Hyde, J. S. (1996) Hyperfine Selectivity Using Multiquantum Electron-Nuclear-Electron Triple Resonance. *J. Chem. Phys.* **104**, 9644-9646.

Closs, G. L., Evanochko, W. T., and Norris, J. R. (1982) Structure and Dynamics of the Trimethylcyclopropenyl Radical as Determined by Electron and Nuclear Magnetic Resonance. *J. Am. Chem. Soc.* **104**, 350-2.

Dalton, L. A., Monge, J. L., Dalton, L. R., and Kwiram, A. L. (1974) Molecular and Applied Modulation Effects in Electron-Electron Double Resonance. III. Bloch Equation Analysis for Inhomogeneous Broadening. *Chem. Phys.* **6**, 166-82.

Dammers, A. J., Levine, Y. K., and Tjon, J. A. (1982) Computation of Magnetic Resonance Spectra From the Stochastic Liouville Equation with Pade Approximants. *Chem. Phys. Lett.* **88**, 198-201.

Doi, Y., and Kuwata, K. (1979a) Electron-Electron Double Resonance Study of Nuclear-Spin Exchange in Quinone Anions Due to Cation Migration. *Mol. Phys.* **37**, 1159-74.

Doi, Y., and Kuwata, K. (1979) Electron-Electron Double-Resonance Study of Heisenberg Spin Exchange of Quinone Anions in Alcohols. *J. Magn. Reson.* **34**, 617-29.

Dorio, M. M. (1979) Polymer Studies, in *Mult. Electron Reson. Spectrosc.,* Dorio, M. M. and Freed, J. H. eds., Plenum Press, N. Y., p. 393-408.

Dorio, M. M., and Chien, J. C. W. (1975) Field-Swept and Frequency-Swept ELDOR [Electron-Electron Double Resonance' and the Manifestation of Rapid-Passage Effects. *J. Chem. Phys.* **62(10)**, 3963-7.

Dorio, M. M., and Chien, J. C. W. (1975a) Spin Diffusion in Electron-Electron Double Resonance. *J. Magn. Reson.* **20**, 114-23.

Dorio, M. M., and Chien, J. C. W. (1975b) Study of Molecular Motion in Polymeric Solids by Electron-Electron Double Resonance. *Macromolecules* **8**, 734-9.

Dorio, M. M., and Chien, J. C. W. (1976) Matrix Signals in Electron-Electron Double Resonance. *J. Magn. Reson.* **21**, 491-8.

Dorio, M., and Freed, J. H. eds. (1979). *Multiple Electron Resonance Spectroscopy.* Plenum, New York.

Dubinskii, A. A., Maresch, G. G., and Spiess, H. W. (1994) Two-Dimensional Electron Paramagnetic Resonance Spectroscopy of Nitroxides: Elucidation of Restricted Molecular Motions in Glassy Solids. *J. Chem. Phys.* **100**, 2437-48.

Dulcic, A., and Poric, M. (1982) Spin Magnetizations in a Coupled Electron-Proton Spin System. *J. Phys. C.* **15**, 7193-8.

Dzuba, S. A., and Tsvetkov, Y. D. (1982) Slow Motion of Nitroxide Radicals Studied by Pulse EPR Spectroscopy. *Khim. Fiz.* **1**, 1197.

Dzuba, S. A., Maryasov, A. G., Salikhov, K. M., and Tsvetkov, Yu. D. (1984). Superslow Rotations of Nitroxide Radicals Studied by Pulse EPR Spectroscopy. *J. Magn. Reson.* **58**, 95-117.

Eastman, M. P., Bruno, G. V., and Freed, J. H. (1970) ESR Studies of Heisenberg Spin Exchange. III. ELDOR [Electron-Electron Double Resonance] Study. *J. Chem. Phys.* **52**, 132-7.

Enemark, J. H., Codd, R., Astashkin, A. V., Raitsimring, A. M., and Pacheco, A. (2000) Preparation and Pulsed ELDOR Spectroscopy of the Mo(V)/Fe(III) State of Sulfite Oxidase. *Abstr. Pap.-Am. Chem. Soc.* 220th INOR-008.

Evans, J. C., and Rowlands, C. C. (1984) ENDOR and ELDOR. *Electron Spin Reson.* **8**, 61-85.

Feix, J. B., Popp, C. A., Venkataramu, S. D., Beth, A. H., Park, J. H., and Hyde, J. S. (1984) An Electron-Electron Double-Resonance Study of Interactions Between [14N]- and [15N]Stearic Acid Spin-Label Pairs: Lateral Diffusion and Vertical Fluctuations in Dimyristoylphosphatidylcholine. *Biochemistry* **23**, 2293-9.

Feix, J. B., Yin, J. J., and Hyde, J. S. (1987) Interactions of Nitrogen-14:Nitrogen-15 Stearic Acid Spin-Label Pairs: Effects of Host Lipid Alkyl Chain Length and Unsaturation. *Biochemistry* **26** 3850-5.

Freed, J. H. (1974) Theory of Saturation and Double Resonance in Electron Spin Resonance Spectra. VI. Saturation Recovery. *J. Phys. Chem.* **78**, 1155-67.

Freed, J. H. (1979) Theory of Multiple Resonance and EPR Saturation in Liquids and Related Media in *Multiple Electron Resonance Spectroscopy,* Dorio, M. M., and Freed, J. H.,eds., Plenum Press, New York, pp. 73-142.

Freed, J. H. (1979a). Theory of ESR Saturation Recovery in Liquids and Related Media in *Time Domain Electron Spin Resonance*, Kevan, L and Scwartz, R. N., eds., Wiley (Interscience), New York. Chapter 2.

Froncisz, W., Camenisch, T. G., Ratke, J. J., and Hyde, J. S. (2001) Pulse Saturation Recovery, Pulse ELDOR, and Free Induction Decay Electron Paramagnetic Resonance Detection Using Time-Locked Subsampling. *Rev. Sci. Instrum.* **72**, 1837-1842.

Gamliel, D., and Freed, J. H. (1990) Theory of Two-Dimensional ESR With Nuclear Modulation. *J. Magn. Reson.* **89**, 60-93.

Geoffroy, M., Kispert, L. D., and Hwang, J. S. (1979) An ESR, ENDOR, and ELDOR Study of Tunneling Rotation of a Hindered Methyl Group in X-Irradiated 2,2,5-Trimethyl-1,3-Dioxane-4,6-dione Crystals. *J. Chem. Phys.* **70**, 4238-42.

Gorcester, J., and Freed, J. H. (1986). Two-dimensional Fourier Transform ESR Spectroscopy. *J. Chem. Phys.* **85**, 5375-5377.

Gorcester, J., and Freed, J. H. (1988). Linear Prediction and Projection of Pure Absorption Lineshapes in Two-Dimensional FT ESR Correlation Spectroscopy. *J. Magn. Reson.* **78**, 292-301.

Gorcester, J., and Freed, J. H. (1988a) Two-Dimensional Fourier Transform ESR Correlation Spectroscopy. *J. Chem. Phys.* **88**, 4678-93.

Gorcester, J., Millhauser, G. L., and Freed, J. H. (1990). Two-Dimensional Electron Spin Resonance, in *Modern Pulsed and Continuous-Wave Electron Spin Resonance*, Kevan, L. and Bowman, M. K. eds., Wiley, New York, Ch 3.

Gorcester, J., Rananavare, B., and Freed, J. H. (1989) Two-Dimensional Electron-Electron Double Resonance and Electron Spin-Echo Study of Solute Dynamics in Smectics. *J. Chem. Phys.* **90**, 5764-86.

Haas, D. A. (1993) Analysis of the Power Saturation Method and Using SR-ELDOR and Global Analysis to Extract Multiple Rates From Recovery Curves. *Diss. Abstr. Int. B* **54**, 4176.

Haas, D. A., Mailer, C., Sugano, T., and Robinson, B. H. (1992) New Developments in Pulsed Electron Paramagnetic Resonance: Direct Measurement of Rotational Correlation Times From Decay Curves. *Bull. Magn. Reson.* **14** 35-41.

Haas, D. A., Sugano, T., Mailer, C., and Robinson, B. H. (1993) Motion in Nitroxide Spin Labels: Direct Measurement of Rotational Correlation Times by Pulsed Electron Double Resonance. *J. Phys. Chem.* **97**, 2914-21.

Hara, H., Kawamori, A., Astashkin, A. V., and Ono, T. (1996) The Distances From Tyrosine D to Redox-Active Components on the Donor Side of Photosystem II Determined by Pulsed Electron-Electron Double Resonance. *Biochim. Biophys. Acta* **127**,140-146.

Hecht, J. L., Honig, B., Shin, Y-K., and Hubbell, W. L. (1995) Electrostatic Potentials Near the Surface of DNA: Comparing Theory and Experiment. *J. Phys. Chem.* **99**, 7782-6.

Hornak, J. P., and Freed, J. H. (1983) ELDOR Spin Echoes and Slow Motions. *Chem. Phys. Lett.* **101**, 115-19.

Huber, M., van Amsterdam, I. M. C., Ubbink, M., Canters, G. W. (2002) Dipolar Distances Between Two Copper Centers in Azurin Using DEER Measurements, *Biophys. J.* **82**, 2328.

Hwang, J. S., Dickinson, A. C., and Kispert L. D. (1979) ENDOR and ELDOR Studies of X-Irradiated Polycrystalline Dipeptides, Myosin and Actomyosin. *J. Phys. Chem.* **83**, 3381-7.

Hwang, J. S., Wang, H. C., Andersson, B., Kispert, L. D., and Geoffroy, M. (1981) An Electron Spin Double Resonance Study of X-Ray Irradiated Phenacyl Chloride Single Crystals. *J. Chem. Phys.* **75**, 3758-64.

Hyde, J. S. (1974) Paramagnetic Relaxation. *Annu. Rev. Phys. Chem.*, **25**, 407-35.

Hyde, J. S. (1998) ELDOR Spectroscopy in *Foundations of Modern EPR*, Eaton, S. S., Eaton, G. R., and Salekhov, K. M., eds. World Scientific Publishing, pp. 607-618.

Hyde, J. S., and Feix, J. B. (1988). Electron-Electron Double Resonance in *Biological Magnetic Resonance: Spin Labeling, Theory and Applications*, Berliner, L. J., and Reuben, J., eds., Plenum Press, New York, P. 305-337.

Hyde, J. S., and Feix, J. B. (1989) Biological Magnetic Resonance: Spin Labeling Theory and Applications, Vol. 8. Edt. L. J. Berliner and J. Reuben, New York, 305-337.

Hyde, J. S., Chien, J. C. W., and Freed, J. H. (1968) Electron Electron Double Resonance of Free Radicals in Solution, *J. Chem. Phys.* **48**, 4211-4226.

Hyde, J. S., Froncisz, W., and Mottley, C. (1984) Pulsed ELDOR Measurement of Nitrogen T1 in Spin Labels. *Chem. Phys. Lett.* **110**, 621-5.

Hyde, J. S., Kispert, L. D., Sneed, R. C., and Chien, J. W. (1968a) Frequency Swept Electron-Electron Double Resonance: Separation of Overlapping Spectra in Irradiated Malonic Acid, *J. Chem. Phys.* **48**, 3824-3825.

Hyde, J. S., Mchaourab, H. S., Strangeway, R. A., and Luglio, J. R. (1995) Multiquantum ESR: Physics, Technology and Applications to Bioradicals. *Bioradicals Detected ESR Spectrosc.* 31-47.

Hyde, J. S., Smigel, M. D., Dalton, L. R., and Dalton, L. A. (1975) Molecular and Applied Modulation Effects in Electron Electron Double Resonance]. IV. Stationary ELDOR [Electron-Electrode Double Resonance] of Very Slowly Tumbling Spin Labels. *J. Chem. Phys.* **62(5)**, 1655-67.

Hyde, J. S., Sneed, R. C., Jr., and Rist, G. H. (1969) Frequency-Swept Electron-Electron Double Resonance. DPPH [1,1-Diphenyl-2-Picrylhydrazyl] in Liquid and Frozen Solution. *J. Chem. Phys.* **51**, 1404-16.

Hyde, J. S., Yin, J. J., Feix, J. B., and Hubbell, W. L. (1990) Advances in Spin Label Oximetry. *Pure Appl. Chem.* **62**, 255-60.

Hyde, J. S., Yin, J. J., Froncisz, W., and Feix, J. B. (1985) Electron-Electron Double Resonance (ELDOR) With a Loop-Gap Resonator. *J. Magn. Reson.* **63**, 142-50.

Iwasaki, M., Toriyama, K., and Nunome, K. (1974) ELDOR Enhancement Arising From Admixture of Nuclear Spin States. *J. Chem. Phys.* **61**, 106-14.

Jeschke, G., Koch, A., Jonas, U. and Godt, A. (2002). Direct Conversion of EPR Dipolar Time Evolution Data to Distance Distributions. *J. Magn. Reson.* **155**, 72-82.

Jeschke, G., Pannier, M., and Spiess, H. W. (2000). Double Electron-Electron Resonancc in *Biological Magnetic Resonance: Distance Measurements in Biological Systems by EPR.* Berliner, J. L., Eaton, S. S., and Eaton, G. R. eds., Kluwer Academic, N.Y., Ch. 11.

Jeschke, G., Pannier, M., Godt, A., and Spiess, H. W. (2000a) Dipolar Spectroscopy and Spin Alignment in Electron Paramagnetic Resonance. *Chem. Phys. Lett.* **331**, 243-252.

Kawamori, A., Hard, H., Shigemori, K., and Astashkin, A. V. (1998) Pulsed Electron-Electron Double Resonance to Determine Distances Between Paramagnetic Species in Photosystems. *Mod. Appl. EPR/ESR, Proc Asia-Pac. EPR/ESR Symp., 1st,* 226-332.

Kevan, L. and Narayana, P. A. (1979) ENDOR and ELDOR in disordered Matrixes, in *Mult. Electron Reson. Spectrosc.,* Dorio, M. M. and Freed, J. H. eds., Plenum Press, N. Y., p. 229-59.

Kevan, L., and Bowman, M. K. eds. (1990). *Modern Pulsed and Continuous-Wave Electron Spin Resonance.* John Wiley & Sons, New York.

Kevan, L., and Kispert, L. D. (1976). *Electron Spin Double Resonance Spectroscopy.* John Wiley and Sons, N.Y.

Kirillov, S. T., Kozhushner, M. A., and Stryukov, V. B. (1976). Diffusion of Nitroxide Radicals in Plastic Solids as Studied by Electron-Electron Double Resonance. *Chem. Phys.* **17**, 243-8.

Kispert, L. D. (1979) Spectroscopic Study on Crystalline Systems, in *Mult. Electron Reson. Spectrosc.,* Dorio, M. M. and Freed, J. H. eds., Plenum Press, N. Y., p. 261-95.

Kispert, L. D., and Wang, P. S. (1974) Electron-Electron Double Resonance Investigations of Irradiated Malonic Acid-d_4 Single Crystals. *J. Phys. Chem.* **78**, 1839-44.

Kispert, L. D., Chang, K., and Bogan, C. M. (1972) Electron-Electron Double Resonance [ELDOR] of Irradiated Dichlorofluoroacetamide. Dominant Role the Chlorine Quadrupole. *Chem. Phys. Lett.* **17**, 592-7.

Kispert, L. D., Chang, K., and Bogan, C. M. (1973) Electron-Electron Double Resonance of Irradiated Dimethylmalonic Acid, α-Aminoisobutyric Acid, and L-Alanine Single Crystals. Role of Methyl Substitutents. *J. Chem. Phys.* **58**, 2164-76.

Kispert, L. D., Chang, K., and Bogan, C. M. (1973b) Electron-Electron Double Resonance of Irradiated Single Crystals of Zinc Acetate and Malonic Acid. Influence of Nuclear Spin Exchange. *J. Phys. Chem.* **77**, 629-33.

Kispert, L. D., Chang, K., and Chen, T. C. S. (1976), ELDOR Study of Hydrogen-Deuterium Exchange Reactions in Irradiated Glycine-d_2 Crystals. *Chem. Phys. Lett.* **44**, 269-72.

Kispert, L. D., Chang, K., and Mottley, C. (1975) Electron-Electron Double Resonance (ELDOR) Spectroscopy. New Approach to Studying Relaxation and Radiation Processes in Irradiated Organic Crystals. *Magn. Reson. Relat. Phenom,* **1**, 279-80.

Kispert, L. D., Chang, K., and Wang, P. S. (1974) Electron-Electron Double-Resonance Investigations of Irradiated Organic Crystals. Influence of Deuterium Substitution. *J. Magn. Resonance,* **14**, 339-47.

Kispert, L. D., Chen, T. C. S., Hill, J. R., and Clough, S. (1978) ELDOR Detection of ENDOR Transitions in Irradiated Organic Crystals. *J. Chem. Phys.* **69**, 1876-80.

Kispert, L. D., Hill, J. R., and Mottley, C. (1979) A Frequency-Swept ELDOR Study of Irradiated Ammonium Acetate Crystals at 77 K: Resolution of Spin Packets. *J. Magn. Reson.* **33**, 379-87.

Kispert, L., and Chang, K. (1973) Electron-Electron Double Resonance of Irradiated Monofluoroacetamide Single Crystals. Role of Fluorine p-Orbital Anisotropy. *J. Magn. Resonance*, **10**, 162-9.

Koptyug, I. V., Bossmann, S. H., and Turro, N. J. (1996) Inversion-Recovery of Nitroxide Spin Labels in Solution and Microheterogeneous Environments. *J. Am. Chem. Soc.* **118**, 1435-45.

Kuroiwa, S., Tonaka, M., Kawamori, A., and Akabori, K. (2000) The Position of Cytochrome b559 Relative to QA in Photosystem II Studied by Electron-Electron Double Resonance (ELDOR). *Biochim. Biophys. Acta,* **1460**, 330-337.

Kurshev, V. M., Raitsimring, A. M., and Tsvetkov, Yu D. (1989). Selection of Dipolar Interaction by the "2+1" Pulse Train ESE. *J. Magn. Reson. 81*, 441-454.

Kurshev, V. V., Raitsimring, A. M., and Salikhov, K. M. (1988). Angular Dependence of the Dipole Broadening Function of the ESR Line of Single Crystals Containing Paramagnetic Centers with an Anisotropic g Tensor: Er^{3+} in $CaWO_4$. *Sov. Phys. Solid State* **30**, 239-242.

Lai, C. S., Wirt, M. D., Yin, J. J., Froncisz, W., Feix, J. B., Kunicki, T. J., and Hyde, J. S. (1986) Lateral Diffusion of Lipid Probes in the Surface Membrane of Human Platelets. An Electron-Electron Double Resonance (ELDOR) Study. *Biophys. J.* **50**, 503-6.

Larsen, R. G., and Singel, D. J. (1993) Double Electron-Electron Resonance Spin-Echo Modulation: Spectroscopic Measurement of Electron Spin Pair Separations in Orientationally Disordered Solids. *J. Chem. Phys.* **98**, 5134-46.

Lee, S., Budil, D. E., and Freed, J. H. (1994) Theory of Two-Dimensional Fourier Transform Electron Spin Resonance for Ordered and Viscous Fluids. *J. Chem. Phys.* **101**, 5529-58.

Lee, S., Patyal, B. R., and Freed, J. H. (1993) A Two-Dimensional Fourier-Transform, Electron-Spin-Resonance (ESR) Study of Nuclear Modulation and Spin Relaxation in Irradiated Malonic Acid. *J. Chem. Phys.* **98**, 3665-89.

Leniart, D. S. (1979) Instrumentation and Experimental Methods in *Mult. Electron Reson. Spectrosc.,* Dorio, M. M. and Freed, J. H. eds., Plenum Press, N. Y., p. 5-72.

Lin, D-P., and Kevan, L. (1977) Mechanism of Spin Diffusion in Electron Spin Resonance Spectra of Trapped Electrons in Aqueous Glasses. Electron-Electron Double Resonance Studies. *J. Phys. Chem.* **81**, 966-9.

Lin, D. P., Feng, D. F., Ngo, F. Q. H., and Kevan, L. (1976) Electron-Electron Double Resonance Study of Magnetic Energy Transfer Between Trapped Electrons and Radicals in Organic Glasses: Relation Between Dipolar Cross Relaxation Times and Dipolar Interaction Distances. *J. Chem. Phys.* **65**, 3994-4000.

Lund, A., Gillbro, T., Feng, D-F., and Kevan, L. (1975) EPR and ELDOR Studies of the Dichloromethyl Radical in g-Irradiated Single Crystal Dichloromethane. *Chem. Phys.* **7**, 414-23.

Mailer, C., Robinson, B. H., and Haas, D. A. (1992) New Developments in Pulsed Electron Paramagnetic Resonance: Relaxation Mechanisms of Nitroxide Spin Labels. *Bull. Magn. Reson.* **14** 30-4.

Maresch, G. G., Weber, M., Dubinskii, A. A., and Spiess, H. W. (1992) 2D-ELDOR Detection of Magnetization Transfer of Nitroxides in Disordered Solid Polymers. *Chem. Phys. Lett.* **193**, 134-40.

Marsh, D. (1992) Influence of Nuclear Relaxation on the Measurement of Exchange Frequencies in CW Saturation EPR Studies. *J. Magn. Reson.* **99**, 332-7.

Martin, R. E., Pannier, M., Diederich, F., Gramlich, V., Hubrich, M., and Spiess, H. W. (1998) Determination of End-to-End Distances in a Series of TEMPO Diradicals of up to 2.8 nm Length With a New Four-Pulse Double Electron Electron Resonance Experiment. *Angew. Chem., Int. Ed.*, **37**, 2834-2837.

Maryasov, A. G., Tsvetkov, Y. D., and Raap, J. (1998) Weakly Coupled Radical Pairs in Solids. ELDOR in ESE Structure Studies. *Appl. Magn. Reson.* **14**, 101-113.

Mchaourab, H. S. (1994) Theory and Methodology of Multiquantum EPR, ELDOR and ENDOR With Application to a Spin-Labeled Analog of the Ion-Channel Peptide Cecropin AD. *Diss. Abstr. Int., B* **57**, 4927.

Mchaourab, H. S., Christidis, T. C., Froncisz, W., Sczanieck, P. B., and Hyde, J. S. (1991). I. Multiple-Quantum Electron-Electron Double Resonance. *J. Magn. Reson.* **92**, 429-433.

Mehdizadeh, M., Ishii, T. K., Hyde, J. S., and Froncisz, W. (1983). Doubly Tuned Resonators, *IEEE Trans. Microwave Theory Tech.* **MTT-31**, 1059.

Mehlkopf, A. F., Kuiper, F. G., Smidt, J., and Tiggelman, T. A. (1983) Modulation Scheme for Electron-Electron Double Resonance Spectroscopy. *Rev. Sci. Instrum.* **54**, 695-6.

Miick, S. M., and Millhauser, G. L. (1994) Measuring Heisenberg Spin Exchange Between Spin-Labeled Peptides with 2D-ELDOR. *J. Magn. Reson.. B* **104**, 81-4.

Milov, A. D., and Tsvetkov, Y. D. (2000) Charge Effect on Relative Distance Distribution of Fremy's Radical Ions in Frozen Glassy Solution Studied by PELDOR. *Appl. Magn. Reson.* **18** 217-226.

Milov, A. D., Mar'yasov, A. G., Samoilova, R. I., Tsvetkov, Y. D., Raap, J., Monaco, V., Formaggio, F., Crisma, M., and Toniolo, C. (2000a) Pulsed Electron Double Resonance of Spin-Labeled Peptides: Data on the Secondary Structure of the Peptide Chain. *Dokl. Akad. Nauk,* **370**, 265-268.

Milov, A. D., Maryasov, A. G., and Tsvetkov, Y. D. (1998) Pulsed Electron Double Resonance (PELDOR) and Its Applications in Free-Radicals Research. *Appl. Magn. Reson.* **15**, 107-143.

Milov, A. D., Maryasov, A. G., Tsvetkov, Y. D., and Raap, J. (1999) Pulsed ELDOR in Spin-Labeled Polypeptides. *Chem. Phys. Lett.* **303**, 135-143.

Milov, A. D., Ponomarev, A. B., and Tsvetkov, Y. D. (1984). Electron-Electron Double Resonance in Electron-Spin Echo Model Biradical Systems and the Sensitized Photolysis of Decalin. *Chem. Phys. Lett.* **110**, 67-72.

Milov, A. D., Salikhov, K. M., and Schirov, M. D. (1981). Application of Double Resonance Method to Electron Spin Echo in a Study of the Spatial Distribution of Paramagnetic Centers in Solids. *Sov. Phys. Solid State*, 23, 565.

Milov, A. D., Tsvetkov, Y. D., Formaggio, F., Crisma, M., Toniolo, C., and Raap, J. (2000a) Self-Assembling Properties of Membrane-Modifying Peptides Studied by PELDOR and CW-ESR Spectroscopies. *J. Am. Chem. Soc.* **122**, 3843-3848.

Milov, A. D., Tsvetkov, Y. D., Formaggio, F., Crisma, M., Toniolo, C., and Raap, J. (2001) The Secondary Structure of a Membrane-Modifying Peptide in a Supramolecular Assembly Studied by PELDOR and CW-ESR Spectroscopies. *J. Am. Chem. Soc.* **123**, 3784-3789.

Milov, A. F., Tsvetkov, Y. D., and Raap, J. (2000) Aggregation of Trichogin Analogs in Weakly Polar Solvents: PELDOR and ESR Studies. *Appl. Magn. Reson.* **19**, 215-226.

Mino, H., Kawamori, A., and Ono, T. (2000) Pulsed EPR Studies of Doublet Signal and Singlet-Like Signal in Oriented Ca^{2+}-depleted PS II Membranes: Location of the Doublet Signal Center in PS II. *Biochemistry* **39**, 11034-40.

Moebius, K. (1977) ENDOR and ELDOR. *Electron Spin Reson.*, **4**, 16-29.

Moebius, K. (1979). ENDOR and ELDOR. *Electron Spin Reson.* **5**, 52-65.

Moebius, K. (1981) ENDOR and ELDOR. *Electron Spin Reson.* **6**, 32-42.

Moran, P. (1964). Electron Spin Double Resonance Studies of *F* Centers in KCl. *Phys. Rev.* **135**, A247-A265.

Mottley, C., Chang, K., and Kispert, L. (1975a) Exploitation of Molecular Motion in the Analysis of Complex ESR Spectra by ELDOR Spectroscopy. *J. Magn. Reson.* **19**, 130--43.

Mottley, C., Kispert, L. D., and Clough, S. (1975) Electron-Electron Double Resonance Study of Coherent and Random Rotational Motion of Methyl Groups. *J. Chem. Phys.* **63**, 4405-11.

Mottley, C., Kispert, L. D., and Lin, I. B. (1979) An ELDOR Study at 77 K of the Proton Environment Surrounding the Methyl Radical in Irradiated Ammonium Acetate Crystals. *J. Magn. Reson.* **36** 239-48.

Mottley, C., Kispert, L. D., and Wang, P. S. (1976) An ELDOR Study of Methyl Radical Production at 77 K in Irradiated Acetate Powders as a Function of Metal Cation. *J. Phys. Chem.* **80**, 1885-91.

Mottley, C., Kispert, L., and Wang, P. S. (1975b) Electron-Electron Double Resonance Studies of a Methyl Group Undergoing Tunneling Rotation. *Magn. Reson. Relat. Phenom.* **2**, 385-6.

Mukai, K., Shikata, H., Azuma, N., and Kuwata, K. (1979) ELDOR Studies of 1,3,5-Triphenylverdazyl and 1,5-diphenyl-3-(4-Chlorophenyl)verdazyl Radicals. *J. Magn. Reson.* **35** 133-7.

Nechtschein, M., and Hyde, J. S. (1970). Pulsed Electron-Electron Double Resonance in an S = $^1/2$, I = $^1/2$ System. *Phys. Rev. Lett.* **24**, 672-674.

Nordio, P. L., and Segre, U. (1980) Lineshape Simulations in Nonlinear EPR Spectroscopy for Spin Probes Undergoing Jump Diffusion. *J. Magn. Reson.* **39**, 229-35.

Pace, M. D. (1979) An Investigation of Free Radical Formation in Irradiated Halogen Substituted Organic Single Crystals Using Scanning Electron Microscopy, ESR and ELDOR Spectroscopy. *Diss. Abstr. Int. B* **41**, 218.

Pannier, M., Schaedler, V., Schoeps, M., Wiesner, U., Jeschke, G., and Spiess, H. W. (2000) Determination of Ion Cluster Sizes and Cluster-to-Cluster Distances in Ionomers by Four-Pulse Double Electron Electron Resonance Spectroscopy. *Macromolecules* **33**, 7812-7818.

Pannier, M., Schoeps, M., Schaedler, V., Wiesner, U., Jeschke, G., and Spiess, H. W. (2001) Characterization of Ionic Clusters in Different Ionically Functionalized Diblock Copolymers by CW EPR and Four-Pulse Double Electron-Electron Resonance. *Macromolecules* **34**, 5555-5560.

Pannier, M., Veit, S., Godt, A., Jeschke, G., and Spiess, H. W. (2000a) Dead-Time Free Measurement Of Dipole-Dipole Interactions Between Electron Spins. *J. Magn. Reson.* **142**, 331-340.

Pannier, M., Veit, S., Godt, A., Jeschke, G., and Spiess, H. W. (2000). Dead Time Free Measurement of Dipole-Dipole Interactions Between Electron Spins. *J. Magn. Reson.* **142**, 331-340.

Patyal, B. R., Crepeau, R. H., Gamliel, D., and Freed, J. H. (1990) Two-Dimensional Fourier-Transform ESR in the Slow-Motional and Rigid Limits: 2D-ELDOR. *Chem. Phys. Lett.* **175**, 453-60.

Percival, P. W., Hyde, J. S., Dalton, L. A., and Dalton, L. R. (1975) Molecular and Applied Modulation Effects in Electron Electron Double Resonance. V. Passage Effects in High Resolution Frequency and Field Swept ELDOR. *J. Chem. Phys.* **62**, 4332-42.

Perkins, R. C. Jr., Dalton, L. R., and Kispert, L. D. (1977) Molecular and Applied Modulation Effects in Electron-Electron Double Resonance. VII. Modulation Frequency Effects for the C·H_2COO-Radical in Irradiated Zinc Acetate. *J. Magn. Reson.* **26**, 25-33.

Perkins, R. C., Jr. (1977) Saturation Transfer Electron Paramagnetic Resonance Spectroscopy: Theoretical Passage Signal Sensitivity of Slowly Tumbling Nitroxide Radicals and Electron-Electron Double Resonance Spectral Dependence on Zeeman Modulation Frequency for the CH_2COO-Radical. *Diss. Abstr. Int. B*, **38**, 3714.

Persson, M., Harbridge, J. R., Hammarstrom, P., Mitri, R., Martensson, L.-G., Carlsson, U., Eaton, G. R., Eaton, S. S. (2001) Comparison of Electron Paramagnetic Resonance Methods to Determine Distances Between Spin Labels on Human Carbonic Anhydrase II, *Biophys. J.* **80**, 2886-97.

Pfannebecker, V., Klos, H., Hubrich, M., Volkmer, T., Heuer, A., Wiesner, U., and Speiss, H. W. (1996). Determination of end-to-end Distances in Oligomers by Pulsed EPR, *J. Phys. Chem.* **100**, 13428-13432.

Piasecki, W., Froncisz, W., and Hyde, J. S. (1996) Bimodal Loop-Gap Resonator. *Rev. Sci. Instrum.* **67**, 1896-1904.

Piven, N. P., and Benderskii, V. A. (1984) Study of Rotations of Peroxide Radicals in Polystyrene Using Electron-Electron Double Resonance. *Khim. Fiz.* **3**, 386-92.

Popp, C. A., and Hyde, J. S. (1981) Effects of Oxygen on EPR Spectra of Nitroxide Spin-Label Probes of Model Membranes. *J. Magn. Reson.* **43**, 249-58.

Popp, C. A., and Hyde, J. S. (1982) Electron-Electron Double Resonance and Saturation-Recovery Studies of Nitroxide Electron and Nuclear Spin-Lattice Relaxation Times and Heisenberg Exchange Rates: Lateral Diffusion in Dimyristoylphosphatidylcholine. *Proc. Natl. Acad. Sci.* **79**, 2559-63.

Raitsimring, A. (2000). "2 + 1" Pulse Sequence as Applied for Distance and Spatial Distribution Measurements of Paramagnetic Centers in *Biological Magnetic Resonance: Distance Measurements in Biological Systems by EPR.* Berliner, J. L., Eaton, S. S. and Eaton, G. R. eds., Kluwer Academic, N.Y., Ch. 10.

Rengan, S. K., Bhagat, V. R., Sastry, V. S. S., and Venkataraman, B. (1979) Magnetic Field-Pulsed ELDOR Spectroscopy *J. Magn. Reson.* **33**, 227-40.

Renk, G. E., Crouch, R. K., and Feix J. B. (1988) Lack of Interaction of Rhodopsin Chromophore With Membrane Lipids. An Electron-Electron Double Resonance Study Using Nitrogen-14:Nitrogen-15 Pairs. *Biophys. J.* **53**, 361-5.

Rinard, G. A., Quine, R. W., Ghim, T. B., Eaton, S. S. and Eaton, G. R. (1996) Easily Tunable Crossed-Loop (Bimodal) EPR Resonator. *J. Magn. Res. A* **122**, 50-57.

Rinard, G. A., Quine, R. W., and Eaton, G. R. (2000) An L-band Crossed-loop (Bimodal) EPR Resonator. *J. Magn. Res.* **144**, 85-88.

Robinson, B. H., Haas, D. A., and Mailer, C. (1994) Molecular Dynamics in Liquids: Spin-Lattice Relaxation of Nitroxide Spin Labels. *Science* **263**, 490-3.

Robinson, B. H., Monge, J. L., Dalton, L. A., Dalton, L. R., and Kwiram, A. L. (1974) Theory of Modulation Effects in Electron Electron Double Resonance. *Chem. Phys. Lett.* **28**, 169-75.

Saalmueller, J. W., Long, H. W., Maresch, G. G., and Spiess, H. W. (1995) Two-Dimensional Field-Step ELDOR. A Method for Characterizing the Motion of Spin Probes and Spin Labels in Glassy Solids. *J. Magn. Reson., Ser. A* **117**, 193-208.

Saalmueller, J. W., Long, H. W., Volkmer, T., Wiesner, U., Maresch, G. G., and Spiess, H. W. (1996) Characterization of the Motion of Spin Probes and Spin Labels in Amorphous Polymers with Two-Dimensional Field-Step ELDOR. *Polym. Phys.* **34**, 1093-104.

Sands, R. H. (1979) ENDOR and ELDOR on Iron-Sulfur Proteins, in *Mult. Electron Reson. Spectrosc.*, Dorio, M. M. and Freed, J. H. eds., Plenum Press, N. Y., p. 331-74.

Sarna, T. (1979) Application of Double Resonance Methods (ENDOR and ELDOR). *Inst. Biol. Mol.*, **4**, 49-63.

Sastry, V. S. S., Polimeno, A., Crepeau, R. H., and Freed, J. H. (1996) Studies of Spin Relaxation and Molecular Dynamics in Liquid Crystals by Two-Dimensional Fourier Transform Electron Spin Resonance. II. Perdeuterated-Tempone in Butoxy Benzylidene Octylaniline and Dynamic Cage Effects. *J. Chem. Phys.* **105**, 5773-5791.

Sastry, V. S. S., Polimeno, A., Crepeau, R. H., and Freed, J. H. (1996a) Studies of Spin Relaxation and Molecular Dynamics in Liquid Crystals by Two-Dimensional Fourier Transform Electron Spin Resonance. I. Cholestane in Butoxy Benzylidene-Octylaniline and Dynamic Cage Effects. *J. Chem. Phys.* **105**, 5753-5772.

Saxena, S., and Freed, J. H. (1997) Absorption Lineshapes in Two-Dimensional Electron Spin Resonance and the Effects of Slow Motions in Complex Fluids. *J. Magn. Reson.* **124**, 439-54.

Saxena, S., and Freed, J. H. (1997) Two-Dimensional Electron Spin Resonance and Slow Motions. *J. Phys. Chem. A* **101**, 7998-8008.

Schosseler, P., Wacker, T., and Schweiger, A. (1994) Pulsed ELDOR Detected NMR. *Chem. Phys. Lett.* **224**, 319-24.

Schweiger, A., and Jeschke, G. (2001). *Principles of Pulse Electron Paramagnetic Resonance*, Oxford University Press, Oxford, Chpt. 13.

Shin, Y. K., and Hubbell, W. L. (1992) Determination of Electrostatic Potentials at Biological Interfaces Using Electron-Electron Double Resonance. *Biophys. J.* **61**, 1443-53.

Smigel, M. D., Dalton, L. A., Dalton, L. R., and Kwiram, A. L. (1974) Very Slowly Tumbling Spin Labels. Saturation Recovery. *Chem. Phys.* **6**, 183-92.

Smigel, M. D., Dalton, L. R., Hyde, J. S., and Dalton, L. A. (1974a) Investigation of Very Slowly Tumbling Spin Labels by Nonlinear Spin Response Techniques. Theory and Experiment for Stationary Electron Double Resonance. *Proc. Nat. Acad. Sci.* **71**, 1925-9.

Sorokin, P. P., Lasher, G. J., and Gelles, I. L. (1960) Cross Relaxation Studies in Diamond. *Phys. Rev.* **118**, 939-945.

Stetter, E., Vieth, H. M., and Hausser, K. H. (1976) ELDOR Studies of Nitroxide Radicals: Discrimination Between Rotational and Transitional Correlation Times in Liquids. *J. Magn. Reson.* **23**, 493-504.

Tsapin, A. I., Hyde, J. S., and Froncisz, W. (1992) Bimodal Loop-Gap Resonator. *J. Magn. Reson* **100** 484-90.

Unruh, W. W., and Culvahouse, J. W. (1963). Experimental Study of Relaxation Processes for Divalent Cobalt Ions. *Phys. Rev.* **129**, 2441-2453.

Van Der Struijf C., and Levine, Y. K. (1998) On the Use of the Phase Memory Time T2 for the Quantitative Characterization of the Rotational Motions of Proteins in Lipid Bilayer Systems. *J. Magn. Reson.* **130**, 244-252.

Van der Drift E., Dammers, A. J., and Smidt, J. (1981) Structural Information From Proton ELDOR on Radicals in Solution. *J. Phys. Chem.* **85**, 829-34.

Van der Drift E., Mehilkopf, A. F., and Smidt, J. (1975) Comparison of Different Sweep Methods in ELDOR Spectroscopy. *Chem. Phys. Lett.* **36**, 385-9.

Van der Drift, E., and Smidt, J. (1982) EPR, ELDOR, and ENDOR Studies of Alkali Metal-o-Dimesitoylbenzene Radical Complexes in Solution. II. The Lithium and Sodium Complexes. *J. Magn. Reson.* **46**, 9-22.

Van der Drift, E., Dammers, A. J., Smidt, J., Plato, M., and Moebius, K. (1980) EPR, ELDOR, and ENDOR Studies of Alkali Metal-*o*-Dimesitoylbenzene Radical Complexes in Solution. I. Rubidium and Cesium Complexes. *J. Magn. Reson.* **40**, 551-79.

Van der Drift, E., Rousseeuw, B. A. C., and Smidt, J. (1984) EPR and ELDOR Studies on Spin Relaxation in Perdeuterated 2,2,6,6-Tetramethyl-4-Piperidone N-Oxide in Liquid

Solutions. The Slowly Relaxing Local Structure Mechanism. *J. Phys. Chem.* **88**, 2275-84.

Veksli, Z., and Rakvin, B. (1997) New Methods of Electron Paramagnetic Resonance for the Determination of Slow-Motion Dynamics in Polymers. *Polimeri (Zagreb)* **18**, 163-170.

Wu, G., and Hubbell, W. L. (1993) Phospholipid Asymmetry and Transmembrane Diffusion in Photoreceptor Disc Membranes. *Biochemistry* **32**, 879-88.

Xu, D., Crepeau, R. H., Ober, C. K., and Freed, J. H. (1996) Molecular Dynamics of a Liquid Crystalline Polymer studied by Two-Dimensional Fourier Transform and CW ESR. *J. Phys. Chem.* **100**, 15873-15885.

Yang, H. W. H., and Chien, J. C. W. (1978) Electron-Electron Double Resonance Study of Polymer Chain Motions. *J. Polym. Sci., Polym. Symp.* **63**, 263-9.

Yang, H. W. H., and Chien, J. C. W. (1978a) Electron-Electron Double-Resonance Study of the Molecular Motion in End-Labeled Poly(Styrene). *Macromolecules* **11**, 759-63.

Yin, J. J., Feix, J. B., and Hyde, J. S. (1987) The Effects of Cholesterol on Lateral Diffusion and Vertical Fluctuations in Lipid Bilayers. An Electron-Electron Double Resonance (ELDOR) Study. *Biophysical*, **52**, 1031-8.

Yin, J. J., Feix, J. B., and Hyde, J. S. (1988) Solution of the Nitroxide Spin-Label Spectral Overlap Problem Using Pulse Electron Spin Resonance. *Biophysical J.* **53**, 525-31.

Yin, J.-J., Feix, J. B., and Hyde, J. S. (1988) Solutions of the Nitroxide Spin Label Overlap Problems Using Pulse Electron Spin Resonance. *Biophys. J.* **53**, 525-531.

Yin, Y. J., and Hyde, J. S. (1987) Application of Rate Equations to ELDOR and Saturation Recovery Experiments on Nitrogen-14:Nitrogen-15 Spin-Label Pairs. *J. Magn. Reson.* **74**, 82-93.

Yin, Y. J., and Hyde, J. S. (1989) Use of High Observing Power in Electron Spin Resonance Saturation-Recovery Experiments in Spin-Labeled Membranes. *J. Chem. Phys.* **91**, 6029-35.

Yoshida, H., Feng, D.-F., and Kevan, L. (1972) Electron-Electron Double Resonance Demonstration of Cross Saturation Between Trapped Electrons and Radicals in g-Irradiated 2-Methyltetrahydrofuran Glass. *J. Amer. Chem. Soc.* **94**, 8922-4.

Yoshida, H., Feng, D.-F., and Kevan, L. (1973) Electron-Electron Double Resonance Study of Trapped Electrons in g-Irradiated 2-Methyl-trahydrofuran Glass. Magnetic Energy Transfer Between Two Different Spin Systems. *J. Chem. Phys.* **58**, 4924-9.

Yoshida, H., Feng, D.-F., and Kevan, L. (1973a) Electron-Electron Double Resonance Study of Trapped Electrons in 10 M Sodium Hydroxide Alkaline Ice Glass. *J. Chem. Phys.* **58**, 3411-19.

Chapter 7

Digital Detection by Time-Locked Sampling in EPR

James S. Hyde[1], Theodore G. Camenisch[1], Joseph J. Ratke[1], Robert A. Strangeway[1,2], Wojciech Froncisz[1,3]
[1]*Department of Biophysics, Medical College of Wisconsin, Milwaukee, WI, USA*
[2]*Milwaukee School of Engineering, Milwaukee, WI, USA*
[3]*Jagiellonian University, Krakow, Poland*

Abstract: All frequencies in a magnetic resonance spectrometer should be phase-locked to a single master oscillator. Departure from this principle leads to degraded instrument performance. The use of digital technology is making superheterodyne detection increasingly attractive, relative to homodyne detection, which has been used in most "modern" EPR spectrometers. The signal modulation frequency, the sampling, frequency, and the intermediate frequency (from the signal down-converter) are all locked to the same clock, so the method is called "time-locked." The sampling of the analog signal to digitize it is done four times in an odd number of cycles, typically 3, 5, or 7, so this is "sub-sampling" relative to the Nyquist criterion. Hence, the name time-locked subsampling (TLSS). An essential feature of TLSS is broad-bandedness followed by digital filtering with internal consistency between the two quadrature detection channels. This type of broad-band acquisition followed by digital analysis permits, for example, study of multiple harmonics of the field modulated signal.

1. INTRODUCTION

This article seeks to provide a background for digital detection methods in EPR spectroscopy that will be useful in future spectrometer-design initiatives. Both continuous wave (CW) and pulse EPR are considered. We propose here a system of classification of these methods as illustrated in Fig. 1. In the detection systems of this figure, the microwave EPR signal-of-interest is represented by $f_0(t)$. Generally it will be a periodic time series, even in the case of pulse EPR since pulse experiments are usually repeated periodically to increase the signal-to-noise ratio (SNR). The harmonic content of the signal can be quite high. Each circuit employs a low noise amplifier (LNA), based on our view that an LNA should always be incorporated in high performance spectrometers. In addition, each circuit also shows an anti-aliasing (aa) filter, designated LP for lowpass or BP for bandpass. This filter is an essential element whenever an analog-to-digital (A/D) converter is used.

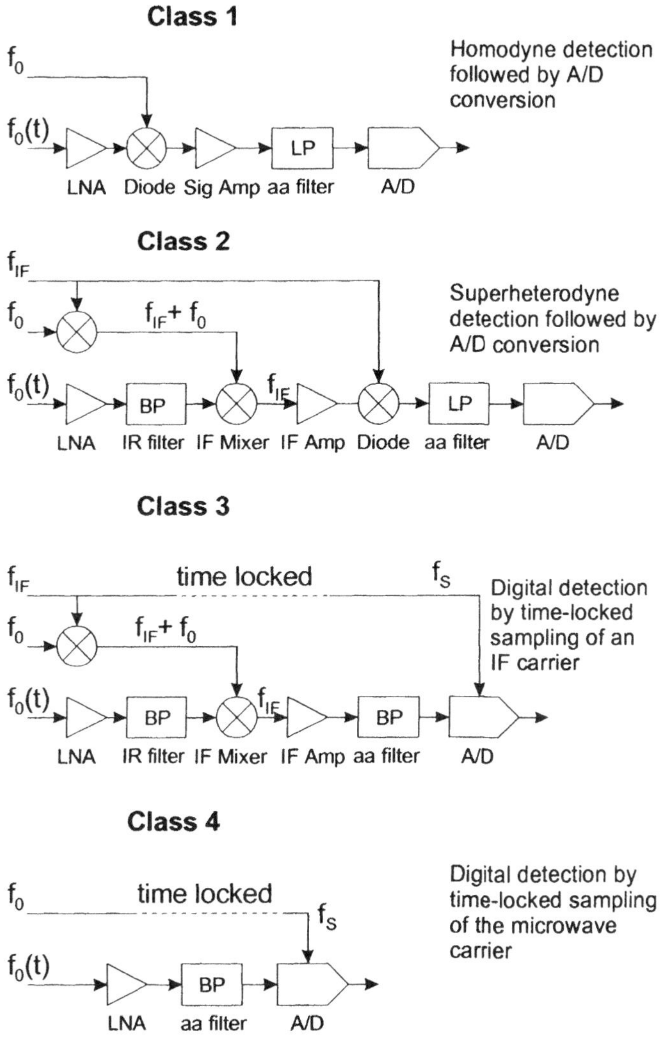

Figure 1. Classes of digital detection methods for EPR spectroscopy. Notation is defined in the text.

Class 1, Fig. 1, is homodyne detection of the microwave signal followed by A/D conversion. See King (1978) for a discussion of homodyne detection. One can detect either dispersion or absorption, but not both, in the circuit shown by changing the phase of the reference arm microwaves at frequency f_0. However, since the LNA establishes the noise floor, quadrature detection could be used without an SNR penalty. Signal processing, including phase sensitive detection if it is needed, is carried out in a computer after the A/D converter. The signal amplifier (Sig Amp) should have level gain and constant time delay over the frequencies described by $FT[f_0(t)]$ where FT is a Fourier Transform of the output of the diode detector. Class 1 detection has been used extensively by us for saturation recovery EPR. The greatest technical difficulty has been design of the Sig Amp and the aa lowpass filter. They both should have a flat amplitude response and a linear phase response for many octaves. The Sig Amp is AC coupled and attenuates frequencies below the repetition rate of the saturating pulses. There can be a tendency for signal "droop" to occur in the time domain display from attenuation and phase shift of low frequencies. The lowpass response of the aa filter is determined by the sampling frequency f_S and must attenuate frequencies beyond the Nyquist limit to prevent aliasing effects.

Class 2, Fig. 1, is superheterodyne detection. As in the case of homodyne detection, additional circuitry is required to detect both dispersion and absorption. The primary benefit of this scheme, relative to homodyne detection, is the flexibility gained in distributing gains between the LNA, IF amplifier, and signal amplifier. Level response across the chain of amplifiers and filters can be an engineering challenge. Selection of dispersion or absorption can be made in two ways, changing the phase of $f_{IF} + f_O$ or of f_{IF}, which are functionally equivalent. A central rationale for the Class 2 circuit is reduced sensitivity to low frequency noise originating in the diode detector, compared with homodyne detection. This was a more serious problem, historically, prior to the introduction of LNAs. The image reject (IR) bandpass filter rejects noise at the IF image frequency. Use of an IR IF mixer is an *alternative* to IR filtering. Both methods can be used simultaneously. The choice between these alternatives involves tradeoffs that are determined by center frequency, bandwidth and required rejection of the image frequency band (see Section 5 of this chapter for additional details). The aa bandpass filter cuts off at the Nyquist limit ($f_S/2$), as is also the case for Class 1.

Class 3, Fig. 1, is time-locked A/D detection of an intermediate frequency carrier, the primary subject of this article. It might be considered to be a modern version of superheterodyne detection. In this method, the information-of-interest is centered about the IF carrier. Since the A/D sampling frequency, f_s, is time locked to the carrier, changes in IF phase can be detected and both dispersion and absorption can be detected by digital signal processing. Both time-locked subsampling (TLSS) (i.e., four times in an odd number of IF cycles – say five) and time-locked oversampling (TLOS) (i.e., sampling many times per IF cycle) are possible. A significant benefit of this method is that the aa filter is bandpass and centered on the IF. If the IF is appropriately chosen, this filter can be relatively broad in absolute units but nevertheless narrow – say a Q of 10 – relative to the IF. This is a favorable situation for filter design.

A design that is closely related to Class 3 has been published by Murugesan *et al.* (1998) and by Subramanian *et al.* (1999) for use at 300 MHz. Data is converted to an IF of 50 MHz and oversampled using a high speed A/D converter that is not time-locked.

Class 4, Fig. 1, is direct digital detection of the microwave EPR signal, almost certainly using TLSS. It would appear to be technically feasible with current technology for a microwave frequency of 1 GHz and less, which might be practical for small animal *in vivo* EPR studies. EPR detection based on the Class 4 scheme has not yet been demonstrated. It can be expected to be more robust and less costly than the other classes of digital detection because fewer synthesizers and mixers are required.

A central rationale of each of these four schemes is broadband digital detection followed by digital filtering. The amplitude is uniform and the phase shift is linear across the bandwidth of interest. The aa filters for Class 3 and 4 may perform better than those for Class 1 and 2. Nevertheless, the homodyne scheme of Class 1 is better than the older analog methods employing phase-sensitive detection when using 100 kHz field modulation for experiments involving EPR of transient species. This is because filtering for adequate SNR is done digitally after A/D conversion.

Since nearly all EPR experiments can be characterized as producing a number of signals at the fundamental of a basic repetition rate and the harmonics of this rate, and the bandwidth can be large in Class 3 and Class 4 schemes, simultaneous capture of all harmonics is feasible. If the number of expected frequencies is small, for example, when using field modulation, it may be more convenient to process the digitized time series by cross correlation with the known sinusoidal waveform and a few of its harmonics, performing this operation at each field point, to arrive eventually at spectra corresponding to the first, second, third, etc. harmonics. Filtering by Gaussian smoothing, for example, provides a digital equivalent for each spectrum of the "time constant" in conventional phase-sensitive detection. This would seem to be an ideal way to investigate phase shifts arising from passage effects when using field modulation. If the number of harmonics is large, it may be more convenient to process the digitized time series by taking a Fourier transform, which is, in fact, the cross correlation of the signal with cosines and sines at the fundamental and all harmonics. This is not unlike having thousands of phase-sensitive detectors available with consistent relative gains and perfect phasing. However, because all data are digitized, optimization of the digital filter for each spectrum can be done retrospectively. There are four possible spectra for each harmonic and two relevant phases: microwave phase corresponding to dispersion and absorption, and field modulation phase. Information regarding both phases is captured simultaneously for all harmonics when using Class 3 and Class 4 schemes.

2. TIME LOCKING AND SUPERHETERODYNE DETECTION – EPR INSTRUMENT-DESIGN BACKGROUND

All frequencies in a magnetic resonance instrument, whether NMR, MRI, or EPR, should in principle be phase locked to a single master oscillator. Departure from this rule can give rise to some degree of degraded instrument performance. The rationale for this design rule is that our instruments have extremely low noise and are becoming increasingly complex. If two frequencies in the system are not locked, they (or their harmonics) can beat as well as mix. The only general way to avoid this without analyzing every detail of instrument design and use is to lock all frequencies.

In the case of superheterodyne spectrometers, the microwave oscillator itself serves as the master oscillator. For detection, the RF is converted to an intermediate frequency. Since the phase and frequency of the IF are known, the IF becomes the surrogate "master oscillator." If all frequencies are time locked to the IF, including for example, the sampling rate and the field modulation frequency or, in the case of pulse EPR, pulse timing frequencies, the rule is satisfied.

This technique works extremely well for CW EPR when used with low microwave power incident on the sample cavity. At higher power, superheterodyne EPR spectrometers become difficult to use. This stability problem seems to have its origin primarily in the fact that as the microwave power is increased, it is necessary to balance the microwave bridge ever more closely in order not only to avoid saturation of the IF amplifier, but also to satisfy the condition that the EPR microwave signal and its microwave carrier be much less than the local oscillator level at the IF mixer. (See Fig 1, Class 2.) This results in enhanced sensitivity to: i) microphonics in the low audio frequency range, noting that field modulation frequencies used in the early instruments were in this range, ii) klystron phase noise, noting that this was poor in the klystrons used in the immediate post-WWII period, and iii) thermal instability of the cavity and microwave circuit.

Homodyne detection using 100 kHz field modulation became the dominant EPR instrumental configuration beginning about 1960. Success was based on several technological advances: i) Microwave detector diodes exhibit noise with a dependence on frequency that varies approximately as 1/f. A rationale for superheterodyne detection had been that the intermediate frequency was high enough that this noise source was insignificant. Development of improved detector diodes permitted use of field modulation at frequencies as low as 100 kHz without serious impact from 1/f noise. ii) A second generation of klystrons had been developed with reduced phase noise, and iii) most mechanical structures, including microwave circuits at X-band, are acoustically "insensitive" at 100 kHz, eliminating much of the microphonics problem.

The hypothesis of this article, and indeed much of the work of the authors for the past five years, is that the pendulum for EPR spectrometer design is now swinging back to superheterodyne detection. This is based on rapid developments in digital technology including readily available high performance personal computers, advances in microwave synthesizers and high speed A/D converters with high vertical resolution. It is also based on an increased awareness in magnetic resonance system design of the need for locking every frequency in the system to a single master oscillator. We are convinced that this hypothesis is correct and that digital detection will eventually become dominant, both for CW and for pulse EPR. The shift from homodyne detection will be driven by increasing awareness of the benefits of digital detection.

3. TIME-LOCKED SUBSAMPLING DETECTION FOR CW EPR

Our paper, Electron Paramagnetic Resonance Detection by Time-Locked Sub-Sampling (TLSS) (Hyde et al., 1998), establishes feasibility for application to CW EPR using sinusoidal field modulation. This paper is reviewed here. Time-locked subsampling detection is in use in MRI scanners manufactured by GE Medical Systems as described in their patent (Stormont et al., 1991), and commercial reliability motivated the work that led to Hyde et al., 1998. Our paper was the first to use TLSS in EPR and the first to use it for detection of periodically modulated signals in any application, to the best of our knowledge. Another significant distinguishing characteristic is that the MRI application is pulsed, i.e., the incident RF is off during detection; in CW EPR it is on, which introduces a number of complications. Our paper is also the first paper using TLSS in magnetic resonance, since no paper was written by GE scientists on this subject.

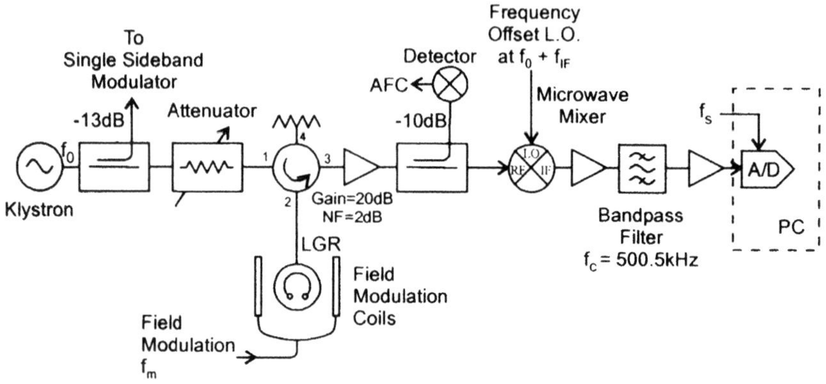

Figure 2. Microwave bridge with TLSS detection (see Hyde et al., 1998).

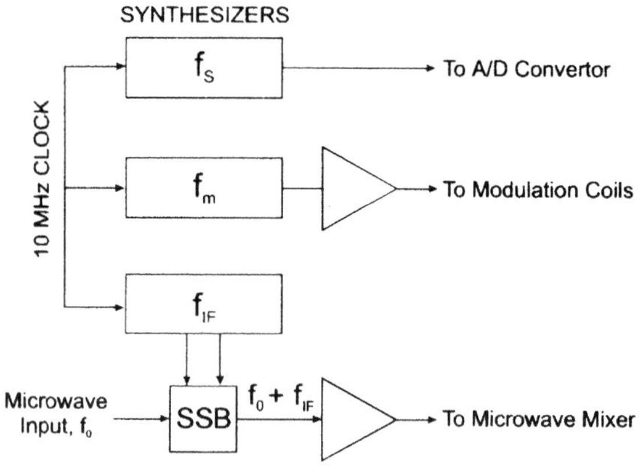

Figure 3. Block diagram of the frequency synthesizer array.

Figures 2 and 3 are schematics of the method used in Hyde et al., 1998. The microwave circuit diagram (Fig. 2) shows four signal inputs: f_0 (the microwave frequency), f_{IF} (the intermediate frequency), f_s (the A/D converter sampling frequency), and f_m (the field modulation frequency). Three of these, f_{IF}, f_s and f_m are time locked, but the fourth frequency, f_0, is free running as illustrated in Fig. 3.

The novel aspect of TLSS is the use of a time-locked sampling rate that is four times in an odd number of cycles, typically 3, 5 or 7. Figure 4 indicates this main idea for three cycles (although seven were used in Hyde et al., 1998): the solid sinusoidal waveform is the real part of the IF carrier and the dotted waveform is the imaginary part. Every dot, both open and filled, is a sample point. One can see that there are four samples in three cycles. The filled dots sample the real waveform and the open dots the imaginary waveform. These digitized data go to a PC (Fig. 2), where the data points corresponding to the filled dots are separated from the points corresponding to the open dots. Thus I and Q, or absorption and dispersion, are in separate memory locations. The sign of every other data point in I is changed, and similarly for Q. We call this process of I and Q formation and sign reversal the "word shuffle." It constitutes detection of the envelope of the IF carrier. To summarize, I and Q EPR signals at the field modulation frequency started out as periodic modulations of a microwave carrier that are translated to an IF carrier, and then separated and detected using time-locked sampling with subsequent manipulation by a PC.

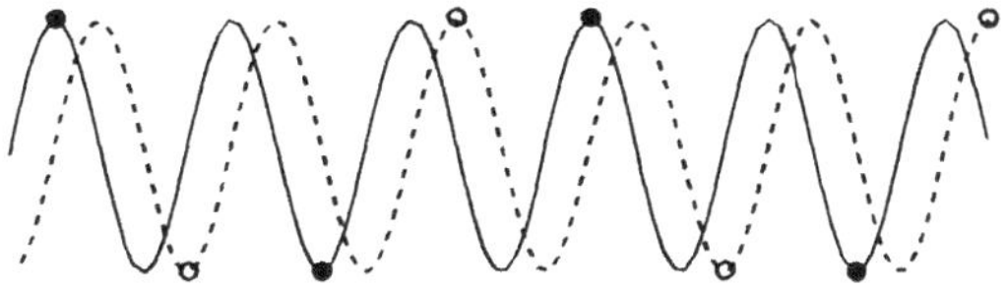

Figure 4. TLSS detection, sampling four times in three cycles.

A carefully designed bandpass filter (Fig. 2) is required. This filter must have a bandwidth less than $f_S/2$, consistent with the Nyquist condition. The usual filter for an A/D converter is a *low pass* filter cutting off at half the sampling rate. For TLSS, the filter is centered at the IF frequency and has a *bandpass* width of less than half the sampling rate. An appropriate A/D converter board for TLSS must have an input frequency response that is consistent with the desired IF and bandpass width.

There are a number of constraints on the choice of f_m, f_s and f_{IF}. If the A/D conversion rate and either the number of harmonics or the field modulation frequency are fixed, other parameters are predetermined. Table 1 shows the frequencies that were used in Hyde et al., 1998. Frequencies must be carefully selected such that ratios are terminated fractions, i.e., $f_s/f_{IF} = 4/7$, $f_s/f_m = 20$.

Table 1. Frequencies

$f_{IF} = 500.5$ kHz	the intermediate frequency
$f_s = 286$ kHz	the sampling frequency
$f_m = 14.3$ kHz	the field modulation frequency
$f_0 = 8.9$ GHz	the microwave frequency
BW = 143 kHz	filter bandwidth
$f_s/f_{IF} = 4/7$	the subsampling ratio
$f_s/f_m = 20$	10 samples in U and 10 in V

The essence of TLSS detection is broadbandedness followed by digital filtering with essentially perfect internal consistency between I and Q (which is the main claim in the GE patent (2). All information that can pass the input bandpass filter is preserved in broadband form until the process of cross-correlation is performed. All harmonics of field modulation are in essentially perfect internal consistency with respect to amplitude and phase (Fig. 5). If information is time varying because of, for example, signal decay, time constants can be extracted by suitable analysis in the PC if this information can pass the input bandpass filter. Digital filtering can be optimized retrospectively.

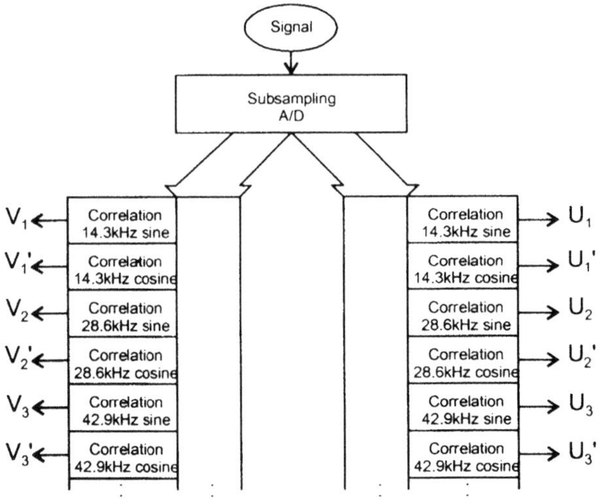

Figure 5. Schematic of information flow when using TLSS detection for SW EPR with field modulation.

Figure 6 shows eight spin label spectra that were acquired simultaneously in a single sweep of the magnetic field using TLSS detection: dispersion and absorption, first four harmonics in phase. A field modulation amplitude of 1.5 G at 14.4 kHz was used along with an estimated microwave field intensity at the sample in the rotating frame of 0.2 G. The incident microwave power on the loop gap resonator was 1 mW. The peak-to-peak line width for the sample (10^{-4} M TEMPO in H_2O at room temperature) using a low field modulation amplitude is 1.2 G. Under the conditions used to obtain Fig. 6, the modulation amplitude is close to the value that yields the largest possible first harmonic signals, V_1, U_1. The noise is about the same for both dispersion and absorption, and was independent of incident microwave power under the conditions used to obtain Fig. 6. The limiting noise source was not firmly established in Hyde et al., 1998, but it was felt that noise was determined by the noise figure of the low noise microwave amplifier (Fig. 2).

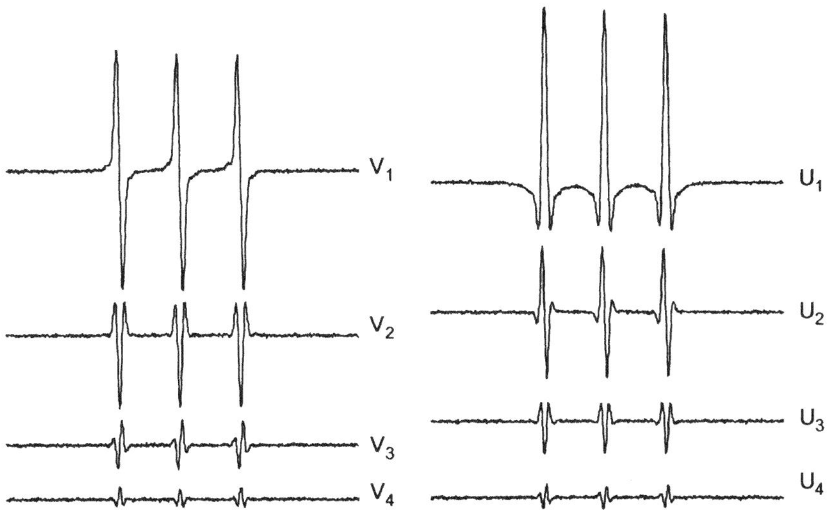

Figure 6. Example of eight spectra of a nitroxide radical spin label produced using TLSS detection in a single sweep of the magnetic field. The notation is defined in Fig. 4.

The overall SNR compared with conventional homodyne detection using 14.3 kHz field modulation frequency was about the same, extrapolated to estimates of the same effective integrating time constants. In unpublished work done at the time the work in Hyde et al., 1998 was carried out, the field modulation amplitude was increased. The amplitude of the higher harmonics (Fig. 7) increased, as expected. In the past, extreme overmodulation such as in Fig. 7 had always been avoided. Since pseudomodulation (Hyde et al., 1990, 1992) can be applied to a simulated spectrum to simulate these harmonics, and since they can now be collected experimentally using TLSS detection, data such as shown in Fig. 7 may turn out to be useful. These spectra are rich in information content.

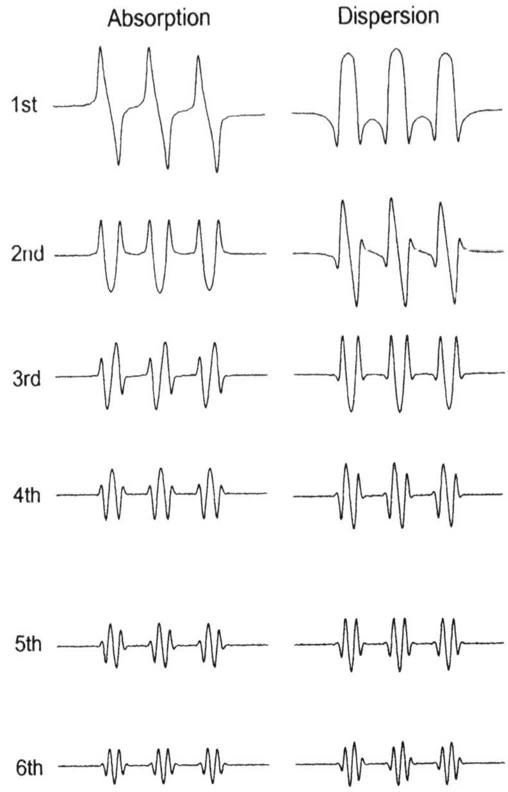

Figure 7. Detected harmonics from a spin label using TLSS detection and severe overmodulation.

4. PULSE SATURATION RECOVERY USING TIME-LOCKED SUBSAMPLING

Pulse EPR experiments can be divided into two broad classes: "driven" and "free precession." The driven category is characterized by detection of the EPR pulse response using an observing CW microwave source. Saturation recovery (SR) is the most familiar type of driven pulse EPR. It involves measurement of the recovery of saturation arising from an intense microwave irradiating pulse using a weaker observing pulse. This subject is reviewed in the chapter 1 of this volume by Eaton and Eaton. See also Hyde (1979, 1998). There are other kinds of driven pulse EPR including: i) the response of the spin system when the observing power cannot truly be characterized as "weak," ii) the response when the incident power is stepped up rather than down, iii) the response to a temperature jump as observed by an unchanging incident microwave power (so-called T-jump), iv) pulse electron-electron double resonance (ELDOR) where the transient response is induced by irradiating one transition and is observed by inspection of another transition, v) jumps in other experimental conditions such as pH, ionic strength, irradiating light level, potential across the sample, and more. The driven category can, alternatively, be labeled step-recovery. As a class, these experiments detect changes in M_z indirectly through changes in M_x or M_y. The time scale for these experiments is characterized by T_1.

Free-precession experiments include free induction decay (FID) and the many variants of spin echo (SE) EPR. The time scale for these experiments is characterized by transverse relaxation, T_2, which is always shorter than T_1. Data collection in the ideal free-precession experiment is carried out in the absence of any microwave power incident on the sample, which eliminates several sources of noise or instability in the detection process. As a class, these experiments directly detect the time evolution of M_x or M_y. Free-precession effects can occur in driven experiments, but can be suppressed by microwave phase modulation techniques (Huisjen and Hyde, 1974). The chapter by Freed in this volume gives additional information on pulse EPR. Berliner et al., 2000 contains extensive information on pulse EPR – see particularly Chapter 2, Relaxation Times of Organic Radicals and Transition Metal Ions, by Eaton and Eaton, in that volume. The recent monograph on pulse EPR by Schwieger and Jeschke (2001) provides a foundation for future progress in that field.

Froncisz et al. (2001) describes the first application of TLSS detection to pulse EPR. Two driven pulse experiments – saturation recovery and pulse ELDOR, and one free precession experiment – free induction decay, were described. The intermediate frequency, f_{IF}, was 187.5 MHz, the sampling frequency, f_s, was 250 MHz (i.e., three samples in four cycles), the overall bandwidth was 125 MHz, and the bandwidths for the separate I and Q channels were each 62.5 MHz. The apparatus employed four frequency synthesizers locked to a common 10 MHz clock. Reference 6 provides extensive technical detail. Experiments were conducted on nitroxide radical spin labels.

Figure 8a is a direct representation of the recovery of magnetization from strong saturation using a weak observing microwave power for 3.3 mM Tempo in 30% glycerol/water, not deoxygenated. It shows the temporal change of the microwave signal that is reflected from the resonator, noting that the microwave signal after translation to the subsampled intermediate frequency is actually displayed. Figure 8b is the conventional saturation-recovery display after signal processing. Namely, the sign of the extreme negative points in Fig. 8a is changed to positive and the resulting envelope is displayed. At this sample concentration, Heisenberg exchange is strong, and single exponential decay is expected. The data are the result of 3.28×10^6 repetitions at a rate of 28×10^3 repetitions per second (about 2 min).

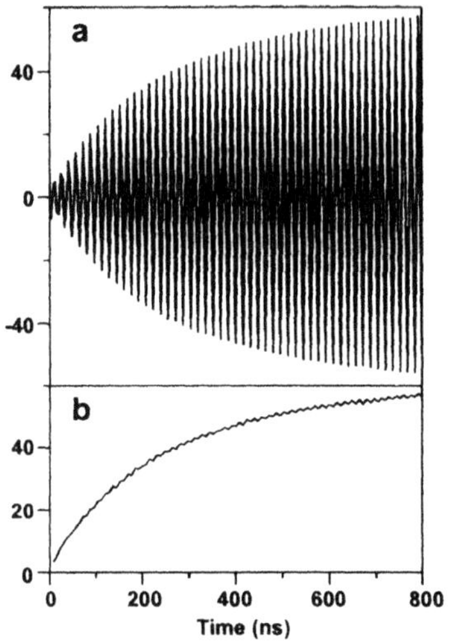

Figure 8. Saturation-recovery data: (a) raw IF recovery; (b) SR after the word shuffle.

The primary benefits of TLSS detection of SR data based on the experiments conducted thus far include simultaneous detection of saturation-recovery dispersion and absorption and elimination of low frequency distortions that cause droop of the tail of the exponential decay. In data not shown in Hyde (1998), the pulse ELDOR experiment of Hyde et al. (1984) was replicated using TLSS detection. The pulse TLSS spectrometer that was constructed is convenient for pulse ELDOR experiments. One simply translates the SR pump arm to a different frequency such that the frequency difference between the pump and observed microwaves matches an interval of interest in the EPR spectrum.

Figure 9 shows an FID signal of one line of a rapidly tumbling nitroxide radical spin label that was obtained at the output of the time-locked A/D converter. The magnetic field was not precisely centered on the EPR line, being shifted slightly down field. The resulting FID signal is a decaying microwave signal with a frequency that differs from the frequency of the excitation by an amount that depends on the offset of the magnetic field from the center of the line. In a conventional FID display, mixing of the FID microwave frequency with a microwave reference at the excitation frequency results in decaying oscillations *at the offset frequency*. This is *not* what one sees in Fig. 9. This figure shows the decaying FID signal directly as translated to an intermediate frequency. The oscillations in Fig. 9 are not at the offset frequency, but rather at the excitation minus the offset frequency – translated, of course, to an intermediate frequency.

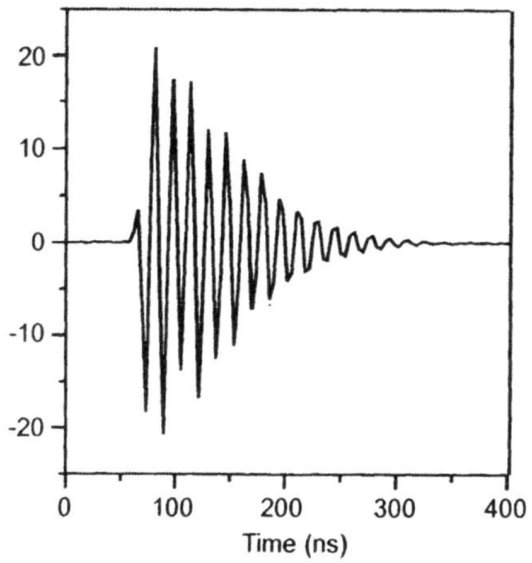

Figure 9. FID signal using TLSS detection.

5. SELECTED ENGINEERING CONSIDERATIONS

There are two goals in this section: to give to the EPR spectroscopist an overview of key engineering aspects of digital detection, and to provide the EPR spectrometer designer with technical detail and access to the literature that will be of practical value. There are five parts in this section of the chapter: i) frequency synthesizers, ii) image reject filters, iii) anti-aliasing filters, iv) A/D converters and v) data processing.

5.1 Frequency Synthesizers

Frequency synthesizers are required in Class 3 digital detectors (Fig. 1) as well as to generate various excitation irradiation patterns. Synthesizer-based excitation methods can introduce distinctive noise effects into digital detection systems. In this section, we review our experience in the use of synthesizers both for excitation and detection.

Various techniques for generating the frequencies in the microwave bridge arms have been considered (Strangeway et al., 1995). The double sideband/ fixed filter technique has been pursued extensively (Berliner et al., 2000; Hyde et al., 1995) and utilizes synthesizers to drive the mixers in a direct translation approach. The synthesizers are time-locked through the use of a common reference clock, typically a low-noise crystal oscillator. A block diagram of a generic synthesizer-based microwave bridge is shown in Fig. 10 (Schweiger and Jeschke, 2001). The bridge can be used for either conventional field modulation or multiquantum (MQ) EPR. ELDOR and

MQ-ELDOR modes of operation are also available if a sufficient number of excitation (main) arms are present.

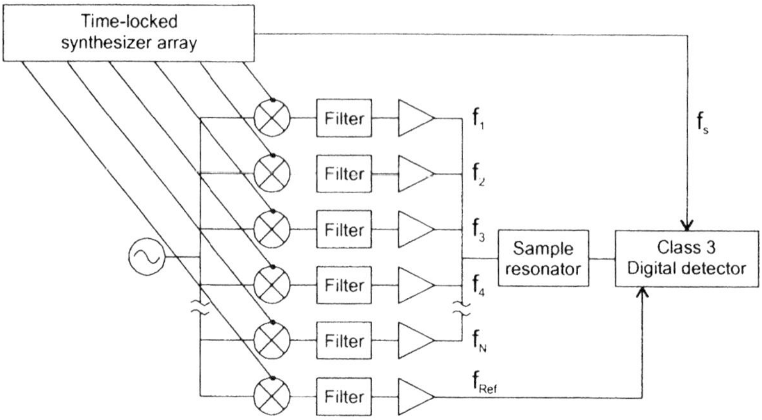

Figure 10. EPR bridge with multiple time-locked microwave signals incident on the sample followed by Class 3 digital detection.

Synthesizers are utilized for three primary functions in this microwave bridge circuit:

- Drive the translation mixer(s) in the main arm(s) that feed into the sample resonator.
- Drive the translation mixer in the reference arm at a frequency appropriate to the desired bridge detection mode. The reference arm frequency is f_{REF} in Fig. 10 and is (f_{IF} + f_o) in Fig. 1, Class 3.
- Provide the A/D converter clock (f_s).

Incorporation of synthesizers into bridge designs introduces another set of characteristics that must be considered. The following parameters are relevant (beyond the normal parameters of output frequency, microwave output power, and reference clock frequency):

- spurious content
- amplitude noise
- absolute phase noise
- residual phase noise
- residual phase drift
- frequency resolution

Knowledge of the spurious content of the synthesizer output spectrum is important, because a spurious frequency could interfere with a frequency component of an EPR signal. For example, a spur at 100 kHz would be detrimental to detecting 100 kHz EPR signals in field modulation. The modulation frequency could be adjusted to place the spur in the rejection band of eventual filtering. In MQ EPR, one should check that none of the spurs align with the MQ EPR frequencies (if the main arms are separated by

10 kHz, then the frequencies of concern are 15, 25, 35, etc. kHz from the nominal mean frequency). The maximum spurious level of a synthesizer is normally specified by the manufacturer, but the frequencies at which the spurs occur are not specified and generally must be measured.

Amplitude noise, often stated through a noise floor specification, is usually less than phase noise in modern synthesizers, but should be checked for a given synthesizer. Phase noise is considered from two aspects: absolute phase noise and residual phase noise. Absolute phase noise is represented by the phase noise power density-to-carrier ratio vs. offset frequency under the small angle condition (Fantanas, 1992). In the context of synthesizers, residual phase noise is the uncorrelated phase noise between two synthesizers when they are locked to a common reference clock. The distinction is complicated in synthesizers because, for example in indirect synthesizers (Goldberg, 1999), there are frequency offset regions where either the absolute phase noise or the residual phase noise is dominant and regions where both are significant. Furthermore, these frequency offset regions are dependent on the particular synthesizer design. Impact of synthesizer phase noise on the spectrometer performance is dependent on the bridge operating mode. In general, a lower absolute phase noise vs. frequency offset for a synthesizer produces lower system noise levels. Hence, low phase noise synthesizers are clearly desirable. Absolute phase noise is normally specified for modern synthesizers, but residual phase noise specifications are rare and must generally be measured by the user.

Residual phase drift is the drift of the phase between two synthesizers with a common reference clock. The outputs of two synthesizers set to the same frequency have a nominal phase shift between them. If the synthesizer frequencies are the same, residual phase drift is the change of phase shift between the two sinusoidal signals and is easily measured (Bates, 1999). If the synthesizer frequencies are different, it is the change of phase shift from what the phase ought to be vs. time. The latter is more difficult to measure. Synthesizer frequencies must be mixed and then phase detected against a reference source that is also time-locked to the synthesizer reference clock. The reference source must have a phase drift that is known to be significantly less than the residual phase drift being measured. Often, the reference is at a lower frequency (the difference frequency between the two synthesizers) and satisfies this condition.

Residual phase drift is an unspecified parameter. It is important because, if significant, it can change the absorption/dispersion proportion while the scan is in progress. One must insure that the synthesizers are fully warmed up and their temperature is stable. It is insufficient that the synthesizers are connected to the wall power source – the reference clock may be warmed-up and stable, but not the remainder of the circuits. All synthesizers should be turned on and set to the desired frequencies. A minimum warm-up period is typically one day. Many users leave the synthesizers on at all times.

Resolution, i.e., frequency resolution, is the minimum frequency increment that can be set in the synthesizer. It is not a measure of absolute frequency accuracy (Yates, 1982). Resolution is significant to the extent that one wishes to resolve absolute nominal frequencies. For example, if an ideal sampling frequency of $4/7f_{IF}$ is desired, a higher resolution allows one to set the actual sampling frequency closer to the ideal. Frequency resolution is a standard synthesizer specification.

Other parameters, such as switching speed and modulation capability, may become significant as EPR applications with synthesizer-based bridges evolve.

Synthesizer performance has improved significantly over the past few decades. The original intent for synthesizers was to use them to generate several frequencies, often from one stable source, although sets of stable sources such as crystal oscillators, have been used (Smith, 1998). Generally, synthesizers are classified as either *direct* synthesizers, based on arithmetic generation of the output frequency from a reference frequency, or *indirect* synthesizers, based on the phase-locked loop (PLL). Analog and digital versions of both classes exist. Direct analog synthesizers are based on frequency mixing, division, multiplication or combinations thereof (Galani and Campbell, 1991). The direct digital synthesizer (DDS) has grown in prominence in the last decade with the ever-increasing capacity of digital circuits. It is based on digital circuits and digital-to-analog converters to generate the output signal (Pozar, 2001). Impressive phase noise performance of DDS synthesizers with good spurious specifications are now available: for a 0.01 to 3.0 GHz frequency range with at least 0.1 Hz resolution, the phase noise is -112, -132, and -155 dBc/Hz at 1, 10, and 100 kHz offsets, respectively, with –80 dBc non-harmonic spurious suppression (Stavenick, 2002).

Indirect analog synthesizers have a long history, but indirect digital synthesizers now dominate with the evolution of digital circuit technology. Indirect digital synthesizers with divide-by-N PLLs traditionally have elevated phase noise levels when fine frequency resolution is required (Dell'Aera and Riley, 2002). Multi-loop architectures exist to reduce the impact of the divide-by-N on phase noise (Goldberg, 1999), but the advent of the fractional-N PLL has resulted in significant phase noise reductions while maintaining fine frequency resolutions. The main difficulty with fractional-N PLL has been spurs, which have been significantly reduced recently through digital correction techniques (Owen, 2001). Dell'Aera and Riley present a succinct overview of integer-N and fractional-N synthesizers in Smith, 1998. Impressive phase noise performance of indirect digital synthesizers with good spurious specifications is now available: for a 0.675

to 1.35 GHz frequency range with a 0.1 Hz resolution, the phase noise is -115 dBc/Hz at 1 kHz offset and is –140 dBc/Hz at a 20 kHz offset and above, with –90 dBc non-harmonic spurious suppression (see IFR Systems Product Brochure).

These recent synthesizer improvements are providing direct system performance improvement in synthesizer-based microwave bridges. For example, previous synthesizer phase noise exceeded the phase noise of the microwave oscillator in the microwave bridge. Synthesizer phase noise is now equal to or lower than the phase noise of many fundamental microwave oscillators. Improved spur suppression and frequency resolution further promote the usefulness of synthesizers in microwave bridges. Synthesizer performance has improved and costs have decreased to the point where they are viable for routine incorporation into microwave bridges.

5.2 Image Rejection in Class 2 and 3 Receivers

Any superheterodyne receiver including Classes 2 and 3 receivers, Fig. 1, must take into consideration noise introduced by the image frequency band into the IF output of the signal mixer. The image frequency band consists of frequencies that, when mixed with the $f_O + f_{IF}$, will produce an output within the IF bandwidth at the mixer IF output port (Stremler, 1979). Noise and spurious signals in this band will elevate the noise at the output of the signal mixer, and should be suppressed. Increased noise degradation may occur in an EPR bridge due to reactive effects of the sample resonator on frequencies outside its bandwidth. In other words, noise originating in the main arm source around the image frequency will be reflected off the sample resonator into the signal LNA and mixed to the IF bandwidth by the signal mixer.

This additional noise can be diminished by bandpass IR filtering of the signal incident on the RF port of the signal mixer, or through the use of an IR mixer as the IF mixer. This type of mixer inherently rejects frequencies either above or below $f_{IF} + f_0$, depending on the internal phasing of quadrature hybrids used in its construction. It will operate over relatively wide RF and IF bandwidths (up to one octave), with typically 20 dB of image signal suppression (Maas, 1993).

The choice of suppression method depends on the relationship of the image frequency band to the incoming RF, desired IF frequency and bandwidth. For example, if the IF is very low compared to the incoming RF (say 10 MHz IF and 10 GHz RF), it may not be practical to construct a tunable bandpass filter that can remove the image frequency band over the full 1 GHz tuning range of the bridge. In this case, an IR mixer may be used to reduce at least 20 dB of noise in the receiver resulting from conversion of the image band into the IF. Where an appropriate filter can be constructed, considerably more rejection of image noise can be achieved. A double conversion scheme using conversion to a high and then to a lower IF may be used to improve image rejection over a wide bridge tuning range. The final IF can be relatively low, which may be mandated by analog input bandwidth considerations of the A/D converter.

5.3 Anti-Aliasing (aa) Filters

The anti-aliasing filter for the A/D converter in an IF sampling system such as Class 3 or 4 receivers, Fig. 1, must have a bandpass response. The maximum allowable signal bandwidth is limited by the sampling freqeuncy ($BW < f_S/2$). Frequencies outside this bandwidth must be rejected or they will alias into the passband in the A/D converter. This may be particularly troublesome if the aliasing frequencies are harmonically related to the desired signals, as would be the case in a greatly over-modulated EPR line. Out-of-band noise not attenuated by filtering will fold over into the passband, increasing the apparent noise level of the receiver. An engineering guideline is that the undesired frequencies and spurious signals should be attenuated to a level of less than one quantization level of the A/D converter, but greater attenuation may be desirable. The bandpass filter must not distort the phase and amplitude of the IF signal over the desired bandwidth in order to allow accurate digital detection of I and Q (absorption and dispersion).

The need for a sharp filter cutoff characteristic, and for uniform amplitude and group delay (implying linear phase response) may result in an unrealizable or very high order filter design (Zverev, 1967). There are several solutions to this problem. The sampling frequency of the A/D converter may be increased, resulting in oversampling of the desired signal bandwidth. This allows the band-reject specifications of the filter to be relaxed because the filter cutoff frequencies may be moved further from the center frequency, and less steep attenuation characteristics will be required to reject out-of-band signals to a level below the final resolution. Signals outside of the desired IF passband will not be aliased because they are actually within the wider Nyquist bandwidth implied by the higher sampling frequency. Secondly, amplitude and phase distortion can be corrected by signal processing after A/D conversion if the filter is adequately characterized over the passband.

5.4 A/D Converters

Table 2 lists A/D converters that have been considered by us for possible use for digital detection in EPR spectrometers. Of the entries in the table, we have considerable practical experience with the Harris/Intersil HI1276 and the Analog Devices AD6644 converters. The table provides a snapshot in late 2002 of a rapidly changing technology.

The Nyquist ratio of the bandwidth-to-sampling frequency of about ½ is used only in the HI1276 chip. All other entries in the table could be used for subsampling. Several of the entries have high ratios: about 3.5 for the Analog Devices products and 5 for the Texas Instruments ADS5422. These same devices also have the highest vertical resolution, 14 to 16 bits.

Table 2: A/D Converters for Digital Detection in EPR*

Manufacturer	Part #	Bits	Analog BW	f_S:MSPS	Use
Harris/Intersil	HI1276	8	300	500	baseband
Harris/Intersil	HI3276	8	250	160	subsampling
Maxim	MAX108	8	2200	1500	subsampling
Analog Devices	AD6644	14	250	65	subsampling
Analog Devices	AD6645	14	270	105	subsampling
Analog Devices	AD10677	16	200	65	subsampling
Fairchild/SPT	SPT7760	8	900	1000	baseband
Fairchild/SPT	SPT7871	10	180	100	subsampling
Texas Instruments	ADS5422	14	300	62	subsampling

* BW-bandwidth (3 db), f_S-sampling frequency, MSPS-megasamples per second

The Maxim MAX108 device has by far the highest analog bandwidth. A 2.2 GHz microwave signal can be introduced directly to the device and sampled at 1500 MSPS. For example, the sampling of the 2.2 GHz microwave frequency could be four samples in seven cycles, which would be convenient for Class 4 digital TLSS detection, Fig. 1. One can also imagine the use of this device for Class 3 digital TLSS detection based on an intermediate frequency of 2.2 GHz. This would result in the fastest possible time response for pulse EPR at X-band and higher microwave frequencies within the constraints of the entries in the Table.

We were attracted to the AD6644 chip in part because of its high vertical resolution. For digital detection in CW EPR, there is a potential risk of over-ranging the A/D converter in Class 3 designs because the IF carrier is too high. This device minimizes the risk. The fairly large analog bandwidth, 250 MHz, permits reasonable temporal response for pulse EPR using a relatively high intermediate frequency, noting that high IF facilitates design of the aa filter, Fig. 1. For subsampling with I-Q detection, one can collect a data point in each channel every 60 nanoseconds. This is 1/4 of f_s^{-1}. With the AD6645, this interval drops to 40 nanoseconds, and with the recently announced AD10677, the vertical resolution increases from 14 to 16 bits.

The number of A/D converters with sampling rates over 1 MSPS – including 67 devices from Texas Instruments alone – indicates their growth in popularity. Of the more than 40 chips with sample rates above 10 MSPS, only a handful have analog bandwidths below the sample rate, indicating that the target market for most of the chips is subsampling.

5.5 Data Processing

A number of factors combine to make archiving and processing possible. Subsampling (bandpass sampling) is a way of reducing data flow from the A/D to subsequent processing stages. The rates and bandwidths need only be sufficient to capture the desired information. Decimation, the process of sample rate reduction, can also be used for reducing data rates to a value high enough to capture the bandwidth required and low enough to be processed by the attached computer.

Device clock rates, particularly central processing units (CPU) and memory, have increased. The advent of GHz PCs, for example, pushed the development of memory technology as the CPUs became starved for data. Data buses also increased in clock speed. Peripheral Component Interconnect (PCI) bus clock speed, for example, doubled from 33 MHz to 66 MHz with PCI-eXtended (PCI-X) enhancements. At the same time, the width of the bus – the number of bits it carries in parallel – also doubled from 32 to 64 bits.

Storage rates of disk systems also increased due to increased bus speeds and Redundant Array of Inexpensive Disk (RAID) arrays. There are several ways in which RAID arrays can be used. The one of interest in our application writes a stream of data across several disks to reduce access delays. The RAID system we have in-house streams data to disk at 50 to 60 MB/sec. Raw data streaming is now near 200 MB/sec.

When the bus is wider, in terms of data bits, than the data word we are storing (14 bits for the AD6644), only clock speed and not bus width increases the rate at which our samples can be acquired. This is true in the typical PC architecture where the PCI bus and the path to disk are 32 bits or larger. Additional hardware or software is needed to pack the 8-16 bits output data of the A/D onto a wider bus. Most manufacturers have added multiple A/Ds operating in parallel to fill the bus width. These give very impressive throughput rates – hundreds of MB/sec – but this does not help our single channel application. As switched fabrics are used in place of conventional backplanes, throughput rates will rise because multiple transfers can take place in parallel. Switched fabrics are used in switches and routes for network applications. But unless the data is multiplexed, single channel acquisition speeds are not improved.

Several devices have been optimized for the communications industry that are applicable to our work. They include the subsampling A/Ds mentioned above as well as Digital Down Converters on a single chip. These are digital logic integrated circuits, several of which are made by both Intersil and Graychip, designed to connect directly to the A/D. They include a sine generator, quadrature mixer and filters, as well as data handling and formatting hardware. The Digital Signal Processor (DSP) chip is another such device. DSPs were designed to use wide data buses and have multiple processor units on the chip. They are optimized for the multiply-accumulate instructions needed in fast Fourier transforms. The C6xxx family from Texas Instruments, for example, has eight processors in parallel. This provides instruction rates above one Giga Floating Point Operation per second (GFLOP), with only a 167 MHz clock. With new products, DSP clock speeds are also rising.

Field Programmable Gate Arrays (FPGAs) are standard parts that can be programmed for specific functions. Their clock speeds and ability to handle complex algorithms allow them to replace DSPs in some applications. Pentek is providing an environment to make use of FPGAs in a development setting. They will even provide an FPGA programmed to perform FFTs. Xilink, a manufacturer of FPGA chips, has an intellectual property (ip) core that programs the FPGA to provide the functions of a Digital Down Converter. FPGAs can also perform data packing and improve data acquisition rates. Hardware and software tradeoffs are again blurred by the FPGAs.

Platforms designed for development of digital radio include most of the pieces mentioned above. These development stations designed for the commercial market have made it easier and less expensive for us to obtain a system useful for our objectives. Evaluation modules also make creating digital receiver systems possible while avoiding concerns with other issues such as power supply noise, ground planes and clock skew.

As hardware improves, software is following. Higher level programming is now easier to accomplish. For efficient implementations, DSPs previously required extensive hand coding in assembly language. The C language compiler in Texas Instruments (TI) Code Composer developer software is reported to generate assembly language instructions that not only make use of the parallel processors available in the DSP chip, but also produce code that is within 90% of the performance of what can be done by hand. National Instruments is developing an interface for their high level graphical program (LabVIEW) to allow control of the DSP and its integration into LabVIEW-based software systems. MATLAB from Mathworks, as well as similar software from other companies, allows creation of DSP algorithms and integrates them with the DSP chip.

6. CONCLUSION

Digital detection in EPR spectroscopy will become increasingly common in the years ahead. The benefits are primarily broadband detection, digital filtering and increased opportunities for data analysis. Real-time displays of data will become increasingly elaborate. For example, in CW EPR using field modulation, all harmonics for both I and Q could be displayed almost instantly. Raw data will be stored, permitting sophisticated off-line analysis. Specialized software for this analysis will be developed with plug-in capabilities that facilitate sharing of analysis programs.

At the present level of digital technology, Class 3 digital receivers are a practical approach: namely conversion to an intermediate frequency carrier that can be sampled in a time-locked manner. State-of-the-art A/D chips are often available on so-called "development boards" that are very flexible and well-suited for use in EPR spectrometers. A possible disadvantage for Class 3 digital detection is the cost of frequency synthesizers as well as subtle issues of noise generated during the numerous multiply and divide operations that are necessary to create a specific frequency. However, the quality of frequency synthesizers is steadily improving and the costs are dropping.

Sampling rates and bandwidths of A/D converters are increasing. Sampling rates of 10 GSPS (gigasamples per second) at 8-bit resolution are within reach, if not already realized. In a recent search of the Internet, we found a report of a proof-of-concept device with 8 GSPS. Direct Class 4 EPR detection at X-band will be a reality, it is predicted, within the next five to ten years. This will eliminate the need to generate the intermediate and sampling frequencies, thereby reducing the need for frequency synthesizers, increasing the bandwidth, lowering the cost, and possibly reducing noise levels.

Direct Class 4 digital detection at higher microwave frequencies, Q-band and above, probably will remain out of reach for a decade or more. Class 3 digital detection will be required for spectrometers operating at these frequencies, possibly using an intermediate frequency as high as X-band.

7. REFERENCES

Bates, P.C. (1999) A Simple Method Measures the Coherence of Phase Locked Devices, *RF Design*, July 1999, p. 70.

Berliner, L.J., Eaton, G.R. and Eaton, S.S. (2000). *Distance Measurements in Biological Systems by EPR , Biol. Magn. Reson.* **19**

Dell'Aera, S. and Riley, T. (2002) A Hybrid Fractional-N Synthesizer for Direct Modulation Applications, *Appl. Microwave & Wireless*, **14**:7, 34-39.

Fantanas, C. (1992) Introduction to Phase Noise, *RF Design*, August 1992, pp. 50-57.

Froncisz, W., Camenisch, T.G., Ratke, J.J. and Hyde, J.S. (2001) Pulse Saturation Recovery, Pulse ELDOR and Free Induction Decay EPR Detection Using Time-Locked Subsampling. *Rev. Sci. Instrum.* **72,** 1837-1842.

Galani, Z. and Campbell, R.A. (1991) An Overview of Frequency Synthesizers for Radars, *IEEE Trans. MTT*, **39**, 782-790.

Goldberg, B.-G. (1999) Analog and Digital Fractional-n PLL Frequency Synthesis: A Survey and Update, *Applied Microwave and Wireless*, **11**:6,32-42.

Hyde, J.S. (1979) Saturation Recovery Methodology. In Kevan, L. and Schwartz, R. N. (eds), Time Domain Electron Spin Resonance, pp. 1-30. Wiley & Sons, New York.

Hyde, J.S. (1998) Saturation Recovery. In Eaton, S. S., Eaton, G. R., and Salikhov, K. M. (eds), Foundations of Modern EPR, pp. 607-618. World Scientific Publ., New York.
Hyde, J.S., Froncisz, W. and Mottley, C. (1984) Pulsed ELDOR Measurement of Nitrogen T_1 in Spin Labels. *Chem. Phys. Lett.* **110**, 621-625.
Hyde, J.S., Pasenkiewicz-Gierula, M., Jesmanowicz, A., and Antholine, W. E. (1990). Pseudo Field Modulation in EPR Spectroscopy. *Appl. Magn. Reson.* 1, 483-496.
Hyde, J. S., Jesmanowicz, A., Ratke, J. J., and Antholine, W. E. (1992) Pseudomodulation: A Computer-Based Strategy for Resolution Enhancement. *J. Magn. Reson.* **96**, 1-13.
Hyde, J.S., Strangeway, R.A., Luglio, J., Mchaourab, H.S. and Froncisz, W. (1995) Noise in EPR Bridges with Multiple Time-Locked Microwave Frequencies, *Bull. Magn. Reson.*, **17**, 54-60.
Hyde, J.S., Mchaourab, H.S., Camenisch, T.G., Ratke, J J., Cox, R.W. and Froncisz, W. (1998). Electron Paramagnetic Resonance Detection by Time-Locked Subsampling. *Rev. Sci. Instrum.* **69**, 2622-2628.
IFR Systems Product Brochure (2001) 2040, 2041, 2042 Low Noise Signal Generator www.ifrsys.com, **4**, July 2001.
King, R.J. (1978) Microwave Homodyne Systems, Peregrinus Ltd. on Behalf of the Institution of Electrical Engineers, Herts, England.
Maas, S.A. (1993) Microwave Mixers, 2nd Ed., Artech House, Boston, p. 280.
Murugesan, R., Afeworki, M., Cook, J.A., Devasahayam, N., Tschudin, R., Mitchell J.B., Subramanian,, S. and Krishna M.C. (1998). A Broadband Pulsed Radio Frequency Electron Paramagnetic Resonance Spectrometer For Biological Applications. *Rev. Sci. Instrum.* **69**, 1869-1876.
Owen, D. (2001) Fractional-N Synthesizers, *Microwave Journal*, **44**:10, pp. 110-121.
Pozar, D.M. (2001) Microwave and RF Wireless Systems, Wiley & Sons, New York, pp. 268-271.
Schwieger, A. and Jeschke, G. (2001). *Principles Of Pulse Electron Paramagnetic Resonance,* Oxford University Press, New York.
Smith, J.R. (1998) Frequency Synthesizers, in *Modern Communication Circuits*, 2nd ed., WCB McGraw-Hill, Boston, pp. 407-411.
Stavenick, P. (2002) Synthesizers Offer Submicrosecond Switching, *Microwaves and RF*, June 2002, pp. 98-102.
Stormont, R.S., Anas, M.C., Pelc, N.J. (1991) U.S. Patent No. 4,992,736, Serial No. 289456, issued Feb. 12, 1991.
Strangeway, R.A., Mchaourab, H.S., Luglio, J., Froncisz, W. and Hyde, J.S. (1995) A General Purpose Multiquantum Electronic Paramagnetic Resonance Spectrometer, *Rev. Sci. Instrum.*, **66**, 4516-4528.
Stremler, F.G. (1979) Introduction to Communication Systems, Addison-Weseley Publishing Co., Reading, MA, p. 225.
Subramanian, S., Murugesan. R., Devasahayam, N. Cook, J.A., Afeworki, M., Pohida, T., Tschudin, R.G., Mitchell, J.B. and Krishna, M.C. (1999). *J. Magn. Reson.* **137**, 379-388.
Yates, W. (1982) Meeting Today's Stringent Requirements: Synthesized Signal Generators, *Electronics Products*, Oct. 25,1982, pp. 71-74.
Zverev, A.I. (1967) Handbook of Filter Synthesis. Wiley & Sons, New York.

Chapter 8

Measurement of Distances Between Electron Spins Using Pulsed EPR

Sandra S. Eaton and Gareth R. Eaton
Department of Chemistry and Biochemistry, University of Denver, Denver, Colorado 80208

Abstract: Distances between unpaired electrons ranging from ca. 15 Å to > 50 Å can be measured by pulsed electron paramagnetic resonance (EPR) techniques. Techniques are available to measure distances between two slowly relaxing centers or between a rapidly relaxing center and a slowly relaxing center. An overview of these methods is provided with an emphasis on recent examples that demonstrate the power of these techniques.

1. INTRODUCTION

Fundamental to current thinking about molecular biology is the relationship between structure and function, and the time-dependence of both. How do proteins fold, how do molecules assemble into multi-molecular units, how do species cross membranes? Answers to these, and similar questions, require measurement of distances between sites in a protein and the time dependence of those distances. EPR provides a unique insight into distances between locations in biological systems and can be applied to any system with two paramagnetic centers, provided the distance falls within the ranges discussed below. There is no need to have single crystals, as for x-ray crystallography, and no limit on molar mass, as for NMR spectroscopy. Since the electron magnetic moment is much larger than the nuclear magnetic moment, much larger distances can be probed by EPR than by NMR. Samples do not need to be in solution; they may be suspensions or films. Some systems have two intrinsic paramagnetic centers. Alternatively, site-directed spin labeling can be used to place nitroxyl radicals at the locations one wants to study on a single molecule, or in an assembly of molecules, and the distance between the spin labels can be measured as a function of relevant parameters. Metal sites also can be introduced by site directed mutagenesis (Regan, 1993; Voss *et al.*, 1995; Lu and Valentine, 1997; Lu *et al.*, 2001). Some disadvantages of EPR for distance measurements are that distances must be studied one-by-one and that the probes have significant size and flexibility that may cause uncertainty in relating spin-spin distance to the distance between protein backbone atoms. However, molecular modeling and increased understanding of probe conformations hold promise for interpreting the interspin distances (Borbat *et al.*, 2002). Also, the pulse methods discussed in this chapter require immobilization of the sample both to prevent motional averaging of small dipolar couplings and to make spin echo dephasing times as long as possible.

Many CW and pulsed EPR methods for measuring distances between unpaired electrons are reviewed in *Biological Magnetic Resonance*, vol 19 (Berliner *et al.*, 2000). As discussed in the introductory chapter in that volume (Eaton and Eaton, 2000a), a range of CW methods can be used to determine distances up to about 20 Å, depending upon the linewidths in the spectra. Pulse methods are required to measure longer distances. Volume 19 includes discussions, with references to the literature through 1999, of distance measurements by saturation recovery (Eaton and Eaton, 2000c; Lakshmi and Brudvig, 2000), double quantum ESR (Borbat and Freed, 2000), the 2+1 pulse sequence (Raitsimring, 2000), and the out-of-phase echo (Dzuba and Hoff, 2000). This chapter provides an overview of distance measurements by pulse methods (Table 1), with an emphasis on recent examples that demonstrate the power of these techniques.

2. FUNDAMENTAL PRINCIPLES OF INTERACTION BETWEEN ELECTRON SPINS

Measurement of the distance between paramagnetic centers relies upon determination of the dipole-dipole interaction, which is a through-space interaction. A complete description of the dipolar interaction requires inclusion of g anisotropy. However, if it is assumed that the g values are isotropic, the dipolar splitting of an EPR signal can be expressed in terms of the parameter D, where D is the splitting of the signal, in the limit of strong exchange interaction, when the interspin vector is perpendicular to the external magnetic field and –2D is the splitting when the interspin vector is parallel to the external magnetic field (Luckhurst, 1976). In a randomly oriented sample, this results in a classic "Pake pattern" where, in the ideal case, the value of D can be read from the splitting between the intense perpendicular turning points.

Table 1. Summary of Pulsed EPR Methods for Determining Electron-Electron Distances[a]

Method	R_{exp} [b] (Å)	R_{max} [c] (Å)	comments	reference
2+1 pulse sequence	30-60	60		Raitsimring, 2000
double-quantum coherence	20-47	60		Borbat and Freed, 2000; Borbat *et al.*, 2002
SIFTER	28-50	50		Jeschke, 2000b
3-pulse DEER (PELDOR)	15-80[d]	80		Milov *et al.*, 1998
4-pulse DEER	15-66	80		Jeschke *et al.*, 2000a; Jeschke, 2002
out-of-phase echo	25-40	40	pair must be spin-correlated	Bittl and Zech, 2001
change in T_1	12-40	40	longer distances for faster relaxing metals	Eaton and Eaton 2000c; Lakshmi and Brudvig, 2000
change in T_m	12-30	40	characteristic temperature dependence	Eaton and Eaton, 2000c
selective hole burning	25-50			Dzuba and Kawamori, 1996

[a]In this table, as in the text, slow and fast relaxation times are relative to the dipolar splitting expressed in frequency units. [b] Values of r that have been measured with a particular technique. [c] Maximum values of r that have been proposed as measurable by a particular technique. [d] Some early measurements claimed to measure distances as long as 100 Å, but these values are judged to be unreliable.

$$D\,(\text{erg}) = \frac{3g^2\beta^2}{2r^3} \text{ and r is in cm.} \qquad (1)$$

Conversion of units from erg to gauss gives

$$D\,(\text{gauss}) = \frac{3g\beta}{2r^3} = 1.39\text{x}10^4\,\frac{g}{r^3} \qquad (2)$$

where r is in Å, the g value for one of the unpaired electrons is assumed to be 2.00 and the g value for the second unpaired electron is entered explicitly.

D can also be given in the following units.

$$D\,(\text{MHz}) = \frac{-78000}{r^3}\text{, where r is in Å, and g is assumed to be 2,} \qquad (3)$$

$$\text{and } D\,(\text{mT}) = \frac{-2786}{r^3}\text{, and r is in Å, and g is assumed to be 2.} \qquad (4)$$

To calibrate our thinking about the ways in which dipolar couplings are measured, it is useful to consider the magnitude of D for various interspin distances as shown in Table 2.

Table 2. Magnitude of D as a function of interspin distance

r (Å)	D (gauss)	2D/3 (gauss)
5	222	148
10	28	19
15	8.2	5.5
20	3.5	2.3
40	0.43	0.29
60	0.13	0.09

When the exchange interaction is less than the separation between the resonance frequencies for the two paramagnetic centers (the weak exchange limit), the dipolar splitting that is observed for a Pake pattern is 2D/3, which is designated by some authors as d (Jeschke, 2002). When d is expressed in frequency units it is called the dipolar frequency. As discussed below, the weak exchange limit pertains at the distances currently measured by pulsed methods.

Electron-electron interaction also has an exchange contribution that depends upon the overlap of the orbitals that contain the unpaired electron (Coffman and Buettner, 1979; Eaton and Eaton, 1988). Measurement of a distance requires separation of the exchange and dipolar contributions. Since electron delocalization is strongly dependent on the electronic structure of the paramagnetic center, it is difficult to make generalized predictions concerning the distance dependence of the exchange interaction. Coffman and Buettner (1979) proposed a "limiting" function that predicted the longest distance at which exchange interaction of a particular magnitude would be observed. Other limiting functions for exchange interaction are discussed by Jeschke (2002). Depending upon the data that are considered, these models indicate that exchange interactions become much smaller than dipolar interactions at distances greater than 10 to 15 Å. As discussed below, pulse measurements currently are used for distances greater than about 15 Å so exchange interactions usually can be neglected in analyzing the spin-spin interactions measured by pulsed techniques. Two types of systems that may not fit these generalizations are unpaired electrons in delocalized molecular orbitals (Eaton and Eaton, 2000a) and proteins that are optimized for electron transfer.

Nitroxyl radicals with normal isotopic abundance typically have frozen-solution (powder) linewidths of 6-8 gauss. Replacement of hydrogen with deuterium narrows the lines by about a factor of two. Comparison of these widths with values of 2D/3 (Table 1) indicates that for distances greater than about 20 Å the dipolar interaction is small compared with typical spin label linewidths, which puts an upper limit on distances that can be measured by CW EPR. In contrast, typical values of the spin-echo dephasing time constant, T_m, for nitroxyl radicals at temperature below about 80 K are about 2 μs, which corresponds to a spin packet linewidth of about 30 mG. Dipolar interactions that are small compared to CW linewidths are significant compared with spin-packet linewidth, which makes pulse techniques advantageous for measurements of longer interspin distances. In addition, the spin-lattice relaxation time, T_1, is much longer than T_m in typical immobilized samples of nitroxyl radicals. Consequently, changes in either T_m or T_1 can be sensitive indicators of spin-spin interactions in distance regimes where CW lineshape changes are too small to detect.

3. DISTANCE BETWEEN TWO SLOWLY RELAXING CENTERS

A variety of pulse sequences have been developed to measure dipolar interactions between slowly relaxing centers. The dipolar coupling is proportional to r^{-3} (eq. 1) so the dipolar frequencies observed in these experiments vary as r^{-3}. Experiments that use a single microwave frequency include the "2+1" sequence (Raitsimring, 2000), single-quantum coherence (Borbat and Freed, 2000), and the SIFTER sequence (Jeschke, 2000b). A comparison of these techniques was provided by Jeschke (2002). To separate the excitation of the "observed" and "neighboring" coupled spins, pulse sequences have been developed that use two microwave frequencies. The terms 3-pulse DEER (double electron-electron resonance) and PELDOR (pulsed electron-electron resonance) are used by different groups to refer to the same experiment in which the amplitude of a 2-pulse spin echo is perturbed to varying extents by an additional pulse at a second microwave frequency, for which the timing relative to the other pulses is varied. These 3-pulse experiments (as in the 2+1 sequence) have an experimental deadtime that prevents observation of the rapid loss of coherence that is characteristic of short interspin distances. This deadtime is avoided in the 4-pulse DEER experiments (Jeschke *et al.*, 2000a). Spiess and co-workers (Jeschke, 2002) have applied this tool to studies of synthetic polymers, but their results are directly transferable to biopolymers and assemblies of biomolecules. An important advantage of the DEER and double quantum coherence methods is that the modulation arises only from dipolar coupled spins so these measurements are less susceptible to interference from singly-labeled protein than CW measurements. Another very important aspect of the analysis of the data obtained by these pulse measurements is the development of methods to determine not just distances, but also distributions in these distances (Jeschke *et al.*, 2002; Pannier *et al.*, 2000).

A special case occurs for spin-correlated pairs as in light-induced radical pairs, for which a phase shifted "out-of-phase" spin echo signal can be observed (Bittl and Zech, 2002). The echo exhibits intensity modulation as a function of the time between the pulses. The frequency of this modulation is characteristic of the distance between the two centers.

4. DISTANCE BETWEEN A SLOWLY RELAXING CENTER AND A RAPIDLY RELAXING CENTER

Dipolar coupling between a slowly relaxing center and a more rapidly relaxing center, typically a metal ion, enhances the spin lattice relaxation for the slowly relaxing center (Eaton and Eaton, 2000c; Lakshmi and Brudvig, 2000). Long-pulse saturation recovery (see chapter 1) is the method of choice for measuring spin-lattice relaxation rates for the slowly relaxing center in the absence and presence of a rapidly relaxing metal, because the long pulses can mitigate the effects of competing spectral diffusion processes including nuclear spin relaxation and cross relaxation (Eaton and Eaton, 2000b; Harbridge *et al.*, 2003). When the relaxation rate for the slowly relaxing center in the absence of interaction, and the relaxation rate for the metal are known, the interspin distance and relative orientations of the magnetic axes for the interacting spins are the only adjustable parameters and the effect of the metal on the saturation recovery curves for the radical can be simulated to determine the interspin distance (Zhou *et al.*, 2000). When the relaxation rate for the metal is not known, distances can be determined by comparison of the relaxation enhancement with that for similar systems for which the metal-radical distance is known. The perturbation of the electron spin relaxation of the radical by the metal depends on the square of the dipolar matrix elements and therefore varies as r^{-6}.

When the metal relaxation rate is comparable to the dipolar splitting expressed in frequency units, the electron spin relaxation of the metal is an effective spin-echo dephasing mechanism for the slowly relaxing center, analogous to intermediate exchange in NMR (Eaton and Eaton, 2000c). The rate of 2-pulse spin echo dephasing is enhanced and echo amplitude at constant pulse spacing is decreased. These effects can be analyzed to determine the interspin distance.

5. SOME PRACTICAL CONSIDERATIONS

For most of the pulse techniques discussed in this chapter, spin concentrations in the range of 0.2 to 0.5 mM are optimal for distances up to about 50 Å. Lower concentrations can be used, but may require extensive signal averaging. As the target distance becomes longer, it becomes necessary to use increasingly low sample concentrations to ensure that the intramolecular spin-spin distance of interest is significantly shorter than the average random distance between spins in the sample. A fundamental limit will be the spectrometer sensitivity at these low spin concentrations. The spin echo dephasing time, T_m, also contributes to limitations. As the interspin distance increases, the dipolar frequency decreases, which means that the period for the corresponding oscillation increases. The longer the period of the oscillation, the longer the time that is required between the pulses that form the echo or detect the coherence. The shorter the value of T_m the more the echo intensity decreases as the time between the pulses is increased. Thus the echo intensity becomes smaller for the same long interspin distances that require low concentrations to minimize intermolecular interactions. In favorable situations, however, it seems feasible to measure distances as large as 80 Å. Measurements of shorter distances (closer to 20 Å) can be made more readily at lower concentrations than measurements of long distances, because the higher dipolar frequencies can be adequately defined with shorter interpulse spacings.

When a solvent crystallizes there is a tendency to form regions with locally-high solute concentration, which enhances intermolecular spin-spin interaction. To minimize solvent crystallization, solvents or solvent mixtures are selected that form glasses, or cryoprotectants such as sucrose or glycerol are added to water to decrease the tendency to crystallize.

Although each of the pulse methods was developed in a research lab using locally-available instrumentation, many of the methods now can be performed on the commercially-available Bruker Elexsys spectrometers. Realistically, a person who is exploring the use of one or more of these pulse methods for measuring distances would be well advised to take samples to a lab that has some experience with the method to obtain their initial results.

6. RECENT EXAMPLES FOR DISTANCES BETWEEN TWO SLOWLY RELAXING RADICALS

6.1 Spin-Labeled T4 Lysozyme (Borbat *et al.*, 2002)

Eight double-cysteine mutants of T4L were spin-labeled with methanethiosulfonate spin label (MTSSL). Sucrose was added to the solutions as a cryoprotectant. Interspin distances were measured with the 6-pulse double quantum coherence sequence, which is essentially dead-time free. Due to limitations on the accessible temperature range for the spectrometer, experiments at Ku-band (17.35 GHz) were performed at ca. 200 K which is a local maximum in the temperature dependence of nitroxide T_m's. The short T_m's (300 to 350 ns) at this temperature limited the Ku-band experiments to samples with shorter distances. For the double quantum coherence experiments, optimal results are obtained when B_1 excites the full dipolar coupled spectrum, which requires high-power microwave amplifiers and resonators with large B_1 per square root of watt incident on the resonator. The higher B_1 that was available on the Ku-band spectrometer (about 30 G), relative to the B_1 available at X-band, was important for characterization of the shorter interspin distances. Experiments at X-band (9.2 GHz) were performed at 77 K where T_m is about 4 μs. The available B_1 on the X-band system (about 11 G) was adequate for samples with interspin distances greater than 30 Å. Fourier transformation of the double quantum coherence signals gave the dipolar spectrum. Distinctive features in the shapes of the dipolar spectra permitted characterization of distribution widths and characterization of multiple conformations of the spin labels. Average distances between 29 and 47 Å with distribution widths of 1.0 to 2.7 Å were observed. It was shown that a relatively small number of these long distance constraints could define the three-dimensional conformation of the protein.

6.2 Conformation of Doubly Spin-Labeled Peptide (Milov *et al.*, 2001)

An analog of the antibiotic trichogin GA IV was prepared that contained two spin labels, 7 amino acids apart. The spin label was TOAC, which is an analog of the amino acid α-aminoisobutyric acid, and contains less flexible linkages than the more commonly-used MTSSL label. Three-pulse DEER was used to characterize the conformation of the peptide in glassy toluene-chloroform solutions at 77 K. Experiments were performed at X-band with a bimodal resonator, using microwave frequencies separated by about 100

MHz. For the DEER experiments B_1 typically is selected such that there is minimal overlap between the bandwidth of spins observed at ν_1 and pumped at ν_2. To permit distinctions between intramolecular and intermolecular electron-electron spin-spin interaction, the spin-labeled peptide was diluted with varying ratios of unlabeled protein and the total peptide concentration was varied. At high dilutions with unlabeled protein, characteristic DEER oscillations were observed that correspond to an interspin distance of 15.7 Å, which is consistent with a 3_{10}-helix conformation of the peptide. Comparison of the amplitudes of rapidly decaying and oscillating contributions to the DEER signal for peptide aggregates indicated that about 19% of the peptides adopted the conformation with this interspin distance. The DEER results were consistent with a model in which four of the amphiphilic peptides form a structured aggregate.

6.3 Characterization of Clusters in Ionomers (Pannier *et al.*, 2000)

The 4-pulse DEER experiment was used to characterize the size of clusters and distance between clusters in poly(isoprene) polymers with sulfonate end groups and for poly(styrene)-poly(isoprene) diblock polymers with sulfonate end groups on the poly(isoprene) chains. The 4-pulse DEER experiment was important for this application because the absence of a deadtime permitted characterization of the short interspin distances within the clusters. The potassium salt of 4-carboxy-tempo (K-tempo) was added in a ratio of two spin labels per 15 ionic end groups. The immobilization of the spin label shown by CW EPR indicated that the charged nitroxyls were associated with the charged endgroups of the polymers. The DEER experiments were performed at X-band at 15 K on a Bruker E380E spectrometer using an ENDOR resonator that was overcoupled to Q ~ 100 to give a bandwidth that was large enough to accommodate two microwave frequencies that differed by 50 – 70 MHz. The DEER signals for poly(isoprene) exhibited three components that could be modeled by two Gaussian distance distributions with mean values of 15 – 20 Å and 40 to 70 Å, and a uniformly distributed background. The 15 – 20 Å distribution was attributed to interacting spins within a cluster, which therefore defined the size of the clusters. The distribution with longer distances defined the intercluster distances. The intercluster distance was consistent with prior results from small angle X-ray scattering (SAXS) and the distance within a cluster was consistent with molecular modeling calculations. The DEER method also gave plausible results for the diblock polymer, for which attempts to measure the intercluster distance by SAXS were unsuccessful.

This study demonstrates the utility of 4-pulse DEER in characterizing heterogeneous systems.

6.4 Separation between spins radicals pairs of the Photosystem I (Bittl and Zech, 2001)

Modulation of the out-of-phase echo is a powerful technique to determine the distance within the charge-separated pair, $P_{700}^{\bullet+}A_1^{\bullet-}$. Detection of the spin echo is synchronized with laser pulses that generate the radical pair. Subsequent X-ray crystallographic results validated the distance of 28.4 Å obtained by this technique. This is a convenient technique for monitoring structural changes that occur (or do not occur) due to site-directed mutagenesis or quinone substitution. A detailed discussion of this technique is given in Dzuba and Hoff (2000).

7. RECENT EXAMPLES FOR DISTANCES BETWEEN A RAPIDLY RELAXING AND A SLOWLY RELAXING SPIN

7.1 Spin-Labeled High-Spin Metmyoglobin Studied by Saturation Recovery (Zhou *et al.*, 2000)

Calculations of the distance between a rapidly relaxing metal ion and a neighboring organic radical based on enhancement of spin lattice relaxation rates have used the Bloembergen equation and its modifications (Eaton and Eaton, 2000c; Lakshmi and Brudvig, 2000). However, the Bloembergen equation was derived for nuclear spins, so the zero-field splitting (ZFS) that is present for metals with $S > \frac{1}{2}$ was not included. An approach analogous to the derivation of the Bloembergen equation was performed, including the additional energy splittings that arise from the zero-field splitting, for a metal with $S = 5/2$ and ZFS much greater than the EPR quantum. An additional complication arises for Kramers' ions with $S > \frac{1}{2}$ and large ZFS in that the observable EPR transitions are between the $m_s = \pm \frac{1}{2}$ spin states. Since the metal spin-lattice relaxation rates are expected to be different for different values of m_s, measurement of relaxation rates for the observed transitions will not be representative of all the metal spin states interacting with the slowly relaxing spin. The approach that was taken in this study was to fix the interspin distance for one spin-labeled high-spin met-myoglobin variant at the value obtained for the low-spin analog and treat the metal relaxation rate as the adjustable parameter in simulating long-pulse saturation recovery curves for the interacting spin label. The resulting values of the iron relaxation rates were systematically faster than values obtained by simulation of the temperature dependent contribution to the CW lineshapes for the high-spin Fe(III) EPR spectra. When these calculated iron relaxation rates were used to obtain interspin distances by analysis of long-pulse saturation recovery curves for a set of 12 spin-labeled metmyoglobin variants, preliminary results indicate reasonable agreement in distances between high-spin and low-spin variants for distances between about 16 and 30 Å (Ulyanov *et al.*, unpublished).

7.2 Spin-Labeled High-Spin Metmyoglobin Studied by Spin Echo (Ulyanov *et al.*, unpublished)

When the relaxation rate for the rapidly relaxing spin is comparable to the magnitude of the electron-electron dipolar coupling in frequency units, the metal relaxation is an effective dephasing mechanism for the two-pulse spin echo of the slowly relaxing center (Rakowsky *et al.*, 1998; Eaton and Eaton, 2000c). These dramatic effects on spin echo decays have been demonstrated (Seiter *et al.*, 1998), but the analysis of the full decay curves to determine the interspin distance is tedious. The enhanced rate of dephasing results in decreased echo intensity at each point along the decay curve. For a fixed timing of the two-pulse echo sequence, as the rate of relaxation for the fast relaxing spin increases with increasing temperature, the echo intensity for the slowly relaxing spin goes through a minimum. Calculations predict that the minimum echo intensity observed as a function of temperature decreases as the interspin distance decreases (Eaton and Eaton, 2000c). For a series of spin-labeled variants of low-spin cyano-metmyoglobin, preliminary results show that the minimum nitroxyl echo intensity correlates well with the interspin distance obtained by analysis of long-pulse saturation recovery curves (Figure 1). The advantage of the spin echo measurements relative to the saturation recovery measurements is that values of the iron relaxation rates as a function of temperature are not required to calculate the interspin distance from the spin echo intensity. This method shows considerable promise for distance measurements.

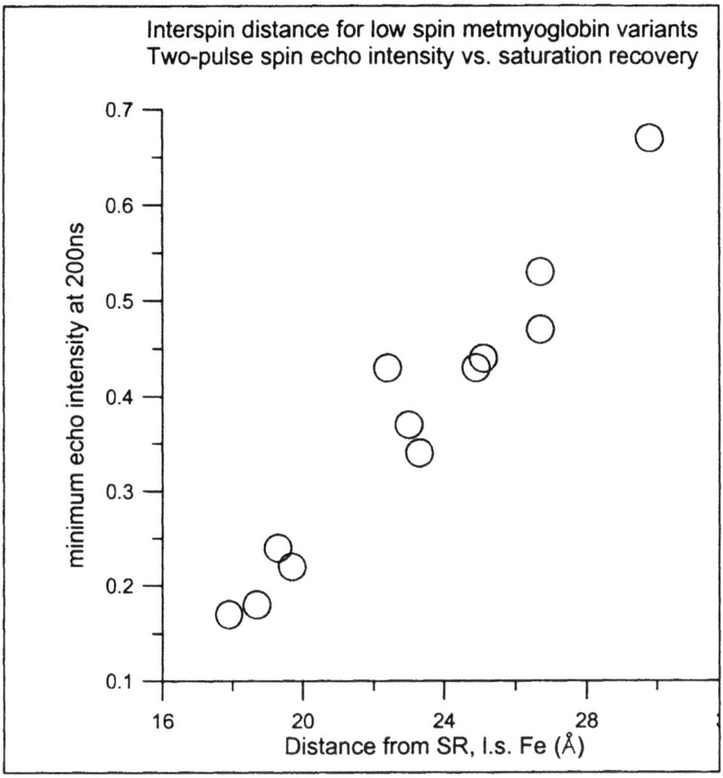

Figure 1. Correlation between minimum nitroxyl two-pulse spin echo intensity as a function of temperature for a pulse spacing of 200 ns, and interspin distance determined by analysis of long-pulse saturation recovery curves (Ulyanov *et al.*, unpublished).

8. PROGNOSIS

Interspin distances in the range of 20 to about 50 Å determined by pulsed EPR provide important constraints on structural models in a particularly useful distance range. As computational techniques become more powerful in predicting secondary structure, it may become possible to distinguish between postulated tertiary structures based on a few long-distance constraints. One area in which we expect many applications of EPR distance measurements is assemblies of subunits. For example, with appropriate spin labeling, it will be possible to determine whether one or another portion of a molecule associates with a particular region of another molecule in an assembly. Similarly, structural changes could be measured during successive steps in a biological process.

9. ACKNOWLEDGMENTS

Our studies of distances in biomolecules are supported by NIH grant GM21156.

10. REFERENCES

Berliner, L. J., Eaton, S. S., and Eaton, G. R., eds. (2000). Distance Measurements in Biological Systems by EPR, *Biol. Magn. Reson.* **19**.

Bittl, R., and Zech, S. G. (2001). Pulsed EPR Spectroscopy On Short-Lived Intermediates In Photosystem I, *Biochim. Biophys. Acta* **1507**, 194-211.

Borbat, P. P. and Freed, J. H. (2000). Double-Quantum ESE and Distance Measurements Distance Measurements in Biological Systems by EPR, L. J. Berliner, S. S. Eaton, and G. R. Eaton, eds., *Biol. Magn. Reson.* **19**, 383-459.

Borbat, P. P., Mchaourab, H. S., and Freed, J. H. (2002). Protein Structure Determination Using Long-Distance Constraints from Double-Quantum Coherence ESR: Study of T4 Lysozyme, *J. Am. Chem. Soc.* **124**, 5304-5314.

Coffman, R. E., and Buettner, G. R. (1979). A Limit Function for Long-Range Ferromagnetic and Antiferromagnetic Superexchange. *J. Phys. Chem.* **83**, 2387-2392.

Dzuba, S. A. and Hoff, A. J. (2000). Photo-Induced Radical Pairs Investigated using Out-of-Phase Electron Spin Echo, *Biol. Magn. Reson.* **19**, 569-596.

Dzuba, S. A., and Kawamori, A. (1993). Selective Hole Burning: Spectral Diffusion and Dipolar Broadening, *Concepts in Magnetic Resonance* **8**, 49-61.

Eaton, G. R. and Eaton, S. S. (1998). EPR Studies of Long-Range Intramolecular Electron-Electron Exchange Interaction, *Accts. Chem. Res.* **21**, 107-113.

Eaton, S. S. and Eaton, G. R. (2000a). Distance Measurements by CW and Pulsed EPR in Distance Measurements in Biological Systems by EPR, L. J. Berliner, S. S. Eaton, and G. R. Eaton, eds., *Biol. Magn. Reson.* **19**, 2-21.

Eaton, S. S. and Eaton, G. R. (2000b). Relaxation Times of Organic Radicals and Transition Metal Ions in Distance Measurements in Biological Systems by EPR, L. J. Berliner, S. S. Eaton, and G. R. Eaton, eds., *Biol. Magn. Reson.* **19**, 29-154.

Eaton, S. S., and Eaton, G. R. (2000c). Determination of Distances Based on T_1 and T_m. in Distance Measurements in Biological Systems by EPR, L. J. Berliner, S. S. Eaton, and G. R. Eaton, eds., *Biol. Magn. Reson.* **19**, 347-381.

Harbridge, J. R., Eaton, S. S., Eaton, G. R. (2003). Electron Spin-Lattice Relaxation Processes of Radicals in Irradiated Crystalline Organic Compounds, *J. Phys. Chem. A* **107**, 598-610.

Jeschke, G. (2002) Determination of the Nanostructure of Polymer Materials by Electron Paramagnetic Resonance Spectroscopy, *Macromol. Rapid Commun.* **23**, 227-246.

Jeschke, G., Pannier, M., Spiess, H. W. (2000a). Double Electron-Electron Resonance in Distance Measurements in Biological Systems by EPR, L. J. Berliner, S. S. Eaton, and G. R. Eaton, eds., *Biol. Magn. Reson.* **19**, 493-512.

Jeschke, G., Pannier, M., Godt, A., and Spiess, H. W. (2000b). Dipolar Spectroscopy and Spin Alignment in Electron Paramagnetic Resonance, *Chem. Phys. Lett.* **331**, 243-252.

Jeschke, G., Koch, A., Jonas, U., Godt, A. (2002). Direct Conversion of EPR Dipolar Time Evolution Data to Dipolar Distances, *J. Magn. Reson.* **155**, 72-82.

Lakshmi, K. V., and Brudvig, G. W. (2000). Electron Paramagnetic Resonance Distance Measurements in Photosynthetic Reaction Centers in Distance Measurements in Biological Systems by EPR, L. J. Berliner, S. S. Eaton, and G. R. Eaton, eds., *Biol. Magn. Reson.* **19**, 493-512.

Lu, Y., and Valentine, J. S. (1997). Engineering Metal-Binding Sites in Proteins, *Curr. Opin. Struct. Biol.* **7**, 495-500.

Lu, Y. S., Berg, S. M., and Pfister, T. D. (2001). Engineering Novel Metalloproteins: Design of Metal-Binding Sites into Native Protein Scaffolds, *Chem. Rev.* **101**, 3047-3080.

Luckhurst, G. R. (1976). Biradicals as Spin Probes in Spin Labeling: Theory and Applications, L. J. Berliner, ed., Academic Press, N. Y., ch. 4.

Milov, A. D., Maryasov, A. G., and Tsvetkov, Y. D. (1998). Pulsed Electron Double Resonance (PELDOR) and its Applications in Free Radicals Research. *Appl. Magn. Reson.* **15**, 107-143.

Milov, A. D., Tsvetkov, Yu. D., Formaggio, F., Crisma, M., Toniolo, C., Raap, J. (2001). The Secondary Structure of a Membrane-Modifying Peptide in a Supramolecular Assembly Studied by PELDOR and CW-ESR Spectroscopies, *J. Am. Chem. Soc.* **123**, 3784-3789.

Pannier, M., Schädler, V., Schöps, M., Wiesner, U., Jeschke, G., Spiess, W. (2000). Determination of Ion Cluster Size and Cluster-to-Cluster Distances in Ionomers by Four-Pulse Double Electron Electron Resonance Spectroscopy, *Macromol.* **33**, 7812-7818.

Rakowsky, M. H., Zecevic, A., Eaton, G. R., and Eaton, S. S. (1998). Determination of High-Spin Iron(III)-Nitroxyl Distances in Spin-Labeled Porphyrins by Time-Domain EPR. *J. Magn. Reson.* **131**, 97-110.

Ratsimring, A. (2000). "2+1" Pulse Sequence as Applied for Distance and Spatial Distribution Measurements of Paramagnetic Centers, in Distance Measurements in Biological Systems by EPR, L. J. Berliner, S. S. Eaton, and G. R. Eaton, eds., *Biol. Magn. Reson.* **19**, 461-491.

Regan, L. (1993). The Design of Metal-Binding Sites in Proteins. *Ann. Rev. Biophys. Biomol. Struct.* **22**, 257-281.

Seiter, M., Budker, V., Du, J.-L., Eaton, G. R., and Eaton, S. S. (1998). Interspin Distances Determined by Time Domain EPR of Spin-Labeled High-Spin Methemoglobin. *Inorg. Chim. Acta* **273**, 354-366.

Ulyanov, D., Bowler, B. E., Eaton, G. R., and Eaton, S. S. (to be published).

Voss, J. L., Hubbell, W. L., and Kaback, H. R. (1995). Distance Determination in Proteins using Designed Metal Binding Sites and Site-Directed Spin Labeling: Application to the Lactose Permease of *Escherichia Coli*, *Biochemistry* **34**, 6272-6277.

Zhou, Y., Bowler, B. E., Lynch, K., Eaton, S. S. and Eaton, G. R. (2000). Interspin Distances in Spin-Labeled Metmyoglobin Variants Determined by Saturation Recovery EPR. *Biophys. J.* **79**, 1039-1052.

II

Motion, Proteins, and Membranes

Chapter 9

ESR and Molecular Dynamics

Jack H. Freed
Department of Chemistry, and Chemical Biology, Baker Laborator, Cornell University, Ithaca, New York 14853-1301

Abstract: The development of ESR for the study of spin-relaxation and molecular dynamics of organic radicals and spin labels in fluids is reviewed from a historical perspective.

1. MOTIONAL NARROWING AND ORGANIC RADICALS

My interest in electron spin relaxation and molecular dynamics began when I was a graduate student with George Fraenkel at Columbia University from 1958-1962. His laboratory was teeming with interest and activity in the area of spin-relaxation, mainly of organic free radicals in liquid solution. Fraenkel had developed a new theory (Stephen and Fraenkel, 1960) which could successfully account for the fact that the measured T_1's from each hyperfine line of semiquinone spectra obtained in his lab were different in magnitude (Schreurs and Fraenkel 1961). Kivelson was completing his theory of unsaturated linewidths, (Kivelson, 1960) that he developed from the seminal Kubo and Tomita theory of lineshapes, (Kubo and Tomita 1954). The Stephen-Fraenkel theory deriving more from the Wangness-Bloch (1953) and Redfield (1957) theories, (more commonly known as Redfield theory today) also incorporated components of Kubo and Tomita theory. In fact, the Stephen-Fraenkel theory of ESR (electron-spin resonance) saturation and the Kivelson theory of unsaturated linewidths were complementary in the insight and understanding they provided into spin-relaxation of organic radicals in solution.

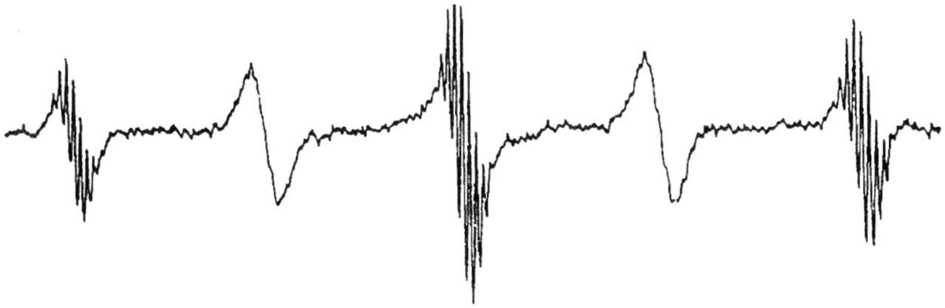

Figure 1. ESR spectrum of *p*-dinitrotetramethylbenzene radical at 20°C showing the alternating linewidth effect. (Magnetic field increases to the right). From Freed and Fraenkel (1962).

In Fraenkel's lab my interest was piqued by the "anomalous alternating linewidth" effect. The electrochemically generated ESR spectrum of the p-dinitrotetramethylbenzene anion showed features that had not been seen before: well-resolved proton shfs appeared on the 1st, 3rd, and 5th lines of the hf splitting from the two equivalent ^{14}N nuclei; but the 2nd and 4th lines were so broad that the proton shfs was completely masked (see Fig. 1). Neither the Kivelson theory of linewidths nor the Stephen-Fraenkel theory of spin-relaxation could explain such a phenomenon. I found that the problem with the earlier theories rested in their improper treatment of multiple or degenerate hf lines which are commonplace for organic radicals. A simple extension of the Kubo-Tomita theory led to the viewpoint that such a multiple hf line must be an "average Lorentzian". I was able to show rigorously from the Redfield theory, that such a multiple hf line must, in general, be a superposition of Lorentzians. For the specific case of the alternating-linewidths of p-dinitrotetramethylbenzene, a particular molecular motional model was needed to complete the explanation. The new theory required out-of-phase correlation of the two ^{14}N hfs, which are assumed to be fluctuating in time, (Freed and Fraenkel, 1962). (One such model would be a rotation of one nitro-group into the benzene plane, thereby increasing its spin-density, while the other is forced to rotate out of the plane, thereby decreasing its spin density, possibly assisted by counterion motions). Such a process would broaden all the hf components except for those arising from nuclear spin configurations in which the two ^{14}N nuclear spin quantum numbers were equal. Work by Bolton and Carrington (1962) at that time on alternating linewidths in durosemiquinone using modified Bloch equations was also consistent with this analysis.

As is often the case in science, the resolution of an "anomaly" led to the formulation of a more generally inclusive theory, in this case the theory of linewidths for organic free radicals. It is, with pleasure, that I note this theory, published in 1963, (Freed and Fraenkel, 1963) is still accepted today as valid for spectra in the motional narrowing regime. The important improvements to the Freed-Fraenkel theory since then have largely to do with the incorporation of more precise and detailed models of the molecular dynamics into the formulation. One important example of this, was the incorporation of Perrin's model (Perrin, 1934) of anisotropic rotational diffusion into the linewidth theory, and its illustration by reinterpreting a linewidth study on p-dinitrobenzene, (Freed, 1964). This work made clear the utility of ESR for the study of molecular dynamics in liquids.

Another improvement was Fraenkel's (1965) extension of the linewidth theory to include the effects of dynamic frequency shifts, which accompany the fast motional linewidths but are smaller except in special cases, such as for the alternating linewidth effect.

2. DOUBLE RESONANCE AND MOLECULAR DYNAMICS

In 1964, Jim Hyde and Gus Maki first observed ENDOR for organic radicals in liquids, (Hyde and Maki, 1964) (Note ENDOR stands for electron-nuclear double resonance, wherein the ESR signal is partially saturated and the nuclear spins on the radical are irradiated at their NMR frequency). At that time there was no theory for explaining why ENDOR can occur in liquids, and the reason for their successful observation was a mystery. Did it involve spin relaxation and therefore would be relevant for studies of molecular dynamics? Thus it seemed appropriate at that time to undertake a reformulation and generalization of the theory of ESR saturation by analogy to what Fraenkel and I had done with ESR linewidths, and to see if a complete theory would allow for a satisfactory explanation of the Hyde-Maki experiment. This ultimately led to a very general theory of ESR saturation and double resonance which appeared in 1965, (Freed, 1965). This theory showed that any ENDOR effects must be small. [At that time, Jim Hyde came to Cornell to give a lecture. I explained to Jim how I had looked everywhere in "spin-relaxation space", and I still couldn't find effects of more than about a percent. Jim promptly assured me that was about the magnitude of the effects he was seeing! By chopping the NMR frequency, and detecting at the chopping frequency, Jim could get just the difference signal due to the ENDOR]. This formulation and its later extensions have served as the basis of interpreting ESR saturation and ENDOR experiments for motionally-narrowed spectra up to today, (Dorio and Freed, 1979; Kurreck et al, 1988; Möbius et al, 1989).

At Cornell we quantitatively tested the ENDOR theory on several semiquinones, (Leniart et al, 1975). We showed that aside from explaining the ENDOR enhancements in terms of the spin-relaxation processes and molecular dynamics, a very important feature was the observed linewidths for the NMR transitions of the free radicals. They very nicely complement the information obtained from the ESR linewidths. This is most important for the case of concentration-dependent linewidths. The sources are Heisenberg spin exchange and electron-electron dipolar interactions between colliding radicals. Their very similar effects on ESR linewidths make it very difficult to separate them out, but it is important to do so in order to utilize these interactions to study microscopic molecular diffusion in liquids. However, these two mechanisms have very different relative effects on the NMR vs. the ESR linewidths, so we were able to successfully separate them from our ENDOR studies.

Jim Hyde and I collaborated on aspects of ENDOR. However, our most important collaboration was undoubtedly the development of ELDOR (electron-electron double resonance) in liquids, (Hyde et al, 1968). This came about during a visit to Varian in the spring of 1967, when Jim showed me the very exciting ELDOR spectra he and Jimmy Chien had obtained with a bimodal cavity that Jim had developed. Using the saturation and double resonance theory that I had developed, I could come up with the appropriate theory for the ELDOR experiment. ELDOR, today, has taken on many new configurations, but in all of them it serves as a powerful means of studying spin-relaxation for purposes of exploring both rotational and translational motions in liquids.

3. SLOW MOTIONAL ESR AND MOLECULAR DYNAMICS

With the theory of ESR linewidths, spin relaxation, saturation, and double resonance well-established by the mid to late 1960's, it could have appeared that all that remained was to apply it to a wide variety of experiments to study molecular dynamics. However, I personally was concerned with some of the theoretical foundations of Wangness-Bloch-Redfield and Kubo-Tomita theory. It appeared to work so well in the motional narrowing region, but it was essentially a perturbation theory, which was ill-defined in the sense that one kept the first non-trivial term in a perturbation expansion, yet there was no way of generating the higher order terms in the expansion. The question remained how to extend Redfield theory to all orders in a systematic fashion as motional rates slow. Using Kubo's method of generalized cumulant expansions in the methodology of statistical mechanics, (Kubo, 1962, 1963), I was able to provide a formal, general solution. Ordinary cumulant expansions arise in probability theory when one examines statistical averages of exponential functions. Spin relaxation, however, involves ensemble averages of exponentiated spin Hamiltonian super-operators, so one must generalize the conventional probability techniques. Thus, it proved possible to utilize Kubo's generalized cumulant method to develop general theorems about spin-relaxation. For example, one could in principle, calculate the relaxation matrix to all orders in perturbation theory, and I could show that the nth order term involved a particular form of the nth order time correlation function of the molecular motion treated as a random function. Relaxation theory is a long-time limiting theory, but the generalized cumulant method even allowed one to get the finite time corrections. This yielded a general formulation of relaxation theory valid to all orders, (Freed, 1968). Redfield theory provides just the leading or 2nd order term. However, generalized cumulant expansions suffer very severely from the problems of perturbation expansions, viz. each higher order term is much harder to compute than the previous term.

The slow motional ESR problem becomes important when one studies spin-labeled macromolecules, as McConnell was doing at that time. Other efforts on this problem, included that of Kivelson, who had another method that also supplied perturbative corrections to Redfield theory, (Sillescu and Kivelson, 1968), as well as the early work of Korst and Khazanovitch (1964), who had solved the slow-motional problem for the simple case of only a secular perturbation, i.e. a perturbation term in the spin-Hamiltonian that commutes with the main term. With all its complexities, at least the generalized cumulant theory was appropriate for all types of perturbation and all types of motions.

In 1969, I spent a sabbatic with Kubo in Tokyo. He called my attention to two of his papers, wherein he had congealed, in a largely heuristic sense, his ideas on stochastic Liouville equations, (Kubo, 1969a,b). I realized how I could incorporate an approach based on the stochastic Liouville equation (SLE) to develop a complete analysis of the ESR slow-motional problem. In order to use the SLE one had to assume that the motional dynamics could be described statistically by a Markov process. More serious in practical terms was the need to take the matrix representations of the SLE and solve for the ESR spectrum on a computer, which could not be done in Tokyo at that time and had to be done at Cornell. The theory was worked out for all relevant cases of g-tensor and hyperfine anisotropy, and it was extended to include saturation phenomena, (Freed et al, 1971a).

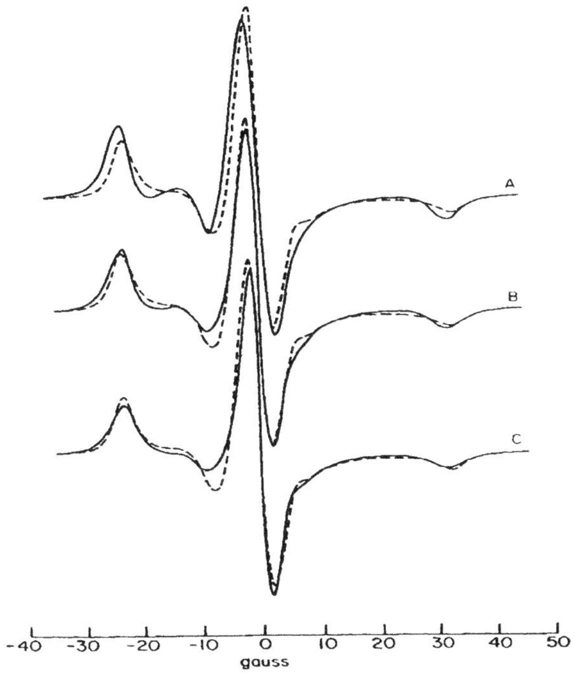

Figure 2. Slow motional ESR spectrum from peroxylamine disulfonate anion in frozen D_2O at T = -60°C. The dashed line is the experimental spectrum, and the solid lines are calculated for a particular microscopic model of rotational reorientation but with differing anisotropy of the rotational diffusion tensor: A) isotropic; B) moderate anisotropy; C) larger anisotropy. From Goldman et al (1972).

At Cornell, we successfully completed experiments using peroxylamine-disulfonate (PADS) in ice. These gave lovely slow-motional spectra with none of the inhomogeneous broadening of typical spin-labels (see Fig. 2). An important discovery was that we could fit these spectra significantly better if we assumed jump-type reorientations rather than simple Brownian motion, (Goldman et al, 1972). Thus, the slow motional spectra were proving to be more sensitive to the microscopic molecular dynamics than are the fast motional spectra. It was also possible to extend the SLE approach to a complete solution of slow-tumbling triplets, (Freed et al, 1971b) generalizing earlier work of Norris and Weissman (1969). In addition both Gordon and Lynden-Bell made early contributions to the slow motional problem, (Gordon and Messenger, 1972).

Work continued on slow-motional ESR through the 1970's utilizing the spectra to obtain new insights into molecular rotational motions in ordinary isotropic fluids, in liquid crystals, and in model membranes, (Hwang et al, 1975; Polnaszek and Freed, 1975). The SLE approach was also utilized to provide a quantitative theory for the then new phenomena of chemically-induced dynamic spin polarization: CIDEP and CIDNP (Freed and Pederson, 1976). The great challenge was in carrying out the rather tedious slow-motional simulations.

Also, in that period, Dalton and Robinson, (Hyde and Dalton, 1979; Beth and Robinson, 1989) managed to employ the SLE to provide a theory for Hyde's new saturation transfer technique, which is useful for studying very slow motions.

During that period it was also possible to improve on the formulation of the SLE, to study more thoroughly its range of validity, and to extend its range of applicability. Nevertheless, these initial efforts have successfully withstood the tests of time. Clearly, the most important accomplishment was the development in the 1980's of a very efficient method of computing solutions to the SLE that drastically reduced the computation time and storage requirements and ultimately led to versions that could be made generally available, (Moro and Freed, 1981).

Working with Giorgio Moro, we uncovered material about the Method of Moments. This is a formal procedure for projecting out a sub-space, known as a Krylov space, starting from a real symmetric or Hermitian matrix and an initial vector, each of dimension n. This sub-space will, in general, be of dimension m lower than n (i.e. $m < n$), and in the representation of the basis vectors obtained, the m-dimensional approximation to the original matrix will be in tridiagonal form. The practical implementation for computation involved a specific algorithm, known as the Lanczos Algorithm (LA), which, however was known to suffer badly from computer round-off error. Another problem confronted us in that the LA had been developed for real-symmetric or Hermitian matrices, but not for the complex symmetric matrices (which are non-Hermitian) that one generates with the SLE. The available theorems no longer necessarily applied, including a guarantee that the matrix is diagonalizable. We were not especially troubled by the latter fact, since we could diagonalize SLE matrices by standard, but slow, methods. However, the complex symmetric arithmetic could (and does) increase the problem of computer round-off error.

Despite these concerns, we found the LA succeeded admirably for numerically solving the SLE, (Moro and Freed, 1981). It reduced computation time by at least an order of magnitude, and it also greatly reduced storage requirements. Why does the LA work so well? First of all, it takes full advantage of the sparsity of the SLE matrix. Secondly, after just a few Lanczos projections it produces a sub-space that very effectively includes what is important for the ESR experiments. This is partly because the initial vector is a kind of statement of the physics of the ESR experiment. It essentially represents the ESR transition moments. Thirdly, the ESR experiment is dominated by the slowly decaying eigenvalues of the SLE, and these are accurately obtained from the small sub-space approximation. But what about the eigenvalues that are poorly represented? They are automatically projected out of the solution when one calculates the specific ESR observable, viz. the lineshape. And what about the round-off error? Since we only need a small sub-space generated by a relatively small number of Lanczos projections, the calculation is terminated before the round-off error becomes serious. Important for the execution of these programs, (Schneider and Freed 1989a) are very powerful methods we developed for selecting the minimum basis set to represent the SLE, and for reliably determining when sufficient Lanczos projections have been utilized, (Schneider and Freed, 1989b).

Thus the Lanczos algorithm provides both a conceptually insightful approach as well as an extremely powerful computational algorithm for the SLE. It has since been possible to show the connection of the LA with other mathematical methodologies, but none other lends itself so effectively to a computational algorithm, (Schneider and Freed, 1989b). The most instructive connection is that to the Mori method, well-known in statistical mechanics.

4. HIGH FIELD ESR AND MOLECULAR DYNAMICS

With the advent of high field far-infrared (FIR) ESR, with its enhanced resolution to motional dynamics, we have found it desirable to use enhanced models for molecular reorientation to fit these spectra. We can now dispense with the old and worn jump models of diffusion. Instead we approximate the many-body problem of dealing with the microscopic details of the fluid by a set of collective degrees of freedom that represent the main effects of the solvent on a rotating solute. These collective variables are taken as a loose solvent "cage" that is slowly relaxing. The solute is then reorienting more rapidly in this cage. This is called the slowly relaxing local structure (SRLS) model. Since we approximate the combined system of solute plus cage by Markovian equations, the SLE remains valid in this augmented form. Then the Lanczos projections effectively determine the extent to which the cage variables are needed to interpret the ESR spectrum, (Polimeno and Freed, 1995).

In addition to the greater sensitivity of FIR-ESR (e.g. 250 GHz) to the details of the molecular motions in fluids, another virtue of FIR-ESR (e.g. 250 GHz) over ESR at conventional microwave frequencies is the excellent orientational resolution it provides for studies utilizing nitroxide spin labels

(Budil et al, 1989; Earle et al, 1993, 1997, 1998). As a result, at 250 GHz, once motion is discernible in the spectrum, one can discern about which axis (or axes) the motion occurs, (Earle et al, 1993).

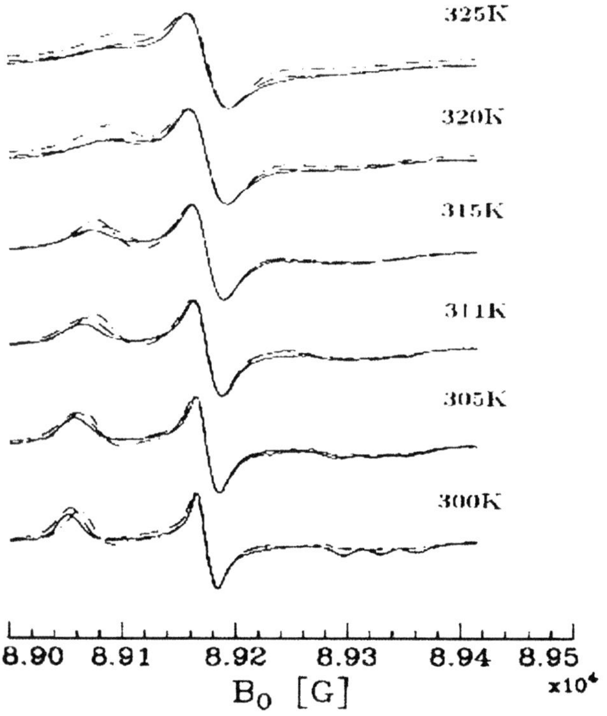

Figure 3. Comparison of two models for fitting effects of rotational diffusion on 250 GHz electron spin resonance spectra of spin probe of a cholesterol-like nitroxide (CSL) in ortho-terphenyl solvent. (*Solid line*) Experiment, (*dashed line*) the SRLS model, and (*dashed-dotted line*) simple Brownian diffusion (Earle et al, 1997).

In a 250-GHz ESR study of the dynamics of several nitroxide spin probes dissolved in the glass-forming solvent ortho-terphenyl (OTP), we demonstrated how the enhanced sensitivity to rotational dynamics of the slow-motional spectra could be utilized to explore details of the dynamic solvent cage, (Earle et al, 1997). The SRLS model adequately fits the model-sensitive regions of the 250-GHz spectra (cf. Figure 3) and leads to a coherent picture of the dynamics: The rotational diffusion tensors of the various probes exhibit simple behavior such that the smaller the probe is the larger the diffusion coefficient. The cage relaxation rate is the slowest, but it is independent of the particular probe. This interesting observation appears reasonable when one considers that the cage relaxation involves just the movement of the OTP solvent molecules. In addition, the magnitude and directionality of the cage-orienting potential could be obtained. As expected, only probes comparable to or larger than the OTP molecules experience substantial potentials, of 2-4 kT. It was possible to show that the nonlinear way in which the dynamics affects the slow-motional ESR spectra allows one to distinguish between two limiting cases. The first is that of a homogeneous liquid, but with a complex motional dynamics, (e.g. the SRLS model that was used). The second is that of an inhomogeneous liquid with a distribution of simple relaxation times (e.g. Brownian tumbling). The latter was shown to be incompatible with the 250-GHz spectra.

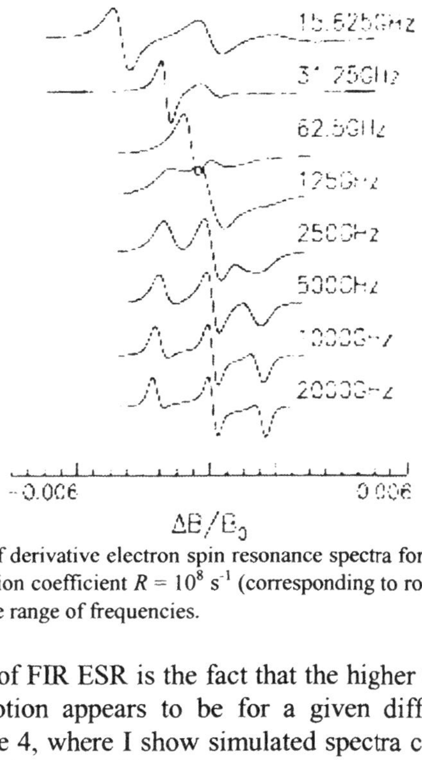

Figure 4. Simulation of derivative electron spin resonance spectra for a nitroxide, reorienting with a rotational diffusion coefficient $R = 10^8$ s^{-1} (corresponding to rotational correlation time $\tau_R = 1.67$ ns) for a wide range of frequencies.

Another virtue of FIR ESR is the fact that the higher the ESR frequency, the slower the motion appears to be for a given diffusion rate. This is illustrated in Figure 4, where I show simulated spectra corresponding to the same motional rate but for different ESR frequencies, ranging from 15 GHz to 2 THz. At the low frequency end, one observes simple motionally narrowed spectra, whereas at the high frequency end, the spectra are very slow motional, almost at the rigid limit. Thus we see that the higher-frequency ESR spectra act as a faster "snapshot" of the dynamic, (Earle et al, 1993, 1997). This is because of the increased role of the g-tensor term, which is linear in magnetic field, B_o in the spin-Hamiltonian. As the orientation-dependent part of the spin-Hamiltonian, $H_1(\Omega)$ increases in magnitude with increasing frequency, ω_o and B_o, the motional-narrowing condition $|H_1(\Omega)|^2 \tau_R^{\,2} \ll 1$ fails, (where τ_R is the rotational relaxation time) and the spectra become slow motional.

This snapshot feature suggests a multifrequency ESR approach to the study of the dynamics of complex fluids, such as glass-forming fluids and liquid crystals, as well as to the complex modes of motion of proteins and DNA, which should enable one to decompose the different modes according to their different timescales (Liang and Freed, 1999). For example, in the case of proteins, the higher frequency ESR spectra should "freeze-out" the slow overall tumbling motions, leaving only the faster internal modes of motion, whereas ESR performed at lower frequencies is sensitive to the motions on a slower timescale. In glass-forming fluids, as we have seen, the faster motions consist of reorientations of probe molecules, whereas the slower motions relate to the dynamics of the solvent cage.

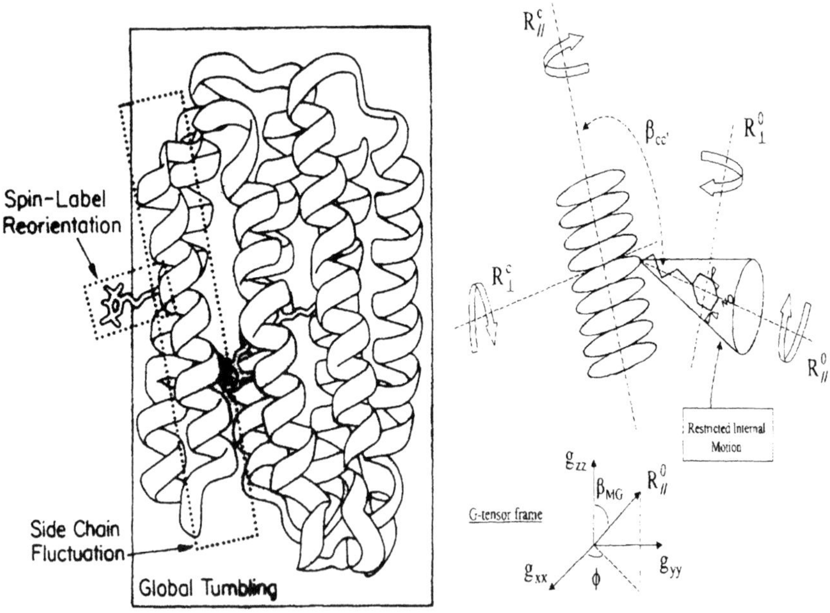

Figure 5. (left) Protein Dynamics of Spin-labeled Protein: There are three kind of motions, spin-label reorientation, side chain fluctuations and global tumbling. (right) The SRLS model is illustrated including relevant motional parameters (Liang and Freed, 1999; Liang et al, 2000).

The virtues of such a multifrequency approach were demonstrated in a study, using 9- and 250-GHz spectrometers, on spin-labeled mutants of the soluble protein T4 lysozyme in aqueous solution (Barnes et al, 1999). In the fast timescale of the 250-GHz ESR experiment, the overall rotation was too slow to significantly affect the spectrum, so that it could satisfactorily be described by the simple MOMD (microscopic order but macroscopic disorder) model (Meirovitch et al 1984), wherein the overall motion is so slow that it corresponds to the rigid limit (see below), which yielded good spectral resolution for the internal dynamics. Then, by fixing the internal motional parameters at the values obtained from the 250-GHz data, the SRLS fits to the 9-GHz line shapes successfully yielded the rates for the global dynamics. Thus the two types of motion were separated, and spectral resolution to these motions was significantly enhanced. The SRLS model as it applies to protein dynamics is shown in Figure 5.

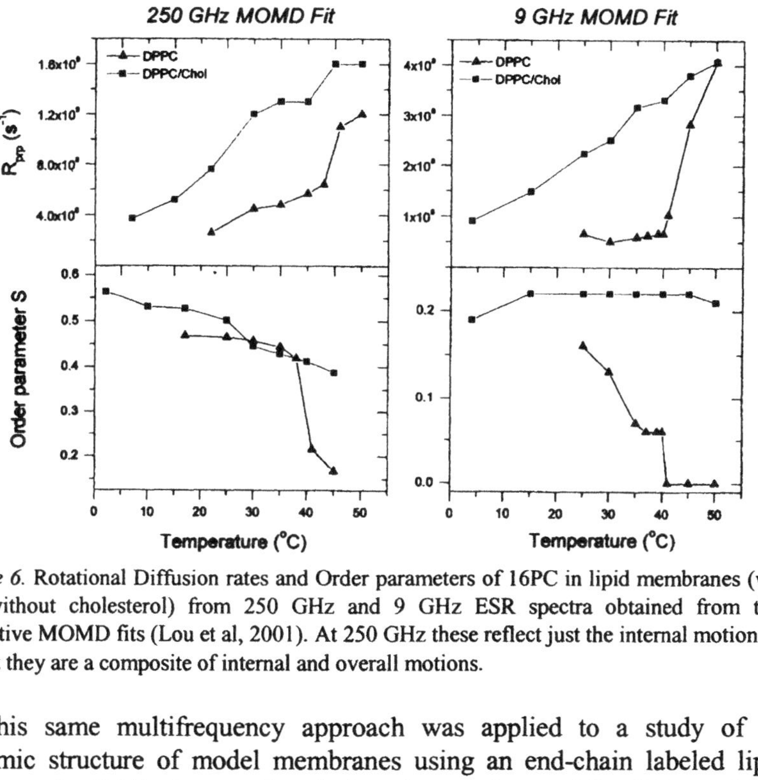

Figure 6. Rotational Diffusion rates and Order parameters of 16PC in lipid membranes (with and without cholesterol) from 250 GHz and 9 GHz ESR spectra obtained from their respective MOMD fits (Lou et al, 2001). At 250 GHz these reflect just the internal motions; at 9 GHz they are a composite of internal and overall motions.

This same multifrequency approach was applied to a study of the dynamic structure of model membranes using an end-chain labeled lipid, (Lou et al, 2001). It was found that the results at 250 GHz could be interpreted in terms of the MOMD model relating to just the internal dynamics and ordering of the ends of the acyl-chains, with the slower overall lipid dynamics frozen-out on the time-scale of the 250 GHz experiment, (cf Figure 6). The 9 GHz spectra, however, are affected by both the internal and overall motions, so they were analyzed in terms of the SRLS model, which explicitly includes both types of motion, using the parameters for the internal dynamics obtained from the analysis of the 250 GHz spectra. It is worth noting, however, that if the 250 GHz spectra are ignored, then the 9 GHz spectra, with their limited resolution to dynamics, could be fit to a simple

MOMD model, but the dynamic and ordering parameters obtained must be interpreted as a composite of both the internal and overall motions, (cf Figure 6) with no obvious way of separating them.

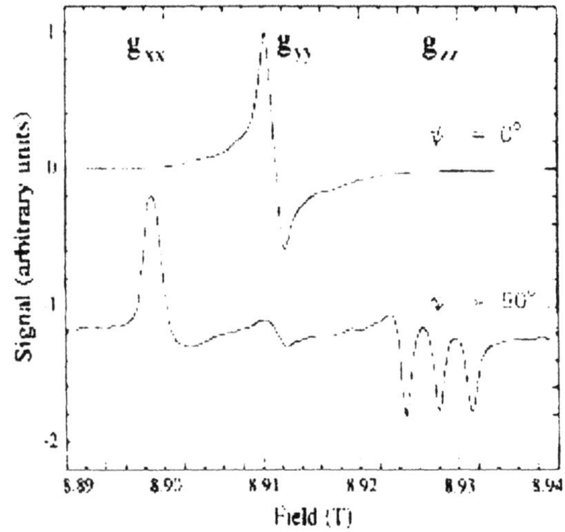

Figure 7. 250-GHz derivative electron spin resonance spectra from cholesterol-like nitroxide in aligned PC-rich membrane with the membrane normal parallel (Ψ = 0°) and perpendicular (Ψ = 90°) to the magnetic field, (Barnes and Freed, 1998).

A striking demonstration of the excellent orientational resolution at 250 GHz in studies utilizing nitroxide spin labels was provided by a study on macroscopically aligned membranes containing a mixture of headgroups: zwitterionic phosphatidylcholine (PC) and negatively charged phosphatidylserine (PS) using the cholesterol-like spin label CSL, (Barnes and Freed, 1998). The macroscopic alignment further enhanced the orientational resolution at 250 GHz and permitted an orientation-dependent study, (cf Figure 7).

5. SPIN-ECHOES AND MOLECULAR DYNAMICS

A major weakness of cw ESR for relaxation studies is the problem of extracting reliable homogeneous line broadening from inhomogeneously broadened ESR spectra such as from nitroxide spin labels. This homogeneous line broadening is the contribution to the linewidth that arises from the motional modulation of the hyperfine and g-tensors as well as the other spin-relaxation processes. It is obscured by the inhomogeneous broadening, which is due, for example, to the unresolved proton superhyperfine interactions. This was a particular problem for our work on macroscopically aligned samples of liquid crystals, because small amounts of misalignment could appear as extra inhomogeneous broadening, which varies for the different ^{14}N hf lines, thereby being easily mistaken for the homogeneous broadening with its well-known variation with hf line. However, by means of electron-spin echoes, one can cancel inhomogeneous broadening and obtain the homogeneous widths, which are the inverse of T_2.

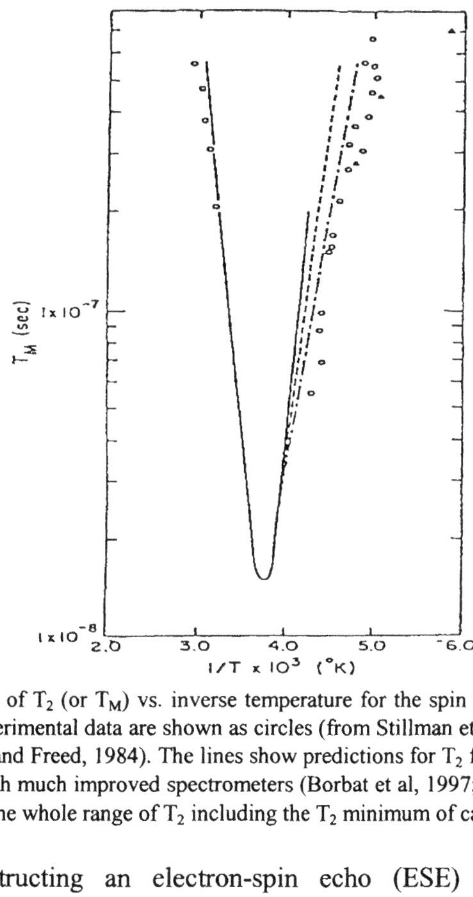

Figure 8. A graph of T_2 (or T_M) vs. inverse temperature for the spin probe tempone in 85% glycerol/H_20. Experimental data are shown as circles (from Stillman et al, 1980) and triangles (from Millhauser and Freed, 1984). The lines show predictions for T_2 for Brownian and Jump models. Today with much improved spectrometers (Borbat et al, 1997; Freed, 2000) it is now possible to cover the whole range of T_2 including the T_2 minimum of ca. 14 ns.

After constructing an electron-spin echo (ESE) spectrometer, we conducted initial ESR experiments on fluids. In particular, we were able to explore T_2's for fast through slow motions for the system of PD-Tempone in glycerol-water solvent (see Fig. 8), (Stillman et al 1980). For fast motion, T_2 has the well-known inverse dependence on correlation time, but for slow motion the homogeneous T_2 depends on the correlation time to a positive, usually fractional, power. Thus there is a T_2 minimum not generally appreciated (but also observed by Ian Brown, 1974). We were able to offer a coherent explanation in terms of the SLE, and this led to a comprehensive theory for spin-relaxation and ESE, (Schwartz et al, 1982). What then is the interpretation for T_2 in the slow-motional regime? In the limit of strong jump reorientation each jump leads to a large change in resonant frequency thereby leading to an uncertainty in lifetime broadening. Thus, in this limit, T_2 equals the correlation time (as noted earlier by Mason and myself in Mason and Freed, 1974). In the limit of simple Brownian motion, T_2 is roughly proportional to the half power of the correlation time. A heuristic interpretation of this is due to Kivelson and Lee, (1982).

A disturbing limitation of ESR measurements of T_2 is that one just obtains a single parameter from which to extract information on motional dynamics. Of course, in the fast motional regime one may study the variation of T_2 with hf line. For the slow motional regime, initial theory and experience did show that the measured T_2 displays some variation across the spectrum. It seemed reasonable to suppose that by studying this variation, sufficient information could be obtained from which to infer details of motional models. This would be analogous to studying the full lineshape of a slow-motional ESR experiment, but it would have the big advantage that the homogeneous T_2 relates solely to the dynamical processes. This advantage is, however, limited by the fact that when molecular motions are slow enough, then solid-state relaxation processes, such as spin diffusion, take over. However, one can explore slower processes by studying the homogeneous T_2 rather than the near-rigid-limit cw-ESR spectra.

Initial ESE experiments of this type were performed by sweeping the magnetic field and collecting the spin echo from weak, or highly selective, microwave pulses, (Millhauser and Freed, 1984). When Fourier-Transformed in the echo delay time, τ this led to a 2D-ESR spectrum in which the homogeneous lineshape is plotted along the frequency axis, and essentially the ESR lineshape appears along the field axis (see Fig. 9). This 2D spectrum thus effectively supplies the homogeneous T_2 variation across the ESR spectrum. For the case of Tempone in glycerol-water, we found a substantial variation in the T_2. Our theoretical analysis showed that a Brownian reorientational model could quite successfully explain this variation. This is because, given the $\cos^2\theta$ type of angular dependence of the hf and g-tensor interaction terms, the variation in these terms with a small change in angle, θ depends significantly on the value of θ, hence on the position in the very slow-motional spectrum. On the other hand, a strong jump diffusion model leads to a uniform T_2 across the spectrum, since for this case $T_2 = \tau_R$, the mean rotational jump time, as already mentioned. In fact, we found that the patterns of T_2 variation across the spectrum, plotted in a normalized contour fashion could themselves be utilized to distinguish the model of motion, and the degree of rotational anisotropy (see Fig. 9). This method was then extended to spin labels in oriented model membranes and to labeled proteins, to slow motions on surfaces.

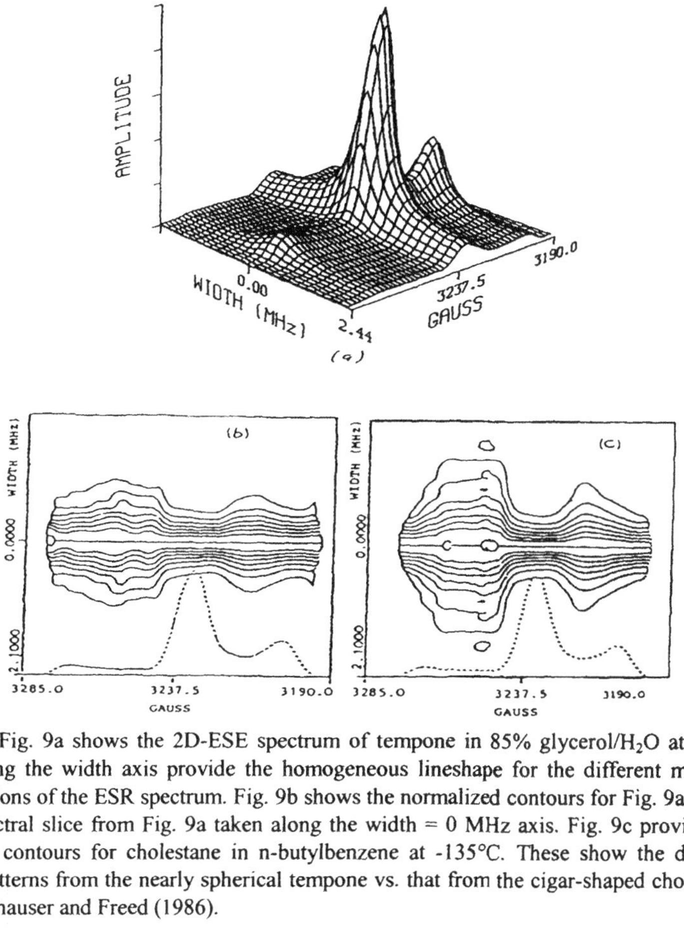

Figure 9. Fig. 9a shows the 2D-ESE spectrum of tempone in 85% glycerol/H_2O at -75°C. Slices along the width axis provide the homogeneous lineshape for the different magnetic field positions of the ESR spectrum. Fig. 9b shows the normalized contours for Fig. 9a as well as the spectral slice from Fig. 9a taken along the width = 0 MHz axis. Fig. 9c provides the analogous contours for cholestane in n-butylbenzene at -135°C. These show the different contour patterns from the nearly spherical tempone vs. that from the cigar-shaped cholestane. From Millhauser and Freed (1986).

Next, a field-swept 2D-ESE experiment, from which one obtains the magnetization transfer rates across the ESR spectrum, was performed in a manner analogous to the T_2-type 2D-ESE experiment, except that a stimulated echo sequence: $\pi/2$ - $\pi/2$ - $\pi/2$, replaces the spin-echo sequence $\pi/2$ - π, and one steps out the time T between the second and third pulses, (Schwartz et al, 1986). Here theory showed that as a function of T there are at least two exponential decays: one is in T_1 and the second (to a reasonable approximation) is in T_A. an effective magnetization transfer time (for the relevant case of $T_A \ll T_1$). The slow rotational reorientations shift spin-bearing molecules irradiated by the first two $\pi/2$ pulses to frequencies outside the irradiated region. Thus they are not detected after the third $\pi/2$ pulse. This magnetization transfer process, thus leads to a more rapid decay of the stimulated echo as a function of T. A Brownian rotation model will also give a T_A variation across the spectrum, because the effectiveness of rotation taking the spins out of the irradiated region depends upon angle, θ through the $\cos^2\theta$ dependence of the magnetic tensor terms. We obtained dramatic variation of T_A across the spectrum for NO_2 adsorbed on crushed vycor, which could be attributed to very anisotropic rotational motion on the surface. For this case there is an enhanced T_A for the spectral regimes corresponding to x and z molecular axes being parallel to the magnetic field, which clearly implies more rapid rotation about the y-axis (which is parallel to the line connecting the two Oxygen atoms). This motional anisotropy is clearly visible from the 2D-contours without the need for detailed spectral analysis (see Fig. 10).

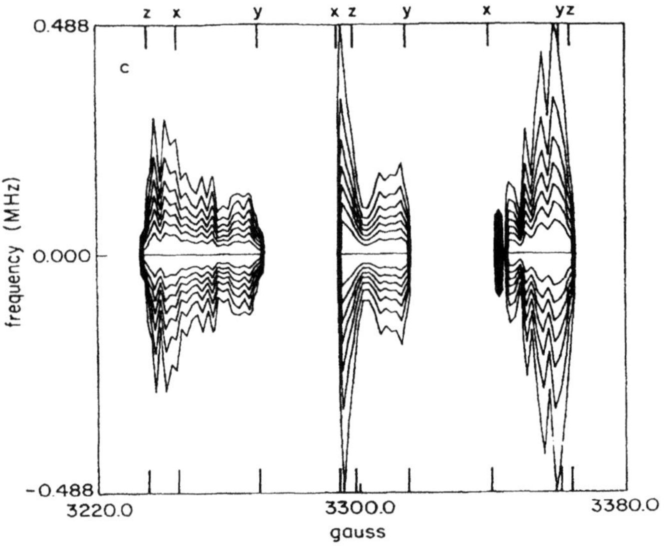

Figure 10. 2D-ESE contours from the stimulated echo sequence for NO_2 adsorbed on vycor at 35°K showing rates of magnetization transfer. It shows relatively rapid rotation about the molecular y axis (i.e. the axis parallel to the oxygen-oxygen internuclear vector). From Schwartz et al (1986).

Both longitudinal and cross-relaxation in liquids were included in the comprehensive theory of spin relaxation in ESE for fast and slow motions, (Schwartz, 1984; Schwartz et al, 1986). A major motivation for the analysis of cross-relaxation was the spin-echo ELDOR experiment we performed in the very viscous regime for PD-Tempone in glycerol/water, (Hornak and Freed, 1983). Instead of using two microwave frequencies, the magnetic field was stepped out during the time between the first inverting π pulse and the detecting π/2 - π spin echo sequence. [This technique had been independently developed by Tsvetkov and co-workers (Dzuba et al, 1984)]. Using the theory we showed that a substantial orientation-independent nuclear-spin-flip rate could explain the ELDOR experiment, (Schwartz, 1984; Schwartz et al, 1986).

Clearly the most informative method of studying magnetization transfer is by ELDOR. One observes not only the transitions out of a certain spectral region but also the spectral region to which the transition is made. This was the basic idea of the stepped field spin-echo ELDOR experiment. One could attempt, by a combination of sweeping one or both frequencies of an ELDOR experiment and/or sweeping the field and the field jump, to perform a 2D experiment as a function of the pumping and observing frequencies. (Tsvetkov and his co-workers did, in fact, develop the use of two microwave sources, wherein they swept one of them, (Dzuba and Tsvetkov, 1988). This requires a resonator with a low enough Q that it could sustain two separated frequencies. But once this is the case, another more general and more elegant method suggests itself, which removes the need for field sweeping and stepping, and it only requires one microwave source. It does require collection of the free-induction decay or the echo decay after the last pulse, but it could be obtained very rapidly.

6. TWO-DIMENSIONAL FOURIER TRANSFORM ESR

In 1976 Richard Ernst and co-workers published the first 2D-NMR experiments, (Aue et al, 1976) that used Fourier Transform (FT) methods with their multiplex advantage for collecting the whole spectrum simultaneously. This also means the successful irradiation of the whole spectrum with a single non-selective rf pulse, and the ability to collect data shortly after such a pulse. Further, the non-selective pulse from a single rf source introduces coherence simultaneously to all spectral components enabling the observation of coherence transfer between these components. In 1979, Ernst and Jeener showed how magnetization transfer could also be studied in this manner, (Jeener et al, 1979).

Why were these ideas not incorporated into ESR until 1986? Clearly the ESR experiment is much more difficult than the comparable NMR one. In ESR we use microwave rather than rf technology. Relaxation times are orders of magnitude faster, pulse widths need to be orders of magnitude shorter, and spectral bandwidths are orders of magnitude wider.

Clearly, it was necessary to develop modern FT techniques in ESR as a prerequisite to developing the ESR analogues to 2D-NMR. Modern FT techniques appeared almost the same time in Bowman's lab in Argonne, (Angerhofer et al, 1988) Dinse's lab in Dortmund, Germany, (Dobbert et al, 1986) Lebedev's lab in Moscow, (Panferov et al, 1984) and my own lab in the 1984-1986 period. My motivation was partly the hope of performing modern 2D-FT-ESR experiments and partly the hope of studying the spectra from transient radicals. In fact, in 1984, we succeeded in obtaining the free induction decay (FID) of a transient photogenerated electron from Rb/THF solutions, thereby distinguishing its spectrum from the stable solvated electron, (Eliav et al, 1984). In another experiment, we showed that the B_1 microwave fields in the rotating frame need not be much larger than the spectral bandwidth to obtain reasonable coverage in an FT-ESR experiment, (Hornak and Freed, 1986). One merely has to accept a rotation of the spins into the rotating x′-y′ plane instead of precisely along the x′ axis (for a B_1, along the y′ axis). Then quadrature detection plus standard phase corrections yielded the pure absorption from the FID. Also we showed the advantages of utilizing a loop-gap resonator which can supply large B_1 fields, but with low Q's to reduce resonator ringing and thereby spectrometer dead time after the pulse.

After introducing a digitizing oscilloscope and a home-built quadrature detector, we were able to obtain good FID's and FT spectra from fast motional nitroxides with a total spectral width of 90 MHz, and this immediately led to the first two-dimensional FT-ESR experiments on the fast motional nitroxide system. They consisted of a 2D-ESE experiment, appropriately called a SECSY (spin-echo correlated spectroscopy) experiment and an FID-based 2D-exchange experiment, which we now call 2D-ELDOR, (Gorcester and Freed, 1986). This first FT-based 2D-ELDOR experiment showed cross-peak development resulting from Heisenberg spin-exchange (see Fig. 11). The SECSY experiment showed how the homogeneous T_2 values from all the hf lines could be obtained simultaneously from an inhomogeneously broadened ESR signal. Thus 2D-FT-ESR became a reality.

There were still a number of major challenges to make 2D-FT-ESR generally applicable. Sophisticated phase cycling was added and a full theoretical analysis for the fast motional 2D spectra in terms of how Heisenberg exchange (HE) and electron-nuclear dipolar (END) terms generate the cross peaks, was developed. We showed how their respective contributions could be readily distinguished. This led to quantitative measurements of HE in an isotropic fluid, (Gorcester and Freed, 1988) and of END terms in a liquid crystal, (Gorcester et al, 1989). This latter study could be utilized to provide sophisticated insights on molecular dynamics in ordered fluids in a way that cw-ESR linewidths could not. Also, Bowman showed how 2D-ELDOR could be used to measure rates of chemical exchange in a semi-quinone system, (Angerhofer et al, 1988).

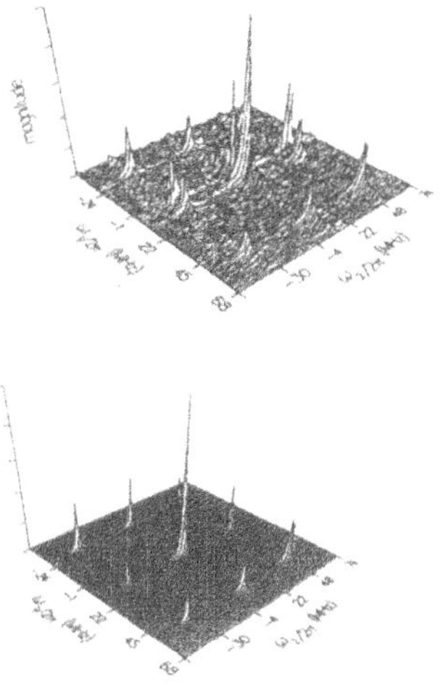

Figure 11. 2D- ELDOR spectrum of solution of PD-Tempone in toluene-d_8 at 21°C, a) raw data; b) after analysis by linear predictive methods. The cross-peaks are due to Heisenberg spin exchange. From Gorcester and Freed, (1988).

With additional improvements to the 2D-FT-ESR spectrometer, which increased the spectral coverage to about 250 MHz, and greatly increased our data acquisition rates and significantly reduced spectrometer dead-times, it became possible to extend 2D-FT-ESR to the slow motional regime, (Patyal et al, 1990). These developments necessitated more subtle instrumental and filtering improvements before we could fully benefit from the increased signal-to-noise that was achieved. Also, a general theory for these experiments was needed for their interpretation. When developed, we could demonstrate the good agreement between theory and experiment.

It then became possible to perform detailed studies on complex fluids, (Lee et al, 1994; Crepeau et al, 1994). These include phospholipid membrane vesicles (cf. Figure 12) (Lee et al, 1994a; Crepeau et al, 1994), liquid crystalline solutions (Sastry et al, 1996a, b) and liquid crystalline polymers, (Xu et al, 1996). A key feature was dead times of ca 50-60 ns. The detailed theory (Lee et al, 1994b) enabled quantitative analysis of these 2D spectra. In the case of 2D-ELDOR, simultaneous fits of experiments at several mixing times, T_m, provided in effect, a third dimension. One can watch how the cross peaks grow in relative to the auto peaks with increasing mixing time, (cf. Figures 13 and 14). This supplies quantitative information on the nuclear-spin-flip-inducing processes of both HE, which reports on translational diffusion, and the intramolecular electron-nuclear dipolar interaction, which reports on the tumbling motions.

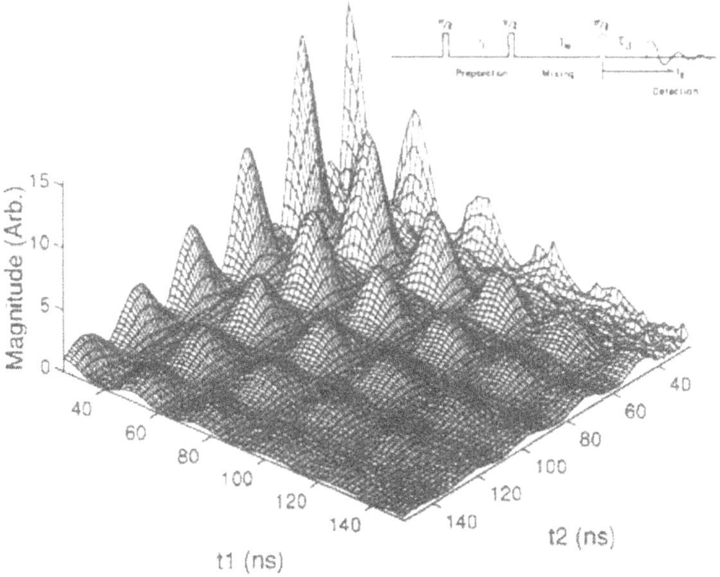

Figure 12. 2D-ELDOR at 17.3 GHz. The time domain S_{c-} spectrum showing ESR timescale; from phospholipid that is end chain labeled with nitroxide (16-PC) in lipid vesicles, (Borbat et al, 1997). (*Inset*) Pulse sequence.

In addition, the line shapes of the auto and cross peaks are particularly informative. In fact, there are two types of line shapes provided by the COSY (correlation spectroscopy) and 2D-ELDOR experiments. They arise because the experiment provides two types of 2D spectrum, depending on the coherence pathway: One is FID-like (sometimes referred to as the anti-echo) and the other is echo-like, i.e. there is a refocusing of the inhomogeneous broadening (IB) terms in the spin-Hamiltonian leading to their cancellation in the echo formation. The echo-like (or S_{c-}) 2D signal can in fact be transformed to provide just the homogeneous broadening (HB) along one frequency dimension, ω_1, whereas the other frequency dimension, ω_2, provides essentially the cw spectrum. This transformation takes one from the COSY to the SECSY format. In the 2D-ELDOR spectrum, this same transformation will yield the HB for the auto peaks, but the cross-peaks will be affected by any differences in the IB existing between the two spectral lines connected by that cross peak. Thus, the 2D-ELDOR S_{c-} spectrum provides detailed information on spin relaxation via the cross-peak development and the HB of the auto peaks, whereas the differences in IB show up in the cross peaks. The FID-like S_{c+} 2D spectra include the full effects of inhomogeneous broadening. The 2D-SECSY format is particularly useful for ultraslow motions, for example macromolecules in viscous media, (Saxena and Freed, 1997).

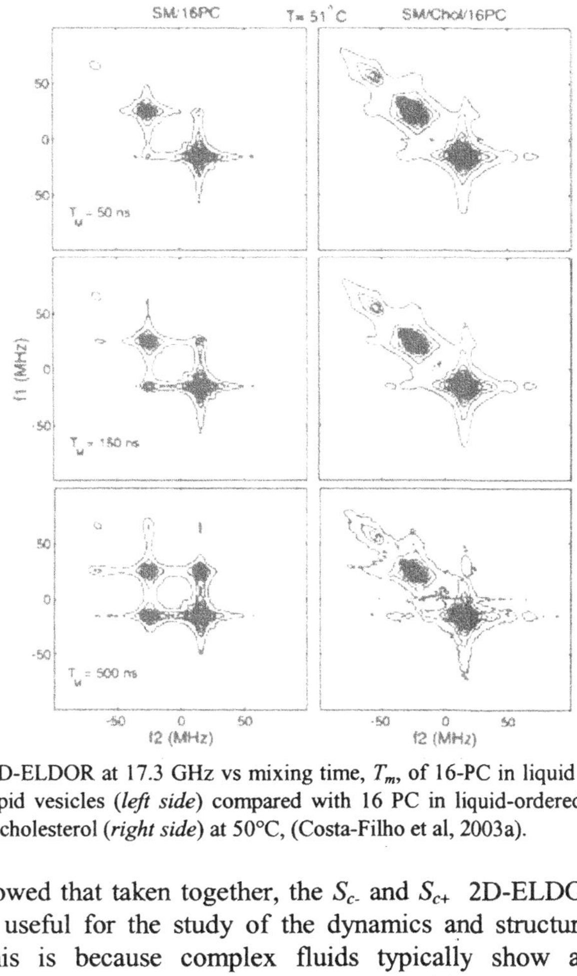

Figure 13. 2D-ELDOR at 17.3 GHz vs mixing time, T_m, of 16-PC in liquid crystalline phase from pure lipid vesicles (*left side*) compared with 16 PC in liquid-ordered phase from 1:1 ratio lipid to cholesterol (*right side*) at 50°C, (Costa-Filho et al, 2003a).

We showed that taken together, the S_{c-} and S_{c+} 2D-ELDOR spectra are especially useful for the study of the dynamics and structure of complex fluids. This is because complex fluids typically show a microscopic structure, such that molecular tumbling occurs with respect to this structure, which provides the local orientational alignment. This can be readily appreciated in the case of lipid membranes. If they are macroscopically aligned, then one would observe the different "single-crystal-like" spectra obtained for each orientation of the membrane normal with respect to the constant magnetic field. Membrane vesicles, however, simultaneously have membrane components at all angles with respect to the magnetic field, and they thereby provide "powder-like" spectra that is referred to as microscopically ordered, but macroscopically disordered. The extent of the local ordering is thus reflected in the IB, and details about the aligning fields can be obtained from the differences of the IB for the different hyperfine (hf) lines. At the same time, the S_{c-} spectra permit one to obtain dynamics from the homogeneous T_2's and the development of the cross peaks with mixing time. Such a program was carried out (Patyal et al, 1994; Crepeau et al, 1994) using several different nitroxide spin labels in phospholipid membrane vesicles to obtain accurate dynamics and ordering parameters.

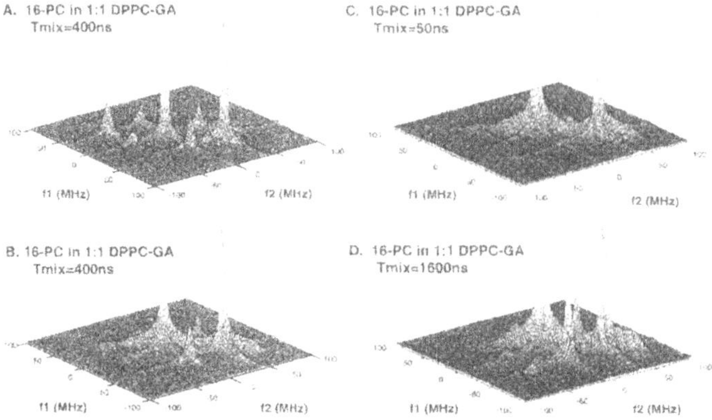

Figure 14. 2D-ELDOR at 17.3 GHz showing effect of peptide gramicidin A (GA) on dynamic structure of lipid membrane containing (end chain) nitroxide labeled lipid (16-PC) at 75°C. (*A*) Pure lipid, mixing time, T_m = 400 ns. (*B, C, D*) 1:1 lipid to GA with T_m = 400 ns, 50 ns, and 1.6 μs, respectively (Costa-Filho et al, 2003b).

In general, one finds that the 2D-ELDOR spectra from membrane vesicles show more dramatic changes as the membrane properties are varied. This can even enable simple interpretations of these spectra just in terms of pattern recognition. For example, in Figure 13, 2D-ELDOR contour plots as a function of mixing time, T_m, are shown for the spin-labeled lipid, 1-palmitoyl-2-(16-doxyl stearoyl) phosphatidylcholine (16-PC) in pure lipid vesicles and for a lipid-cholesterol mixture in the ratio 1:1. The former is in the standard liquid crystalline phase, whereas the latter is in a "liquid-ordered" LO phase, (Ge et al, 1999). The spectra are qualitatively different, emphasizing that the LO phase exhibits significantly greater ordering than the liquid crystalline phase at 51°C. The increased microscopic ordering leads to increased IB affecting the spectra from the LO phase. In addition, the restriction of the range of orientational motion, due to the microscopic ordering in the LO phase, shows up as a much slower development of cross peaks vs T_m, (Costa-Filho et al, 2003a).

In addition to the microscopic ordering but macroscopic disorder (MOMD) (Meirovitch et al, 1984) that gives rise to complex inhomogeneous line shapes, these spectra are often in the slow-motional regime, i.e. the motions are too slow to provide complete averaging of the rigid-limit line shapes. This is another source of IB that is effectively dealt with in the theory for MOMD spectra. These slow motional spectra provide more insight into the microscopic details of the molecular dynamics because their timescales are comparable. It was found that for complex fluids, a more sophisticated model than the MOMD model, i.e. the SRLS model referred to above, was needed to analyze the 2D-ELDOR spectra in order to achieve reasonably good agreement with experiment.

We used studies on a macroscopically aligned liquid crystal solvent called 4O,8 to test the applicability of the SRLS model, (Sastry et al 1996a, b). This is a liquid crystal that exhibits many phases as a function of temperature, including isotropic, nematic, liquid-like smectic A, solid-like smectic B, and crystalline phases. We found consistently better fits using a SRLS model (in addition to the macroscopic liquid crystalline orienting potential) than with the standard simpler model that does not include any local structure. These studies demonstrated the very extensive relaxation, dynamic, and structural information that one can obtain from 2D-ELDOR experiments performed as a function of mixing time. In all, 10 such parameters could be effectively extracted. They include the two-term (asymmetric) macroscopic orienting potential in the liquid crystalline phases, the axially symmetric diffusion tensor for the probe, its two-term orienting potential in the local structure or cage, the relaxation rate for the cage, the residual homogeneous T_2^{-1} due to processes other than the reorientational modulation of the ^{14}N dipolar and g-tensors, the residual (Gaussian) inhomogeneous broadening not due to the specific slow-motional contributions from the ^{14}N hf and g-tensors, and the overall T_1 for the electron-spins. These constitute virtually all the parameters that one can hope to obtain from any ESR experiment(s) on spin relaxation in a complex fluid!

The virtues of the improved 2D-FT-ESR technology were further demonstrated in studies of the effect of the peptide, gramicidin A (GA), on the dynamic structure of model membranes. Earlier studies that showed that the changes in the 2D-ELDOR spectra on adding GA were much more dramatic than the changes in the cw-ESR spectra, emphasizing the much greater sensitivity of the former to molecular dynamics, (Patyal et al, 1997). However, these studies, performed at 9.3 GHz with a τ_d of 50-60 ns, related just to the bulk lipids. They showed no clear indications of the so-called boundary lipids that coat the peptide. Evidence for the boundary lipid exists in cw-ESR spectra but is of very limited resolution. More recently using 17.3-GHz 2D-ELDOR with its increased SNR and decreased dead times (τ_d ~ 25-30 ns), we have been able to obtain 2D-ELDOR spectra (Costa-Filho et al, 2003b) that show the presence of two components, viz the bulk lipid component, previously seen by Patyal et al, (1997) which shows relatively fast dynamics, and a second, the presumed boundary lipid, which grows in as GA is added (cf. Figure 14). Its 2D-ELDOR spectrum is clearly that of a more slowly reorienting lipid, as expected. In addition, simulations of these spectra are consistent with a dynamic bending of the end-chain of the lipid as it coats the GA. This level of detail of the dynamic structure of complex membrane systems is not likely achievable by other means.

7. PROSPECTUS

At present the modern methods of ESR have rendered it a powerful technique for studying molecular dynamics in a wide variety of chemical, physical, and biological systems. In NMR, molecular motions in fluids lead to nearly complete averaging of the motion-dependent terms in the spin Hamiltonian, so only their residual effects, reflected in the T_1 and T_2, report on dynamics. In ESR, however, there are often dramatic lineshape variations resulting from the molecular motions, which are particularly sensitive to the microscopic details of the dynamics. This feature is significantly enhanced in the multi-frequency approach, as we have seen. 2D-ELDOR provides unique features in resolving homogeneous from inhomogeneous broadening, clearly distinguishing cross-relaxation processes, as well as T_1's, all of which are valuable for studying molecular dynamics.

A key future development would be to extend 2D-ELDOR to higher frequencies and then to perform multifrequency studies of molecular dynamics by this powerful method. A coherent pulsed high-power spectrometer at 95 GHz has recently been developed to address this objective, (Hofbauer et al, 2003). Another challenge continues to be the development of spin labels with more limited flexibility and well-defined conformations, especially with regard to the study of protein dynamics, (Columbus and Hubbell, 2002). This would reduce effects of the internal motions of the spin label's tether that otherwise can interfere with extracting the more relevant features of the molecular dynamics.

Additional related reviews may be found elsewhere, (Freed, 1998, 2000, 2002; Borbat et al, 2001).

8. ACKNOWLEDGEMENTS:

I am grateful to my co-workers at ACERT. This work was supported by grants from NIH/NCRR, NIH/GMS, and NSF/CHE.

9. GLOSSARY OF ABBREVIATIONS

CIDEP: Chemically-induced Dynamic Electron Spin Polarization
CIDNP: Chemically-induced Dynamic Nuclear Spin Polarization
COSY: Correlation Spectroscopy
CSL: Cholestane Spin Label
ELDOR: Electron-Electron Double Resonance
END: Electron-Nuclear Dipolar Interaction
ENDOR: Electron-Nuclear Double Resonance
ESE: Electron Spin Echoes
ESR: Electron Spin Resonance
FID: Free Induction Decay
FIR: Far Infrared
FT: Fourier Transform
GA: Gramicidin A
HB: Homogeneous Broadening
HE: Heisenberg Spin Exchange
IB: Inhomogeneous Broadening
LA: Lanczos Algorithm
LO: Liquid Ordered
MOMD: Microscopic Order with Macroscopic Disorder
NMR: Nuclear Magnetic Resonance
OTP: Ortho-Terphenyl
SECSY: Spin Echo Correlation Spectroscopy
SLE: Stochastic Liouville Equation
SNR: Signal-to-Noise Ratio
SRLS: Slowly Relaxing Local Structure

10. REFERENCES

Angerhofer, A., Massoth, R.J. and Bowman, M.K. (1988) Fourier transform EPR measurement of homogeneous electron transfer rates. *Israel J. Chem.* **28**, 227-238.

Aue, W.P., Bartholdi and Ernst, R.R. (1976) Two-dimensional spectroscopy application to nuclear magnetic resonance. *J. Chem. Phys.* **64**, 2229-2246.

Barnes, J.P. and Freed, J.H. (1998) Dynamics and ordering in mixed model membranes of DMPC and DMPS: A 250 GHz ESR study. *Biophys. J.* **75**, 2532-2546.

Barnes, J., Liang, Z., Mchaourab, H., Freed, J.H. and Hubbell, W.L. (1999) A multi-frequency ESR study of T4 lysozyme dynamics. *Biophys. J.* **76**, 3298-3306.

Beth, A.H. and Robinson, B.H. (1989) Nitrogen-15 and deuterium substituted spin labels for studies of very slow rotational motion. *Biol. Magn. Res.* **8**, 179-254.

Bloch, F., (1956) Dynamical theory of nuclear induction II. *Phys. Rev.* **102**, 104-135.

Bolton, J.R. and Carrington, A. (1962) Line width alternation in the electron spin resonance spectrum of the durosemiquinone cation. *Mol. Phys.* **5**, 161-167.

Borbat, P.P, Crepeau, R.H., and Freed, J.H. (1997) Multifrequency two-dimensional Fourier transform ESR: An X/Ku band spectrometer. *J. Magn. Res.* **127**, 155-167.

Borbat, P.P., Costa-Filho, A.J., Earle, K.A., Moscicki, J.K. and Freed, J.H. (2001) Electron spin resonance in studies of membranes and proteins. *Science* **291**, 266-269.

Brown, I.M. (1974) Electron spin-echo envelope decays and molecular motion: Rotational and translational diffusion. *J. Chem. Phys.* **60**, 4930-4938.

Budil, D.E., Earle, K.A., Lynch, W.B., and Freed, J.H. (1989) Electron paramagnetic resonance at 1 millimeter wavelengths. In *Advanced EPR Applications in Biology and Biochemistry*, ed. A. Hoff, Elsevier, Amsterdam, **8**, 307-340.

Columbus, L. and Hubbell, W.L. (2002) A new spin on protein dynamics. *Trends in Biochem. Sciences* **27**, 288-295.

Costa-Filho, A.J., Shimoyama, Y. and Freed, J.H. (2003a) A 2D-ELDOR study of the liquid ordered phase in multilamellar vesicle membranes, *Biophys. J.* (in press).

Costa-Filho, A.J., Crepeau, R.H., Borbat, P.P., Ge, M. and Freed, J.H. (2003b) Lipid-Gramicidin interactions: Dynamic structure of the boundary lipid by 2D-ELDOR, *Biophys. J.* (in press).

Crepeau, R.H., Saxena, S.K. Lee, S., Patyal, B.R. and Freed, J.H. (1994) Studies on lipid membranes by two dimensional-fourier transform ESR: Enhancement of resolution to ordering and dynamics. *Biophys. J.* **66**, 1489-1504.

Dobbert, O., Prisner, T. and Dinse, K.P. (1986) Single-channel quadrature FT ESR. *J. Magn. Res.* **70**, 173-175.

Dorio, M. and Freed, J.H. (eds) (1979) *Multiple Electron Resonance Spectroscopy*; Möbius, K., Lubitz, W. and Freed, J.H. (1989) Liquid-state ENDOR and triple resonance. In *Advanced EPR, Applications in Biology and Biochemistry*, (Hoff, A.J., ed.) pp. 441-499, Elsevier, Amsterdam, The Netherlands

Dzuba, S.A., Maryasov, A.G., Salikhov, K.M. and Tsetkov, Yu. D. (1984) Superslow rotations of nitroxide radicals studied by pulse EPR spectroscopy. *J. Magn. Reson.*. **58**, 95-117.

Dzuba , S.A. and Tsvetkov, Yu.D. (1988) Magnetization transfer in pulsed EPR of ^{15}N nitroxides: reorientational motion model of molecules in glassy liquids. *Chem. Phys.* **120**, 291-298.

Earle, K.A., Budil, D.E. and Freed, J.H. (1993) 250 GHz EPR of nitroxides in the slow-motional regime: Models of rotational diffusion. *J. Phys. Chem.* **97**, 13289-13297.

Earle, K.A., Moscicki, J., Polimeno, A. and Freed, J.H. (1997) A 250 GHz ESR study of *o*-terphenyl: Dynamic cage effects above T_c. *J. Chem. Phys.* **106**, 9996-10015.

Earle, K.A., Moscicki, J, Polimeno, A. and Freed, J.H., (1998) Response to "Comment on 'A 250 GHz ESR study of *o*-terphenyl dynamic cage effects above T_c. *J. Chem. Phys.* **109,** 10525-26.

Eliav. U. and Freed, J.H. (1984) The oscillatory nature of polarization evolution in CIDEP. *J. Phys. Chem.* **88**, 1277-1280.

Fraenkel, G.K. (1965) Static and dynamic frequency shifts in electron spin resonance. *J. Chem. Phys.* **42**, 4275-4298.

Freed, J.H. (1964) Anisotropic rotational diffusion and electron spin resonance line-widths. *J. Chem. Phys.* **41**, 2077-2083.

Freed, J.H. (1965) Theory of saturation and double resonance effects in ESR spectra. *J. Chem. Phys.* **43**, 2312-2332.

Freed, J.H. (1968) Generalized cumulant expansions and spin-relaxation theory. *J. Chem. Phys.* **49**, 376-391.

Freed, J.H. (1998) Linewidths, lineshapes, and spin relaxation in the one and two-dimensional ESR of organic radicals and spin-labels. In *Foundations of Modern EPR* (G. Eaton, S. Eaton, and K. Salikhov, eds.) Ch. I.7, pp 658-683. World Scientific, NJ.

Freed, J.H. (2000) New technologies in electron spin resonance. *Ann. Rev. of Phys. Chem.* **51**, 655-689.

Freed, J.H. (2002) Modern ESR methods in studies of the dynamic structure of proteins and membranes. In *EPR in the 21st. Century*, (Kawamori, A., Yamauchi, J. and Ohta, H. eds. Elsevier Science, Amsterdam, Netherlands), 719-730.

Freed, J.H. and Fraenkel, G.K.(1962) Anomalous alternating linewidths in ESR spectra. *J. Chem. Phys.* **37**, 1156-1157.

Freed, J.H. and Fraenkel, G.K. (1963) Theory of linewidths in electron spin resonance spectra. *J. Chem. Phys.* **39**, 326-348.

Freed, J.H., Bruno, G.V. and Polnaszek, C. (1971a) ESR lineshapes and saturation in the slow motional region. *J. Phys. Chem.* **75**, 3385-3399.

Freed, J.H., Bruno, G.V. and Polnaszek, C.F. (1971b) ESR lineshapes for triplets undergoing slow rotational reorientation. *J. Chem. Phys.* **55**, 5270-5281.

Freed, J.H. and Pederson, J.B. (1976) The theory of chemically-induced dynamic spin polarization. *Adv. Magn. Res.* **8**, 1-84.

Ge, M., Field, K.A., Aneja, R, Holowka, D., Baird, B., and Freed, J.H. (1999) ESR characterization of liquid ordered phase of detergent resistant membranes from RBL-2H3 cells. *Biophys. J.*, **77**, 925-933.

Goldman, S.A., Bruno, G.V., Polnaszek, C.F. and Freed, J.H. (1972) An ESR study of anisotropic rotational reorientation and slow tumbling in liquid and frozen media. *J. Chem. Phys.* **56**, 716-735.

Gorcester, J. and Freed, J.H. (1986) Two dimensional fourier transform ESR spectroscopy. *J. Chem. Phys.* **85** 5375-5377.

Gorcester, J. and Freed, J.H. (1988) Two-dimensional fourier transform ESR corrrelation spectroscopy. *J. Chem. Phys.* **88**, 4678-4693.

Gorcester, J., Rananavare, S.R. and Freed, J.H. (1989) Two dimensional ELDOR and electron spin echo study of solute dynamics in smectics. *J. Chem. Phys.* **90**, 5764-5786.

Gordon, R.G. and Messenger, T. (1972). Magnetic resonance line shapes in slowly tumbling molecules. In *Electron Spin Relaxation in Liquids*, (Muus, L.T. and Atkins, P.W. Eds.) Plenum, New York.

Hofbauer, W., Earle, K.A., Dunnam, C. and Freed, J.H. *High-Power Pulsed Spectrometer at 95 GHz for 2D-FT-ESR* (in preparation).

Hornak, J.P. and Freed, J.H. (1983) ELDOR spin echoes and slow motions. *Chem. Phys. Lett.* **101**, 115-119.

Hornak, J.P. and Freed, J.H. (1986) Spectral rotation in pulsed ESR spectroscopy. *J. Magn. Reson.* **67**, 501-518.

Hwang, J.S., Mason, R.P., Hwang, L.P. and Freed, J.H. ((1975) ESR studies of anisotropic rotational reorientation and slow tumbling in liquid and frozen media III: PD-tempone and an analysis of fluctuating torques. *J. Phys. Chem.* **79**, 489-511.

Hyde, J.S. and Maki, A.H. (1964) ENDOR of a free radical in solution. *J. Chem. Phys.* **40**, 3117-3118.

Hyde, J.S., Chien, J.C.W. and Freed, J.H. (1968) Electron-electron double resonance of free radicals in solution. *J. Chem. Phys.* **48**, 4211-4226.

Hyde, J.S. and Dalton, L.R. (1979) Saturation-Transfer spectroscopy. In *Spin Labeling II. Theory and Applications*, (Berliner, L.J. ed.) pp 3-70, Academic Press, NY.

Jeener, J., Meier, B.H., Bachmann, P. and Ernst, R.R. (1979) Investigation of exchange processes by two-dimensional NMR spectroscopy. *J. Chem. Phys.* **71**, 4546-4593.

Kivelson, D. (1960) Theory of ESR linewidths of free radicals. *J. Chem. Phys.* **33**, 1094-1106.

Kivelson, D. and Lee, S. (1982) Theory of ESR parallel-edge lines of slowly tumbling molecules. *J. Chem. Phys.* **76**, 5746-5754.

Korst, N.N. and Khazanovitch, T.N. (1964) Relaxation and shape of paramagnetic resonance lines in highly viscous media. *Soviet Phys. JETP*, **18**, 1049-1055.

Kubo, R. (1962) Generalized cumulant expansion method. *J. Phys. Soc. Japan*, **17**, 1100-1120.

Kubo, R. (1963) Stochastic Liouville equations. *J. Math.Phys.***4,** 174-183.

Kubo, R. (1969a). A stochastic theory of line shape. In *Advan. Chem. Phys.*, (Schuler, K. E., ed). **15**, 101-127, Wiley, NY.

Kubo, R. (1969b) Stochastic theories of randomly modulated systems. *J. Phys. Soc. Japan*, **26,** Supplement 1-5.

Kubo, R. and Tomita, K. (1954) A general theory of magnetic resonance absorption. *J. Phys. Soc. Japan* **9**, 888-919.

Kurreck, H., Kirste, B. and Lubitz, W. (1988) Electron nuclear double resonance spectroscopy of radicals in solution. VCH Verlagsgesellschaft, Weinheim, Germany.

Lee, S., Patyal, B.R., Saxena, S.H., Crepeau, R.H., and Freed, J.H. (1994a) Two-dimensional fourier transform ESR in complex fluids. *Chem. Phys. Lett.* **221**, 397-406.

Lee, S., Budil, D.E., and Freed, J.H. (1994b) Theory of Two-Dimensional Fourier Transform ESR for Ordered and Viscous Fluids. *J. Chem. Phys.* **101**, 5529-5558.

Leniart, D.S., Connor, H.D. and Freed, J.H. (1975) An ESR and ENDOR study of spin-relaxation of semi quinones in liquid solution. *J. Chem. Phys.* **63**, 165-199.

Liang, Z.C. and Freed, J.H. (1999) As assessment of the applicability of multifrequency ESR to study the complex dynamics of biomolecules. *J. Phys. Chem.B* **103**, 6384-6396.

Lou, Y., Ge, M. and Freed, J.H., (2001) A Multifrequency ESR Study of the Complex Dynamics of Membranes, *J. Phys. Chem. B* **105**, 11053-11056.

Mason, R.P. and Freed, J.H. (1974) Estimating microsecond rotational correlation times from lifetime broadening of nitroxide ESR spectra. *J. Phys. Chem.* **78**, 1321-1323.

Meirovitch, E., Nayeem, A. and Freed, J.H. (1984) An analysis of protein-lipid interactions based on model simulations of ESR spectra. *J. Phys. Chem.* **88,** 3454-3465.

Millhauser, G.L. and Freed, J.H. (1984) Two dimensional electron-spin-echoes and slow motions. *J. Chem. Phys.* **81**, 37-48.

Möbius, K., Lubitz, W. and Plato, M., (1989) Liquid State ENDOR and triple resonance. In *Advanced EPR, Applications in Biology and Biochemistry* (Hoff, A.J. ed.), pp. 441-449, Elsevier, Amsterdam.

Moro, G. and Freed, J.H. (1981) The efficient computation of magnetic resonance spectra and related correlation functions from stochastic Liouville Equation. *J. Phys. Chem.* **84**, 2837-2840.

Norris, J.R. and Weissman, S.I. (1969) Studies of rotational diffusion through the electron-electron dipolar interaction. *J. Phys. Chem.* **73**, 3119-3124.

Panferov, Y.F., Grinberg, O.Y., Dubinskii, A.A. and Lebedev, Y.S. (1984) ESR fourier transform spectroscopy with correction for nonuniformity of phase distortion. *Dokl. Phys. Chem.* **278**, 888-890.

Patyal, B.R., Crepeau, R.H., Gamliel, D. and Freed, J.H. (1990) Two dimensional fourier transform ESR in the slow motional and rigid limits: SECSY:ESR, Two-dimensional fourier transform ESR in the slow motional and rigid limits: 2D-ELDOR. *Chem. Phys. Lett.* **175**, 445-452 ,453-460.

Patyal, B., Crepeau, R.H. and Freed, J.H. (1997) Lipid-Gramicidin interactions using two-dimensional fourier-transform electron spin resonance. *Biophys. J.*, **73**, 2201-2220.

Perrin, F. (1934) Mouvement brownien d'un ellipsoïde, 1: Dispersion diélectrique pour des molécules ellipsoïde. *J. Phys. Radium* **5**, 497-511.

Polimeno, A., and Freed, J.H. (1995) Slow motional ESR in complex fluids: The slowly relaxing local structure model of solvent cage effects. *J. Phys. Chem.* **99**, 10995-11006

Polnaszek, C.F. and Freed, J.H. (1975) ESR Studies of anisotropic ordering, spin relaxation, and slow tumbling in liquid crystalline solvents. *J. Phys. Chem.* **79**, 2283-2306.

Redfield, A.G. (1965) The theory of relaxation processes. *Adv. Magn. Res.* **1**, 1-32.

Sastry, V.S.S., Polimeno, A., Crepeau, R.H. and Freed, J.H. (1996a) Studies of spin relaxation and molecular dynamics in liquid crystals by two dimensional fourier transform ESR: I. Cholestane in butoxy benzylidene-octylaniline and dynamic cage effects. *J. Chem. Phys.* **105**, 5753-5772.

Sastry, V.S.S., Polimeno, A., Crepeau, R.H. and Freed, J.H. (1996b) Studies of spin relaxation and molecular dynamics in liquid crystals by two dimensional fourier transform ESR: II PD-tempone in butoxy benzylidene-octylaniline and dynamic cage effects. *J. Chem. Phys.* **105**, 5773-5791.

Saxena, S.K. and Freed, J.H. (1997) Two dimensional ESR and slow motions. *J. Phys. Chem. A*, **101**, 7998-8008.

Schreurs, J.W.H. and Fraenkel, G.K. (1961) Anomalous relaxation of hyperfine components in electron spin resonance. *J. Chem. Phys.* **34**, 756-768 .

Schneider, D.J. and Freed, J.H. (1989a) Calculating Slow Motional Magnetic Resonance Spectra: A User's Guide, in *Spin Labeling: Theory and Applications, Vol. III, Biol. Magn. Res,*.**8**, 1-76

Schneider, D.J. and Freed, J.H. (1989b) Spin relaxation and motional dynamics, in *Lasers, Molecules and Methods,* J.O. Hirschfelder, R.E. Wyatt, and R.D. Coalson, Eds. *Adv. Chem. Phys.* **73**, 387-528.

Schwartz, L.J., (1984). *Molecular Reorientation and Time Domain ESR*, Ph.D. Thesis, Cornell University.

Schwartz, L.J., Stillman, A.J. and Freed, J.H. (1982) Analysis of electron-spin echoes by spectral representation of the Stochastic-Liouville equation. *J. Chem. Phys.*77, 5410-5425.

Schwartz, L.J., Millhauser, G.L. and Freed, J.H. (1986) Two dimensional electron spin echoes: Magnetization transfer and molecular dynamics. *Chem. Phys. Lett.* **127**, 60-66.

Sillescu, H. and Kivelson, D. (1968) Theory of spin-lattice relaxation in classical liquids. *J. Chem. Phys.* **48**, 3493-3505.

Stephen, M.J. and Fraenkel, G.K. (1960) Theory of saturation in electron spin resonance spectra. *J. Chem. Phys.* **32**, 1435-1444.

Stillman, A.E., Schwartz, L.J. and Freed, J.H. (1980) Direct determination of rotational correlation time by electron-spin echoes. *J. Chem. Phys.* **73**, 3502-3503.

Wangness, R.K. and Bloch, F. (1953) The dynamical theory of nuclear induction. *Phys. Rev.*, **102**, 728-739

Xu, D., Crepeau, R.H., Ober, C.K. and Freed, J.H. (1996) Molecular dynamics of a liquid crystalline polymer studied by two-dimensional fourier transform and cw-ESR. *J. Phys. Chem.* **100**, 15873-15885.

Chapter 10

SDSL: A Survey of Biological Applications

Candice S. Klug and Jimmy B. Feix
Department of Biophysics, Medical College of Wisconsin, Milwaukee, WI 53226

Abstract: Site-directed spin labelling (SDSL) is a powerful method for investigating protein structure, function, and dynamics. SDSL involves adding a nitroxyl spin label at a specifically-placed amino acid residue in a peptide or protein. In most cases a sulfhydryl-specific spin label is bound to a cysteine side chain. Such specific placement monitors a local environment via the usual parameters of a nitroxyl spin label such as sensitivity to motion, spin-spin interaction and accessibility to other species that impact the relaxation time.

1. INTRODUCTION

The applications of site-directed spin labeling (SDSL) electron paramagnetic resonance (EPR) spectroscopy to biological systems have grown rapidly in the years since its introduction. SDSL is a powerful method for investigating protein structure, function, and dynamics. The main advantage of this technique is the ability to gain detailed information at a very local site within a protein or peptide, even in complex systems with multiple components.

SDSL typically involves adding a nitroxide spin label to a unique and specifically placed cysteine residue within a protein or peptide. First, if the system contains any native cysteine residues, they need to substituted with a comparable amino acid residue or shown to be unreactive to the spin label (i.e. involved in a disulfide bond or deeply buried within the protein structure). This methodology is typically successful as most proteins do not have an abundance of cysteines and many proteins are remarkably tolerant of substitutions. In the cases where all of the cysteines cannot be removed or removal disrupts function and/or expression, consistent background labeling (though not preferred) can be tolerated. Next, once the reactive-cysteine-free system is created, unique cysteines can be substituted at sites of interest using site-directed mutagenesis techniques. Mutant protein is then purified and checked for retained function. The introduced cysteine residue is then directly reacted with a sulfhydryl-specific nitroxide. The most commonly utilized spin label is 1-oxyl-2,2,5,5-tetramethylpyrroline-3-methylmethanethiosulfonate (MTSL, Figure 1). Since the spin label adds only a small volume to the cysteine side chain, relatively little or no perturbation in the structure or function of most proteins is observed.

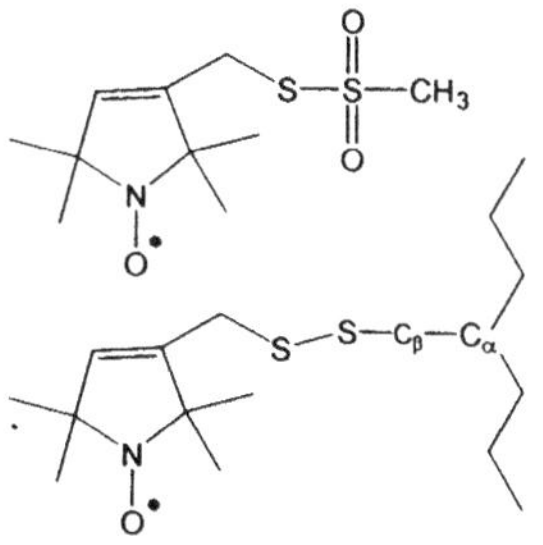

Figure 1. MTSL. Structure of the most commonly used sulfhydryl-specific nitroxide spin label alone and covalently bound to a cysteine side chain.

The advancement of loop-gap resonator (LGR) technology (Hubbell *et al.*, 1987; Froncisz and Hyde, 1982) has greatly enhanced the biological applications of the SDSL EPR spectroscopy technique. Many recombinant proteins of interest are expressed at low levels and the necessary purity required for elimination of background cysteine-containing contaminants often results in a low final yield of protein. In addition, numerous protein preparations may be necessary, depending on the number of sites analyzed within a protein system. Thus, the ability to study relatively small amounts of protein is a key factor enabling biological application of SDSL. LGRs have made this research possible due to i) the small sample size allowed – typically <5μL of 50-100μM spin-labeled protein, ii) the ability to use gas-permeable TPX capillaries that easily allows the introduction and removal of oxygen for accessibility measurements, and iii) the achievement of high microwave field densities with the elimination of heating effects in the sample at higher powers, which are essential for the continuous wave power saturation method of accessibility measurements. Without the LGR, much of the work that has been done on biological systems would not have been possible at such a rapid rate, if at all.

SDSL includes three fundamental classes of experiments that focus on spin label motion, accessibility of the nitroxide side chain, and distance measurements between spin label pairs. The spin label side chain is exceptionally sensitive to local motion of the whole protein, the nitroxide side chain itself, and to motion of the α-carbon (main chain) backbone. In addition, the accessibility of the spin label side chain to polar and nonpolar reagents can be easily measured and gives valuable information on placement of the spin label within the protein, on secondary structure when a series of sites are scanned with the spin label, and on depth of the spin label within a lipid bilayer. Distances between two spin labels in the range of ~ 8-25Å can be determined by continuous wave (CW) SDSL EPR methods, while distances up to nearly 80Å have been determined using pulse methods. Both methods can be applied to a variety of structural and functional studies. The number of spin labeling publications has dramatically increased in recent years and the following is an overview of a representative variety of applications for which SDSL has been useful in studying biological systems.

2. SOLVENT ACCESSIBILITY

The ability to obtain information on the local environment of specifically placed spin labeled side chains within a protein structure is one of the most valuable uses for SDSL. Based on collisions with paramagnetic probes such as oxygen, which is nonpolar and partitions into hydrophobic pockets and into the lipid bilayer, and nickel complexes, which are water soluble probes, a great deal of information can be gained on the environment surrounding the spin label. As a result, it can be determined whether a spin label side chain is exposed to the solvent, buried within the protein core, or membrane-exposed. In addition, by scanning through a region of a protein with the spin label, secondary structure can be resolved. This is especially useful for looking at regions of a protein not found in a crystal structure or that undergo conformational changes upon ligand, protein, or membrane binding. The ability to identify conformational changes in the solution phase is an especially powerful aspect of SDSL. Also, due to the inverse concentration gradients of oxygen and polar nickel reagents into and out of a membrane bilayer, the depth of a spin label side chain in a bilayer can be readily determined. This feature of SDSL is relevant to various applications, including the study of membrane proteins, membrane-associated proteins and peptides, and lipids themselves. The following is an overview of just some of the proteins that have been successfully investigated using this particular aspect of SDSL.

2.1 Secondary structure determination

The ability to determine and resolve local secondary structure within a protein is instrumental in the study of not only its structure, but also its functional dynamics. The SDSL approach has been used successfully to identify secondary structures in soluble proteins for a number of years. The workhorse of this and many of the other SDSL techniques has been T4 lysozyme (T4L), a relatively small, predominantly α-helical protein that is readily expressed and purified.

Secondary structure determination is carried out through the use of paramagnetic probes such as oxygen, nickel complexes (e.g., nickel (II) acetylacetonate, NiAA, and nickel (II) ethylenediaminediacetate, NiEDDA), and chromium oxalate (CROX). The spin label side chain reports on its accessibility to these reagents through changes in its saturation behavior under increasing amounts of microwave power. A plot of microwave power vs. signal amplitude is fit to an equation that yields the parameter $P_{1/2}$, the microwave power at which the height of the center nitroxide line is half what it would be in the absence of saturation (Altenbach *et al.*, 1994). Accessibility of a spin label to a paramagnetic reagent decreases the effective spin-lattice relaxation time, T_1, resulting in an increase in the $P_{1/2}$ value. Thus, the more accessible a spin label is to a given paramagnetic reagent, the higher the $P_{1/2}$ value since the power required for saturation is increased due to collisions of the spin label with the paramagnetic relaxation reagent. Assuming that any change in the spin-spin relaxation time, T_2, upon addition of the relaxation agent is negligible relative to the change in T_1 (a good assumption since T_2 for spin labels in biological systems at ambient temperatures is typically about an order of magnitude shorter than T_1), the *change* in $P_{1/2}$ ($\Delta P_{1/2}$) relative to a standard measured under N_2, for example $P_{1/2}(O_2) - P_{1/2}(N_2) = \Delta P_{1/2}(O_2)$, is directly proportional to the bimolecular collision rate of the spin label with the paramagnetic probe (Altenbach *et al.*, 1989a). To enable comparison between sites with different linewidths (i.e., different T_2s) and between different laboratories, $\Delta P_{1/2}$ values are normalized to the peak-peak width of the center line (ΔH_{pp}) and to the linewidth and saturation properties of a DPPH (diphenylpicrylhydrazyl) standard (Farahbakhsh *et al.*, 1992), generating the accessibility parameter, Π; e.g. $\Pi(O_2) = [\Delta P_{1/2}(O_2)/\Delta H_{pp}]x[\Delta H_{pp}(DPPH)/P_{1/2}(DPPH)]$. This method is termed continuous wave (CW) power saturation, and is by far the most commonly used method to determine spin label accessibility. However, it should be noted that another method involving the use of saturation recovery (SR) EPR has also been utilized that directly measures the spin lattice relaxation time of the nitroxide (e.g. (Altenbach *et al.*, 1989b). The change in T_1 in the presence of a given relaxation agent provides a direct measure of spin label accessibility, circumventing the need to adjust for linewidth or laboratory conditions.

Based solely on accessibility data for a given site, the general environment in which it is located can be established. For example, high accessibility to oxygen indicates a membrane-exposed site and low accessibility could indicate either solvent exposure or burial within the protein core. High accessibility to polar nickel compounds such as NiEDDA or NiAA indicates a solvent exposed site, whereas low or no accessibility would indicate a membrane exposed or a buried site. CROX is a charged polar reagent that, unlike nickel compounds, does not partition into the membrane and can be a good indicator of true solvent exposure. A combination of $P_{1/2}$ values collected for two or three of these reagents will indicate the specific environment and location of a labeled site. For instance, high $\Delta P_{1/2}(O_2)$, very low $\Delta P_{1/2}$(NiEDDA), and no accessibility to CROX indicates a membrane-exposed site; low $\Delta P_{1/2}(O_2)$, very high $\Delta P_{1/2}$(NiEDDA), and high $\Delta P_{1/2}$(CROX) would indicate a solvent-exposed site; and low $\Delta P_{1/2}$ values for all reagents generally indicates a site buried within the protein structure.

Secondary structure can then be identified by scanning through a region of the protein with the nitroxide spin label and plotting the accessibility of each site to either oxygen or nickel compounds against residue number. A periodicity in the data of ~3.6 indicates α-helical structure due to the fact that there are 3.6 residues per turn in an α-helix. Likewise, a periodicity of 2 is observed for β-strands as neighboring residues are positioned on opposite faces of the strand. Oxygen and NiEDDA accessibilities for membrane-embedded secondary structures are out-of-phase (Figure 2), whereas the periodicities of oxygen and NiEDDA accessibility in water-soluble protein and aqueous regions of membrane proteins exhibit in-phase periodicities. Loop regions are identified by the typical lack of any periodicity in a dataset.

The α-helices of the soluble T4L (e.g. (Mchaourab *et al.*, 1996)) and colicin E1 (Todd *et al.*, 1989; Salwinski and Hubbell, 1999; Vogelsang *et al.*, 2001) proteins have been studied extensively, and the transmembrane helical bundles of bacteriorhodopsin (Altenbach *et al.*, 1990; Altenbach *et al.*, 1989a; Altenbach *et al.*, 1994) and rhodopsin (Altenbach *et al.*, 1996; Farahbakhsh *et al.*, 1993; Farrens *et al.*, 1996; Resek *et al.*, 1993) were the first membrane structures studied by SDSL techniques, even prior to the publication of their crystal structures. In fact, SDSL studies were able to resolve loop regions of rhodopsin that the crystal structure was unable to detect (Langen *et al.*, 1999). The ability of this technique to examine local structure in solution, and not just a static structure as is determined by x-ray crystallography, is one of its unique advantages. Crystal structures are excellent starting points for further analysis, but no other technique can so unobtrusively "watch" a protein in solution at such a detailed level. The information on molecular dynamics provided by SDSL is an excellent complement to the high-resolution structural detail provided by crystallographic studies.

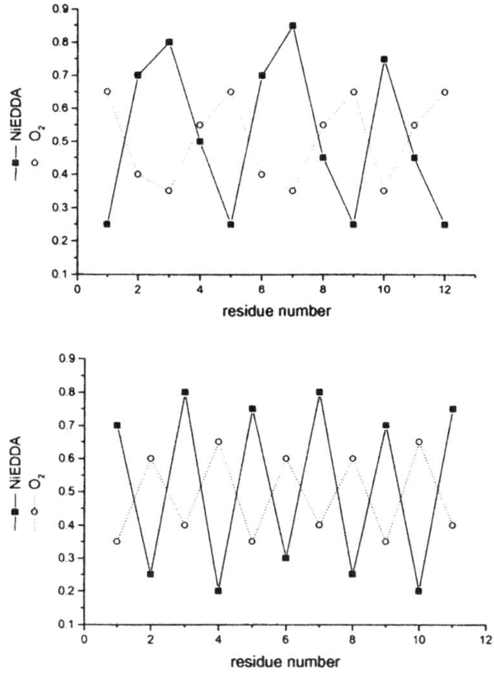

Figure 2. Accessibility periodicity. Examples of out-of phase periodicity in the accessibility parameters, Π(x), for O_2 and NiEDDA observed for a membrane-associated α-helix (top) and *β*-strand (bottom).

Another structure solved by SDSL methods prior to the release of a crystal structure is the *Streptomyces lividans* potassium channel KcsA. The organization of the transmembrane helices was studied by nitroxide scanning and accessibility measurements in the two transmembrane segments that make up each of the four monomers of KcsA (Perozo *et al.*, 1998). It was clearly established that the first transmembrane helix (TM1) was located on the outside of the channel with TM2 lining the interior of the channel at the four-fold axis of symmetry. The publication of the crystal structure of KcsA generally confirmed the SDSL data.

Two helices have been nitroxide scanned in the α-helical annexin XII, a soluble protein that binds to lipid bilayers in the presence of calcium (Isas *et al.*, 2002). The SDSL study confirmed the position and orientation of the helices D and E and the connecting hairpin loop in annexin XII, which is believed to be involved in a calcium-dependent interaction with lipid bilayers. This study also demonstrated a close correlation between accessibility at a given site and the fraction of accessible surface area for the corresponding residue obtained from the crystal structure. Accessibility studies of membrane-bound annexin at pH 4.0 suggested that the helix-loop-helix motif inserted into the bilayer to form a single, membrane-spanning α-helix (Langen *et al.*, 1998a).

Nitroxide scanning accessibility measurements on residues 129–145 of the inhibitory component of the troponin complex TnI, which plays an essential role in the regulation of striated muscle contraction, indicated that TnI was unstructured in a binary complex with the Ca^{2+}-binding component TnC but formed α-helical secondary structure at residues 129–137 upon formation of the ternary TnI-TnC-TnT complex that interacts with tropomyosin (Brown *et al.*, 2002). Helical secondary structure was confirmed by distance measurements between pairs of nitroxides, and changes in spin label motion upon ternary complex formation identified a tertiary contact surface. These studies conflicted with two earlier molecular models of TnI, but were in good agreement with a preliminary, unpublished crystal structure.

Although the α-helical motif was studied first and most extensively, work on β-strands is following closely. Early studies of soluble β-structures were carried out on the soluble proteins cellular retinol binding protein (CRBP) (Hubbell *et al.*, 1998) and α-crystallin (Berengian *et al.*, 1997; Koteiche *et al.*, 1998). These studies established the expected periodicity of 2.0 for β-strand accessibility (Hubbell *et al.*, 1996).

Integral membrane proteins dominated by β-strand structure have also been studied by SDSL. These include the outer membrane protein receptors FepA and BtuB. Investigations of the putative β-barrel structure of the outer membrane protein FepA were the first to identify a transmembrane β-strand by SDSL methods (Klug *et al.*, 1997). One entire strand was structurally characterized and determined to traverse the membrane at a slight angle, consistent with β-barrel structure. The publication of the crystal structure of FepA followed soon after, confirming the SDSL results and supporting the findings that the transmembrane β-strand had continued structure several residues beyond the surface of the membrane. In addition, the vitamin B_{12} transporter, BtuB, was confirmed to have β-structure using SDSL methods (Fanucci *et al.*, 2002). In this study, two consecutive β-barrel strands were characterized and it was found that even numbered sites faced the lipid bilayer and odd numbered sites faced the channel interior. Since the crystal structure of BtuB had not yet been published, a model of the protein was constructed based on the SDSL data collected and compared to the crystal data of FepA and FhuA, another outer membrane transporter. Additional studies on ligand-induced conformational changes in FepA and BtuB are discussed below.

2.2 Protein-Membrane Interactions

One advantage to the study of membrane proteins is that secondary structure information can be combined with depth measurements (described below) to give information on the position and orientation of the α-helix or β-strand relative to the bilayer normal. Oxygen is hydrophobic and therefore has an increasing concentration gradient into the bilayer, with the maximum at the center, whereas nickel complexes are polar and have a decreasing gradient into the bilayer, with a very low concentration in the center of the bilayer and a high concentration on the surface. This reciprocal concentration gradient of commonly used reagents in the membrane allows for the estimation of depth of a spin label within the lipid bilayer.

Spin labeled lipids with the labels at given depths are used to calibrate each bilayer system. The accessibilities of the labels to oxygen and NiEDDA are recorded and converted to a value, Φ, which is defined by the natural log of the ratio of the oxygen and nickel accessibilities: $\Phi = \ln[\Delta P_{1/2}(O_2)/\Delta P_{1/2}(\mathrm{NiEDDA})]$ (Altenbach *et al.*, 1994). Φ is then plotted against the known depths of the nitroxides within the bilayer to give a calibration equation for further depth experiments: $Å = m\Phi + b$.

In addition to the relatively early, seminal studies on the α-helical transmembrane proteins rhodopsin and bacteriorhodopsin mentioned above, SDSL has contributed to the understanding of membrane insertion by channel-forming proteins such as colicin E1 and diphtheria toxin.

The structure of two consecutive helices from the channel forming toxin colicin E1 was studied by SDSL. Twenty three residues were scanned with a spin label and it was found that they formed two separate helices connected by a hairpin loop, a configuration that closely matched that found in the crystal structure of the channel-forming domain (Salwinski and Hubbell, 1999). However, upon binding to negatively charged membranes at low pH (pH4.0), these two colicin E1 helices form one long helix that appears to insert into the lipid bilayer. One face of the helix was found to face a polar environment and the other a hydrophobic environment. In order to distinguish whether the helix was laying across the top of the bilayer or inserted into it, depth measurements were carried out that identified this helix as being transmembrane. Further studies on the structure of neighboring helices reveal that they also extend into one longer helix, but that upon binding to bilayers at low pH the extended helices localize to the membrane/water interface (Vogelsang *et al.*, 2001).

Twenty-one consecutive residues in a putative transmembrane helix (TH9) of diphtheria toxin were analyzed in the presence of phospholipid vesicles at pH 4.6, corresponding to the endosomal pH at which translocation of the toxin occurs (Oh *et al.*, 1996). The pattern of accessibility to O_2 and NiEDDA indicated that this segment was α-helical, with one side of the helix facing the lipid and the other exposed to the aqueous phase. Depth measurements using SDSL on those sites facing the bilayer indicated a transmembrane orientation of the helix.

Accessibility measurements have also played an important role in determining secondary structure and penetration depth for a number of smaller, membrane-active peptides. Although such peptides are amenable to structure determination by NMR and secondary structure analysis by CD spectroscopies, these methods do not provide the detailed information on interaction with the lipid bilayer that is available from SDSL studies. Important examples include alamethicin (Archer *et al.*, 1991; Barranger-Mathys and Cafiso, 1996; Lewis and Cafiso, 1999) and the 25-residue MARCKS (myristoylated alanine-rich protein kinase C substrate) peptide. Depth measurements for twelve single-cysteine analogs indicated that MARCKS is oriented along the bilayer surface with its N-terminus extending into the aqueous phase (Qin and Cafiso, 1996). Additional studies on the interaction of MARCKS with phosphatidylinositol are discussed below. A peptide derived from the protein kinase C/calmodulin binding domain of neuromodulin also bound to negatively-charged bilayers in an extended conformation along the membrane surface with its N- and C-termini extended into the aqueous phase, although the central region of this peptide penetrated more deeply into the hydrophobic phase than MARCKS (Wertz *et al.*, 1996). In contrast, an N-terminal myristoylated peptide derived from Src bound to membranes with its N-terminus close to the bilayer interface and the C-terminal half extended into the aqueous phase (Victor and Cafiso, 1998).

A membrane-binding presequence of yeast cytochrome c oxidase was bound to membranes in an extended conformation, with the nitroxide side chains immersed approximately 13Å below the lipid headgroups (Yu *et al.*, 1994). Accessibility measurements on the N-terminal fusion peptide of influenza virus hemagglutinin indicated that it inserted into the bilayer as an α-helix tilted ~ 25° relative to the membrane surface (Macosko *et al.*, 1997).

An extensive study has recently examined depth of penetration and orientation of the C2 domain of cytosolic phospholipase A2 when bound to membranes in the presence of Ca^{2+} (Frazier *et al.*, 2002). Values of Φ were combined with constraints from the solution NMR structure to generate a model for the protein at the membrane interface. It was suggested that the dependence of Φ on bilayer depth is best represented by a hyperbolic tangent function rather than the linear relationship described above, especially near the bilayer-aqueous interface. This is intuitively consistent with the fact that concentrations of relaxation agents must reach some limiting value in the bulk solution and in an infinitely deep membrane.

Docking of the soluble bee venom phospholipase A2 to small unilamellar vesicles of non-hydrolyzable, anionic phospholipids was studied by measuring the accessibility of several sites to CROX in the presence and absence of membranes (Lin *et al.*, 1998). The concentration of this negatively-charged relaxation agent is expected to decrease near the bilayer surface due to electrostatic repulsion, providing relative distances to the various spin-labeled sites. These relative distances were in turn used to orient the protein, based on its known crystal structure.

In addition to measurement of relative accessibilities to O_2 and NiEDDA, a novel methodology has recently been published for determining the location of residues facing the membrane-aqueous interface (Gross and Hubbell, 2002). The ability to identify residues not only within the membrane bilayer, but also those near the membrane-aqueous interface is invaluable to identification of global protein structure and folding because it restricts the number of possible orientations of the entire protein by confining specific residues to specific surroundings. This technique involves introducing a Ni-chelating lipid (DOGS-NTA; Figure 3) into the membrane and measuring the accessibility of the protein-attached spin label to the paramagnetic nickel ion. Since the metal chelate is attached to the lipid through a spacer, there is an approximately 14Å band near the membrane surface where collisions can occur, which provides an excellent starting point for more accurate positioning. This is especially useful in the absence of a crystal structure.

2.3 Conformational changes in secondary structure

Conformational changes in protein structure are often reflected in both accessibility measurements and in motional changes and therefore cannot strictly belong to one category of SDSL techniques. Illustrated here are two examples of accessibility measurements as a tool for the study of conformational changes; more examples can be found in the sections on motion and distance measurements.

Lactose permease is an α-helical membrane protein involved in the translocation of galactosides and has been extensively studied by SDSL (e.g. (Zhao *et al.*, 1999; Sun *et al.*, 1999; Wang *et al.*, 1998)). Many of the helices have been characterized by nitroxide scanning, and conformational changes have been identified through changes in solvent accessibility (Zhao *et al.*, 1999). Upon nitroxide scanning of helices IV and V, it was found that binding of the ligand, β-galactoside, had no apparent effect on the motion of the labeled side chains. However, an unexpected increase in solvent accessibility at site 137 on helix IV and a decrease in solvent accessibility on one entire face of helix V after ligand binding indicated a translational motion of the entire helix V, yet without perturbation of side chain motion.

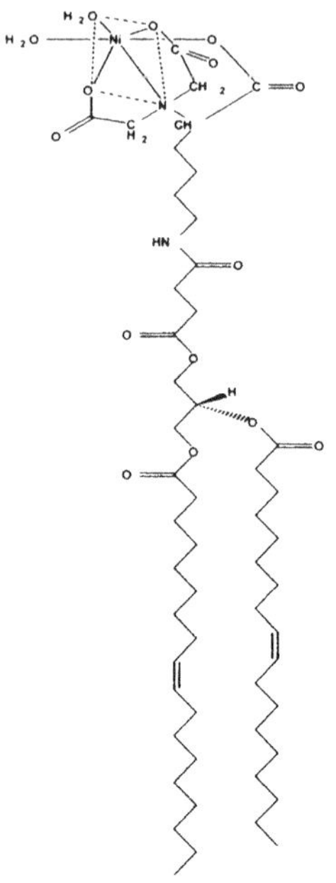

Figure 3. DOGS-NTA.

The mechanosensitive channel, MscL, is a homopentamer of monomers comprised of two transmembrane helices and a cytoplasmic helix that are all thought to be involved in channel gating. It was found by SDSL investigations that a large rearrangement of the monomers occurs upon lipid composition-induced channel opening (Perozo *et al.*, 2002a). Structures of the channel domain in the closed, intermediate, and open states were all determined by SDSL accessibility measurements and the pore size was found to increase from essentially closed to greater than 25Å wide in the open state. Data indicating increases in solvent accessibility and the dynamics of the side chains in both transmembrane helices and the absence of spin-spin interactions between the monomers were combined to map out the nature of the channel rearrangement upon gating.

3. MOTION

The ability to introduce a spin label side chain into a protein at very specifically chosen locations allows the study of the local motion of a particular site within a protein structure. EPR spectra are sensitive to the overall tumbling of the protein, the local motion of the spin label side chain, and the motion of the protein backbone. In order to observe only the motion of interest, it is possible to slow the tumbling of small proteins with viscous solvents (*e.g.*, 30% sucrose, see below), thereby leaving the spectrum sensitive to only side chain and backbone dynamics. In addition, rotation of the bonds within the spin label itself can be dampened to leave the spectrum sensitive to only the backbone motion of the protein. The spectral information gained based on motion alone is another important feature of SDSL and a number of the protein systems that have benefited from this technique are discussed below. Since packing of the nitroxide side chain within the protein structure affects its motion, a good deal of information can be gained on local tertiary and/or quaternary interactions based on motion alone. A particular strength of the spin labeling technique is its ability to report on changes on very localized dynamics. Thus, conformational changes due to substrate binding, membrane binding, secondary, tertiary and protein-protein interactions, denaturation, and other perturbations are all evident in the EPR spectra and can give key information on structural changes at specific sites.

Rotational motion influences the spin label EPR spectrum primarily through modulation of hyperfine and g-tensor anisotropies (Stone *et al.*, 1965; Nordio, 1976), leading to changes in line amplitudes, widths, and positions. The term motion encompasses both the rate and amplitude of changes in the orientation of the spin label in the external magnetic field and rigorous motional analysis requires spectral simulation. Nonetheless, it is often possible to obtain biologically relevant estimates of changes in motion and patterns of motional freedom from readily accessible spectral parameters.

The inverse width of the center line, ΔH_o^{-1}, has proven to be a convenient, semi empirical parameter for the estimation of relative rotational mobility (Hubbell *et al.*, 2000; Hubbell *et al.*, 1998; Hubbell *et al.*, 1996). It is particularly valuable for determination of secondary structure when scanning a continuous region of a protein. For example, ΔH_o^{-1} values for sites in a known α-helix of phage T4L show a periodicity that closely matches the 3.6 residue/turn of an ideal α-helix (Mchaourab *et al.*, 1996). Similar observations have been made for a large number of systems, including colicin E1 (Salwinski and Hubbell, 1999), lactose permease (Voss *et al.*, 1997), and tear lipocalins (Glasgow *et al.*, 1999). Such patterns in rotational mobility generally match quite closely with those obtained from accessibility measurements, but tend not to apply to motionally restricted sites engaged in tertiary contacts.

3.1 Local side chain motion

It has been shown in phage T4L, one of the proteins most thoroughly studied by SDSL, through the study of a large number of mutants by X-ray crystallography (e.g., (Matthews, 1995)) that this protein tolerates single-site mutations without global changes in tertiary structure. In one study (Mchaourab *et al.*, 1996), 30 single cysteine mutants were examined and classified based on the crystal structure of T4L as *1)* solvent exposed helical sites not in the N- or C-cap regions, *2)* solvent-exposed sites near the N- or C-termini of α-helices, *3)* sites with at least some degree of tertiary interaction, *4)* sites in solvent-exposed loops, and *5)* sites buried in the hydrophobic core with no solvent accessibility. Since T4L is a relatively small protein (~17 kDa), rotational tumbling of the entire molecule was suppressed by recording spectra in 30% sucrose (increasing viscosity by a factor of ~ 3) to allow examination of side chain and backbone dynamics.

The first crystal structures of a spin labeled protein were recently published that revealed the orientation of the spin label MTSL attached to various sites on T4L (Langen *et al.*, 2000). Three helix surface sites and one tertiary contact site were presented. This was an important step in linking the side chain dynamics revealed through spectral analysis with a picture of the local structural orientation(s) of the spin label attached to the protein. Most importantly, the preferred conformations of the spin label side chain were correlated with their spectral fingerprints.

In all cases, the electron density of the side chains were clearly resolved, indicating that distinct orientations of the side chain are preferred in these cases. For example, at site 119, an interior helix site with a two-component spectrum, two distinct conformations of the side chain were revealed, accounting for the two spin populations observed in its EPR spectrum. Other sites gave information on molecular and tertiary contact interactions of the side chain that may give rise to its immobilization. Clearly, as additional crystal structures of spin-labeled proteins become available, the relationship between the structure of the protein and side chain mobility will be better understood, allowing correlation of the possible structural interactions with spectral mobility. This will especially benefit future spectral simulations of side chain motion and dynamics in addition to paving the way for the development of novel spin label side chains specifically designed for motional investigations.

The effects of different neighboring side chains on the motion of the spin label in β-sheets was studied in the β-protein CRBP (Lietzow and Hubbell, 1998; Hubbell *et al.*, 1998). It was concluded that nearest-neighbor solvent-accessible side chain interactions strongly influenced spin label motion. This was determined by mutating neighboring side chains and observing the spectral effects on a given spin labeled site. These findings were in contrast to those found in an α-helical model system, T4L, where side chain dynamics of solvent-exposed helical sites are less dependent on nearest neighbors (Mchaourab *et al.*, 1996). Solvent-exposed sites in α-helices appear to be largely dependent on backbone mobility and tertiary, rather than secondary, interactions. The use of spin labels other than MTSL, including the more flexible saturated derivative of MTSL, may also allow better spectral contrast between surface site spin labels and buried labels (Mchaourab *et al.*, 1999).

Changes in mobility and the resultant alteration in signal amplitude at a given spectral position are also the basis for the majority of SDSL studies on time-resolved conformational changes. Often this involves simply "sitting" on a sensitive spectral position such as the maxima of a strongly- or weakly-immobilized peak and observing changes in intensity with time. For example, the kinetics of rhodopsin photoactivation (Farahbakhsh *et al.*, 1993) and the bacteriorhodopsin photocycle (Rink *et al.*, 1997; Mollaaghababa *et al.*, 2000) were followed by optical methods and compared to the kinetics of motional changes at a given spin-labeled site. Agreement not only shows that the relevant biological process is being observed, but also indicates whether the labeled site in question is undergoing a change in local structural environment and to what degree the structural changes and biological processes are coupled. Time-resolved studies have also been used to monitor rigid-body movements associated with the bacteriorhodopsin photocycle (Steinhoff *et al.*, 1994) and membrane insertion of colicin E1 (Shin *et al.*, 1993).

The SDSL approach is also being used in RNA research. Recently, nitroxide derivatives have been introduced at specific backbone locations in an RNA hairpin through a deoxyribo-phosphorothiolate linkage to detect, for the first time, GAAA tetraloop/receptor complex formation (Qin *et al.*, 2001). Here, a method was presented to specifically label internal locations of the RNA backbone independent of the sequence and it was found that the modification and spin labeling of the tetraloop hairpin did not significantly perturb the secondary structure of RNA. The free energy of complex formation and a Mg^{2+}-dependent K_d were determined from EPR motional parameters and laid the foundation for further work on the quantification of weak interactions in nucleic acids and nucleic acid/protein complexes.

In addition, other methods for spin labeling RNA have been published. For example, a 4-thiol was substituted in an unpaired uridine and spin labeled for the detection of long-range RNA/protein interactions by NMR (Ramos and Varani, 1998), an amino-specific spin label was used to label the 2'-amino-modified position of base-paired nucleotides to investigate the trans-activation responsive region of HIV RNA by EPR (Edwards *et al.*, 2001), and Shin and coworkers have used 5' displacement spin labeling by incorporating a guanosine monophosphorothioate at the 5' end of Rev response element, allowing labeling of the 5' end of an RNA molecule (Macosko *et al.*, 1999). Spin labels have also been used in the study of DNA by conjugation to either the sugar-phosphate backbone or a nucleoside base (e.g. (Miller *et al.*, 1995)).

3.2 Protein backbone flexibility

In addition to studying the motion of the protein itself and the motion of the spin label side chain attached to specific sites within a protein, fluctuations in the backbone motion can also be detected by analysis of the spectral dynamics (Columbus and Hubbell, 2002) or with the use of novel modifications to the commonly-used MTSL. Additions to the MTSL nitroxide 4' position (Figure 4) have very recently been shown to give valuable information on the origin of spectral motion (Columbus *et al.*, 2001). The addition of a 4-methyl group restricts motion about the χ_4 and χ_5 bonds of MTSL (Mchaourab *et al.*, 1996; Columbus *et al.*, 2001), while the addition of a 4-phenyl group restricts motion about the remaining bonds, leaving spectral motion sensitive only to backbone fluctuations. This was demonstrated for the first time in the workhorse, T4L, at an external helix site and very nicely illustrated the gradual hindering of spin label motion and the backbone rigidity of the helix (Columbus *et al.*, 2001). A 4-bromo derivative of MTSL was also used to restrict side chain motion and thus improve distance measurements in double-label T4L experiments (Altenbach *et al.*, 2001c). These are unique methods of gathering information about the flexibility of protein backbones in very localized regions and could reveal important information on the function of various secondary structural folds. Although this is a recent advance in SDSL, it likely will become common practice in the near future due to its ability to systematically restrict motion of the spin label side chain and yield direct backbone dynamics.

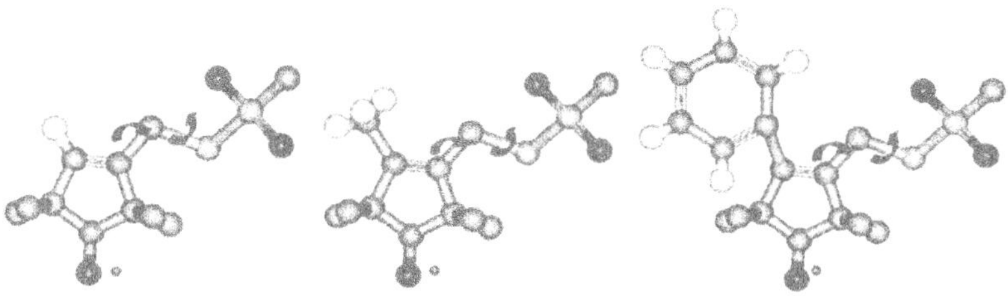

Figure 4. MTSL 4' modifications. MTSL (–H), with -CH_3, and –C_6H_5 modifications. Arrows indicate the χ_4 and χ_5 bond rotations. When the labels are covalently attached to a cysteine residue, χ_4 and χ_5 are the 4th and 5th bonds from the C_α of the peptide backbone (see Figure 1).

3.3 Protein-substrate and protein-protein interactions

The myristoylated alanine-rich protein kinase C substrate (MARCKS) is thought to sequester phosphoinositides within the lipid bilayer. SDSL studies of spin labeled MARCKS peptides indicated that the peptide did indeed sequester phosphatidylinositol 4,5-bisphosphate ($PI(4,5)P_2$) in the membrane, that this mechanism was driven by electrostatic interactions, and that MARCKS did not significantly alter its structure upon binding to $PI(4,5)P_2$ (Rauch *et al.*, 2002). Also interesting in this study was the use of spin labeled $PI(4,5)P_2$. Proxyl-PIP_2 was synthesized, characterized, and used to directly study the sequestration of $PI(4,5)P_2$ in the membrane, which confirmed the hypothesis that MARCKS interacts with multiple $PI(4,5)P_2$ molecules. In both cases, the motional changes in the spectra were followed by a function of the amplitude of the center spectral lineheight and bilayer depth measurements of the MARCKS peptide.

Conformational changes have been seen in the SNARE (soluble NSF acceptor protein receptor) complexes, which appear to be able to switch between helical and random structures either in part or as a whole in a homo- or heterooligomeric complex–dependent manner (Margittai *et al.*, 2001). SNARE complex assembly is necessary for membrane fusion and neurotransmitter release in neurons. SDSL nitroxide scanning studies showed that the membrane/water interfacial domain inserts into the bilayer as the membrane coupling step in membrane fusion (Kweon *et al.*, 2002). Another study on the assembly of the SNARE complex after membrane insertion revealed a possible fusion mechanism first involving the insertion of the membrane domains followed by the assembly of the complex to pull the membranes together for fusion (Kim *et al.*, 2002).

Coupling through trimeric G-proteins is one of the most important signal transduction mechanisms in biology. A Gα protein lacking reactive cysteine residues was engineered and a series of single-cysteine mutations introduced in the functionally-important N-terminus, a region of the protein that is absent in the relevant crystal structures. SDSL showed that this region of the isolated protein is disordered, but adopts an α-helical structure upon interaction with $G\beta\gamma$ (Medkova *et al.*, 2002).

BtuB is an outer membrane protein similar to FepA that is responsible for the uptake of vitamin B_{12} and requires interaction with TonB in order to translocate the bound ligand. SDSL studies were carried out on the TonB box, the N-terminal segment of BtuB that is thought to interact with TonB once ligand is bound to the extracellular loops of the receptor, both before and after addition of cyanocobalamin ligand (Merianos *et al.*, 2000). It was found that this region is a structured helix located within the barrel of the receptor prior to ligand binding, but converts to an extended, disordered segment that likely extends into the periplasm after ligand is bound. In transport-defective mutants of BtuB this region of the protein was unstructured even in the resting state (Coggshall *et al.*, 2001) and showed no evidence of a conformational change upon ligand binding. Both spectral motion and accessibility measurements were utilized in identifying the nature of the ligand-induced conformational change. In addition, experiments were carried out in intact outer membrane preparations rather than purifying and reconstituting the protein. Background labeling was found to be less than 10% of the total signal and the native lipid environment of the receptor was maintained.

Conformational changes near the ferric enterobactin binding loops of FepA showed a dramatic conformational change upon the binding of ligand (Klug *et al.*, 1998) along with a corresponding change in accessibility measurements. In addition, FepA is the receptor for the toxins colicins B and D, for which the mechanism of binding is unknown. Changes in the spectrum were also seen upon the addition of colicin B, but were not as dramatic, indicating that this site is involved in structural rearrangements when both ligands bind, strengthening the theory that distinct binding mechanisms exist for each ligand (unpublished results, C. Klug and J. Feix).

3.4 Protein folding and denaturation

The formation or loss of motional constraints that arise from local tertiary structure provides a sensitive means by which to monitor protein folding or denaturation, respectively. For many systems, chemical denaturation with detergents or compounds that disrupt protein-solvent interactions (e.g., urea or guanidine hydrochloride) is a reversible process, so that determination of the equilibrium between folded and unfolded species as a function of denaturant concentration gives direct insight into the thermodynamic parameters that characterize protein stability. Since the motional properties of an attached spin label will in general be different for the folded and unfolded states, SDSL provides a method for determining these parameters at various sites throughout the protein structure.

Initial studies on FepA indicated that denaturation of a site near the extracellular ligand binding site was well-described by a two-state equilibrium between folded and unfolded species, thus providing a measure of the Gibbs free energy of unfolding (Klug *et al.*, 1995). MTSL bound to this particular site (E280C) was strongly immobilized in the native state, but became freely mobile upon denaturation with both urea and guanidine hydrochloride. Additional studies on the denaturation of FepA have shown this to be generally true for β-strand residues facing the interior, globular domain that fills the channel (Klug and Feix, 1998). It should be noted that free energies of denaturation determined by SDSL are specific for the site examined, i.e. they reflect a local loss of tertiary structure around the nitroxide side chain rather than global unfolding of the protein. The latter can be determined by more non-specific methods such as circular dichroism (CD) or calorimetry. This is also observed with other site-specific methods, for example when hydrogen-deuterium exchange is monitored by NMR or mass spectroscopy, each residue will display a specific free energy of denaturation and these may vary widely within a given protein.

The specificity of denaturation measurements by SDSL for a given labeling site provides the opportunity to compare the stability of different structural elements or domains within a given protein. Similar studies using more global methods such as CD are typically done by expressing each domain independently and measuring their stability in isolation. This can give erroneous results if there are important contacts between domains. In addition, SDSL can be used to examine the effects of mutations on protein stability by maintaining the spin label at a given location and measuring the effects of mutations at other sites of interest. Clearly, there are numerous advantages to the SDSL approach in these types of studies.

Scholes and co-workers have employed time-resolved SDSL to examine the *folding* of cytochrome c (Grigoryants *et al.*, 2000; DeWeerd *et al.*, 2001). Multiple rate constants were observed, demonstrating the complexity of this biological process even for relatively small proteins. Instrumentation was developed and refined allowing observation of kinetic components on the 0.1 msec time scale. This methodology has tremendous potential for examining early events in protein folding and denaturation.

3.5 Peptide-Membrane Interactions

A change in the mobility of a peptide-bound spin label upon membrane binding forms the basis for a sensitive method to measure peptide-membrane binding affinity. In the aqueous phase, peptides and small proteins are generally unstructured and/or rapidly tumbling, leading to a nearly isotropic spectrum with rotational motion on the subnanosecond time scale. Peptide binding to a liposome or cell surface restricts mobility, producing a two-component spectrum that is a superposition of signals arising from free and bound peptide. Because the high-field (M_I = -1) line of the bound peptide is relatively broad (Figure 5), it contributes little to the peak-peak amplitude of the high-field line, so that simple measurement of this amplitude allows rapid determination of the fractions of free and bound peptide. This leads to ready determination of membrane binding affinities and partition coefficients that are essential parameters for understanding the physical basis of peptide-membrane interactions.

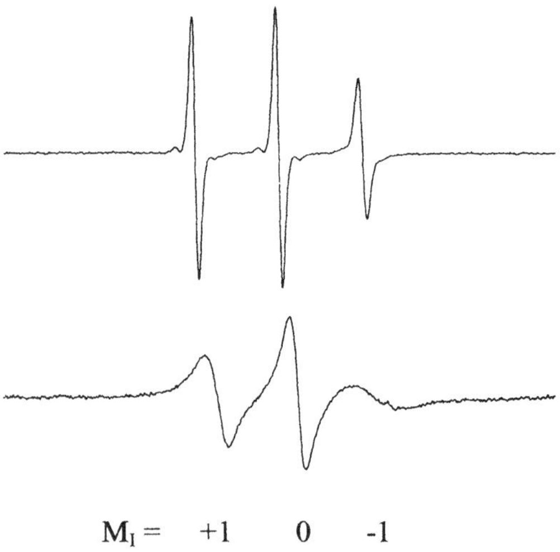

Figure 5. EPR spectra of peptides. Spin labeled peptides free in solution (top) and bound to membranes (bottom).

An excellent example of this methodology is a recent study of five- and six-residue model peptides containing lysine and phenylalanine that were analyzed to determine the relative contributions of hydrophobic and electrostatic interactions to membrane association (Victor and Cafiso, 2001). These peptides bind membranes with partition coefficients that vary from < 10^1 to > 10^3 M^{-1}, depending on peptide composition and the mol fraction of negatively charged lipid. In contrast, the hydrophobic fungal peptide antibiotic alamethicin binds even to neutral bilayers with partition coefficients on the order of 10^4 - 10^5 M^{-1} (Archer *et al.*, 1991; Lewis and Cafiso, 1999). A spin-labeled analog of the 33-residue insect peptide antibiotic cecropin binds to membranes containing 30 mol % PG with a partition coefficient of ~ 5 x 10^3 M^{-1} (Mchaourab *et al.*, 1994), and use of this methodology has allowed us to rapidly evaluate binding affinities for a number of modified cecropin analogs (J. Feix, unpublished data).

The ability to design membrane-binding peptides and the development of insights into the forces governing insertion of integral membrane proteins requires an understanding of the relative affinities of the individual amino acids for the hydrophobic phase of the bilayer. A number of relative thermodynamic scales have been developed, such as those of Wimley and White (Wimley and White, 1996) and Li and Deber (Li and Deber, 1994). Shin and co-workers have used SDSL to develop a similar scale for hydrophobic propensity (Thorgeirsson *et al.*, 1996). They employed a host-guest system based on the membrane binding presequence of yeast cytochrome c oxidase, with each peptide containing a spin-labeled site and a second site for the guest amino acid, to determine the free energy of transfer (ΔG_t) from aqueous solution to the membrane. Changes in ΔG_t relative to glycine were in good agreement with those determined from octanol:water partition coefficients of the isolated N-acetyl amino acid amides. A study on the temperature-dependence of membrane partitioning for four of these peptides indicated that membrane binding was largely entropy-driven (Russell *et al.*, 1996). A subsequent study that employed a designed α-helical peptide as the host suggested that hydrophobic propensity measured in this manner was not dependent on the site at which the guest peptide was introduced (Russell *et al.*, 1999). Thus, SDSL allowed analysis of this fundamental physical property in a context much more biologically relevant than simple partitioning between aqueous solution and an organic solvent.

3.6 Additional approaches

In addition to the standard approach of site-specific cysteine incorporation followed by specific labeling of the -SH group with MTSL, peptides prepared by solid-phase peptide synthesis (SPPS) can incorporate the nitroxide amino acid TOAC (2,2,6,6-tetramethylpiperidine-1-oxyl-4-amino-4-carboxylic acid; Figure 6) (e.g. Marchetto *et al.*, 1993, Hanson *et al.*, 1996; McNulty *et al.*, 2000; Victor and Cafiso, 2001), reviewed by (McNulty and Millhauser, 2000)). The TOAC label is tightly coupled to the peptide backbone, and the nitroxide moiety lies only ~ 2.4Å from Cα. TOAC incorporation appears to stabilize α- and 3_{10}-helices (Hanson *et al.*, 1996). Recently, a method was described for coupling a carboxylate spin label to a diaminopropionic acid (Dap) residue during SPPS (McNulty *et al.*, 2002). The Dap-spin label side chain (Figure 7) is expected to be less perturbing than the TOAC label, especially at non-helical sites. Amino group-specific succinimide spin labels can also be used to label small peptides and proteins, binding to lysine residues or the N-terminus (Altenbach *et al.*, 1989b; Archer *et al.*, 1991), however most proteins have far too many reactive amino groups to make this approach very useful.

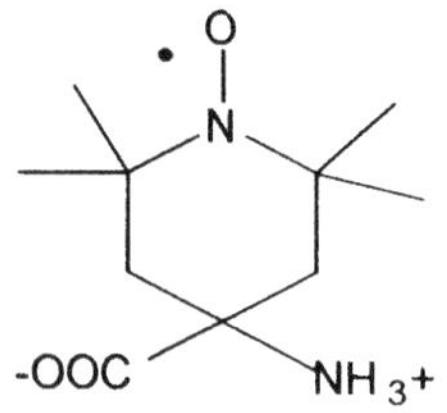

Figure 6. TOAC. Spin label used in peptide synthesis.

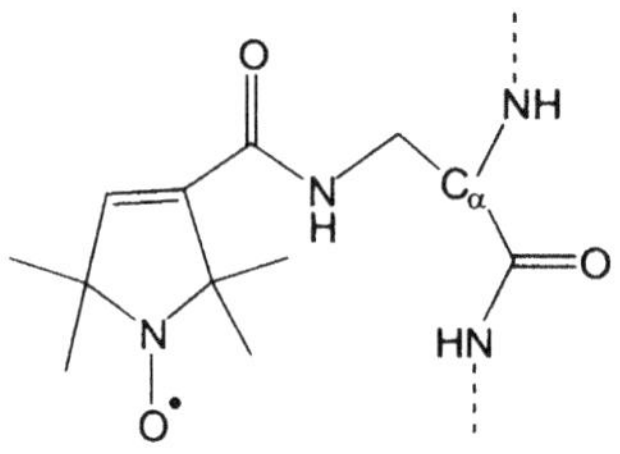

Figure 7. Dap-Spin Label.

Additional spin labels other than MTSL and its analogs can also be used to label the cysteine sulfhydryl group. Traditionally, maleimide (e.g. (Singh *et al.*, 1995; Hustedt and Beth, 1996; Blackman *et al.*, 2001) and iodoacetamide (e.g. Panse *et al.*, 2001; Kersten *et al.*, 2000) labels have been used with good success. Although their absolute chemical specificity for cysteine is not as high as the methanethiosulfonates, they do have the advantage of being chemically stable in the presence of reducing agents. It is also likely that there will continue to be numerous modifications and adaptations of MTSL (some of which are discussed above), such as the addition of charged or polar groups that more accurately mimic the native amino acid side chain being replaced.

It is becoming apparent that a multifrequency approach, including the use of Q- and W-band as well as even higher frequencies, will provide a more complete characterization of a given labeling site (Borbat *et al.*, 2001). At frequencies of 94GHz and higher, the spin label spectrum is more sensitive to the faster motional dynamics range and shows better resolution of the g and A tensors for anisotropic motion. These benefits have already been useful for the study of membranes and membrane proteins (Mangels *et al.*, 2001; Smirnov *et al.*, 1995; Steinhoff *et al.*, 2000; Barnes *et al.*, 1999; McNulty *et al.*, 2000; Bennati *et al.*, 1999; Borbat *et al.*, 2002). In addition, dipolar couplings between sets of two spin labels are more accurately analyzed at higher frequencies (Hustedt *et al.*, 1997; Hustedt and Beth, 1999; McNulty *et al.*, 2000).

Also, pulsed and double quantum EPR methods that provide more direct insights into relaxation properties are likely to play an increasingly important role in future SDSL studies (Borbat *et al.*, 2001; Eaton *et al.*, 2000). These approaches are particularly powerful for making nitroxide-metal and nitroxide-nitroxide distance measurements (discussed below).

4. DISTANCE MEASUREMENTS

Distances between two nitroxides, or a nitroxide and a metal ion, can provide information on both protein structure and functional dynamics. SDSL EPR can give information on distances between spin labels within the range of about 8-25Å using CW methods, and of 50Å or greater using pulse methods (extensively reviewed in Volume 19 of this series). Various methodologies have been developed over the years to study the interaction of spin labels in biological systems. For example, methods exist for data acquisition in frozen solution or at room temperature, and various programs exist to analyze and quantitate the spectra that result from spin–spin interactions. The ability to monitor conformational changes within a protein due to structural rearrangement is a unique benefit of this technique.

4.1 Acquisition and Analysis Methodologies

A growing application of SDSL is the measurement and analysis of magnetic dipolar interactions between two spin labels to determine interspin distance. Following initial mapping of α-helices, loops, and β-strands, distance measurements can provide insight into how these secondary structural elements pack in the tertiary structure of the protein, and the observation of alterations in distance between sites upon ligand binding or protein-protein interaction is a powerful approach to characterizing conformational changes. Determination of spin-spin distances between different monomeric components can provide insights into the quaternary organization of a macromolecular complex. Some of the more commonly used approaches for measuring interspin distance are discussed below in representative applications.

Rabenstein and Shin introduced a method for determination of interspin distance based on Fourier transform deconvolution of dipolar-coupled spectra (Rabenstein and Shin, 1995; Xiao and Shin, 2000). The EPR spectra of interacting spins were treated as a convolution of non-interacting powder pattern spectra with a dipolar broadening function as described by Pake

(Pake, 1948). This approach was validated using a series of lysine-containing polyalanine peptides with the general sequence Ac-$(A_4K)_4$A-$CONH_2$ that folds into a well-defined α-helix. Pairs of cysteines were substituted for alanine residues at various distances apart, labeled with MTSL, and their spectra obtained in frozen solution. Excellent agreement between the measured distances and those based on a molecular model was obtained for sites spaced between 7 and 25Å (Rabenstein and Shin, 1995). Modeling necessarily must take into consideration the size of the nitroxide side chain, and this study found the best fit for an arm length (from the Cα carbon to the nitroxide) of 6.7Å, in good agreement with theoretical expectation. One drawback of this method is the requirement to have rigid-limit spectra, which will usually require freezing of the sample. A strength, however, is the ability to accurately determine distances even in the presence of singly-labeled species. This is of particular importance since it is often difficult to attain stoichiometric labeling.

The Rabenstein and Shin approach has been utilized in a large number of double-labeling applications including studies of HIV gp41 peptides (Rabenstein and Shin, 1996), the KcsA potassium channel (Perozo *et al.*, 1998; Perozo *et al.*, 1999; Liu *et al.*, 2001; Gross *et al.*, 1999), the mechanosensitive MscL channel (Perozo *et al.*, 2002b), the neuronal SNARE complex (Kim *et al.*, 2002), and the inhibitory component of cardiac muscle troponin (Brown *et al.*, 2002).

Hustedt and Beth have developed rigorous simulation methodologies that provide the most precise assessment of distance and relative orientation between two nitroxides currently available (Hustedt *et al.*, 1997; Hustedt and Beth, 2000). Their approach determines all of the independent variables describing the spatial relationship of the two nitroxides, including the interspin distance and five independent angles. The accuracy of the fit is significantly enhanced by obtaining spectra at multiple frequencies, and this study utilized data at X-, Q-, and W-band. Resolution was also enhanced by using perdeuterated spin labels to reduce inhomogeneous broadening. The distance between two spin-labeled NAD^+ analogs bound to glyceraldehyde-3-phosphate dehydrogenase was measured as 12.85Å with 99% confidence levels on the order of 0.1 – 0.2Å. It should be noted that the resolution attained in this approach depends on having two labels at a fixed distance and in a fixed relative orientation. These conditions are not often achieved in SDSL, except in the case of buried, strongly-immobilized sites. Further development to accommodate the distribution of distances and angles normally found with spin labeled cysteine residues is in progress (Hustedt and Beth, 2000).

An important goal in SDSL is the accurate measurement of interspin distances at ambient temperature. This would allow the study of biomolecules in their native state (*i.e.*, the liquid phase), and open up numerous possibilities for the study of protein dynamics. In the rapid-tumbling limit, Redfield relaxation theory has been used to determine interspin distances in T4L (Mchaourab *et al.*, 1997). For this relatively small protein, rotational modulation of the dipolar interaction provides a relaxation mechanism leading to homogeneous line broadening. Double-mutant spectra were fit by convolution of a Lorentzian broadening function with the sum of the corresponding single mutant spectra. The width at half-height of the Lorentzian determined in this manner correlated well with interspin distances from a molecular model, and had a distance dependence of $1/r^6$ that is consistent with theory for dynamic modulation of the dipolar coupling (in contrast to the $1/r^3$ dependence for static dipolar coupling). Distance measurements in the presence and absence of substrate indicated a large domain displacement, supportive of a proposed mechanism for T4L catalysis (Mchaourab *et al.*, 1997).

Altenbach *et al.* have described a method based on static dipole-dipole coupling that is applicable to larger proteins at ambient temperature (Altenbach *et al.*, 2001c). This interactive approach uses Fourier deconvolution of dipolar-coupled spectra as introduced by Rabenstein and Shin (Rabenstein and Shin, 1995) to yield a broadening function that is then compared to a simulated broadening function generated from a user-selected algebraic sum of Pake functions. A distribution of spin-spin distances (or in some cases a single distance) is derived from the set of Pake functions used to simulate the experimentally-derived broadening function. Recombination of the simulated broadening function with the non-interacting spectra and comparison to the original dipolar-coupled spectrum provides a self-consistent validation of the result. Deconvolution operations are performed on the experimentally-obtained first derivative spectra (processed to remove baseline and phasing artifacts), and a smoothing bias or low-pass filter is used to improve the Fourier analysis.

This interactive approach was first tested on a set of T4L mutants analyzed in frozen solution, and at ambient temperature with sucrose added to decrease the tumbling rate of the whole protein (Altenbach *et al.*, 2001c). Cysteines were labeled with MTSL or with an MTSL analog modified at the 4'-ring position with bromine to reduce the rotational mobility of the probe relative to the protein backbone. Non-interacting spectra were taken either as the sum of the single mutants or from the double mutant labeled with a mixture of MTSL and its diamagnetic N-acetylated analog. Good agreement was found between distances obtained at ambient temperature and in frozen solution. Although spectra of the brominated pyrroline derivative were fit somewhat better than those obtained with MTSL, overall results indicated that residual motion of the nitroxide at ambient temperature had little effect on the estimated distances. Furthermore, distance distributions obtained with this method in several cases showed more than one maximum, consistent with the rotameric positions on the spin label side chain relative to the protein backbone observed in crystal structures of MTSL-labeled T4L (Langen *et al.*, 2000).

This interactive approach was also used to examine light-dependent structural changes in rhodopsin (Altenbach *et al.*, 2001b; Altenbach *et al.*, 2001a). In the first study, a reference nitroxide was placed at the cytoplasmic end of a transmembrane helix (TM1) and distances determined to a series of spin labels that spanned the cytoplasmic end of transmembrane helix 7 and a helix that lies along the cytoplasmic surface (helix 8) near the C-terminal (Altenbach *et al.*, 2001b). In the second study, the reference site was placed in helix 8 and distances measured to a series of sites spanning the cytoplasmic ends of two helices and an intervening loop (Altenbach *et al.*, 2001a). In both cases, displacements upon light-activation could be measured with a resolution in the 2–4Å range. These studies illustrate the general strategy of measuring coupling for a series of residues to a single reference site to generate a pattern of distances that reflect local structure (Altenbach *et al.*, 2001b; Altenbach *et al.*, 2001a; Altenbach *et al.*, 2001c). This at least partially eliminates errors due to packing of the nitroxide side chain, and can be especially useful in identifying rigid-body movements of secondary structure elements such as the twisting and tilting of helices.

Dipolar coupling between a spin label and a protein-bound metal ion has also been used to make distance measurements. In one study, a poly-histidine binding site for Cu^{2+} was inserted into T4L and distances measured to a series of spin labels along an adjacent helix (Voss *et al.*, 1995b). Dipolar broadening theory as described by Leigh (Leigh, 1970) gave distances in good agreement with the known structure. This approach was then applied to the characterization of a transmembrane α-helix in lactose permease (Voss *et al.*, 1995a). In another study, a sulfhydryl-reactive gadolinium (III) complex was selectively attached to a cysteine in lac permease that could be protected by substrate during spin labeling and used for distance measurements to nearby helices (Voss *et al.*, 2001). Although it was possible to extract information on helix packing in this study, the large size of the Gd^{3+} chelator added an extra degree of uncertainty to the measured distances. In addition to these studies, Eaton and Eaton have thoroughly developed the field of metal-nitroxide distance measurements. Their studies have been recently reviewed (Eaton *et al.*, 2000) and are discussed in another chapter of this book.

Recent developments in pulsed EPR methods for distance measurements, most notably the double electron-electron resonance (DEER) (Pfannebecker *et al.*, 1996; Larsen and Singel, 1993; Jeschke *et al.*, 2000) and double-quantum coherence (Borbat *et al.*, 2001; Borbat *et al.*, 2002; Borbat and Freed, 2000) techniques, hold significant promise for SDSL applications. Interspin distances of ~ 50Å have been reported (Borbat *et al.*, 2002), and in principle, these approaches are capable of detecting spin-spin coupling at distances as long as 80Å (Borbat *et al.*, 2001; Pannier *et al.*, 2000), greatly expanding the number of experiments that can be envisioned for elucidating structure, dynamics, and intermolecular interactions.

4.2 Monitoring structural changes

The conformational change that occurs in rhodopsin upon the absorption of a photon of light was first mapped out by SDSL EPR (Farrens *et al.*, 1996; Altenbach *et al.*, 1996). Of the seven transmembrane helices of rhodopsin, it was found that a residue on helix III, site 139, remained fairly fixed upon light activation. Double mutants between helix site 139 and various sites on helix VI were constructed. The distances were measured both in the dark, resting state, and then after light activation. Through the distance data collected on various double pairs, the nature of the conformational change that occurred in rhodopsin was mapped out and illustrated an upward twist and outward movement of the entire helix VI. In a related non-EPR study, this movement was confirmed by showing that disulfide bonds between the pairs prevented the conformational change and thus the activation of rhodopsin (Yang *et al.*, 1996).

KcsA has also been studied using EPR distance measurements. Since KcsA is comprised of four monomers, distances were measured between the same sites on different monomers. In order to eliminate multiple spin-spin interactions (i.e. between >2 spin labels in close proximity), tandem dimers were constructed that contained only one cysteine each, leaving only two cysteines to interact in the tetramer and giving only one distance measurement. The rearrangement that occurs at the bottom of the channel between open and closed states was mapped out using a set of ten mutant pairs (Liu *et al.*, 2001).

A second approach used to circumvent problems caused by multiple (i.e. >2) labels in close proximity in the tetrameric KcsA channel was the use of diamagnetic (N-acetylated) labels (Gross *et al.*, 1999). Underlabeling, (i.e. labeling of the protein at a spin label concentration of < 4 labels/ tetramer) has also been used as a technique to study oligomers. However, underlabeling can be problematic given the variability in reactivity for different sites. For labeling sites in close proximity, binding of the first spin label can modify the reactivity of nearby cysteines due to steric hindrance, altering the distribution of labels among oligomers from that expected based on the labeling stoichiometry. This can make it difficult, if not impossible, to obtain reliable quantitative results. Use of a diamagnetic spin label analog such as that shown in Figure 8 overcomes these problems by allowing one to label with an excess of reagent, saturating the available labeling sites at a known ratio of paramagnetic to diamagnetic labels and to more accurately mimic the packing state of two labels per tetramer without the broadening due to spin-spin interaction seen with two paramagnetic labels.

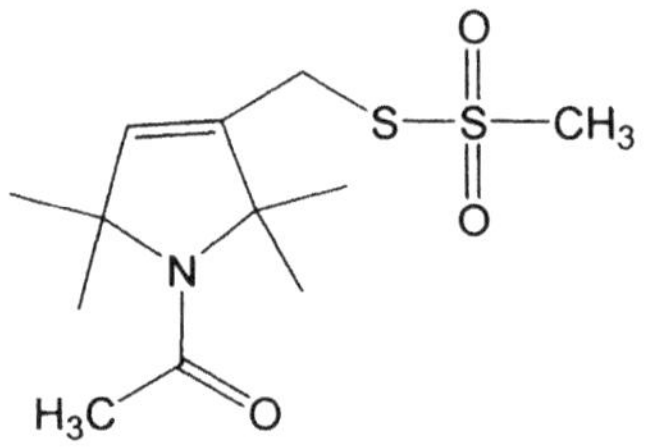

Figure 8. Diamagnetic label. MTSL is N-acetylated to remove the free electron.

In addition to providing distance measurements within a protein or between subunits of a multimeric protein, spin-spin interactions also provide an approach for determination of the number of subunits in an oligomeric assembly. An elegant example of this is the elucidation of the oligomeric state of membrane-bound annexin XII (Langen *et al.*, 1998b). Although annexins exist as monomers in solution, they crystallize in a variety of quaternary states. For the membrane-bound state, electron microscopy images were interpreted as trimers, and chemical cross-linking indicated the presence of both trimers and hexamers. SDSL studies were undertaken to determine the oligomeric state of membrane-bound annexin under more physiologically-relevant conditions. Labeling sites were selected based on the crystal structure so that, for one single mutant (K132C) and two double mutants, the labeling sites would be far apart in the monomer but close (< 5Å) in the trimer or hexamer, thus producing strong spin-spin interactions only in the higher oligomeric states. Second, an additional single mutant and double mutant were constructed to similarly distinguish between trimer and hexamer. In the absence of Ca^{2+} all of the constructs gave well-resolved spectra indicating nanosecond scale rotational motion. Upon addition of Ca^{2+} and subsequent membrane binding, extensive spin-spin broadening gave a clear indication of oligomer formation indicative of trimer, but not hexamer, formation. To further verify the association state of membrane-bound annexin, spin-dilution experiments were performed. The EPR spectrum of spin-labeled K132C mixed with a large (9-fold) molar excess of unlabeled, cysteine-less annexin indicated relatively free rotational mobility (consistent with its location in the crystal structure) and no indication of spin-spin interaction. As the fraction of spin-labeled K132C was increased, line broadening and the loss of signal amplitude characteristic of spin-spin interactions increased. The dependence of the normalized signal amplitude on the mol fraction of labeled protein was modeled for dimers, trimers, and hexamers according to a binomial distribution, and fit very well with the dependence expected for trimers.

Spin-spin interactions have also been exploited to examine protein association for the cardiac peptide phospholambin (Karim *et al.*, 1998), fibril formation by prion protein (Lundberg *et al.*, 1997), a conserved sequence in α-crystallin and small heat-shock proteins (Berengian *et al.*, 1999), and α-helical coiled-coil alignment for vimentin intermediate filaments (Hess *et al.*, 2002).

SDSL has also recently been used to examine the organization of protomers within amyloid fibrils formed by the protein, transthyretin (Serag *et al.*, 2001; Serag *et al.*, 2002). These insoluble fibrils are considered to be models for the deposits formed by β-amyloid in Alzheimer's disease, α-synuclein in Parkinson's disease, and in prion diseases. Transthyretin (TTR) normally exists as a soluble tetramer, however a number of clinically-relevant mutations resulting in fibril formation are known and the native protein can be induced to form fibrils under acidic, partially denaturing conditions.

In an initial study, distance measurements in the native dimer interface were used to identify sites that were in close proximity (Serag *et al.*, 2001). To obtain spectra corresponding to each site in the absence of spin-spin interactions, magnetically-dilute samples were prepared by labeling with a mixture of MTSL and an excess of its diamagnetic analog. This strategy was developed earlier to study tetramers of the potassium channel, KcsA (Gross *et al.*, 1999). Distance estimates were obtained from room temperature spectra by treating dipolar-coupled spectra as a convolution of the spectrum in the absence of spin-spin interactions with a broadening function composed of a weighted average of Pake functions, as described above (Altenbach *et al.*, 2001c). Positions of close contact in soluble TTR were identified and found to be relatively constant upon transformation to the fibrillar state, providing evidence that strands making up the native dimer interface remained in close proximity in the fiber.

A subsequent study (Serag *et al.*, 2002) used the same basic approach to identify a conformational change and formation of a new, non-native interface upon fibrillization of transthyretin. Fiber formation was accompanied by displacement of β-strands C and C', eliminating a strong spin-spin interaction observed in soluble TTR, and formation of a new interface between strands B and B' that produced dipolar couplings not observed in the soluble state. A model for fiber elongation was proposed based on these and the earlier results. The insolubility and non-crystalline nature of amyloid fibrils make them difficult to study by other physical techniques, and it is likely that many additional SDSL studies on these highly medically-relevant systems will be forthcoming.

4.3 Substrate-protein interactions

Distance measurements between spin-labeled protein and spin-labeled substrate is the next obvious step in distance methodologies. These studies require both an effective labeled substrate that retains high binding affinity and a labeling site on the protein that does not block or perturb binding of the substrate and yet is within the range of accurate distance measurement, and these can be significant obstacles depending on the system under study. Substrates may either be chemically synthesized containing a spin label or modified to contain a reactive site such that one of the commonly used labels can be specifically attached. Both have been done successfully; for example, proxylPIP, used in the study of MARCKS (Rauch *et al.*, 2002), and spin labeled galactosides for binding to lactose permease (Zhao *et al.*, 2000) have been chemically synthesized, as have spin labeled high-affinity inhibitors of the erythrocyte anion channel, band 3 (Hustedt and Beth, 1996). Spin labeled γS-ATP has been made by modifying the third phosphate to a sulfur group and spin labeling with MTSL (Koteiche *et al.*, 1995). However, it was found in the latter case that once labeled, the ATP analog more closely resembled and biologically mimicked NADH rather than ATP due to its larger structure. Spin labeled NADH and CoA analogs have also been used to study cofactor binding sites (Hustedt *et al.*, 1997; Panse *et al.*, 2001; Kersten *et al.*, 2000).

Although these examples exist and many more are likely in progress, they have all been carried out on unlabeled protein systems and do not involve distance measurements. In the case of double label experiments, not only does the labeled substrate have to be present and functionally effective, the labeled site on the protein must be within about 25Å (for CW measurements) or 80Å (for pulse methods) of the spin labeled substrate. Although that is a fairly broad range to work within, the flexibility of the linker arm of the spin label and local packing of two closely positioned labels are not necessarily conducive to a successful experiment. Nonetheless, it is expected that these hurdles can be overcome and that successful examples of this method for mapping out protein contact interfaces will be published in the years to come.

Another possibility for distance measurements between substrate and protein is to detect metal to spin label distances (reviewed in (Eaton and Eaton, 2000)). This methodology has been successfully carried out on several systems, including FepA. Specifically, a distance estimate of about 20-30Å was determined between the Fe^{3+}-containing ligand and a spin-labeled site in the loop region of FepA using electron spin echo (ESE) EPR spectroscopy (Klug *et al.*, 1998). For those systems in which the protein or ligand contain a metal (that is either paramagnetic or can be substituted with one that is), this technique is less intrusive than double-cysteine labeling.

5. METHODOLOGY

Although nearly every journal article on SDSL contains a descriptive methods section on how to spin label a protein, the following is a brief summary of general techniques and pitfalls.

1. Remove any native cysteines or determine to be unreactive.

The first step in SDSL is to assure that the native peptide or protein to be studied is unreactive toward labeling. If native cysteines are present, they must be removed by replacement with serine or another appropriate residue, or they must be shown to be inaccessible to labeling due to disulfide bonding or burial within the protein structure. Very often serine is the residue of choice for substituting out cysteines, however other residues such as alanine have been used in order to retain native function and folding.

2. Introduce new cysteine site(s) using site-directed mutagenesis.

Once the protein to be studied is free of reactive cysteines, unique cysteine residues can be introduced at selected sites of interest. Cysteines are introduced by site-directed mutagenesis of the plasmid-encoded gene for expressed proteins or via SPPS for peptides.

3. Purify the mutant protein and check for retained activity.

Once the cysteine mutation has been verified by sequencing of the gene, the protein is expressed and purified under the same conditions as the native protein. Peptides are typically purified by reverse-phase HPLC and their mass verified by mass spectrometry. Protein purification is an important step for SDSL studies, as contaminating proteins likely contain cysteine residues that will readily label and lead to background labeling problems. It is best to check the purity of the purification by spin labeling the reactive-cysteine-free protein preparation. Ideally, there should be no labeling of the preparation.

4. Spin label introduced unique cysteine(s).

The most common spin labels, including MTSL, are sulfhydryl-specific labels. Therefore, only the introduced cysteine residues in the protein or peptide will be modified with the spin label. Typically, the spin label comes as a dried powder that may be dissolved in 100% acetonitrile to make a stock solution that can be kept at -20°C. MTSL reacts with itself to form dimers in aqueous solutions, thus the use of neat acetonitrile is necessary for long-term storage. For labeling of the protein in solution, a more dilute stock (<10%

acetonitrile) in an appropriate buffer is made up from the acetonitrile stock and added to the protein solution at a 10:1 concentration ratio for overlabeling, or a 1:1 ratio for stoichiometric labeling. The addition of acetonitrile to protein solutions often denatures the protein and spin label dimers form upon storage in aqueous solutions, thus the need for a second stock solution to be made up just prior to spin labeling.

Labeling for exposed cysteines should be complete within minutes of adding spin label, whereas overnight labeling is common for more buried sites.

5. Remove excess spin label and record EPR spectrum

Excess spin label can be removed by various methods. Dialysis of the labeling reaction against buffer, the use of small desalting columns, and successive dilution and concentration in centrifugal filters are all commonly used for removal of excess label. It is important to remove excess free spin label as even small (low μM) amounts are readily observed in the EPR spectrum.

6. Spin concentration

The concentration of the spin label within a sample can be readily calibrated using a spin standard of known concentration. A spectrum of a spin labeled protein is taken under the same conditions and instrument settings as a sample of spin label alone of known concentration. Since the area under the spectral lines is equal to the number of spins in the sample, the first-derivative spectra are double-integrated to get the area under the spectrum. This integrated number is compared to the known spin concentration and then used as a calibration for determining the spin concentration in the protein sample. (If the EPR spectrum of the labeled protein is broad, it is often useful to denature the protein by addition of an equal volume of 8M guanidine hydrochloride or urea in order to obtain narrower lines that can be more accurately integrated.) The labeling stoichiometry can be determined if both the spin label and the protein concentrations are known.

7. Sample considerations

For most SDSL studies, a loop-gap resonator (LGR) is used and therefore the sample size is 2-5μL. Spin concentrations of 25-200μM are routinely used and depend on the shape of the spectrum, the instrument sensitivity and settings, and the number of signal averages recorded. Other techniques may require sample sizes of up to 500μL and 1mM in concentration.

6. CONCLUSION

In conclusion, a large amount of progress in the SDSL field has been made in the last ten years, and other recent reviews on this subject have been published (Columbus and Hubbell, 2002; Mchaourab and Perozo, 2000; Hubbell *et al.*, 2000; Feix and Klug, 1998; Hubbell *et al.*, 1998). The technique has moved from new and experimental to routine in many laboratories. Even non-EPR spectroscopists are realizing the value of this method, further increasing its breadth of use. In addition, the technique is continuing to expand with constant innovations and ideas. The ability to use very small amounts of sample, gas-permeable TPX tubes, and continued successes in the methodology has greatly increased the number of researchers able to carry out SDSL EPR on a great variety of biological systems. Of course, there are more studies utilizing the SDSL technique that have been carried out recently than have been mentioned here and we hope that the progress continues to advance at such a rapid rate.

7. REFERENCES

Altenbach, C., Cai, K., Klein-Seetharaman, J., Khorana, H. G., and Hubbell, W. L. (2001a) Structure and function in rhodopsin: mapping light-dependent changes in distance between residue 65 in helix TM1 and residues in the sequence 306-319 at the cytoplasmic end of helix TM7 and in helix H8. *Biochemistry 40*: 15483-15492.

Altenbach, C., Flitsch, S. L., Khorana, H. G., and Hubbell, W. L. (1989a) Structural studies on transmembrane proteins. 2. Spin labeling of bacteriorhodopsin mutants at unique cysteines. *Biochemistry 28*: 7806-7812.

Altenbach, C., Froncisz, W., Hyde, J. S., and Hubbell, W. L. (1989b) Conformation of spin-labeled melittin at membrane surfaces investigated by pulse saturation recovery and continuous wave power saturation electron paramagnetic resonance. *Biophys. J. 56*: 1183-1191.

Altenbach, C., Greenhalgh, D. A., Khorana, H. G., and Hubbell, W. L. (1994) A collision gradient method to determine the immersion depth of nitroxides in lipid bilayers: application to spin-labeled mutants of bacteriorhodopsin. *Proc. Natl. Acad. Sci. USA 91*: 1667-1671.

Altenbach, C., Klein-Seetharaman, J., Cai, K., Khorana, H. G., and Hubbell, W. L. (2001b) Structure and function in rhodopsin: mapping light-dependent changes in distance between residue 316 in helix 8 and residues in the sequence 60-75, covering the cytoplasmic end of helices TM1 and TM2 and their connection loop CL1. *Biochemistry 40*: 15493-15500.

Altenbach, C., Marti, T., Khorana, H. G., and Hubbell, W. L. (1990) Transmembrane protein structure: spin labeling of bacteriorhodopsin mutants. [Review]. *Science 248*: 1088-1092.

Altenbach, C., Oh, K. J., Trabanino, R. J., Hideg, K., and Hubbell, W. L. (2001c) Estimation of inter-residue distances in spin labeled proteins at physiological temperatures: experimental strategies and practical limitations. *Biochemistry 40*: 15471-15482.

Altenbach, C., Yang, K., Farrens, D. L., Farahbakhsh, Z. T., Khorana, H. G., and Hubbell, W. L. (1996) Structural features and light-dependent changes in the cytoplasmic interhelical

E-F loop region of rhodopsin: a site-directed spin-labeling study. *Biochemistry 35*: 12470-12478.

Archer, S. J., Ellena, J. F., and Cafiso, D. S. (1991) Dynamics and aggregation of the peptide ion channel alamethicin. Measurements using spin-labeled peptides. *Biophys. J. 60*: 389-398.

Barnes, J. P., Liang, Z., Mchaourab, H. S., Freed, J. H., and Hubbell, W. L. (1999) A multifrequency electron spin resonance study of T4 lysozyme dynamics. *Biophys. J. 76*: 3298-3306.

Barranger-Mathys, M. and Cafiso, D. S. (1996) Membrane structure of voltage-gated channel forming peptides by site-directed spin-labeling. *Biochemistry 35*: 498-505.

Bennati, M., Gerfen, G. J., Martinez, G. V., Griffin, R. G., Singel, D. J., and Millhauser, G. L. (1999) Nitroxide side-chain dynamics in a spin-labeled helix-forming peptide revealed by high-frequency (139.5-GHz) EPR spectroscopy. *J. Magn. Reson. 139*: 281-286.

Berengian, A. R., Bova, M. P., and Mchaourab, H. S. (1997) Structure and function of the conserved domain in alphaA-crystallin. Site-directed spin labeling identifies a beta-strand located near a subunit interface. *Biochemistry 36*: 9951-9957.

Berengian, A. R., Parfenova, M., and Mchaourab, H. S. (1999) Site-directed spin labeling study of subunit interactions in the alpha-crystallin domain of small heat-shock proteins. Comparison of the oligomer symmetry in alphaA-crystallin, HSP 27, and HSP 16.3. *J. Biol. Chem. 274*: 6305-6314.

Blackman, S. M., Hustedt, E. J., Cobb, C. E., and Beth, A. H. (2001) Flexibility of the cytoplasmic domain of the anion exchange protein, band 3, in human erythrocytes. *Biophys. J. 81*: 3363-3376.

Borbat, P. P., Costa-Filho, A. J., Earle, K. A., Moscicki, J. K., and Freed, J. H. (2001) Electron spin resonance in studies of membranes and proteins. [Review]. *Science 291*: 266-269.

Borbat, P. P. and Freed, J. H. (2000) Double-Quantum ESR and Distance Measurements in *Biological Magnetic Resonance* (Berliner, L. J., Eaton, S. S., and Eaton, G. R., Eds.) pp 383-459, Kluwer Academic/Plenum Publishers, New York.

Borbat, P. P., Mchaourab, H. S., and Freed, J. H. (2002) Protein structure determination using long-distance constraints from double-quantum coherence ESR: study of T4 lysozyme. *J. Am. Chem. Soc. 124*: 5304-5314.

Brown, L. J., Sale, K. L., Hills, R., Rouviere, C., Song, L., Zhang, X., and Fajer, P. G. (2002) Structure of the inhibitory region of troponin by site directed spin labeling electron paramagnetic resonance. *Proc. Natl. Acad. Sci. USA 99*: 12765-12770.

Coggshall, K. A., Cadieux, N., Piedmont, C., Kadner, R. J., and Cafiso, D. S. (2001) Transport-defective mutations alter the conformation of the energy-coupling motif of an outer membrane transporter. *Biochemistry 40*: 13964-13971.

Columbus, L. and Hubbell, W. L. (2002) A new spin on protein dynamics. *Trends Biochem. Sci. 27*: 288-295.

Columbus, L., Kalai, T., Jeko, J., Hideg, K., and Hubbell, W. L. (2001) Molecular motion of spin labeled side chains in alpha-helices: analysis by variation of side chain structure. *Biochemistry 40*: 3828-3846.

DeWeerd, K., Grigoryants, V. M., Sun, Y., Fetrow, J. S., and Scholes, C. P. (2001) EPR-Detected Folding Kinetics of Externally Located Cysteine-Directed Spin-Labeled Mutants of Iso-1-cytochrome *c*. *Biochemistry 40*: 15846-15855.

Eaton, G. R., Eaton, S. S., and Berliner, L. J. (2000) *Distance Measurements in Biological Systems by EPR, Volume 19 of Biological Magnetic Resonance* Kluwer, New York.

Eaton, S. S. and Eaton, G. R. (2000) Determination of Distances Based on T_1 and T_m Effects in *Biological Magnetic Resonance* (Berliner, L. J., Eaton, S. S., and Eaton, G. R., Eds.) pp 347-381, Kluwer Academic/Plenum Publishers, New York.

Edwards, T. E., Okonogi, T. M., Robinson, B. H., and Singh, R. J. (2001) Site-Specific Incorporation of Nitroxide Spin-Labels into Internal Sites of the TAR RNA; Structure-Dependent Dynamics of RNA by EPR Spectroscopy. *J. Am. Chem. Soc. 123*: 1527-1528.

Fanucci, G. E., Cadieux, N., Piedmont, C. A., Kadner, R. J., and Cafiso, D. S. (2002) Structure and Dynamics of the beta-Barrel of the Membrane Transporter BtuB by Site-Directed Spin Labeling. *Biochemistry 41*: 11543-11551.

Farahbakhsh, Z. T., Altenbach, C., and Hubbell, W. L. (1992) Spin labeled cysteines as sensors for protein-lipid interaction and conformation in rhodopsin. *Photochem. Photobiol. 56*: 1019-1033.

Farahbakhsh, Z. T., Hideg, K., and Hubbell, W. L. (1993) Photoactivated conformational changes in rhodopsin: a time-resolved spin label study. *Science 262*: 1416-1419.

Farrens, D. L., Altenbach, C., Yang, K., Hubbell, W. L., and Khorana, H. G. (1996) Requirement of rigid-body motion of transmembrane helices for light activation of rhodopsin. *Science 274*: 768-770.

Feix, J. B. and Klug, C. S. (1998) Site-directed spin labeling of membrane proteins and peptide-membrane interactions in *Biological Magnetic Resonance, Volume 14: Spin Labeling: The Next Millennium* (Berliner, L. J., Ed.) pp 252-281, Plenum Press, New York.

Frazier, A. A., Wisner, M. A., Malmberg, N. J., Victor, K. G., Fanucci, G. E., Nalefski, E. A., Falke, J. J., and Cafiso, D. S. (2002) Membrane orientation and position of the C2 domain from cPLA2 by site-directed spin labeling. *Biochemistry 41*: 6282-6292.

Froncisz, W. and Hyde, J. S. (1982) The loop-gap resonator: a new microwave lumped circuit ESR sample structure. *J. Magn. Reson. 47*: 515-521.

Glasgow, B. J., Gasymov, O. K., Abduragimov, A. R., Yusifov, T. N., Altenbach, C., and Hubbell, W. L. (1999) Side chain mobility and ligand interactions of the G strand of tear lipocalins by site-directed spin labeling. *Biochemistry 38*: 13707-13716.

Grigoryants, V. M., Veselov, A. V., and Scholes, C. P. (2000) Variable velocity liquid flow EPR applied to submillisecond protein folding. *Biophys. J. 78*: 2702-2708.

Gross, A., Columbus, L., Hideg, K., Altenbach, C., and Hubbell, W. L. (1999) Structure of the KcsA potassium channel from Streptomyces lividans: a site-directed spin labeling study of the second transmembrane segment. *Biochemistry 38*: 10324-10335.

Gross, A. and Hubbell, W. L. (2002) Identification of protein side chains near the membrane-aqueous interface: a site-directed spin labeling study of KcsA. *Biochemistry 41*: 1123-1128.

Hanson, P., Millhauser, G., Formaggio, F., Crisma, M., and Toniolo, C. (1996) ESR Characterization of Hexameric, Helical Peptides Using Double TOAC Spin Labeling. *J. Am. Chem. Soc. 118*: 7618-7625.

Hess, J. F., Voss, J. C., and FitzGerald, P. G. (2002) Real-time observation of coiled-coil domains and subunit assembly in intermediate filaments. *J. Biol. Chem. 277*: 35516-35522.

Hubbell, W. L., Cafiso, D. S., and Altenbach, C. (2000) Identifying conformational changes with site-directed spin labeling. [Review]. *Nat. Struct. Biol. 7*: 735-739.

Hubbell, W. L., Froncisz, W., and Hyde, J. S. (1987) Continuous and Stopped Flow EPR Spectrometer Based on a Loop Gap Resonator. *Rev. Sci. Instr. 58*: 1879-1886.

Hubbell, W. L., Gross, A., Langen, R., and Lietzow, M. A. (1998) Recent advances in site-directed spin labeling of proteins. [Review]. *Curr. Opin. Struct. Biol. 8*: 649-656.

Hubbell, W. L., Mchaourab, H. S., Altenbach, C., and Lietzow, M. A. (1996) Watching proteins move using site-directed spin labeling. *Structure 4*: 779-783.

Hustedt, E. J. and Beth, A. H. (1996) Determination of the orientation of a band 3 affinity spin-label relative to the membrane normal axis of the human erythrocyte. *Biochemistry 35*: 6944-6954.

Hustedt, E. J. and Beth, A. H. (1999) Nitroxide spin-spin interactions: applications to protein structure and dynamics. [Review]. *Annual Review of Biophysics & Biomolecular Structure 28*: 129-153.

Hustedt, E. J. and Beth, A. H. (2000) Structural Information from CW-EPR Spectra of Dipolar Coupled Nitroxide Spin Labels in *Biological Magnetic Resonance* (Berliner, L. J., Eaton, S. S., and Eaton, G. R., Eds.) pp 155-184, Kluwer Academic/Plenum Publishers, New York.

Hustedt, E. J., Smirnov, A. I., Laub, C. F., Cobb, C. E., and Beth, A. H. (1997) Molecular distances from dipolar coupled spin-labels: the global analysis of multifrequency continuous wave electron paramagnetic resonance data. *Biophys. J. 72*: 1861-1877.

Isas, J. M., Langen, R., Haigler, H. T., and Hubbell, W. L. (2002) Structure and dynamics of a helical hairpin and loop region in annexin 12: a site-directed spin labeling study. *Biochemistry 41*: 1464-1473.

Jeschke, G., Pannier, M., and Spiess, H. W. (2000) Double Electron-Electron Resonance: Methodical Advances and Application to Disordered Systems in *Biological Magnetic Resonance* (Berliner, L. J., Eaton, S. S., and Eaton, G. R., Eds.) pp 493-412, Kluwer Academic/Plenum Publishers, New York.

Karim, C. B., Stamm, J. D., Karim, J., Jones, L. R., and Thomas, D. D. (1998) Cysteine reactivity and oligomeric structures of phospholamban and its mutants. *Biochemistry 37*: 12074-12081.

Kersten, M. V., Dunn, S. D., Wise, J. G., and Vogel, P. D. (2000) Site-directed spin-labeling of the catalytic sites yields insight into structural changes within the F0F1-ATP synthase of Escherichia coli. *Biochemistry 39*: 3856-3860.

Kim, C. S., Kweon, D. H., and Shin, Y. K. (2002) Membrane Topologies of Neuronal SNARE Folding Intermediates. *Biochemistry 41*: 10928-10933.

Klug, C. S., Eaton, S. S., Eaton, G. R., and Feix, J. B. (1998) Ligand-induced conformational change in the ferric enterobactin receptor FepA as studied by site-directed spin labeling and time-domain ESR. *Biochemistry 37*: 9016-9023.

Klug, C. S. and Feix, J. B. (1998) Guanidine hydrochloride unfolding of a transmembrane beta-strand in FepA using site-directed spin labeling. *Protein Sci. 7*: 1469-1476.

Klug, C. S., Su, W., and Feix, J. B. (1997) Mapping of the residues involved in a proposed beta-strand located in the ferric enterobactin receptor FepA using site-directed spin-labeling. *Biochemistry 36*: 13027-13033.

Klug, C. S., Su, W., Liu, J., Klebba, P. E., and Feix, J. B. (1995) Denaturant unfolding of the ferric enterobactin receptor and ligand-induced stabilization studied by site-directed spin labeling. *Biochemistry 34*: 14230-14236.

Koteiche, H. A., Berengian, A. R., and Mchaourab, H. S. (1998) Identification of protein folding patterns using site-directed spin labeling. Structural characterization of a beta-sheet and putative substrate binding regions in the conserved domain of alpha A-crystallin. *Biochemistry 37*: 12681-12688.

Koteiche, H. A., Narasimhan, C., Runquist, J. A., and Miziorko, H. M. (1995) Utility of a novel spin-labeled nucleotide in investigation of the substrate and effector sites of phosphoribulokinase. *Biochemistry 34*: 15068-15074.

Kweon, D. H., Kim, C. S., and Shin, Y. K. (2002) The membrane-dipped neuronal SNARE complex: a site-directed spin labeling electron paramagnetic resonance study. *Biochemistry 41*: 9264-9268.

Langen, R., Cai, K., Altenbach, C., Khorana, H. G., and Hubbell, W. L. (1999) Structural features of the C-terminal domain of bovine rhodopsin: a site-directed spin-labeling study. *Biochemistry 38*: 7918-7924.

Langen, R., Isas, J. M., Hubbell, W. L., and Haigler, H. T. (1998a) A transmembrane form of annexin XII detected by site-directed spin labeling. *Proc. Natl. Acad. Sci. USA 95*: 14060-14065.

Langen, R., Isas, J. M., Luecke, H., Haigler, H. T., and Hubbell, W. L. (1998b) Membrane-mediated assembly of annexins studied by site-directed spin labeling. *J. Biol. Chem. 273*: 22453-22457.

Langen, R., Oh, K. J., Cascio, D., and Hubbell, W. L. (2000) Crystal structures of spin labeled T4 lysozyme mutants: implications for the interpretation of EPR spectra in terms of structure. *Biochemistry 39*: 8396-8405.

Larsen, R. G. and Singel, D. J. (1993) Double electron-electron resonance spin-echo modulation: Spectroscopic measurement of electron spin pair separations in orientationally disordered solids. *Journal of Chemical Physics 98*: 5134-5146.

Leigh, Jr. J. S. (1970) ESR Rigid-Lattice Line Shape in a System of Two Interacting Spins. *The Journal of Chemical Physics 52*: 2608-2612.

Lewis, J. R. and Cafiso, D. S. (1999) Correlation between the free energy of a channel-forming voltage-gated peptide and the spontaneous curvature of bilayer lipids. *Biochemistry 38*: 5932-5938.

Li, S. C. and Deber, C. M. (1994) A measure of helical propensity for amino acids in membrane environments. *Nat. Struct. Biol. 1*: 558.

Lietzow, M. A. and Hubbell, W. L. (1998) Site-directed spin labeling of cellular retinol-binding protein (CRBP): examination of a b-sheet landscape and its conformational dynamics. *Biophys. J. 74*: A278.

Lin, Y., Nielsen, R., Murray, D., Hubbell, W. L., Mailer, C., Robinson, B. H., and Gelb, M. H. (1998) Docking phospholipase A2 on membranes using electrostatic potential-modulated spin relaxation magnetic resonance. *Science 279*: 1925-1929.

Liu, Y. S., Sompornpisut, P., and Perozo, E. (2001) Structure of the KcsA channel intracellular gate in the open state. *Nat. Struct. Biol. 8*: 883-887.

Lundberg, K. M., Stenland, C. J., Cohen, F. E., Prusiner, S. B., and Millhauser, G. L. (1997) Kinetics and mechanism of amyloid formation by the prion protein H1 peptide as determined by time-dependent ESR. *Chem. Biol. 4*: 345-355.

Macosko, J. C., Kim, C. H., and Shin, Y. K. (1997) The membrane topology of the fusion peptide region of influenza hemagglutinin determined by spin-labeling EPR. *J. Mol. Biol. 267*: 1139-1148.

Macosko, J. C., Pio, M. S., Tinoco, I., Jr., and Shin, Y. K. (1999) A novel 5 displacement spin-labeling technique for electron paramagnetic resonance spectroscopy of RNA. *Rna-A Publication of the Rna Society 5*: 1158-1166.

Mangels, M. L., Harper, A. C., Smirnov, A. I., Howard, K. P., and Lorigan, G. A. (2001) Investigating magnetically aligned phospholipid bilayers with EPR spectroscopy at 94 GHz. *J. Magn. Reson. 151*: 253-259.

Marchetto, R., Schreier, S., and Nakaie, C. R. (1993) A Novel Spin-Labeled Amino Acid Derivative for Use in Peptide Synthesis: (9-Fluorenylmethyloxycarbonyl)-2,2,6,6-tetramethylpiperidine-N-oxyl-4-amino-carboxylic Acid. *J. Am. Chem. Soc. 115*: 11042-11043.

Margittai, M., Fasshauer, D., Pabst, S., Jahn, R., and Langen, R. (2001) Homo- and heterooligomeric SNARE complexes studied by site-directed spin labeling. *J. Biol. Chem. 276*: 13169-13177.

Matthews, B. W. (1995) Studies on protein stability with T4 lysozyme. [Review]. *Adv. Prot. Chem. 46*: 249-278.

Mchaourab, H. S., Hyde, J. S., and Feix, J. B. (1994) Binding and state of aggregation of spin-labeled cecropin AD in phospholipid bilayers: effects of surface charge and fatty acyl chain length. *Biochemistry 33*: 6691-6699.

Mchaourab, H. S., Kalai, T., Hideg, K., and Hubbell, W. L. (1999) Motion of spin-labeled side chains in T4 lysozyme: effect of side chain structure. *Biochemistry 38*: 2947-2955.

Mchaourab, H. S., Lietzow, M. A., Hideg, K., and Hubbell, W. L. (1996) Motion of spin-labeled side chains in T4 lysozyme. Correlation with protein structure and dynamics. *Biochemistry 35*: 7692-7704.

Mchaourab, H. S., Oh, K. J., Fang, C. J., and Hubbell, W. L. (1997) Conformation of T4 lysozyme in solution. Hinge-bending motion and the substrate-induced conformational transition studied by site-directed spin labeling. *Biochemistry 36*: 307-316.

Mchaourab, H. S. and Perozo, E. (2000) Determination of Protein Folds and Conformational Dynamics Using Spin Labeling EPR Spectroscopy in *Biological Magnetic Resonance, Volume 19* (Berliner, L. J., Eaton, S. S., and Eaton, G. R., Eds.) pp 185-247, Kluwer Academic/Plenum Publishers, New York.

McNulty, J., Thompson, D., Carrasco, M., and Millhauser, G. (2002) Dap-SL: a new site-directed nitroxide spin labeling approach for determining structure and motions in synthesized peptides and proteins. *FEBS Lett. 529*: 243.

McNulty, J. C. and Millhauser, G. L. (2000) TOAC: The Rigid Nitroxide Side Chain in *Biological Magnetic Resonance* (Berliner, L. J., Eaton, S. S., and Eaton, G. R., Eds.) pp 277-307, Kluwer Academic/Plenum Publishers, New York.

McNulty, J. C., Silapie, J. L., Carnevali, M., Farrar, C. T., Griffin, R. G., Formaggio, F., Crisma, M., Toniolo, C., and Millhauser, G. L. (2000) Electron spin resonance of TOAC labeled peptides: folding transitions and high frequency spectroscopy. *Biopolymers 55*: 479-485.

Medkova, M., Preininger, A. M., Yu, N.-J., Hubbell, W. L., and Hamm, H. E. (2002) Conformational Changes in the Amino-Terminal Helix of the G-Protein ai1 Following Dissociation From Gbg Subunit and Activation. *Biochemistry 41*: 9962-9972.

Merianos, H. J., Cadieux, N., Lin, C. H., Kadner, R. J., and Cafiso, D. S. (2000) Substrate-induced exposure of an energy-coupling motif of a membrane transporter. *Nat. Struct. Biol. 7*: 205-209.

Miller, T. R., Alley, S. C., Reese, A. W., Solomon, M. S., McCallister, W. V., Mailer, C., Robinson, B. H., and Hopkins, P. B. (1995) A Probe for Sequence-Dependent Nucleic Acid Dynamics. *J. Am. Chem. Soc. 117*: 9377-9378.

Mollaaghababa, R., Steinhoff, H. J., Hubbell, W. L., and Khorana, H. G. (2000) Time-resolved site-directed spin-labeling studies of bacteriorhodopsin: loop-specific conformational changes in M. *Biochemistry 39*: 1120-1127.

Nordio, P. L. (1976) General magnetic resonance theory in *Spin Labeling Theory and Applications* (Berliner, L. J., Ed.) pp 5-52, Academic Press, New York.

Oh, K. J., Zhan, H., Cui, C., Hideg, K., Collier, R. J., and Hubbell, W. L. (1996) Organization of diphtheria toxin T domain in bilayers: a site-directed spin labeling study. *Science 273*: 810-812.

Pake, G. E. (1948) Nuclear Resonance Absorption in Hydrated Crystals: Fine Structure of the Proton Line. *Journal of Chemical Physics 16*: 327-336.

Pannier, M., Veit, S., Godt, A., Jeschke, G., and Spiess, H. W. (2000) Dead-time free measurement of dipole-dipole interactions between electron spins. *J. Magn. Reson. 142*: 331-340.

Panse, V. G., Beena, K., Philipp, R., Trommer, W. E., Vogel, P. D., and Varadarajan, R. (2001) Electron spin resonance and fluorescence studies of the bound-state conformation of a model protein substrate to the chaperone SecB. *J. Biol. Chem. 276*: 33681-33688.

Perozo, E., Cortes, D. M., and Cuello, L. G. (1998) Three-dimensional architecture and gating mechanism of a K+ channel studied by EPR spectroscopy. *Nat. Struct. Biol. 5*: 459-469.

Perozo, E., Cortes, D. M., and Cuello, L. G. (1999) Structural rearrangements underlying K+-channel activation gating. *Science 285*: 73-78.

Perozo, E., Cortes, D. M., Sompornpisut, P., Kloda, A., and Martinac, B. (2002a) Open channel structure of MscL and the gating mechanism of mechanosensitive channels. *Nature 418*: 942-948.

Perozo, E., Kloda, A., Cortes, D. M., and Martinac, B. (2002b) Physical principles underlying the transduction of bilayer deformation forces during mechanosensitive channel gating. *Nat. Struct. Biol. 9*: 696-703.

Pfannebecker, V., Klos, H., Hubrich, M., Volkmer, T., Heuer, A., Wiesner, U., and Spiess, H. W. (1996) Determination of End-to-End Distances in Oligomers by Pulsed EPR. *J. Phys. Chem. 100*: 13428-13532.

Qin, P. Z., Butcher, S. E., Feigon, J., and Hubbell, W. L. (2001) Quantitative analysis of the isolated GAAA tetraloop/receptor interaction in solution: a site-directed spin labeling study. *Biochemistry 40*: 6929-6936.

Qin, Z. and Cafiso, D. S. (1996) Membrane structure of protein kinase C and calmodulin binding domain of myristoylated alanine rich C kinase substrate determined by site-directed spin labeling. *Biochemistry 35*: 2917-2925.

Rabenstein, M. D. and Shin, Y. K. (1995) Determination of the distance between two spin labels attached to a macromolecule. *Proc. Natl. Acad. Sci. USA 92*: 8239-8243.

Rabenstein, M. D. and Shin, Y. K. (1996) HIV-1 gp41 tertiary structure studied by EPR spectroscopy. *Biochemistry 35*: 13922-13928.

Ramos, A. and Varani, G. (1998) A New Method to Detect Long-Range Protein-RNA Contacts: NMR Detection of Electron-Proton Relaxation Induced by Nitroxide Spin-Labeled RNA. *J. Am. Chem. Soc. 120*: 10992-10993.

Rauch, M. E., Ferguson, C. G., Prestwich, G. D., and Cafiso, D. S. (2002) Myristoylated alanine-rich C kinase substrate (MARCKS) sequesters spin-labeled phosphatidylinositol 4,5-bisphosphate in lipid bilayers. *J. Biol. Chem. 277*: 14068-14076.

Resek, J. F., Farahbakhsh, Z. T., Hubbell, W. L., and Khorana, H. G. (1993) Formation of the meta II photointermediate is accompanied by conformational changes in the cytoplasmic surface of rhodopsin. *Biochemistry 32*: 12025-12032.

Rink, T., Riesle, J., Oesterhelt, D., Gerwert, K., and Steinhoff, H. J. (1997) Spin-labeling studies of the conformational changes in the vicinity of D36, D38, T46, and E161 of bacteriorhodopsin during the photocycle. *Biophys. J. 73*: 983-993.

Russell, C. J., Thorgeirsson, T. E., and Shin, Y. K. (1996) Temperature dependence of polypeptide partitioning between water and phospholipid bilayers. *Biochemistry 35*: 9526-9532.

Russell, C. J., Thorgeirsson, T. E., and Shin, Y. K. (1999) The membrane affinities of the aliphatic amino acid side chains in an alpha-helical context are independent of membrane immersion depth. *Biochemistry 38*: 337-346.

Salwinski, L. and Hubbell, W. L. (1999) Structure in the channel forming domain of colicin E1 bound to membranes: the 402-424 sequence. *Protein Sci. 8*: 562-572.

Serag, A. A., Altenbach, C., Gingery, M., Hubbell, W. L., and Yeates, T. O. (2001) Identification of a subunit interface in transthyretin amyloid fibrils: evidence for self-assembly from oligomeric building blocks. *Biochemistry 40*: 9089-9096.

Serag, A. A., Altenbach, C., Gingery, M., Hubbell, W. L., and Yeates, T. O. (2002) Arrangement of subunits and ordering of b-strands in an amyloid sheet. *Nat. Struct. Biol. 9*: 734-739.

Shin, Y. K., Levinthal, C., Levinthal, F., and Hubbell, W. L. (1993) Colicin E1 binding to membranes: time-resolved studies of spin-labeled mutants. *Science 259*: 960-963.

Singh, R. J., Feix, J. B., Mchaourab, H. S., Hogg, N., and Kalyanaraman, B. (1995) Spin-labeling study of the oxidative damage to low-density lipoprotein. *Arch. Biochem. Biophys. 320*: 155-161.

Smirnov, A. I., Smirnova, T. I., and Morse, P. D. (1995) Very high frequency electron paramagnetic resonance of 2,2,6,6-tetramethyl-1-piperidinyloxy in 1,2-dipalmitoyl-sn-glycero-3-phosphatidylcholine liposomes: partitioning and molecular dynamics. *Biophys. J. 68*: 2350-2360.

Steinhoff, H., Savitsky, A., Wegener, C., Pfeiffer, M., Plato, M., and Mobius, K. (2000) High-field EPR studies of the structure and conformational changes of site-directed spin labeled bacteriorhodopsin. *Biochem. Biophys. Acta 1457*: 253-262.

Steinhoff, H. J., Mollaaghababa, R., Altenbach, C., Hideg, K., Krebs, M., Khorana, H. G., and Hubbell, W. L. (1994) Time-resolved detection of structural changes during the photocycle of spin-labeled bacteriorhodopsin. *Science 266*: 105-107.

Stone, T. J., Buckman, T., Nordio, P. L., and McConnell, H. M. (1965) Spin-Labeled Biomolecules. *Proc. Natl. Acad. Sci. USA 54*: 1010-1017.

Sun, J., Voss, J., Hubbell, W. L., and Kaback, H. R. (1999) Proximity between periplasmic loops in the lactose permease of Escherichia coli as determined by site-directed spin labeling. *Biochemistry 38*: 3100-3105.

Thorgeirsson, T. E., Russell, C. J., King, D. S., and Shin, Y. K. (1996) Direct determination of the membrane affinities of individual amino acids. *Biochemistry 35*: 1803-1809.

Todd, A. P., Cong, J., Levinthal, F., Levinthal, C., and Hubbell, W. L. (1989) Site-directed mutagenesis of colicin E1 provides specific attachment sites for spin labels whose spectra are sensitive to local conformation. *Proteins 6*: 294-305.

Victor, K. and Cafiso, D. S. (1998) Structure and position of the N-terminal membrane-binding domain of pp60src at the membrane interface. *Biochemistry 37*: 3402-3410.

Victor, K. G. and Cafiso, D. S. (2001) Location and dynamics of basic peptides at the membrane interface: electron paramagnetic resonance spectroscopy of tetramethyl-piperidine-N-oxyl-4-amino-4-carboxylic acid-labeled peptides. *Biophys. J. 81*: 2241-2250.

Vogelsang, M. S., Salwinski, L., and Hubbell, W. L. (2001) Membrane structure of 425-442 region of colicin E1 pore forming domain: a site-directed spin labeling study. *Biophys. J. 80*: 129a.

Voss, J., Hubbell, W. L., Hernandez-Borrell, J., and Kaback, H. R. (1997) Site-directed spin-labeling of transmembrane domain VII and the 4B1 antibody epitope in the lactose permease of Escherichia coli. *Biochemistry 36*: 15055-15061.

Voss, J., Hubbell, W. L., and Kaback, H. R. (1995a) Distance determination in proteins using designed metal ion binding sites and site-directed spin labeling: application to the lactose permease of Escherichia coli. *Proc. Natl. Acad. Sci. USA 92*: 12300-12303.

Voss, J., Salwinski, L., Kaback, H. R., and Hubbell, W. L. (1995b) A method for distance determination in proteins using a designed metal ion binding site and site-directed spin labeling: evaluation with T4 lysozyme. *Proc. Natl. Acad. Sci. USA 92*: 12295-12299.

Voss, J., Wu, J., Hubbell, W. L., Jacques, V., Meares, C. F., and Kaback, H. R. (2001) Helix packing in the lactose permease of Escherichia coli: distances between site-directed nitroxides and a lanthanide. *Biochemistry 40*: 3184-3188.

Wang, Q., Voss, J., Hubbell, W. L., and Kaback, H. R. (1998) Proximity of helices VIII (Ala273) and IX (Met299) in the lactose permease of Escherichia coli. *Biochemistry 37*: 4910-4915.

Wertz, S. L., Savino, Y., and Cafiso, D. S. (1996) Solution and membrane bound structure of a peptide derived from the protein kinase C substrate domain of neuromodulin. *Biochemistry 35*: 11104-11112.

Wimley, W. C. and White, S. H. (1996) Experimentally determined hydrophobicity scale for proteins at membrane interfaces. [Review]. *Nat. Struct. Biol. 3*: 842-848.

Xiao, W. and Shin, Y. K. (2000) EPR Spectroscopic Ruler: the Method and its Applications in *Biological Magnetic Resonance, Volume 19* (Berliner, L. J., Eaton, S. S., and Eaton, G. R., Eds.) pp 249-276, Kluwer Academic/Plenum Publishers, New York.

Yang, K., Farrens, D. L., Altenbach, C., Farahbakhsh, Z. T., Hubbell, W. L., and Khorana, H. G. (1996) Structure and function in rhodopsin. Cysteines 65 and 316 are in proximity in a rhodopsin mutant as indicated by disulfide formation and interactions between attached spin labels. *Biochemistry 35*: 14040-14046.

Yu, Y. G., Thorgeirsson, T. E., and Shin, Y. K. (1994) Topology of an amphiphilic mitochondrial signal sequence in the membrane-inserted state: a spin labeling study. *Biochemistry 33*: 14221-14226.

Zhao, M., Kalai, T., Hideg, K., Altenbach, C., Hubbell, W. L., and Kaback, H. R. (2000) Binding of spin-labeled galactosides to the lactose permease of Escherichia coli. *Biochemistry 39*: 11381-11388.

Zhao, M., Zen, K. C., Hernandez-Borrell, J., Altenbach, C., Hubbell, W. L., and Kaback, H. R. (1999) Nitroxide scanning electron paramagnetic resonance of helices IV and V and the intervening loop in the lactose permease of Escherichia coli. *Biochemistry 38*: 15970-15977.

Chapter 11

Saturation Transfer Spectroscopy of Biological Membranes

Derek Marsh,[A] László I. Horváth,[B] Tibor Páli[b] And Vsevolod A. Livshits[c]
[a] Max-Planck-Institut für Biophysikalische Chemie, Abteilung Spektroskopie, 37070 Göttingen, Germany; [b] Institute of Biophysics, Biological Research Centre, 6701 Szeged, Hungary and [c]Centre of Photochemistry, Russian Academy of Sciences, 117421 Moscow Russia.

Abstract: Various aspects of the branch of non-linear spectroscopy that is known as saturation transfer (ST) EPR are reviewed, ranging from its inception to the present day. Initial methodological development was by Hyde and Dalton, followed by the introduction into biology by Hyde and Thomas. ST-EPR is a continuous wave spectroscopy, which extends the sensitivity of conventional nitroxide EPR to the microsecond (or submillisecond) correlation time regime of rotational motion, for spin-labelled membranes and biopolymers. Equally, slow exchange processes are accessible to ST-EPR, as are the paramagnetic relaxation enhancements that are essential to site-directed spin-labelling strategies. Central to the latter are the principles of spin-label oximetry, as developed by Hyde and coworkers at the Milwaukee EPR Centre.

1. INTRODUCTION

Saturation transfer EPR (ST-EPR) is a term introduced by Hyde to describe spin-label EPR methods designed to study slow rotational diffusion with correlation times in the microsecond regime or longer (Hyde, 1978). The physics underlying the method is that saturation of an orientationally dispersed powder spectrum is alleviated by rotational motion on a timescale comparable to that of the spin-lattice (T_1) relaxation (see Fig. 1). The effects of transfer of saturation between different spin-label orientations are greatest in those regions of the spectrum at which the change in resonance position with angular orientation is at a maximum. This leads to a sensitivity of spectral lineshape to rotational correlation time in the sub-millisecond regime. Lineshape changes in the saturated conventional EPR spectrum are unremarkable and an essential development of the orientationally resolved method was the introduction of non-linear spectra detected in quadrature phase with the static field modulation (Hyde and Dalton, 1972). Such out-of-phase spectra have appreciable intensity only in the presence of saturation. The result is that the lineshapes of certain of these non-linear displays are exquisitely sensitive to rotational motion on the T_1-timescale.

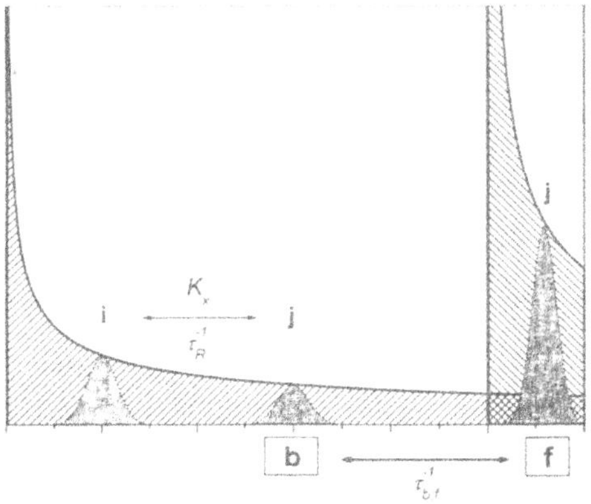

Figure 1. Schematic indication of saturation transfer processes (arrows) in axial powder spectra (heavy lines) for two ensembles of spins, b and f (hatched). Within ensemble b, saturation transfer takes place between orientationally selected component spin packets, i and j (grey), by internal exchange processes: either rotational diffusion (τ_R^{-1}) or Heisenberg spin exchange (K_x). Two-site exchange leads to transfer of saturation between spin ensembles b and f, with characteristic rate constants τ_b^{-1} and τ_f^{-1}.

ST-EPR has come conventionally to be identified with studies of slow rotational motion. It is clear, indeed already to the originators, that this class of experiment can be used to study any type of slow molecular motion that gives rise to transfer of saturation particularly, for instance, two-site exchange (see Fig. 1). Further, the methods are applicable also to the quantitation of any magnetic interaction that alleviates saturation. Especially prominent among these are Heisenberg spin-exchange between nitroxides and interactions with paramagnetic relaxants such as transition metal complexes or molecular oxygen. The latter has been studied intensively by Hyde and coworkers, principally by saturation recovery EPR methods (Hyde and Subczynski, 1989). T_1-relaxation enhancement currently constitutes the most powerful EPR methodology in site-directed spin-labelling applications (Hubbell and Altenbach, 1994). The advantage of saturation-based methods is that they are sensitive to much weaker interactions than are the conventional linewidths which are determined by the T_2-relaxation rate rather than by the slower T_1-relaxation rate.

Here we aim to cover ST-EPR in its widest sense. Appropriate to the dedication of this volume, we begin with a description of the historical development of the subject. At its baldest this simply can be stated as: "nothing would have happened without Jim Hyde". What follows is then devoted to a description of ST-spectroscopy up to its state-of-the-art application. Initial emphasis is placed on rotational diffusion measurements, but in-depth coverage is given also to the less conventional applications for studying exchange processes and paramagnetic relaxation enhancements. The advantage of integral methods is stressed throughout because they are generally insensitive to inhomogeneous broadening and the intensities are additive in multi-component systems.

2. HISTORICAL DEVELOPMENT

Saturation transfer spectroscopy has its origin in the observation by Hyde and collaborators (1970) that the CW saturation behaviour of the conventional low-temperature EPR spectra from flavin free radicals depended on temperature in a way compatible with slow molecular rotation. Following the classic analysis of adiabatic rapid-passage non-linear EPR spectra by Portis (1955) and Weger (1960), Hyde and Dalton (1972) explored the sensitivity to slow rotational diffusion of the first harmonic dispersion spectrum detected 90°-out-of-phase with respect to the field modulation. In the rigid limit, this U_1'-display should have the pure (zeroth harmonic) absorption lineshape, V_o. Rotation-dependent saturation transfer causes angularly selective decreases in intensity of the U_1' ST-EPR lineshape, relative to the absorption spectrum (Fajer and Marsh, 1983a). The predicted sensitivity of the U_1'-spectra to slow motion was found for a small nitroxide spin label in supercooled glasses (Hyde and Dalton, 1972). Additionally, the ST-EPR spectra were found to depend on the field-modulation frequency, ω_m, in a manner expected for rapid passage conditions. Essentially, the U_1'-lineshape is determined by the product $\omega_m\tau_R$, where τ_R is the rotational correlation time. (Throughout this chapter, harmonics (n) refer to a Fourier expansion of the EPR signal with respect to ω_m. Experimentally, the various harmonic spectra are obtained by phase-sensitive detection at frequencies $n\times\omega_m$.)

At the time, dispersion spectra were thought to be unsuitable for biological applications because of problems associated with sensitivity and klystron FM noise. A systematic search was therefore conducted by Hyde and Thomas (1973) to identify the out-of-phase ST-EPR display best suited to biomedical applications. The result – the second-harmonic 90°-out-of-phase absorption display, V_2' – was the next and lasting breakthrough in the development of ST-EPR. Indeed, this is the non-linear display with which saturation transfer spectroscopy has come to be identified. Apart from the adequate intensity of the V_2'-display, its success can be attributed to the richness of the lineshapes. Qualitatively, the V_2'-lineshape appears as an admixture of the out-of-phase first-harmonic dispersion U_1'-spectrum with the in-phase second-harmonic absorption V_2-spectrum.

A comprehensive experimental analysis, including simulations, was presented of the dependence of the V_2'- (and U_1') ST-EPR spectra on the rotational correlation time of spin-labelled haemoglobin in the publication by Thomas, Dalton and Hyde (1976). This was an enormously influential paper and remains the classic reference. Sensitivity of the V_2'-lineshapes to rotational correlation time was demonstrated over almost four decades from 10^{-7} s to 10^{-3} s. Qualitatively, the proportion of U_1'-like contribution to the V_2'-spectrum decreases with decreasing correlation time. For shorter correlation times, contributions from incipient motional narrowing of the V_2-like component contribute also to the motional sensitivity of the V_2'-spectrum. The paper by Thomas, Dalton and Hyde (1976) on spin-labelled haemoglobin represents the first detailed application of ST-EPR to a biomolecular system. It also established the most practical method for analysis of the correlation-time dependence of the V_2'-spectral lineshapes in terms of the diagnostic lineheight ratios L"/L, C'/C and H"/H. These represent the ratios of the V_2'-intensity L", C' and H" in the spectral regions of maximum angular dispersion to those L, C and H at the stationary angular turning points, for the low-field, central and high-field manifolds, respectively. Already, the utility of these lineheight ratios had been anticipated in ST-EPR simulations by Thomas and McConnell (1974) using the diffusion-coupled Bloch equations. Together with elaborations, e.g., for the greater angular dispersion in high-field spectra (Johnson and Hyde, 1981), this remains the most useful empirical approach to analysis of ST-EPR lineshapes.

The initial work on spin-labelled haemoglobin (Hyde and Thomas, 1973) was quickly followed by the application to muscle proteins by Thomas, Hyde, Seidel and Gergely (1975). This represents the first truly biomedical application of ST-EPR and firmly establishes the utility of the method for the study of supramolecular aggregates. The application to biological membranes came somewhat later. Rhodopsin was the first membrane protein to be studied by ST-EPR (Baroin et al., 1977). This again was a landmark study because it represents the other major biological ST-EPR application, in addition to supramolecular assemblies, viz., to the rotational diffusion of integral proteins in the high-viscosity lipid environment of membranes (see also Hidalgo et al., 1978; Kirino et al., 1978; Kusumi et al., 1978; Rousselet and Devaux, 1977). A study of the rotational mobility of spin-labelled lipids in gel-phase membranes (Marsh, 1980) was the first to consider the effects of rotational anisotropy on ST-EPR spectra. These and other earlier applications were reviewed by Hyde and Dalton (1979). An early exposition of the methodology is by Hyde (1978).

In principle, the basic foundations of all further developments in saturation transfer spectroscopy were set by the Hyde and Thomas (1973), and the Thomas, Dalton and Hyde (1976) papers. Robinson and Dalton contributed considerably to the development of theoretical simulations, particularly with respect to anisotropic rotational diffusion (Robinson and Dalton, 1980). Evans (1981) was the first to suggest correlation-time analysis by using the integrated intensity of the V_2'-ST-EPR spectrum. The purpose of this was to remove the effects of contaminating fast tumbling spin labels which contribute little to the integrated ST-EPR intensity but disturb the measurement of lineheight ratios. It was later recognised that this is a viable method for analysing general two-component ST-EPR spectra (Horváth and Marsh, 1983). Use of ST-EPR integrals then led to the discovery of the sensitivity to weak spin-spin interactions (Horváth et al., 1990) and ultimately to establishing out-of-phase intensities as a method to determine relaxation enhancements (Marsh et al., 1998). The final step in this development is the demonstration of the first harmonic out-of-phase absorption V_1' spectrum as a "pure" T_1-sensitive display (Livshits et al., 1998a), as proposed originally by Hyde and Thomas (Hyde and Thomas, 1973).

3. RAPID-PASSAGE SATURATION-TRANSFER-EPR DISPLAYS

The fundamental basis of saturation transfer spectroscopy is to use non-linear CW detection under conditions of partial microwave saturation. In rapid passage experiments, this is invariably done by detecting in phase quadrature with the Zeeman field modulation. However, as a T_1-sensitive CW technique, progressive saturation also belongs generically to the non-linear class of experiments. The second aspect of saturation transfer spectroscopy is spectral resolution: saturation is transferred from one spin packet to a spin packet elsewhere in the spectrum. The rate of transfer must be comparable to the spin-lattice relaxation rate. Spectral intensities are then a measure of the rate of the process that causes the transfer of saturation. Orientational selection of spin packets gives rise to sensitivity of powder lineshapes to slow rotational diffusion. This is the classical saturation transfer experiment. On the other hand, T_1-relaxation enhancement by paramagnetic relaxants cause a decrease in intensity with little effect on lineshape. This is the non-classical saturation transfer experiment that provides certain advantages over progressive saturation experiments.

Out-of-phase detected EPR spectra that have been investigated are the first-harmonic dispersion signal and absorption signals up to the second harmonic (Hyde and Thomas, 1973). With the notation already introduced these are the U_1, U_1', V_1, V_1', V_2 and V_2'-spectra, respectively, where the prime indicates out-of-phase detection. Fig. 2 gives simulated spectra illustrating these six displays. Of the nonlinear (i.e., out-of-phase) displays, the first harmonic dispersion and second-harmonic absorption are sensitive to ultraslow rotational diffusion. This was demonstrated experimentally by Hyde and Thomas (1973) with both the small spin label hydroxy-TEMPO in supercooled *sec*-butyl benzene and maleimide spin-labelled haemoglobin in glycerol. They found that the second harmonic absorption (V_2') out-of-phase display has the greatest sensitivity of lineshape to rotational correlation time. For this reason, V_2'-spectroscopy has become the standard method to study very slow rotation of spin-labelled biomolecular assemblies.

In principle, the first harmonic dispersion, U_1', out-of-phase display contains the same information as the V_2'-spectrum. However, the lineshape changes induced by saturation transfer are not so richly detailed and do not lend themselves so easily to the definition of diagnostic lineheight ratios as do the V_2'-spectra. Fajer and Marsh (1983a) introduced a method of analysis based on the (V_0-U_1') difference spectrum, where the zeroth-harmonic absorption V_0-lineshape is obtained by integration of the unsaturated V_1-spectrum. Rotational correlation time calibrations were produced for (V_0-U_1') by using spin-labelled haemoglobin. This offers a general approach to detecting saturation transfer even in complex or multicomponent systems. Deviations of the U_1'-lineshape from that of the conventional integrated absorption spectrum are a quantitative indication of saturation transfer.

The first-harmonic out-of-phase absorption spectrum (V_1') is insensitive to rotational motion (see Fig. 2). Hyde and Thomas (1973) suggested, however, that this nonlinear rapid passage display might be useful for obtaining information on spin-lattice relaxation times. Indeed, this has formed the basis for development of the non-classical type of saturation transfer experiment that was referred to above. As will be seen later in section 7.2, the V_1'-spectrum turns out to be practically an ideal T_1-sensitive display, with little sensitivity to T_2 or molecular motion.

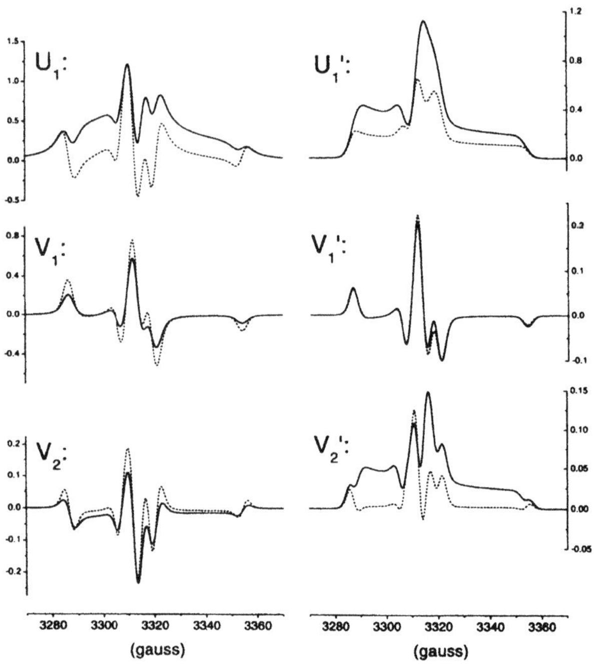

Figure 2. Simulated out-of-phase (right-hand side) and in-phase (left-hand side) nitroxide EPR spectra. *Top row:* first harmonic dispersion, U_1 and U_1'; *middle row*: first harmonic absorption, V_1 and V_1'; *bottom row*: second harmonic absorption, V_2 and V_2'. Spectra are simulated for spin-label rotation rates of $f_R = 0$ (solid line) and 10^6 s^{-1} (dashed line). The simulation procedure is described later in Section 4.

4. MODULATION-COUPLED BLOCH EQUATIONS

Simulation of rapid-passage nonlinear EPR spectra requires explicit inclusion not only of the microwave magnetic field, but also of the Zeeman modulation field. Halbach (1954) provided analytical solutions of the Bloch equations for the first harmonic out-of-phase absorption and dispersion signals, in the low amplitude limit. This was subsequently extended to the second harmonic absorption signal, for studying the dependence of the out-of-phase signal on T_1-relaxation time (Páli et al., 1996). Analysis of slow molecular motion by ST-EPR necessitates use of the diffusion-coupled Bloch equations. This was first done by Thomas and McConnell (1974) with a Brownian diffusion model, and used to validate rotational correlation time calibrations based on diagnostic lineheight ratios (Thomas et al., 1976). Solutions must be obtained numerically when dealing with small-step diffusion. If, however, an uncorrelated jump model is used for the diffusion process, solution of the integral equations for the lineshape is possible in closed form (Livshits, 1976). This latter approach is taken here, together with the adiabatic approximation that limits validity to the slow motional regime.

The Bloch equations that explicitly include the microwave (H_1) and Zeeman modulation (H_m) fields are generalised to take into account molecular rotation by using the random jump diffusion model of Livshits (1976). The resulting equations for the time dependence of the spin magnetisation vector $\vec{\mathbf{X}} = (u, v, z)$ in the rotating frame can be written in matrix form (Livshits et al., 1998b):

$$\dot{\vec{\mathbf{X}}} + (\hat{\mathbf{A}} + f_R \hat{\mathbf{E}}) \cdot \vec{\mathbf{X}} = \vec{\mathbf{M}}_o + f_R \int \vec{\mathbf{X}} d\Omega \tag{1}$$

where $f_R \equiv \tau_R^{-1}$ is the frequency of isotropic rotational reorientation, $\vec{\mathbf{M}}_o = M_o(0,0,T_1^{-1})$, $\Omega \equiv (\theta, \varphi)$ is the orientation of the static magnetic field relative to the magnetic principal axes, and $\hat{\mathbf{E}}$ is the unit matrix. The term containing f_R on the left-hand side of Eq. 1 is the rate of transfer of spin magnetisation to other orientations and that on the right-hand side is the rate of transfer from all other orientations. The Bloch equation matrix $\hat{\mathbf{A}}$ contains both the microwave and modulation fields (as well as the static and resonance fields, and the T_1- and T_2-relaxation times). An expansion is made of the spin magnetisation in Fourier harmonics of the modulation frequency, ω_m:

$$\vec{X} = \sum_{n=-\infty}^{\infty} \vec{X}_n e^{-in\omega_m t} \tag{2}$$

where $\vec{X}_n = (u_n, v_n, z_n)$ are the complex Fourier amplitudes. This gives an infinite system of coupled equations for the amplitudes (Livshits et al., 1998b):

$$\begin{aligned} \hat{\mathbf{A}}_0 \cdot \vec{\mathbf{X}}_0 &= \vec{\mathbf{M}}_0 + f_R \int \vec{\mathbf{X}}_0 d\Omega + \tfrac{1}{2}\hat{\gamma} \cdot H_m(\vec{\mathbf{X}}_1 + \vec{\mathbf{X}}_{-1}) \\ \hat{\mathbf{A}}_n \cdot \vec{\mathbf{X}}_n &= f_R \int \vec{\mathbf{X}}_n d\Omega + \tfrac{1}{2}\hat{\gamma} \cdot H_m(\vec{\mathbf{X}}_{n+1} + \vec{\mathbf{X}}_{n-1}) \end{aligned} \tag{3}$$

where

$$\hat{\mathbf{A}}_n = \begin{pmatrix} T_2^{-1} + f_R - in\omega_m & \gamma_e[H - H_{res}(\theta,\varphi)] & 0 \\ -\gamma_e[H - H_{res}(\theta,\varphi)] & T_2^{-1} + f_R - in\omega_m & -\gamma_e H_1 \\ 0 & \gamma_e H_1 & T_1^{-1} + f_R - in\omega_m \end{pmatrix} \tag{4}$$

is obtained from expansion of the Bloch equation matrix, and the gyromagnetic ratio (γ_e) tensor is:

$$\hat{\gamma} = \begin{pmatrix} 0 & -\gamma_e & 0 \\ \gamma_e & 0 & 0 \\ 0 & 0 & 0 \end{pmatrix} \tag{5}$$

Here H_{res} is the resonance field:

$$H_{res}(\theta,\varphi) = \frac{\hbar\omega_L}{\beta_e g(\theta,\varphi)} - M_I A(\theta,\varphi) \tag{6}$$

with ω_L the microwave frequency and M_I the nuclear magnetic quantum number of the spin label. The g- and hyperfine tensor anisotropies are given by the intermediate field approximation (Van et al., 1974):

$$g(\theta,\varphi) = g_{zz}\cos^2\theta + g_{xx}\cdot\sin^2\theta\cdot\cos^2\varphi + g_{yy}\sin^2\theta\cdot\sin^2\varphi \tag{7}$$

$$A(\theta,\varphi) = (A_{zz}^2\cos^2\theta + A_{xx}^2\sin^2\theta\cos^2\varphi + A_{yy}^2\sin^2\theta\sin^2\varphi)^{\frac{1}{2}} \tag{8}$$

where the principal tensor components are: $\mathbf{g} = (g_{xx}, g_{yy}, g_{zz})$ and $\mathbf{A} = (A_{xx}, A_{yy}, A_{zz})$. Pseudo-secular terms are retained in Eqs. 7,8, and line positions are well reproduced, although non-secular ocuplings are neglected.

A solution is obtained in the small modulation amplitude approximation $\left(h_m = \gamma_e H_m T_2^o < 1\right)$ by expanding the Fourier coefficients in a power series of the dimensionless modulation amplitude, h_m (Halbach, 1954):

$$\vec{X}_n = \sum_{v=0}^{\infty} \vec{X}_{n,v} h_m^v \tag{9}$$

where the coefficients $\vec{X}_{n,\nu} = (u_{n,\nu}, v_{n,\nu}, z_{n,\nu})$ with $\nu<|n|$ disappear because of the symmetry properties of the Bloch equations. Restriction to small modulation amplitudes means that the intensity of the first harmonic signal is linearly dependent on H_m and that of the second harmonic depends on H_m^2. The equations for the vector coefficients $\vec{\mathbf{X}}_{n,n} = (u_{n,n}, v_{n,n}, z_{n,n})$ that define the magnetisation components depending on the nth power of the modulation amplitude are, from Eq. 3:

$$\hat{\mathbf{A}}_0 \cdot \vec{\mathbf{X}}_{0,0} = \vec{\mathbf{M}}_o + f_R \int \vec{\mathbf{X}}_{0,0} d\Omega \tag{10}$$

$$\hat{\mathbf{A}}_n \cdot \vec{\mathbf{X}}_{n,n} = f_R \int \vec{\mathbf{X}}_{n,n} d\Omega + (\hat{\gamma} / \gamma_e) \cdot \vec{\mathbf{X}}_{n-1,n-1} \tag{11}$$

where f_R and the matrix elements of $\hat{\mathbf{A}}_0$ and $\hat{\mathbf{A}}_n$ are now dimensionless, which is obtained by multiplying their initial values by T_2^o.

The solution, $\vec{\mathbf{I}}_1 = \int \vec{\mathbf{X}}_{1,1} d\Omega$, of the integral equation (11) for the first harmonic is given in matrix form by (Livshits et al., 1998b):

$$\vec{\mathbf{I}}_1 = (\hat{\mathbf{E}} - \hat{\mathbf{J}}_1 f_R)^{-1} \cdot \vec{\mathbf{J}}_{10} \tag{12}$$

where $\hat{\mathbf{J}}_1 = \int \hat{\mathbf{A}}_1^{-1} d\Omega$ and $\vec{\mathbf{J}}_{10} = \int \hat{\mathbf{A}}_1^{-1} \cdot \hat{\gamma} \cdot \vec{\mathbf{X}}_{0,0} d\Omega$. The zero-order coefficients $\vec{\mathbf{X}}_{0,0}$ of the spin magnetisation vector that are required for the $\vec{\mathbf{J}}_{10}$ integral are obtained from solution of Eq. 10. These are given by (Livshits et al., 1998b):

$$\vec{\mathbf{X}}_{0,0} = \hat{\mathbf{A}}_0^{-1} \cdot \left[\vec{\mathbf{M}}_o + f_R \left(\hat{\mathbf{E}} - \hat{\mathbf{J}}_0 f_R \right)^{-1} \cdot \hat{\mathbf{J}}_0 \cdot \vec{\mathbf{M}}_o \right] \tag{13}$$

where $\hat{\mathbf{J}}_0 = \int \hat{\mathbf{A}}_0^{-1} \cdot d\Omega$. A similar procedure then yields the solution, $\vec{I}_2 = \int \vec{X}_{2,2} \cdot d\Omega$, for the second harmonic.

The lineshapes of the nth harmonic out-of-phase absorption and dispersion spectra are finally given by:

$$V_n' = \mathrm{Im}\{V_{n,n}\} \tag{14}$$

and $U_n' = \mathrm{Im}\{U_{n,n}\}$ (15)

respectively. The conventional first-harmonic in-phase absorption spectrum is correspondingly given by:

$$V_n = \mathrm{Re}\{V_{n,n}\} \tag{16}$$

where $\vec{\mathbf{I}}_n = (U_{n,n}, V_{n,n}, Z_{n,n})$.

Typical calculated second harmonic out-of-phase absorption V_2'-lineshapes are given in Fig. 3. The characteristic differential loss of intensity at the intermediate field positions, relative to those corresponding to the stationary turning points, is evident with increasing rate, f_R, of rotational diffusion. This behaviour of the lineshape defines the diagnostic lineheight ratios introduced by Thomas, Dalton and Hyde (1976).

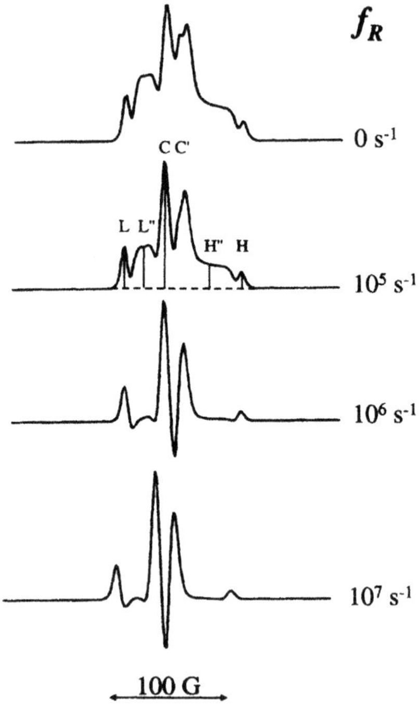

Figure 3. Simulated second-harmonic, 90°-out-of-phase absorption ST-EPR spectra (V_2'-display) for increasing rates ($f_R \equiv 1/\tau_R$) of isotropic rotation. Spectra are calculated for jump diffusion as described in Section 4. Spectra are normalised to the maximum lineheight and do not reflect the decreasing absolute intensity with increasing f_R (cf. Fig. 2). The positions at which the diagnostic lineheight ratios, L''/L, C'/C and H''/H, are measured are shown for one of the spectra.

5. SLOW ROTATIONAL DIFFUSION

The sensitivity of nonlinear ST-EPR spectra to slow rotational motion can be analysed directly by spectral simulation using the diffusion-coupled Bloch equations, as already described and illustrated in Fig. 3. However, an approximate semi-analytical approach based on the formal equivalence between Heisenberg spin exchange and exchange by jump diffusion (Eastman et al., 1969; Marsh, 1992a) has the advantage of containing the basic physical principles of saturation transfer and of giving rise to a very simple expression for parameterising experimental correlation time calibrations (Marsh and Horváth, 1992a). For these reasons, particularly the latter, we outline this simplified treatment here.

The effective spin-lattice relaxation time, $T_1^{eff}(\omega)$, for a given spin packet at resonance position ω depends on the spectral diffusion rate, $\tau_{sd}(\omega)^{-1}$, according to (Eastman et al., 1969; Marsh, 1992a) (and see later in Section 8.1.2):

$$T_1^{eff}(\omega) = T_1^o \frac{1 + Z(\omega)T_1^o \tau_{sd}(\omega)^{-1}}{1 + T_1^o \tau_{sd}(\omega)^{-1}} \tag{17}$$

where T_1^o is the intrinsic spin-lattice relaxation time (in the absence of spectral diffusion) and $Z(\omega)$ is the fractional spin-packet population, or degeneracy (see Section 8.1.2). Thus, $1/Z(\omega)$ is the redistribution factor of the spin packet at position ω, on spectral diffusion over the entire powder lineshape. The spectral diffusion rate at resonance position ω is given by (Fajer et al., 1986):

$$\tau_{sd}(\omega)^{-1} = \left(\frac{8}{3\pi^2}\right)\left(\frac{\partial\omega}{\partial\theta}\right)^2 T_2^2 \tau_R^{-1} \tag{18}$$

where τ_R is the rotational correlation time and $1/T_2$ is the spin packet width by which the resonance position must change in order to alleviate saturation (cf. Fig. 1). This expression (i.e., Eq. 18) was used by Fajer, Hyde and coworkers (1986) to analyse the fast phase in the saturation recovery of spin-labelled haemoglobin, following short microwave pulses. The parameter $(\partial\omega/\partial\theta)$ is the rate at which the resonance position changes with angular orientation, θ, of the spin label with respect to the magnetic field direction (see Hyde and Dalton, 1979).

As will be seen later (in section 7.3), the intensity of the out-of-phase ST-EPR signal, relative to the conventional in-phase EPR, is approximately proportional to T_1^{eff} (Páli et al., 1996). The ST-EPR intensity can therefore be approximated by: $I(\omega) = I_o(\omega)T_1^{eff}(\omega)/T_1^o$, where $I_o(\omega)$ is the intensity in the absence of spectral diffusion (Marsh and Horváth, 1992b). Hence from Eq. 17, the ST-EPR lineshape is given by:

$$I(\omega) = I_o(\omega) \cdot \frac{1 + [I_o(\omega)/T_2]T_1^o \tau_{sd}(\omega)^{-1}}{1 + T_1^o \tau_{sd}(\omega)^{-1}} \tag{19}$$

where the degeneracy factor or fractional population, $Z(\omega)$, for a spectral segment of width $1/T_2$ is given by $I_o(\omega)/T_2$. The following axial orientation dependence of the resonance position contains the essential features of anisotropic powder patterns (cf. Marsh, 1990):

$$\omega = (\omega_{//} - \omega_{\perp})\cos^2\theta + \omega_{\perp} \tag{20}$$

where $\omega_{//}$, $\omega_{\perp}$ are the resonance line positions corresponding to the magnetic field oriented parallel or perpendicular to the principal axis, i.e., $\theta = 0°$ and $\theta = 90°$, respectively. The resulting rate of change of the resonance position with angle is:

$$\frac{\partial\omega}{\partial\theta} = -2[(\omega_{//} - \omega_{\perp})(\omega - \omega_{\perp})]^{1/2} \tag{21}$$

and the normalised lineshape in the absence of rotational diffusion is:

$$I_o(\omega) = \tfrac{1}{2}[(\omega_{//} - \omega_{\perp})(\omega - \omega_{\perp})]^{-1/2} \tag{22}$$

which is valid for the range $\omega_{\perp} < \omega < \omega_{//}$.

Figure 4 gives the model lineshapes predicted from Eqs. 18-22 for various values of the rotational correlation time, τ_R. The calculations reproduce the well-known sensitivity of ST-EPR lineshapes to rotational diffusion at rates comparable to that of the spin-lattice relaxation, which was characterised originally by Thomas, Hyde and coworkers (1976). The lineheight at an intermediate spectral position *P'*, relative to that at the invariant turning point *P*, decreases progressively with decreasing correlation time. The inset to Fig. 4 demonstrates that the relative intensity at a position 1/3 of the way in from the $\omega_{//}$ turning point depends on rotational correlation time in a manner very similar to that found for the experimental diagnostic ST-EPR lineheight ratios L''/L, C'/C and H''/H (see Thomas et al., 1976).

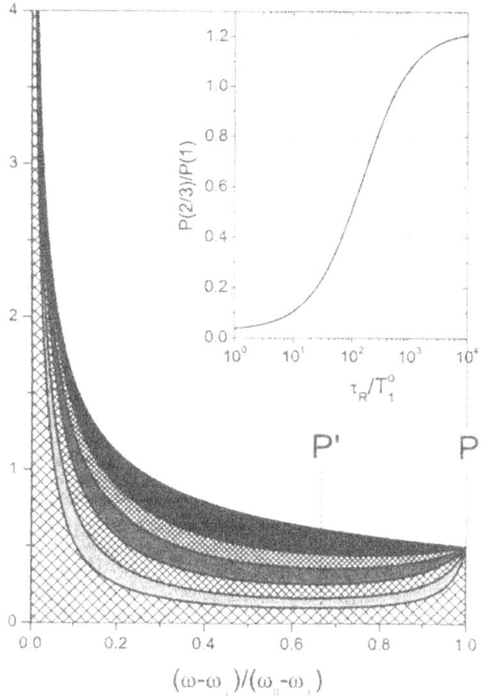

Figure 4. Model axial ST-EPR powder lineshapes (heavy lines) calculated from Eq. 19, together with Eqs. 18, 19, 21 and 22, for increasing rates of rotational diffusion. From top to bottom, $T_1^o\tau_R^{-1}$ = 0, 0.0025, 0.005, 0.01, 0.02 and 0.04; $T_2^{-1} = (\omega_{//}-\omega_{\perp})/25$ in all cases and the integrated intensity is normalised to unity in the absence of rotational diffusion [ordinate in units of $(\omega//-\omega\perp)$-1]. Inset gives the diagnostic ST-EPR lineheight ratio measured at point P', relative to the turning point P at the right-hand extremum of the spectrum (see Marsh and Horváth, 1992a).

The above analysis suggests that spectral lineheight ratios and integrated intensities, R, may have the following dependence on τ_R (cf. Eq. 19):

$$R = R_o \frac{1 + a/\tau_R}{1 + b/\tau_R} \tag{23}$$

where R_o is the value of R in the absence of rotational diffusion, a and b are constants to be fitted that depend only on intrinsic spectral parameters, and the ratio a/b is effectively related to the orientational degeneracy parameter

($Z(\omega_\theta) \approx \sin\theta$) at ω_θ corresponding to the diagnostic spectral position P'. Equation 23 rather well describes the dependence on rotational correlation time of the diagnostic line height ratios and intensities of the saturation transfer EPR spectra from spin-labelled haemoglobin in glycerol-water mixtures (Marsh and Horváth, 1992a). It therefore can be used to give the following simple expression for the correlation time calibrations of the experimental ST-EPR spectra:

$$\tau_R = k/(R_o - R) - b \tag{24}$$

where values of the experimental calibration constants, k, R_0 and b, for the different diagnostic spectral parameters are given in Table 1. This is a much more readily accessible form for the calibrations of rotational correlation time than hitherto was presented (e.g., Horváth and Marsh, 1988) and has the additional advantage of reflecting directly the underlying spectral diffusion process.

Table 1: Parameters for fitting experimental ST-EPR rotational correlation time calibrations (Horváth and Marsh, 1983) from spin-labeled hemoglobin to equation (24) for the diagnostic spectral lineheight ratios (L"/L, H"/H, C'/C) and normalized integrals (I_{ST}). Fits to the calibration curves are given in Marsh and Horváth (1992a).

R	Range fitted	R_o	k (μs)	b (μs)
L"/L[a]	0.2, 2.0	1.825	105.6	63.8
H"/H	0.2, 2.0	2.17	407	210
C'/C	0.2, 1.0	1.01	21.3	21.1
	-0.4, 1.0	0.976	11.9	7.82
I_{ST} (total)	0.15, 1.0	1.07×10^{-2}	0.400	43.6
I_{ST} ($M_I = -1$)[b]	0.06, 0.6	6.97×10^{-4}	2.85×10^{-2}	45.0

[a]Values of L" correspond to the maximum in the low-field diagnostic region. For values corresponding to the position 1/3 of the way in from the low-field turning point (L), as given originally in Horváth and Marsh (1983), see Marsh (1992b).

[b]Integrated intensity for the high-field manifold alone [used to detect anisotropic rotational diffusion - see Horváth and Marsh (1983)].

In principle, slow rotational diffusion also may be studied from the power saturation of the conventional EPR spectra. Squier and Thomas (1986) have done this in terms of saturation factors determined from the ratio of the integrated intensities of the conventional first-derivative EPR spectra recorded at low (subsaturating) and high (partially saturating) microwave powers. Calibrations for this method that are equivalent to those given for ST-EPR by Eq. 24 can be found in Marsh (1995).

6. APPLICATIONS: SLOW ROTATION

This section gives examples of the classical application of saturation transfer EPR to the study of slow rotational diffusion on the submillisecond timescale. Applications are chosen to illustrate specific aspects of rotational diffusion in membranes such as the effects of protein concentration, aqueous viscosity, hydrophobic matching and anisotropic rotation, rather than giving a comprehensive review.

6.1 Dependence on Protein Density

A systematic study of the dependence of the ST-EPR rotational correlation times on lipid/protein ratio, *LP*, was undertaken by Fajer et al. (1989) with spin-labelled cytochrome *c* oxidase reconstituted in bilayer membranes of dimyristoyl phosphatidylcholine. Fig. 5 gives the effective rotational relaxation rate, τ_R^{-1}, deduced from the central ST-EPR lineheight ratio, C'/C, as a function of lipid/protein ratio in the reconstituted membranes. Protein rotational diffusion is drastically reduced in gel-phase membranes at 1°C where the lipid chains are largely frozen. Rotation is much faster in fluid-phase membranes, with effective correlation times in the tens of microsecond regime. The rate at which cytochrome oxidase rotates in the membrane decreases (i.e., the correlation time increases) progressively with increasing protein packing density.

The hindering of cytochrome oxidase rotation by protein crowding can be described with a simple collisional model based on random protein-protein contacts. The observed diffusion coefficient, $\langle D_{R//}\rangle \left(=1/6\tau_{R//}^{eff}\right)$, is a statistical average of that for freely rotating species that do not experience any influence from other proteins ($D_{R,f}$), and that for highly hindered species that have other proteins immediately adjacent ($D_{R,h}$):

$$\langle D_{R//}\rangle = D_{R,f}P_f + D_{R,h}\left(1-P_f\right) \tag{25}$$

where P_f denotes the probability for the freely rotating species. This model assumes that the lifetime of protein-protein contacts is shorter than the rotation period. For a translational diffusion coefficient of 10^{-9} cm^2 s^{-1}, this is likely to be the case (Fajer et al., 1989). The probability P_f is obtained from a lattice model designed to calculate the frequency of lipid-protein contacts in random dispersions (Hoffmann et al., 1981). Each lipid occupies one lattice site and each protein occupies *R* lattice sites, where $R \approx 27$ for the cytochrome *c* oxidase monomer (Deatherage et al., 1982) P_f is then the probability that all *N* lattice sites at the protein perimeter are occupied by lipid molecules:

$$P_f = [1 - R/(LP + R)]^N \tag{26}$$

where LP is the lipid/protein mole ratio (see Fig. 5).

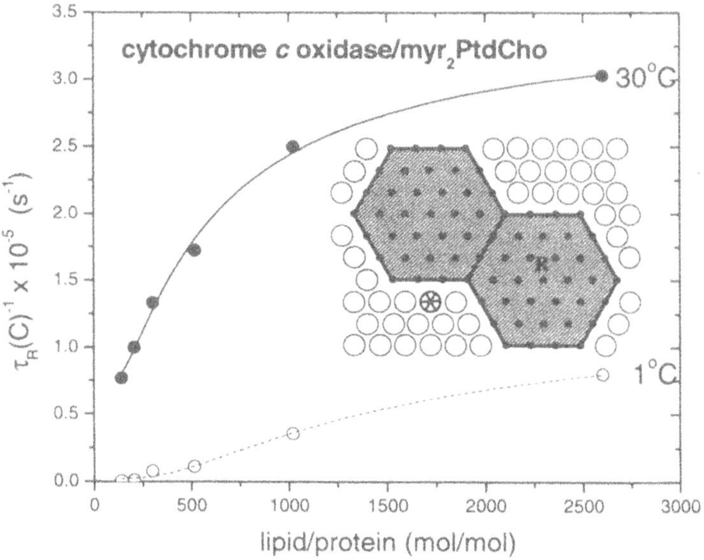

Figure 5. Dependence of the rotational relaxation rate, $\tau_R(0)^{-1}$, on lipid/protein molar ratio for cytochrome *c* oxidase in membranes of dimyristoyl phosphatidylcholine at 30°C (fluid phase, ●) and 1°C (gel phase, ○). Solid line is a non-linear least squares fit of the random collision model (Eqs. 25 and 26) to the data at 30°C with fixed $R = 27$ (see Fajer et al., 1989). The *inset* shows the lattice model used to calculate the probability, P_f, that a protein (given by the hexagons that occupy R lipid lattice sites) does not contact any other protein. The probability that a lipid (circle) does not occupy a site (e.g., the asterisked position) that is adjacent to a protein is given by $R/(LP+R)$ where LP is the lipid/protein mole ratio.

A non-linear least squares fit with R fixed and N as the parameter to be optimised is given by the solid line in Fig. 5. This depicts the time-averaged protein-protein interactions taking place in the fluid phase. The fitted value of $N = 18$ corresponds to 36 first-shell lipid sites at the perimeter of the protein, in both bilayer halves of the membrane. This is somewhat smaller than the number of boundary lipids found by EPR measurements with spin-labelled lipids (Knowles et al., 1979). Presumably, the latter reflects the invaginated nature of the intramembranous surface of the protein, which is effectively smoothed when considering protein-protein contacts.

The effect of lipid/protein ratio on rotational mobility, therefore, can be reasonably described in terms of random protein collisions that occur with increasing protein density. An alternative model of a heterogeneous population of rotating species, with a varying proportion of higher oligomers is unable to explain the dependence on lipid/protein ratio (Fajer et al., 1989). Also, the increase in membrane viscosity with increasing protein concentration is predicted to be insufficient to account for the changes in rotation rate.

6.2 Hydrophobic Matching

Rhodopsin has been reconstituted in bilayer membranes formed from phosphatidylcholines with different acyl chainlengths (Ryba and Marsh, 1992). Rotational diffusion of the spin-labelled protein recorded by ST-EPR was used to follow the extent of dispersal of the protein in the different chainlength lipids. A similar study with equivalent results, was undertaken earlier by Kusumi and Hyde (1982).

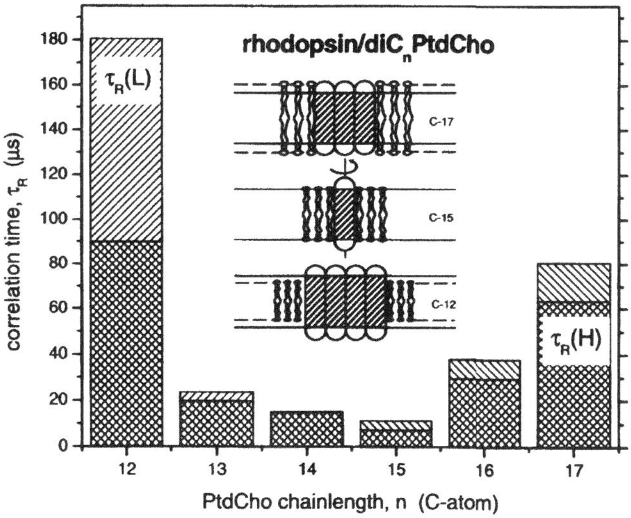

Figure 6. Lipid chainlength (n) dependence of the effective rotational correlation times of maleimide spin-labelled rhodopsin in membranes of different saturated diacyl phosphatidylcholines (diC$_n$PtdCho), at a lipid/protein ratio of 60:1 (mol/mol). Effective correlation times $\tau_R(L)$, $\tau_R(H)$ are deduced from the low-field L''/L (+45° hatching) and high-field, H''/H (-45° hatching) ST-EPR diagnostic lineheight ratios (see Ryba and Marsh, 1992). Matching/mismatching of the hydrophobic lengths of lipid and protein are indicated schematically for lipid chainlengths n = 12, 15 and 17.

Figure 6 gives the effective rotational correlation times of rhodopsin in the lipids of different chainlength that are deduced from the low-field and high-field regions of the ST-EPR spectra. These measurements are all made in the fluid membrane state at equivalent temperatures above the chain-melting transition for the different lipids. Mostly, the correlation times for *L"/L* and *H"/H* are comparable, which is expected because both are reflecting the same anisotropic rotation. The longest effective rotational correlation times are obtained from recombinants with dilauroyl phosphatidylcholine, $diC_{12}PtdCho$, with a steep decrease on increasing the lipid chainlength, through a minimum at chainlengths of C14 to C15, and a subsequent rise on increasing the lipid chainlength to dipalmitoyl and diheptadecanoyl phosphatidylcholines, $diC_{16}PtdCho$ and $diC_{17}PtdCho$.

The pronounced increases in rotational correlation time for rhodopsin in the long and short chainlength lipids can be attributed to protein aggregation. This is driven by hydrophobic mismatch in both cases, as illustrated diagrammatically in the inset to Fig. 6. A lipid that is too short exposes part of the hydrophobic domain of the protein to a polar environment. A lipid that is too long forces contact of the hydrophobic lipid chains with polar groups on the protein. In each case, these energetically unfavourable interactions are alleviated by segregation of the protein from the lipids.

A topic of equal interest is the oligomer state of rhodopsin in the C15 chainlength lipids for which hydrophobic matching is best. The effective rotational correlation time, $\tau_R^{eff}(M_I)$, deduced from the ST-EPR spectra using calibrations from isotropic solutions is related to the diffusion coefficient, $D_{R//}$, for uniaxial rotation by (Robinson and Dalton, 1980; Marsh and Horváth, 1989):

$$\tau_R^{eff}(\pm 1) = \frac{1}{3\left[D_{R//}\sin^2\theta + D_{R\perp}\left(1+\cos^2\theta\right)\right]} \tag{27}$$

where θ is the orientation of the spin-label *z*-axis relative to the membrane normal, and $D_{R\perp} \approx 0$. The rotational diffusion coefficient is related to the cross-sectional dimensions, a_m and b_m, and the intramembrane height, h_m, of the rotating species by the Stokes-Einstein equation (see e.g., Marsh and Horváth, 1989):

$$D_{R//} = \frac{k_B T}{f_{R//}} = \left(\frac{k_B T}{4\pi\eta_m a_m^2 h_m}\right) F_{R//,m} \tag{28}$$

where $f_{R//}$ is the rotational frictional coefficient, η_m is the effective intramembrane viscosity and $F_{R//,m} \leq 1$ is a shape factor that depends weakly on the asymmetry for $a_m/b_m \leq 2$ (see also following section 6.3). The corresponding true rotational correlation time is: $\tau_{R//} = 1/(6D_{R//})$. As usual, k_B is Boltzmann's constant and T is the absolute temperature.

The effective rotational correlation time of rhodopsin in dipentadecanoyl phosphatidylcholine, diC$_{15}$PtdCho, is $\tau_R^{eff} \approx 7$ μs (Fig. 6); an interpolated value for the membrane thickness is $h_m = 4$ nm (Tardieu, 1972); and the effective membrane viscosity is in the region of $\eta_m = 5$ P (Cherry and Godfrey, 1981). This yields a value of $2a_m \approx 4.5$ nm for the intramembranous diameter of the rotating species. This is an upper estimate because it is assumed that $\theta = 90°$ and $F_{R//,m} = 1$. Cross-sectional dimensions of the dimer of frog rhodopsin are $2a_m \approx 7$-8 nm, $2b_m \approx 2.2$-2.5 nm (Corless et al., 1982). Thus rhodopsin is most likely a monomer in diC$_{15}$PtdCho, as it is in rod outer segment disc membranes (Downer, 1985).

6.3 Dependence on Extramembrane Viscosity

As is well known, the intramembrane viscosity that characterises the torque on large integral membrane proteins is much greater than the lipid microviscosity that is determined with small probe molecules by applying the Debye equation (Cherry and Godfrey, 1981). Therefore, because the effective viscosity in the membrane ($\eta_m \sim 2$-5 P) is so much higher than that of water, the rotational diffusion coefficients of membrane proteins are normally determined solely by the intramembranous sections of the proteins (cf., Eq. 28). Only if the extramembrane viscosity is increased considerably, e.g., by addition of sucrose or glycerol, does the rotational diffusion coefficient become dependent on the dimensions of the extramembranous sections of the protein.

The frictional torques exerted on the separate sections of the protein are additive, therefore so are also the individual contributions to the overall frictional coefficient:

$$f_{R//} = \sum_i \frac{f^o_{R//,i}}{F_{R//,i}} \tag{29}$$

where the frictional coefficients of right circular cylinders with volumes, V_i, equal to those of the different sections, i, of the protein are given by:

$$f^o_{R//,i} = 4\eta_i V_i \tag{30}$$

and the corresponding shape factors are:

$$F_{R//,i} = 2\frac{a_i / b_i}{1 + (a_i / b_i)^2} \tag{31}$$

where a_i, b_i are the elliptical semi-axes of the different protein cross-sections (see Fig. 7 inset). The rotational correlation time of the protein therefore is given from Eqs. 28-30 by:

$$\tau_{R//} = \frac{2}{3k_BT}\sum_i \frac{\eta_i V_i}{F_{R//,i}} \tag{32}$$

Specifically, the dependence on external viscosity is given by:

$$\tau_{R//} = \tau_{R,memb} + \frac{2\eta_o}{3k_BT}\sum_j{}' \frac{V_j}{F_{R//,j}} \tag{33}$$

where the summation (as indicated by the prime) now extends only over the extramembrane sections of the protein, and $\tau_{R,memb}$ (= $1/6D_{R,memb}$) is the rotational correlation time given by Eq. 28 above, when external viscosity can be neglected.

Figure 7 gives the dependence of the effective rotational correlation time of membranous spin-labelled Na,K-ATPase on viscosity of the external glycerol-containing medium. From the slope and intercept of the linear viscosity dependence, together with Eqs. 28 and 33, it is concluded that 50-70% of the Na,K-ATPase protein is external to the membrane (Esmann et al., 1994). This conclusion obtained from hydrodynamics is consistent with the results of low-resolution structural studies on this protein (Maunsbach et al., 1989). Fig. 7 also shows that, with polyethylene glycol solutions, a pronouncedly non-linear dependence on the viscosity, η_o, is found that is much larger than the viscosity dependence obtained with glycerol solutions. This greater effect of polyethylene glycol undoubtedly corresponds to a dehydration-induced aggregation of the membrane proteins that may be related to the ability of polyethylene glycol to induce membrane fusion. The rotational correlation time reached at 50% polyethylene glycol corresponds to a degree of aggregation of the membrane proteins between two and five, depending on whether the ethylene glycol polymer is excluded from the membrane surface region (Esmann et al., 1994). Clearing of proteins from areas of apposing membrane, by aggregation, is a prerequisite for the close approach of the lipid bilayers that is needed for effective membrane fusion.

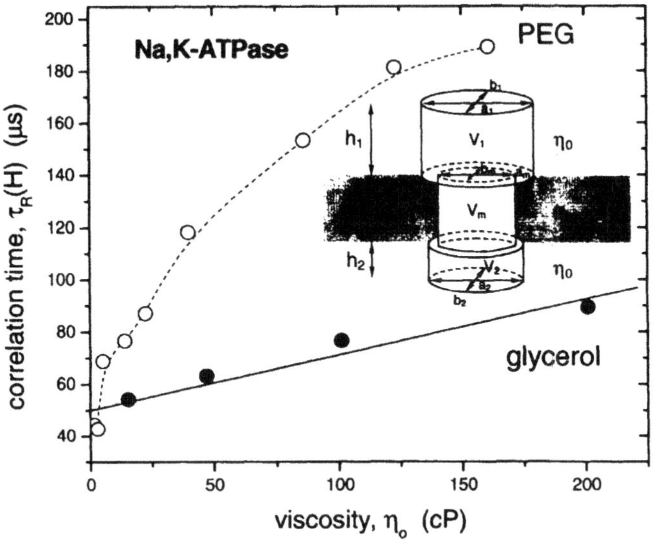

Figure 7. Dependence on extramembrane viscosity, η_o, of the effective rotational correlation time, $\tau_R(H)$ deduced from the high-field diagnostic lineheight ratio in the ST-EPR spectra of Na,K-ATPase spin-labelled with a chloromercuri-reagent. The solid line is a linear regression for the glycerol data with intercept 50 μs and gradient 22 μs P^{-1}. The dashed line for the PEG data is simply to guide the eye (see Esmann et al., 1994). The *inset* illustrates the different dimensions of the extramembrane and intramembrane sections of the protein that experience viscosities η_o and η_m, respectively.

6.4 Anisotropic Rotational Diffusion

Rotational correlation times are routinely deduced from experimental ST-EPR spectra by comparing the diagnostic lineheight ratios in the low-field, central and high-field regions of the spectrum with those obtained from isotropically rotating spin-labelled haemoglobin in solutions of known viscosity (see Table 1). The outer lineheight ratios, L''/L and H''/H, are sensitive to rotation of the nitroxide z-axis, via modulation of the hyperfine interaction, and the central lineheight ratio, C'/C, is sensitive to rotation about all three nitroxide axes, via modulation of the g-value anisotropy (see Eq. 7 and Fajer and Marsh, 1983). For anisotropic rotation, the different lineheight ratios will therefore have differential sensitivities, as illustrated in Table 2 for the rotational diffusion of a spin-labelled phospholipid in gel-phase lipid bilayer membranes (Marsh, 1980). The nitroxide z-axis is oriented along the lipid long molecular axis for this particular spin probe. At low temperatures the effective correlation times, $\tau_R^{eff}(M_I)$, deduced from the different lineheight ratios using the isotropic model system for calibration, are all very similar. Above the bilayer pretransition at 25°C, the effective correlation time deduced from the outer lineheight ratios is relatively unchanged, whilst that deduced from the central spectral region decreases abruptly, indicating the onset of rapid anisotropic rotational diffusion about the long axis of the lipid molecule. A differential response of the different lineheight ratios may therefore be used to diagnose anisotropic rotational diffusion (Fajer and Marsh, 1983b). Comparison of the integral of the high-field region of the saturation transfer spectrum with that of the total ST-EPR spectrum may also be similarly used (see Table 1).

Table 2: Effective rotational correlation times of a spin-labelled phospholipid in gel phase bilayers of dipalmitoyl phosphatidylcholine, deduced from the ST-EPR lineheight ratios using isotropic reference spectra (Marsh, 1980).

	$\tau_R^{eff}(M_I)(s)$			
Parameter	12°C	25°C	30°C	45°C
L''/L ($M_I = +1$)	6×10^{-4}	1×10^{-4}	0.8×10^{-4}	$\approx 10^{-9}$
H''/H ($M_I = -1$)	4×10^{-4}	2×10^{-4}	1×10^{-4}	$\approx 10^{-9}$
C'/C ($M_I = 0$)	0.8×10^{-4}	5×10^{-6}	0.9×10^{-6}	$\approx 10^{-9}$

Simulations of first harmonic phase-quadrature dispersion ST-EPR spectra by Robinson and Dalton (1980) give some guide to the quantitative interpretation of the effective correlation times in terms of the true rotational diffusion parameters. It was found that the effective correlation times deduced from the low-field and high-field regions of the spectrum were very similar, independent of the degree of anisotropy of the motion (cf. Table 2). If the anisotropy of the rotation is great enough and the rotational rates are slow, the dependence of the effective correlation times on the orientation, θ, of the nitroxide *z*-axis with respect to the rotational diffusion axis, is given by Eq. 27 that was introduced above. Precise measurements thus require knowledge of the orientation, θ. However, some estimate of whether θ is close to 0° or close to 90° may be obtained from the relative sizes of $\tau_R^{eff}(\pm1)$ and $\tau_R^{eff}(\pm0)$, or by comparing the high-field and total spectral integrals. In addition, it may be possible to discriminate between protein monomer and oligomer formation on the basis of Eqs. 28-31 without accurate knowledge of θ.

7. T_1-SENSITIVE NONLINEAR EPR DISPLAYS

As will be seen from the examples given later in Section 9, the impetus for exploring the direct sensitivity of non-linear EPR to spin-lattice relaxation and cross-relaxation processes is in the study of paramagnetic relaxation enhancements and slow exchange processes. Both of these aspects have assumed considerable importance in stuctural and dynamic studies of biological membranes. The former is an essential part of site-directed spin-labelling methodology.

Approximate solutions of the Bloch equations that explicitly include the Zeeman modulation field (see section 4 and Páli et al., 1996) yield the following approximate expressions for the different out-of-phase dispersion (U_n') and absorption (V_n') displays, where n is the harmonic with respect to the Zeeman modulation frequency and the prime indicates out-of-phase detection. To illustrate simply the dependence on T_1- and T_2-relaxation times, the effects of molecular motion are neglected.

The amplitude of the first-harmonic dispersion measured at the centre of the resonance is (Marsh et al., 1997):

$$U_1'(0) = \tfrac{1}{2}\gamma_e H_m \frac{\gamma_e H_1 \omega_m T_2^3}{1+\gamma_e^2 H_1^2 T_1 T_2} \qquad (34)$$

where H_m is the amplitude of the modulation field, H_1 that of the microwave field and γ_e is the electron gyromagnetic ratio. The sensitivity of the $U_1'(0)$ amplitude to T_1 therefore does not extend beyond that of the conventional saturation factor $(1+\gamma_e^2 H_1^2 T_1 T_2)^{-1}$. The amplitude of the first-harmonic absorption measured at a distance from the resonance position that is equal to the zeroth harmonic linewidth ($\Delta H = 1/\gamma_e T_2$) is (Marsh et al., 1997):

$$V_1'(\Delta H) = -\tfrac{1}{2}\gamma_e H_m \frac{\gamma_e H_1 \omega_m T_2^3}{1+\frac{1}{2}\gamma_e^2 H_1^2 T_1 T_2} \frac{\gamma_e^2 H_1^2 T_1^2/(1+\omega_m^2 T_1^2)-1}{\left[2+\gamma_e^2 H_1^2 T_1 T_2/(1+\omega_m^2 T_1^2)\right]^2} \qquad (35)$$

where the amplitude at the centre of the line is zero. This displays a considerably greater sensitivity to T_1-relaxation than does U_1' or the conventional in-phase adsorption spectrum, V_1. The amplitude of the second harmonic out-of-phase absorption spectrum is given by (Marsh et al., 1997):

$$V_2'(0) = -\tfrac{1}{4}\gamma_e^2 H_m^2 \frac{\gamma_e H_1 \omega_m T_2^4}{1+\gamma_e^2 H_1^2 T_1 T_2}\left[\frac{\gamma_e^2 H_1^2 T_1(2T_1+5T_2)}{1+2\gamma_e^2 H_1^2 T_1 T_2+4\omega_m^2 T_1^2}-3\right] \qquad (36)$$

This spectral display corresponds to the standard second-harmonic saturation transfer spectroscopy. Again, as for the first-harmonic, the second-harmonic out-of-phase absorption signal possesses an additional sensitivity to T_1, beyond that expressed simply by the saturation factor. The first and second-harmonic non-linear absorption ESR spectra therefore both possess a sensitivity to T_1-relaxation processes superior to that of conventional progressive saturation experiments, which are performed on the in-phase V_1-spectra. As seen from Fig. 8, the first-harmonic out-of-phase V_1'-spectrum has advantages over the second-harmonic out-of-phase V_2' spectrum, for determining T_1-relaxation enhancements. The V_1'-spectrum depends far less on the T_2-relaxation time than does the V_2'-spectrum. As will be seen in Section 7.2, it is also insensitive to molecular motion. This makes it an almost pure T_1-display and, as seen below, this is the nonlinear display of choice for studies of spin-lattice relaxation enhancement.

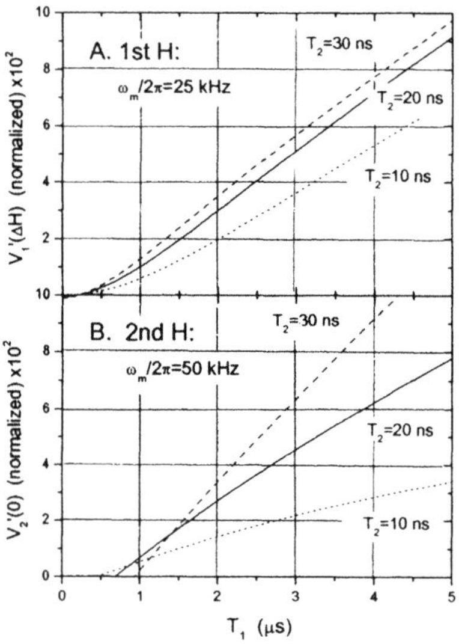

Figure 8. Dependence of the amplitude of: A. the first-harmonic (Eq. 35) and B. the second-harmonic (Eq. 36) out-of-phase EPR absorption signals on T_1-relaxation time, for the values of T_2 indicated. Amplitudes are normalised to the in-phase absorption signal. H_1 = 0.25 G (and H_m = 5 G for V_2') (see Marsh et al., 1997).

Below follows a description of studies of the T_1-dependence of the different nonlinear spin-label EPR displays. Particular attention is paid to the use of integrated intensities in the analysis, and also to the effects of molecular motion in the progressive saturation and out-of-phase first harmonic absorption experiments. For the latter two methods, detailed spin-lattice relaxation time calibrations of the spectral intensities are presented in the Appendix to this chapter.

7.1 Progressive Saturation (V_1) Experiments

The saturation behaviour of the integrated intensity of the zeroth harmonic absorption, V_0, spectrum is independent of the degree of inhomogeneous broadening, because integration over the entire absorption lineshape eliminates the effect of saturation broadening (Páli et al., 1993). Spectral simulation using the methods described in Section 4 confirms that this is also the case in the presence of molecular motion and at 100 kHz Zeeman modulation frequencies (Livshits et al., 1998b). Practically, and in the simulations, the integrated V_0 intensity is obtained by double integration of the conventional first-harmonic V_1-absorption spectrum. The H_1-dependence of the integrated absorption intensity, S, is given by:

$$S(H_1) = \frac{P_1 H_1}{\sqrt{1 + P H_1^2}} \tag{37}$$

where P_1 is a scaling factor and P is the saturation factor that depends upon T_1. With slow passage conditions and in the absence of molecular motion $P = \gamma_e^2 T_1 T_2$. The T_1-dependence of this factor in the presence of molecular motion and Zeeman modulation can thus be expressed to a first approximation in terms of an effective T_2-relaxation time, τ_2^{eff}. More precise calibrations are given in the Appendix.

Figure 9 gives the progressive saturation curves for the integrated absorption intensity of first harmonic spectra as a function of the molecular rotation frequency $f_R \equiv 1/\tau_R$. Spectra were simulated as described in Section 4 with high-frequency Zeeman modulation, $\omega_m/2\pi = 100$ kHz. All saturation curves have the dependence on microwave field intensity, H_1, that is given by Eq. 37, irrespective of rotational frequency, or the presence of inhomogeneous broadening. The saturation behaviour depends strongly on molecular motion in the extreme motional broadening region, for rotational correlation times 10^{-7} s^{-1} > τ_R > 10^{-9} s^{-1}. The effective values of T_2 that are deduced from the saturation curves are given by the inset in Fig. 9. A minimum value of τ_2^{eff} is achieved for $\tau_R \sim 10^8$ s^{-1} that corresponds to the frequency equivalent of the ^{14}N hyperfine anisotropy. In this extreme motional broadening regime the values of τ_2^{eff} are independent of, and much shorter than, the intrinsic T_2^o, and depend only relatively weakly on the rotational correlation time, τ_R. This regime is readily identified from the conventional V_1-spectral lineshapes from which approximate values of τ_R are deduced. Spin-lattice relaxation times can therefore be determined from progressive saturation experiments in this motional regime without a detailed knowledge of τ_2^o (or τ_R).

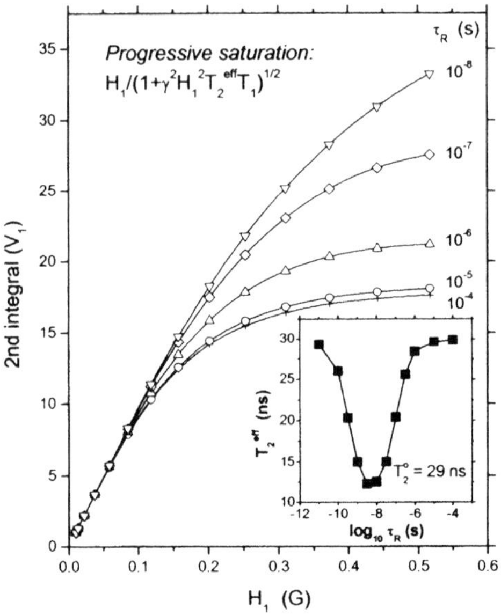

Figure 9. Simulated saturation curves for the double integrated intensity of the V_1-absorption signal at a Zeeman modulation frequency of $\omega_m/2\pi$ = 100 kHz for different values of the inverse frequency of molecular rotation (or rotational correlation time), $\tau_R = 1/f_R$. The intrinsic spin-lattice relaxation time is fixed at T_1^o = 1.8 μs. The *inset* gives the dependence on rotational correlation time τ_R of the effective T_2-value that appears in the saturation factor: $P = \gamma_e^2 T_1 T_2^{eff}$; the intrinsic T_2^o is 29 ns (see Livshits et al., 1998b).

In both the motional narrowing regime ($\tau_R < 10^{-9}$ s^{-1}) and the very slow motion regime ($\tau_R > 10^{-7}$ s^{-1}), the saturation behaviour depends directly on the intrinsic T_2^o, as expected conventionally for progressive saturation experiments. In the extreme motional narrowing regime ($\tau_R < 10^{-10}$ s) and quasi-rigid limit regime ($\tau_R > 10^{-6}$ s), the values of τ_2^{eff} are no longer dependent on rotational correlation time and are equal to the intrinsic values, τ_2^o. More precise calibrations, which are essential in the intermediate regime, are given in the Appendix. The method of extracting the spin-lattice relaxation times from the experimental saturation factors is outlined there.

7.2 First harmonic, out-of-phase absorption (V_1')

As already mentioned, the first-harmonic, out-of-phase absorption spectrum (V_1') has been identified as a non-linear display that is sensitive to T_1, but relatively insensitive to both molecular motion (Hyde and Thomas, 1973) and T_2-relaxation (Livshits et al., 1998a). Fig. 10 gives the dependence of the out-of-phase to in-phase integrated intensity ratio, ρ_1', on spin-lattice relaxation time that is obtained from spectral simulations for different rotational correlation times τ_R ($\equiv 1/f_R$). Clearly the dependence on molecular motion is relatively small, throughout the entire correlation time range τ_R = 10^{-10}-10^{-4} s, especially for a Zeeman modulation frequency of 25 kHz. The dependence on intrinsic τ_2^o is also similarly slight (Livshits et al., 1998a; Livshits and Marsh, 2000).

The best sensitivity to T_1 is obtained at relatively high microwave magnetic field intensities, $H_1 \geq 0.3$ G (Livshits et al., 1998a). The data in Fig. 10 are calculated for $H_1 = 0.5$ G. The dependence of the out-of-phase to in-phase ratio on T_1 can be fitted by an empirical expression of the form (Livshits et al., 1998a):

$$\rho_1' = \frac{\iint V_1' \, \mathrm{d}^2 H}{\iint V_1 \, \mathrm{d}^2 H} = \rho_1^{o}{}' + \frac{a_1' T_1^m}{1 + b_1' T_1^m} \tag{38}$$

where $\rho_1^{o}{}'$, a_1', b_1' and m are fitting parameters. Calibration values are given in the Appendix.

The most striking feature of Fig. 10 is the dependence on Zeeman modulation frequency. For $\omega_m/2\pi$ = 100 kHz, the out-of-phase/in-phase ratio is most sensitive to short spin-lattice relaxation times, $T_1 \leq 2.5$ μs. The lower modulation frequency of 25 kHz extends this range of sensitivity to the T_1 = 2.5-5 μs range, for which the out-of-phase intensities become comparable to those at 100 kHz modulation frequency. Thus the sensitivity of the first-harmonic out-of-phase display can be tuned to different ranges of spin-lattice relaxation time by changing the modulation frequency. Practical examples are given by Livshits et al. (1998a).

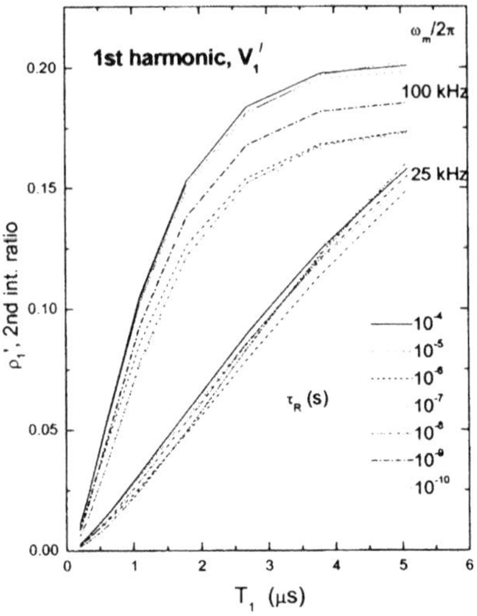

Figure 10. Dependence of the ratio, ρ_1', of out-of-phase to in-phase integrated intensities of the first-harmonic absorption spectra (V_1'/V_1) on spin-lattice relaxation time, T_1. Data are obtained from spectral simulations according to Section 4 for various rotational correlation times, τ_R ($\equiv 1/f_R$), and for Zeeman modulation frequencies: $\omega_m/2\pi$ = 100 kHz and 25 kHz (see Livshits and Marsh, 2000)

7.3 Second Harmonic out-of-phase Absorption (V_2')

Although the second-harmonic out-of-phase absorption spectrum is less optimal for estimation of T_1-relaxation times than is the first harmonic equivalent, it is included here because it is the classical ST-EPR display and to date is also that most used for determining spin-lattice relaxation enhancements (see e.g., Marsh et al., 1998). There are obvious advantages to determining both rotational mobility and spin-lattice relaxation enhancements in one and the same experiment.

Figure 11 illustrates the application to relaxation enhancements induced by spin-spin interactions (i.e., cross relaxation) for a spin-labelled lipid in gel-phase bilayer membranes. Theoretical dependences of the second harmonic V_2' integrated intensity on spin-lattice relaxation time are given in the inset to this figure. The V_2' intensity ratio, ρ_2', is strongly dependent on spin-lattice relaxation time in the range T_1 = 1-5 μs, but also depends strongly on the intrinsic T_2. From the correlation with the effective T_1T_2 relaxation time product obtained from progressive saturation experiments (see Fig. 11), a consistent value of τ_2^o = 8 ns at 25°C (8.5 ns at 15°C and 6.5 ns at 35°C) is obtained to describe the experimental correlation between normalised ST-EPR intensities, I_{ST}, and progressive saturation results. From a linear regression, the numerical calibration of this correlation is (Páli et al., 1996):

$$I_{ST} \times 10^3 \approx 0.17 \times T_1(\mu s) T_2^{eff}(ns) - 1.3 \tag{39}$$

where an effective value of $\tau_2^{eff} \approx 8.5$ ns is appropriate to the experimental data at 15°C and 25°C that essentially correspond to a "no motion" situation.

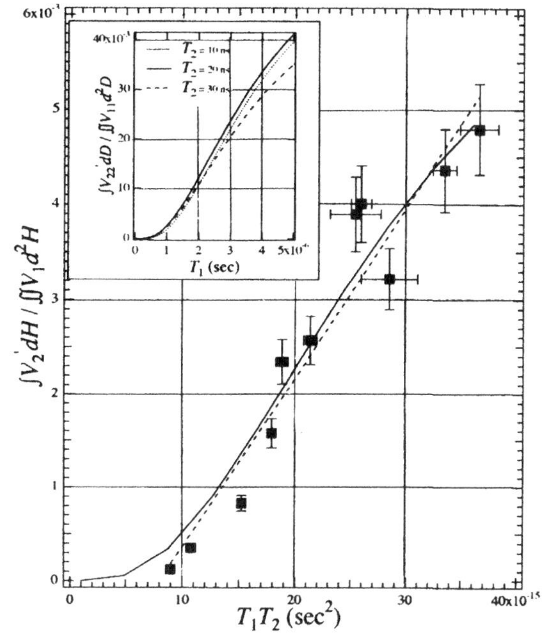

Figure 11. Dependence of the integrated intensities of the second-harmonic, out-of-phase absorption V_2'-spectra of spin-labelled phosphatidylcholine (5-PCSL) at different concentrations in gel-phase (T = 25°C) dipalmitoyl phosphatidylcholine bilayers on the effective T_1T_2 relaxation time products obtained from progressive saturation of the second integral of the conventional V_1-EPR spectra. The solid line is the dependence on T_1 calculated according to Section 4 in the absence of molecular motion with $T_2^o = 7.8 \pm 1.1$ ns. The dashed line is a linear regression. The *inset* gives calculated dependencies of the V_2' integrated intensity on T_1 from no-motion simulations ($f_R = 0$ in Section 4) for different values of T_2 (see Páli et al., 1996).

8. SLOW EXCHANGE AND PARAMAGNETIC ENHANCEMENTS

This section describes the different processes, other than slow rotational diffusion, to which nonlinear spin-label EPR spectroscopy is sensitive. It therefore covers the non-classical aspects of saturation transfer EPR spectroscopy (Marsh et al., 1998). There are three basic areas of application: slow physical or chemical exchange processes; weak spin-spin interactions between spin labels; and weak relaxation enhancements induced by paramagnetic species. In the latter two cases, both Heisenberg spin exchange and magnetic dipole-dipole interactions contribute to the relaxation mechanism.

8.1 Relaxation enhancement by exchange processes

Exchange at rates comparable to that of the spin-lattice relaxation alleviates saturation in nitroxide ESR spectra. This results either from transfer of spin polarisation, in the case of physical or chemical exchange, or by mutual antiparallel spin flips (i.e., cross relaxation), in the case of Heisenberg exchange (see Fig. 1). For exchange between two sites, both cases are equivalent as is seen immediately below. The analysis uses rate equations for spin population differences following Yin and Hyde (1987).

Consider two sites with total spin populations N_i and N_j between which physical or chemical exchange takes place:

$$N_i \underset{\tau_j^{-1}}{\overset{\tau_i^{-1}}{\rightleftarrows}} N_j$$

where τ_i^{-1} is the rate of transfer from site i, and τ_j^{-1} that from site j. These rate constants are related by detailed balance:

$$N_i \tau_i^{-1} = N_j \tau_j^{-1} \tag{40}$$

The rate equation for the spin population difference, $n_i = N_i^- - N_i^+$, at site i is then :

$$\mathrm{d}n_i / \mathrm{d}t = -\left(n_i \tau_i^{-1} - n_j \tau_j^{-1}\right) \tag{41}$$

where N_i^+, N_i^- are the spin populations of the $M_{S,i} = \pm 1/2$ states (with $N_i = N_i^+ + N_i^-$) at site *i*, and similarly for n_j.

Heisenberg spin exchange between the two sites, on the other hand, can be depicted as:

$$N_i^{\pm} \xleftrightarrow{K_x} N_j^{\mp}$$

where K_x is the bimolecular rate constant for spin exchange. The corresponding rate equation for the spin population difference at site *i* is (Yin and Hyde, 1987):

$$\mathrm{d}n_i / \mathrm{d}t = -2K_x\left(N_i^- N_j^+ - N_i^+ N_j^-\right) \tag{42}$$

where the right-hand side can be rewritten in terms of population differences n_i, n_j (Marsh, 1992a):

$$\mathrm{d}n_i / \mathrm{d}t = -K_x\left(N_j n_i - N_i n_j\right) \tag{43}$$

The exchange frequency defined in the usual bimolecular formalism is: $\tau_{ex}^{-1} = K_x\left(N_i + N_j\right)$. Thus physical/chemical exchange and Heisenberg spin exchange are kinetically equivalent, with the following identities obtained from Eqs. 40 to 42: $\tau_i^{-1} = f_j \tau_{ex}^{-1}$ and $\tau_j^{-1} = f_i \tau_{ex}^{-1}$, where the fractional population of state *i* is $f_i = N_i/(N_i+N_j)$ and $f_i + f_j = 1$.

The two-site spin exchange situation is specifically applicable to double labelling experiments (Snel and Marsh, 1994; Páli et al., 1999). Commonly, however, Heisenberg exchange takes place between more than two distinguishable spin label species in a concentration-dependent, single-labelling experiment. Depending on spin-label concentration, the relaxation enhancement can then be considerably more efficient than in the two-site case. This more usual single-labelling case is considered after first treating two-site exchange.

8.1.1 Relaxation by slow two-site exchange

Consider exchange between two sites *b* and *f* (see Fig. 1) where spin transitions are induced at rate $2W$ by H_1-irradiation at site *b*. The steady-state condition for the spin population difference at site *b*, including exchange and spin-lattice relaxation, is:

$$\mathrm{d}n_b / \mathrm{d}t = -2Wn_b + \left(n_b^o - n_b\right)/T_{1,b}^o - n_b \tau_b^{-1} + n_f \tau_f^{-1} = 0 \tag{44}$$

where n_b^o is the spin population difference at Boltzmann equilibrium, and $T_{1,b}^o$ is the spin-lattice relaxation time in the absence of exchange. A corresponding steady-state condition applies to the spins at site f that are not irradiated (i.e., with $W = 0$):

$$\mathrm{d}n_f / \mathrm{d}t = \left(n_f^o - n_f\right)/ T_{1,f}^o - n_f \tau_f^{-1} + n_b \tau_b^{-1} = 0 \tag{45}$$

Solution of rate equations 44 and 45, together with Eq. 40 (that holds also for population differences), yields the standard expression for saturation of the spin system at site b (see Slichter, 1978):

$$n_b = \frac{n_b^o}{1 + 2WT_{1,b}^{eff}} \tag{46}$$

where the effective spin-lattice relaxation time in the presence of exchange is given by (Horváth et al., 1993):

$$T_{1,b}^o / T_{1,b}^{eff} = 1 + \left(1 - f_b\right) T_{1,b}^o \tau_b^{-1} / \left[1 - f_b + f_b T_{1,f}^o \tau_b^{-1}\right] \tag{47}$$

from which the off-rate, τ_b^{-1}, for exchange may be determined. The corresponding expression for Heisenberg spin-exchange between two distinguishable spin-labelled species is (Snel and Marsh, 1994):

$$T_{1,b}^o / T_{1,b}^{eff} = 1 + \left(1 - f_b\right) T_{1,b}^o \tau_{ex}^{-1} / \left(1 + f_b T_{1,f}^o \tau_{ex}^{-1}\right) \tag{48}$$

which can be deduced from the identities between the exchange rates in the two cases, viz., $\tau_b^{-1} = (1 - f_b)\tau_{ex}^{-1}$ (see above).

8.1.2 Concentration dependence of Heisenberg exchange

In the general case, Heisenberg exchange takes place between many spin labels of the same type that are distinguished either by their different ^{14}N-hyperfine states, or by their different orientations with respect to the static magnetic field. Equation 43 must be summed over all states j with which the state i is exchanging. The steady state rate equation for the population difference of the ith transition is then (Marsh, 1992a):

$$\mathrm{d}n_i / \mathrm{d}t = -2Wn_k \delta_{i,k} + \left(n_i^o - n_i\right)/ T_1^o - K_x\left(Nn_i - N_i n\right) \tag{49}$$

where $N = \sum N_j$ and $n = \sum n_j$, and transitions are induced by the H_1-field in the spin system $i = k$. Summing over all i yields the further condition:

$$-2Wn_k + \left(\sum n_i^o - n\right)/T_1^o = 0 \tag{50}$$

Solution of Eqs. 49 and 50 again yields the standard expression for saturation (see Eq. 46) where the effective spin-lattice relaxation time is given by (Marsh, 1992a):

$$T_{1,k}^{eff} = T_1^o\left(1 + Z_k T_1^o \tau_{ex}^{-1}\right)/\left(1 + T_1^o \tau_{ex}^{-1}\right) \tag{51}$$

The exchange frequency is $\tau_{ex}^{-1} = K_x N$, which gives the dependence on spin-label concentration, N, and $Z_k = N_k/N$ is the fractional population (or degeneracy) of the transition being saturated.

For powder spectra, the intensity of the non-linear out-of-phase spectrum is reduced without change in lineshape (Marsh and Horváth, 1992b). The effects of Heisenberg exchange are therefore readily distinguished from those of rotational diffusion. When the degree of degeneracy is high ($Z_k \approx 0$), as in a powder pattern, a large number of states is available for redistribution of saturation. Heisenberg exchange then has the same effect as a true relaxation enhancement, rather than a cross-relaxation. The effective relaxation rate is then simply:

$$1/T_1^{eff} = 1/T_1^o + \tau_{ex}^{-1} \tag{52}$$

which is found to be applicable in many practical situations with low exchange rates.

8.2 Paramagnetic Enhancement by Heisenberg Exchange

Spin-lattice relaxation is induced by Heisenberg spin exchange only if the paramagnetic species comes into direct contact with the spin-label. For strong exchange, the enhancement in spin-lattice relaxation rate is given by the product of the collision rate constant, k_{RL}, and relaxant concentration, c_R, (Molin et al., 1980):

$$T_{1,HE}^{-1} = k_{RL} c_R \tag{53}$$

Under these circumstances, Heisenberg exchange contributes equally to T_2-relaxation, i.e., $T_{2,HE}^{-1} = T_{1,HE}^{-1}$, and is independent of the T_1-relaxation time of the paramagnetic species.

The collision rate constant is given by the Smoluchowski solution of the diffusion equation:

$$k_{RL} = 4\pi\sigma_{RL} r_{RL} D_T \cdot \exp(-z_R e\psi_L / kT) \tag{54}$$

where D_T is the translational diffusion coefficient of the relaxant (that of the spin-labelled system is assumed to be negligible by comparison), r_{RL} is the interaction distance between relaxant and spin label, σ_{RL} is a steric factor, $z_R e$ is the electrical charge on the relaxant, and ψ_L is the electrostatic surface potential of the spin-labelled system. The enhancement in relaxation rate by Heisenberg spin exchange is therefore determined by the diffusion-concentration product, plus any electrostatic interactions. In the case of membranes, the local value of the product $D_T(z)c_R(z)$ at position z along the membrane normal is the determining factor.

Heisenberg spin exchange is the dominant mechanism of spin-label relaxation by paramagnetic molecular oxygen (Hyde and Subczynski, 1989). For paramagnetic ions, as will be seen later, magnetic dipole-dipole interactions - both static and dynamic - can make appreciable contributions depending on the ion spin, magnetic moment and T_1-relaxation time.

8.3 Paramagnetic Enhancement by Magnetic Dipole-Dipole Interaction

Relaxation of the spin label is induced by modulation of the magnetic dipole interaction with the paramagnetic relaxant. Modulation of the dipolar interaction occurs either by rapid spin-lattice relaxation of the paramagnetic species, or by the mutual diffusive motions of the spin label and paramagnetic species. Static dipolar relaxation, i.e., the first case, obtains for T_1-relaxation times, $T_{1,R}$, of the paramagnetic species that are much shorter than the characteristic dipolar correlation time, τ_D, for translational diffusion. Dynamic dipolar relaxation, i.e., the second case, dominates at the opposite extreme, when the dipolar correlation time is much shorter than the spin-lattice relaxation time of the relaxant.

8.3.1 Static Dipolar Relaxation

Relaxation of the spin label by a static magnetic dipole-dipole interaction arises from modulation of the dipolar interaction by the rapid spin-lattice relaxation of the paramagnetic ion. The net enhancement induced by the different ions, k, is given by the Solomon-Bloembergen equation for electron spins (Bloembergen, 1949; Solomon, 1955):

$$\frac{1}{T_{1,dd}(\text{static})} = |\mu_R|^2 \gamma_e^2 \times \sum_k \frac{|F_o(\Omega_k)|^2 J(\omega_L - \omega_R) + |F_1(\Omega_k)|^2 J(\omega_L) + |F_2(\Omega_k)|^2 J(\omega_L + \omega_R)}{r_k^6} \tag{55}$$

where ω_R and ω_L are the Larmor frequencies of the paramagnetic ion and spin label, respectively, r_k is the separation of the spin label from the paramagnetic ion, Ω_k is the angle between the interdipole vector $\mathbf{r}_k$ and magnetic field direction, γ_e is the electron gyromagnetic ratio, and μ_R ($=g_R\beta_e\mathbf{S}_R$) is the magnetic moment operator of the paramagnetic ions. The spectral densities are defined by $J(\omega) = T_{1,R}/(1+\omega^2 T_{1,R})$ where $T_{1,R}$ is the spin-lattice relaxation time of the paramagnetic ion. The latter is assumed to be sufficiently short that it also determines T_2, i.e., $T_{2,R} = T_{1,R}$. The terms F_i involving the angle Ω_k are related to the absolute values of the corresponding spherical harmonics. Summation is over the entire distribution of paramagnetic ions, k.

Volume integration for paramagnetic ions distributed in the aqueous phase, or surface integration for ions adsorbed at the lipid-water interface, yields values of $T_{1,dd}^{-1}$ that depend on the distance of closest approach, R, of the paramagnetic ions to the spin label, and on the angle θ_o between the magnetic field and the membrane normal (Livshits et al., 2001). For macroscopically unoriented membrane dispersions, different parts of the powder pattern will saturate differently. However, the dependence of $T_{1,dd}^{-1}$ (static) on θ_o is much weaker than the initial dependence on Ω. Therefore, for reasonable estimates of the saturation behaviour of the integrated spin-label EPR intensity, one can average $T_{1,dd}^{-1}$(static) over θ_o. The resulting angular-independent effective values of $T_{1,dd}^{-1}$(static) are (Livshits et al., 2001):

$$T_{1,dd}^{-1}(\text{static}) = \frac{\pi}{45}\frac{\mu_R^2\gamma_e^2}{R^3} T_{1,R} \cdot c_R \cdot f_1(\omega_L, \omega_R) \tag{56}$$

$$T_{1,dd}^{-1}(\text{static}) = \frac{\pi}{15}\frac{\mu_R^2\gamma_e^2}{R^4} T_{1,R} \cdot c_{R,s} \cdot f_1(\omega_L, \omega_R) \tag{57}$$

for volume and surface distributions, respectively, where:

$$f_1(\omega_L,\omega_R) = \frac{1}{1+(\omega_L-\omega_R)^2T_{1,R}^2}+\frac{3}{1+\omega_L^{\ 2}T_{1,R}^2}+\frac{6}{1+(\omega_L+\omega_R)^2T_{1,R}^2}, \quad (58)$$

c_R and $c_{R,s}$ are the bulk and surface paramagnetic ion concentrations, and $\mu_R^2 = g_R^2\beta_e^2 S_R(S_R+1)$, with β_e as the Bohr magneton. For Ni^{2+} and Co^{2+} ions, $\omega^2T_{1,R}^2 << 1$ and all three terms in Eq. 58 contribute, whereas for Cu^{2+} and Mn^{2+}, $\omega^2T_{1,R}^2 << 1$ and the first term in $f_1(\omega_L,\omega_R)$ dominates.

Table 3: Predicted enhancements in T_1- and T_2-relaxation rates of spin labels in membranes by static dipolar interactions with 30 mM aqueous paramagnetic ions, and R = 1 nm in Eqs. 56-58 and Eqs. 59-61 for $T_{1,dd}^{-1}$ and $T_{2,dd}^{-1}$, respectively (Livshits et al., 2001).

Ion, R	$\langle g_R \rangle$	S_R	$T_{1,R}$ (s)[a]	$T_{1,dd}^{-1}(s^{-1})$ volume	surface[b]	$T_{2,dd}^{-1}(s^{-1})$ volume	surface[b]
Ni^{2+}	2.25	1	3×10^{-12}	0.8×10^4	3.9×10^4	0.9×10^4	4.3×10^4
			5×10^{-12}	1.1×10^4	5.4×10^4	1.3×10^4	6.6×10^4
Co^{2+}	4.33	1/2	$\approx10^{-12}$	4.6×10^3	2.2×10^4	4.6×10^3	2.2×10^4
Cu^{2+}	2.2	1/2	1×10^{-9}	4×10^3	2.5×10^4	2.4×10^5	1.5×10^6
			3×10^{-9}	1.4×10^3	0.9×10^4	7.3×10^5	4.6×10^6
Mn^{2+}	1.993	5/2	1×10^{-9}	1×10^6	1.1×10^7	2.8×10^6	3.1×10^7
			3.5×10^{-9}	1.5×10^6	1.6×10^7	8.9×10^6	9.6×10^7
Dy^{3+}	1.33	15/2	3.5×10^{-13}	1.3×10^4	1.9×10^5	1.3×10^4	1.9×10^5
			8×10^{-13}	3.0×10^4	4.5×10^5	3.0×10^4	4.5×10^5

[a]For references see Livshits et al. (2001). Differences in T_1 between aqueous and surface-adsorbed ions are neglected.

[b]Surface concentrations calculated from the intrinsic binding constants given in McLaughlin et al. (1978). See also Livshits et al. (2001).

Table 3 gives numerical estimates of the static dipolar T_1-relaxation enhancements for a spin label situated at R = 1 nm in lipid membranes that are immersed in a 30 mM paramagnetic ion solution. For ions with $S > ½$, both the estimates of $T_{1,dd}^{-1}$ and the experimental values of $T_{1,R}$ depend on the way in which zero-field splittings are taken into account (Livshits et al., 2001). The values of $T_{1,dd}^{-1}$ (static) are predicted to be rather small and in the order $Ni^{2+} > Co^{2+} \approx Cu^{2+}$, for paramagnetic ions other than Mn^{2+}. The latter has favourable values of both $T_{1,R}$ and spin, S_R, that give rise to efficient paramagnetic relaxation enhancements, but it has an EPR spectrum at room temperature, which complicates analysis of the saturation behaviour of the spin-label spectrum. For strongly absorbed paramagnetic cations, the values of $T_{1,dd}^{-1}$ (static) may be increased by up to a factor of ten, because of the smaller average distance from the spin label.

Static dipolar interactions modulated by the fast relaxation of the paramagnetic ion also contribute to T_2 relaxation. These contributions influence both progressive saturation studies and also linewidth measurements. In standard Leigh theory (Leigh Jr., 1970), the strong angular dependence of the dipolar relaxation for two isolated dipoles results in practically complete quenching of all resonances other than those for dipole pairs oriented at the magic angle. The much weaker angular dependence resulting from integration over a distribution of paramagnetic ions (cf. above), results in a line broadening rather than an amplitude quenching. The resulting effective transverse relaxation rates, $T_{2,dd}^{-1}$ (static), are given by (Livshits et al., 2001):

$$T_{2,dd}^{-1}(\text{static}) = \frac{\pi \cdot \mu_R^2 \cdot \gamma_e^2}{90R^3} T_{1,R} \cdot c_R \cdot f_2(\omega_L, \omega_R) \tag{59}$$

$$T_{2,dd}^{-1}(\text{static}) = \frac{\pi \cdot \mu_R^2 \cdot \gamma_e^2}{30R^4} T_{1,R} \cdot c_{R,s} \cdot f_2(\omega_L, \omega_R) \tag{60}$$

where

$$f_2(\omega_L, \omega_R) = 4 + \frac{1}{1+(\omega_L - \omega_R)^2 T_{1,R}^2} + \frac{3}{1+\omega_L^2 T_{1,R}^2} + \frac{6}{1+\omega_R^2 T_{1,R}^2} + \frac{6}{1+(\omega_L + \omega_R)^2 T_{1,R}^2} \tag{61}$$

For Cu^{2+} and Mn^{2+} ions ($T_{1,R} \approx 10^{-9}$ s), the first term in $f_2(\omega_L, \omega_R)$ dominates which gives $f_2(\omega_L,\omega_R) \approx 4$. For Ni^{2+}, Co^{2+} and Dy^{3+} ions ($T_{1,R} <<< 10^{-9}$ s), on the other hand, all terms in the spectral density contribute and $f_2(\omega_L,\omega_R) \approx 20$.

Table 3 gives numerical estimates of $T_{2,dd}^{-1}$ (static), again for $R = 1$ nm and 30 mM bulk concentration of paramagnetic ions. Values for Ni^{2+}, Co^{2+} and Dy^{3+} are comparable to those for $T_{1,dd}^{-1}$ (static), whereas for Cu^{2+} (and to a lesser extent for Mn^{2+}), $T_{2,dd}^{-1}$ (static) >> $T_{1,dd}^{-1}$ (static). For Cu^{2+} ions, $T_{2,dd}^{-1}$ (static) is 50 times greater than for Ni^{2+} ions, and in turn is 10-20 times less than for Mn^{2+} ions.

8.3.2 Dynamic Dipolar Relaxation

In the case of rapid translation diffusion, spin-label relaxation can be induced by modulation of the dipolar interaction by the mutual diffusive motions of the spin labels and paramagnetic ions. The criterion that the dynamic mechanism dominates over the static mechanism is $\tau_D^{-1} = r_{RL}^2 / 2D_T >> T_{1,R}^{-1}$, where D_T is the translational diffusion coefficient and r_{RL} is the distancc of closest approach between paramagnetic ion and spin label. Unlike the situation with relaxation by diffusion-controlled Heisenberg spin exchange, the dynamic dipolar interaction is much more significant for T_2-relaxation than for T_1-relaxation, because of the contribution from spectral densities at low frequency. Assuming that spectral densities at the Larmor frequency and above contribute negligibly, the T_2-relaxation enhancement is (Abragam, 1961):

$$T_{2,dd}^{-1}(\text{dynamic}) = \mu_R^2\gamma_e^2\left[\tfrac{1}{6}J^{(0)}(0)+\tfrac{1}{24}J^{(0)}(\omega_R-\omega_L)\right] \tag{62}$$

where the zero-frequency spectral density is $J^{(0)}(0)=(48\pi/15^2)c_R/(D_T r_{RL})$, and for $(\omega_R\text{-}\omega_L)^2\tau_D^2<<1$, $J^{(0)}(\omega_R\text{-}\omega_L)\approx J^{(0)}(0)$, whereas for $(\omega_R\text{-}\omega_L)^2\tau_D^2>>1$, $J^{(0)}(\omega_R\text{-}\omega_L)\approx 0$. The T_2-relaxation enhancement therefore becomes:

$$T_{2,dd}^{-1}(\text{dynamic}) = C.\mu_R^2\gamma_e^2 c_R/(D_T r_{RL}) \tag{63}$$

where $C=2\pi/45$ for $(\omega_R\text{-}\omega_L)^2\tau_D^2<<1$ (i.e., specifically for Mn^{2+}), and $C=8\pi/15^2$ for $(\omega_R\text{-}\omega_L)^2\tau_D^2>>1$ (i.e., for most other ions).

The dynamic dipolar enhancement in T_1-relaxation, on the other hand, is:

$$T_{1,dd}^{-1}(\text{dynamic}) = \tfrac{1}{12}\mu_R^2\gamma_e^2 J^{(0)}(\omega_R-\omega_L) \tag{64}$$

i.e., $T_{1,dd}^{-1} = (2/5)T_{2,dd}^{-1}$ for $(\omega_R\text{-}\omega_L)^2\tau_D^2<<1$, and $T_{1,dd}^{-1} << T_{2,dd}^{-1}$ for $(\omega_R\text{-}\omega_L)^2\tau_D^2>>1$. Therefore, for ions with g-values that differ considerably from that of the spin label ($\Delta g>>0.01$), the dynamic dipolar T_1-relaxation is expected to be small. For ions with g-values close to those of the spin label ($\Delta g<0.01$), viz., Mn^{2+}, the T_1-relaxation enhancement is:

$$T_{1,dd}^{-1}(\text{dynamic}) = \left(4\pi/15^2\right)\mu_R^2\gamma_e^2 c_R/(D_T r_{RL}) \tag{65}$$

i.e., two-fifths that of the dynamic dipolar T_2-relaxation rate. Strictly speaking, the latter condition requires equality of Larmor frequencies.

Therefore, hyperfine structure of the paramagnetic ion can be a complicating factor.

9. APPLICATIONS: RELAXATION ENHANCEMENTS

This section gives examples of the applications of nonlinear EPR to the different mechanisms of saturation transfer that were described in Section 8. They span reasonably well the range of different nonstandard types of ST-EPR experiment.

9.1 Two-site exchange: lipid-protein interactions

The rate of exchange of spin-labelled lipids at the intramembranous perimeter of transmembrane proteins is relatively slow and two-component conventional V_1-EPR spectra are resolved when the rotational mobility differs in the two membrane lipid environments (Horváth et al., 1988). Non-linear EPR methods, both progressive saturation EPR and V_2' ST-EPR, have been used to detect and measure exchange of lipids on and off the protein (Horváth et al., 1993).

Figure 12 gives the integrated V_2' ST-EPR intensity, I_{ST}, as a function of the fraction, f_b, of spin-labelled lipid that is associated with the myelin proteolipid protein in reconstituted membranes. Samples all have the same total lipid/protein ratio; f_b is varied by using spin-labelled lipid species with differing affinities for the protein. The value of f_b is determined by spectral subtraction with the two-component conventional V_1 EPR spectra (Marsh and Horváth, 1998). Below the lipid chain-melting temperature, T_t, any exchange is extremely slow and the normalised saturation transfer intensity, I_{ST}, is simply additive depending linearly on f_b:

$$I_{ST} = (1 - f_b) I^o_{ST,f} + f_b I^o_{ST,b} \tag{66}$$

where $I_{ST,f}{}^o$ and $I_{ST,b}{}^o$ are the values of I_{ST} for lipid-alone and protein-alone samples, respectively. In the fluid lipid phase, above T_t, the dependence of I_{ST} on f_b lies below the straight (dashed) line expected for no exchange. Saturation is partially alleviated by exchange between sites on and off the protein at rates comparable to the spin-lattice relaxation time.

Assuming that I_{ST} is approximately proportional to T_1 (see Fig. 11), the net ST-EPR intensity in the presence of exchange is given by:

$$I_{ST} = (1 - f_b) I^o_{ST,f} \cdot \frac{T_{1,f}}{T^o_{1,f}} + f_b I^o_{ST,b} \frac{T_{1,b}}{T^o_{1,b}} \tag{67}$$

where $T_{1,f}$ and $T_{1,b}$ are the spin-lattice relaxation times at lipid locations respectively off and on the protein, and $T_{1,f}{}^o$, $T_{1,b}{}^o$ are the corresponding values in the absence of exchange. Combining Eq. 67 with Eq. 47 and its equivalent for $T_{1,f}/T_{1,f}{}^o$ gives the predicted dependence of I_{ST} on f_b in the presence of exchange. The non-linear least-squares fit shown in Fig. 12 yields a normalised on-rate constant for lipid exchange of $T_{1,b}{}^o\tau_f{}^{-1} = 2.9$ (at fixed lipid/protein ratio of 37:1 mol/mol and T = 30°C). From this the off-rates $\tau_b{}^{-1}$ for the lipids with different affinities for the protein are determined by the relation $f_b\tau_b^{-1} = (1 - f_b)\tau_f^{-1}$ for detailed balance (see Eq. 40).

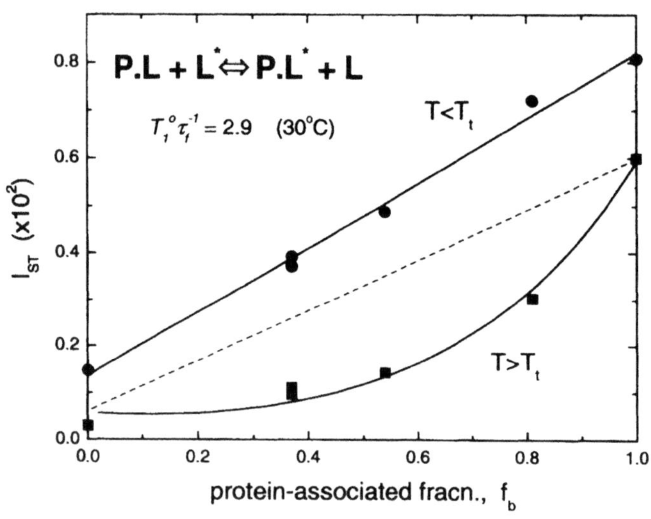

Figure 12. Dependence of the integrated V_2' ST-EPR intensity, I_{ST}, from different spin-labelled lipids on the fraction, f_b, of each lipid species associated with the myelin proteolipid protein in dimyristoyl phosphatidylcholine membranes (lipid/protein = 37:1 mol/mol). Measurements are made in the gel-phase ($T < T_t$) and in the fluid phase ($T > T_t$) at 4°C and 30°C, respectively. Solid lines are fits of Eqs. 67, 47 and equivalents obtained by a linear regression at 4°C and non-linear least squares fit giving a normalised lipid exchange rate of $T_{1,b}{}^o\tau_f{}^{-1} = 2.9$ at 30°C. The dashed line is the dependence expected for no exchange at 30°C (see Horváth et al., 1993).

9.2 Spin-Spin Interactions

This section gives two different examples of the application of ST-EPR techniques to measuring weak spin-spin interactions from the dependence on spin concentration in single-labelling experiments. One example is the use of Heisenberg exchange to determine slow translational diffusion at relatively low label concentrations. The second example is the use of local spin-spin interactions to detect protein oligomer formation. In a third subsection, examples are given of the use of spin-spin interactions detected by nonlinear EPR to determine the membrane location of spin-labelled proteins, relative to spin-labelled lipids, in double-labelling experiments.

9.2.1 Spin exchange: diffusional collisions

A significant example of the use of non-linear EPR to determine low-frequency collision rates from Heisenberg exchange interactions is offered by the translational diffusion of integral membrane proteins (Esmann and Marsh, 1992). The importance of the non-linear spin-label EPR method is that it measures local diffusion coefficients which then may be compared with long-range diffusion detected by such standard techniques as photobleaching (Clegg and Vaz, 1985).

Figure 13 gives the dependence of the reciprocal V_2' ST-EPR intensity, $1/I_{ST}$, on concentration, c_{SL}, of spin-labelled Na,K-ATPase in membranes reconstituted at a constant total lipid/protein ratio equal to that of the native membrane. Given that the integrated second-harmonic out-of-phase absorption intensity, I_{ST}, is approximately proportional to the T_1-relaxation time (see Fig. 11), the relaxation enhancement with increasing spin label concentration found in Fig. 13 is described by the following relation (cf. Eq. 52):

$$1/I_{ST} = \left(1/I_{ST}^{o}\right)\left(1 + k_{ex}T_1^{o}.c_{SL}\right) \qquad (68)$$

where I_{ST}^{o} and T_1^{o} are the values of I_{ST} and T_1 in the absence of spin exchange and k_{ex} is the second-order rate constant for spin exchange between spin-labelled proteins. The latter is related directly to the collision rate constant, k_{coll}:

$$k_{ex} = p_{ex}\sigma k_{coll} \qquad (69)$$

where the probability of exchange on collision is $p_{ex} = ½$ for strong exchange and $\sigma \leq 1$ is the normalised collision cross-section. Thus the gradients with increasing spin-label concentration in Fig. 13 give the normalised exchange rate constant, $k_{ex}T_1^o$, and hence the collision rate constant. Translational diffusion coefficients are extracted from the latter by using specific models.

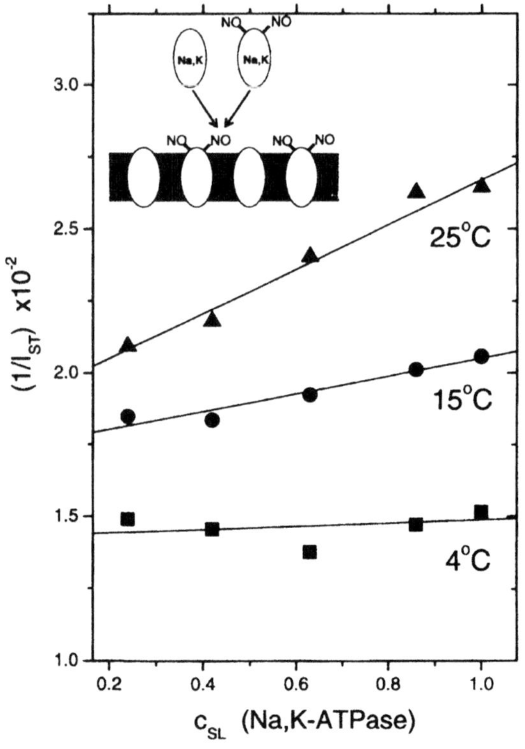

Figure 13. Reciprocal integral intensity, $1/I_{ST}$, of the second-harmonic out-of-phase absorption V_2' ST-EPR spectra as a function of the fractional concentration, c_{SL}, of spin-labelled Na,K-ATPase in reconstituted membranes of fixed lipid/protein ratio (1.2 μmol lipid phosphate/mg protein) at the temperatures indicated. Solid lines are linear regressions. The *inset* indicates the mode of reconstitution by recombining complementary fractions of solubilised spin-labelled and non-spin-labelled protein (see Esmann and Marsh, 1992).

Both a quasi-crystalline lattice model and the two-dimensional diffusion equation with Smoluchowski boundary conditions yield similar values for the translational diffusion coefficients: $D_T \sim 2.4$ and 7.1 $\mu m^2 s^{-1}$ at 15°C and 25°C, respectively (Esmann and Marsh, 1992). These values are comparable to those predicted by the Saffman and Delbrück (1976) hydrodynamic treatment for two-dimensional diffusion with a membrane viscosity of η_m = 1 P (viz., $D_T \sim 3.5$ $\mu m^2.s^{-1}$), and to those determined for various other large transmembrane proteins reconstituted at high dilution in lipid membranes

(Clegg and Vaz, 1985). By contrast, in photobleaching experiments on Na^+,K^+-ATPase from Madin-Darby canine kidney cells, only 50% of the protein exhibited long-range lateral mobility, and that with a diffusion coefficient of 0.05 $\mu m^2.s^{-1}$ (Jesaitis and Yguerabide, 1986). Comparison with the ST-EPR results on reconstituted systems therefore demonstrates the existence of considerable barriers to long-range diffusion of Na^+,K^+-ATPase in this bipolar cellular system.

9.2.2 Spin-spin interactions: oligomer formation

Formation of protein oligomers causes spin-spin interactions between the spin-labelled monomers. The strength of interaction depends on the proximity of spin labels on adjacent monomers, but inevitably will give rise to measurable relaxation enhancement with increasing fraction, F, of monomers that are spin labelled.

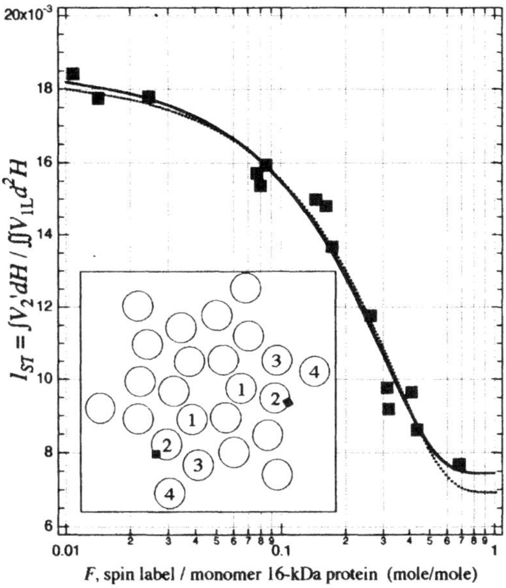

Figure 14. Dependence on spin-labelling level of the integrated V_2' ST-EPR intensity, I_{ST}, from membranous 16-kDa V-ATPase proteolipid subunit spin-labelled with maleimide. I_{ST} is normalised to the low-power second integral of the conventional V_1 EPR spectrum, that is also used to determine the fraction, F, of spin labels per 16-kDa monomer. Solid line is a non-linear least squares fit of Eqs. 70 and 71 yielding a degree of oligomerisation $N = 6.6 \pm 0.7$, and the dotted line is the dependence predicted for $N = 6$ (see Páli et al., 1999). *Inset*: Schematic diagram of the hexamer indicating the site of spin-labelling on Cys54 in transmembrane segment 2.

Figure 14 gives the integrated V_2' ST-EPR intensity (normalised to number of spins) of the vacuolar ATPase proteolipid subunit, as a function of fractional spin labelling with maleimide on the single cysteine (Páli et al., 1999). Alleviation of saturation by spin-spin interaction causes the progressively decreasing intensity with increasing labelling level. If it is assumed that oligomers containing more than one spin-labelled monomer contribute an intensity I'_{ST} that is reduced relative to the value I_{ST}^o for an isolated monomer, then the net normalised ST-EPR intensity is:

$$I_{ST} = \left(I_{ST}^o - I'_{ST}\right)\frac{p_1}{1 - p_0} + I'_{ST} \tag{70}$$

where p_n is the probability that n monomers are labelled in an oligomer composed of N monomers. For a statistical distribution of labelling, the binomial distribution holds and the labelling probabilities are:

$$p_n = \frac{N!}{n!(N-n)!}F^n(1-F)^{N-n} \tag{71}$$

where $\sum_{n=0}^{N} p_n = 1$.

The experimental dependence in Fig. 14 is reasonably well described by Eqs. 70, 71 with $N = 6$. Using N as a non-linear fitting parameter results in a value of $N = 6.6 \pm 0.7$, consistent with a hexameric assembly of the 16-kDa proteolipid subunit. This is the state of oligomerisation that is found in two-dimensional crystalline preparations of the 16-kDa protein (Holzenburg et al., 1993).

In the case of trimer formation in the membrane assembly of annexin XII, spin labels on adjacent monomers are sufficiently close that a binomial distribution is observed in the broadening of the conventional V_1 EPR lineshapes (Langen et al., 1998). Non-linear rapid-passage EPR approaches, as described here, are more versatile in so far as they are sensitive to much weaker spin-spin interactions. This is likely to be the usual case with large integral oligomeric proteins, hence requiring progressive saturation or out-of-phase methods rather than conventional in-phase EPR.

9.2.3 Spin-spin interactions: double labelling

Spin-spin interactions between different spin-labelled species in a double-labelled system are detected as a decrease in the out-of-phase spectral intensity below that predicted from the weighted sum of the two single-labelled systems. The strength of the interaction is a measure of the mutual accessibility of the two labelled species. The technique is most usefully applied to spin-labelled lipids and spin-labelled proteins in membranes. The vertical location in the membrane of the spin-labelled protein residue is determined relative to spin labels systematically stepped down the lipid chain. The depth in the membrane of the DCCD-reactive glutamate that is essential for proton translocation by the 16-kDa V-ATPase proteolipid was mapped in this way (Páli et al., 1999). More importantly, the double-labelling V_2' ST-EPR experiment demonstrates that the essential glutamate is exposed to the lipid, as is required by the rotary mechanism proposed for proton transport by stepwise interaction with the Vph1p subunit of the V-ATPase.

Progressive saturation V_1-EPR was used with double-labelling to demonstrate the deep penetration of the N-terminal section of the haem-less cytochrome *c* precursor, apocytochrome *c*, into negatively charged lipid bilayer membranes (Snel and Marsh, 1994). The spin-labelled holoprotein, cytochrome *c*, by contrast, showed strongest spin-spin interaction with a spin label attached to the phospholipid headgroup, at the membrane surface, as expected for a classical peripheral membrane protein.

9.3 Paramagnetic Relaxation Enhancements

Paramagnetic species used to induce spin-label relaxation enhancement are mostly paramagnetic ions, their complexes and molecular oxygen. Generally, such measurements yield details on spin-label accessibility, penetration profiles of relaxants and, in the case of purely dipolar interactions, molecular separations.

9.3.1 Dipolar relaxation: distance measurements

Figure 15 gives the dependence of the paramagnetic relaxation enhancement by aqueous Ni^{2+} ions on position, *n*, of the spin label in the lipid chain in gel-phase bilayer membranes (Páli et al., 1992). Because the lipid chains are frozen, the Ni^{2+} ions do not penetrate the gel-phase membranes and spin-label relaxation is by distance-dependent magnetic dipole-dipole interaction. As seen from the inset, the reciprocal ST-EPR integral intensity depends linearly on concentration, c_{Ni}, of Ni^{2+} ions. For a static dipolar mechanism (see Eqs. 56, 57), we have:

$$\frac{1}{I_{ST}(c_{Ni})} = \frac{1}{I_{ST}(0)} + k_m\left[\frac{1}{R^m} + \frac{1}{(d_l - R)^m}\right]c_{Ni} \tag{72}$$

where R is the distance of the spin label from the membrane surface, d_l is the membrane thickness and k_m is a constant (Páli et al., 1992). For a three-dimensional, bulk distribution of paramagnetic ions the exponent is $m = 3$, but $m = 4$ for an adsorbed surface layer. It is the term in square brackets of Eq. 72 that depicts the dependence on spin label position, where the second term represents the contribution from Ni^{2+} ions on the opposite face of the membrane from that in which the spin-labelled lipid is located. Because the chain configuration (all-*trans*) is known in the gel-phase it is possible from Fig. 15 to determine both the membrane thickness and the thickness of the lipid polar region from C-atom $n = 1$ to the membrane surface. Results obtained are in good agreement with those from low-angle x-ray diffraction (Páli et al., 1992).

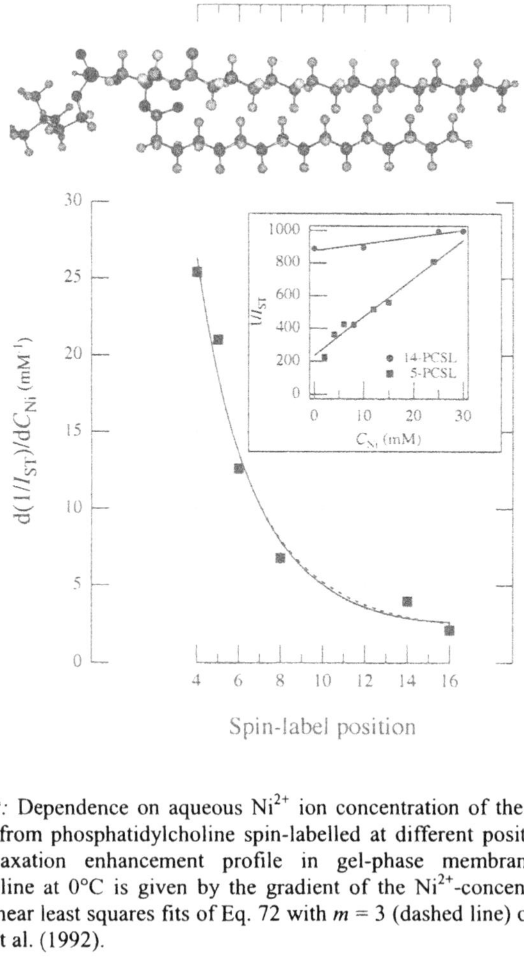

Figure 15. Inset: Dependence on aqueous Ni^{2+} ion concentration of the reciprocal ST-EPR intensity, $1/I_{ST}$, from phosphatidylcholine spin-labelled at different positions, n, in the *sn*-2 chain. The relaxation enhancement profile in gel-phase membranes of dipalmitoyl phosphatidylcholine at 0°C is given by the gradient of the Ni^{2+}-concentration dependence. Lines are non-linear least squares fits of Eq. 72 with $m = 3$ (dashed line) or $m = 4$ (solid line). Data from Páli et al. (1992).

9.3.2 Penetration profiles of paramagnetic ions and relaxation mechanisms

The extent of penetration of paramagnetic ions into fluid lipid membranes depends strongly on the anionic counterion, or complexing ligand. Fig. 16 shows membrane penetration profiles of the sulphate, chloride and perchlorate salts of divalent nickel. The spin-lattice relaxation enhancements of spin labels in the lipid chains that are obtained from progressive saturation V_1-EPR are in the order $ClO_4^- >> Cl^- > SO_4^{2-}$. This is the order of the octanol partition coefficients multiplied by the spin exchange rate constants determined in methanol (Livshits et al., 2001). Heisenberg exchange with ion pairs dissolved in the fluid membrane therefore makes a major contribution to the relaxation enhancement of the spin-labelled lipids.

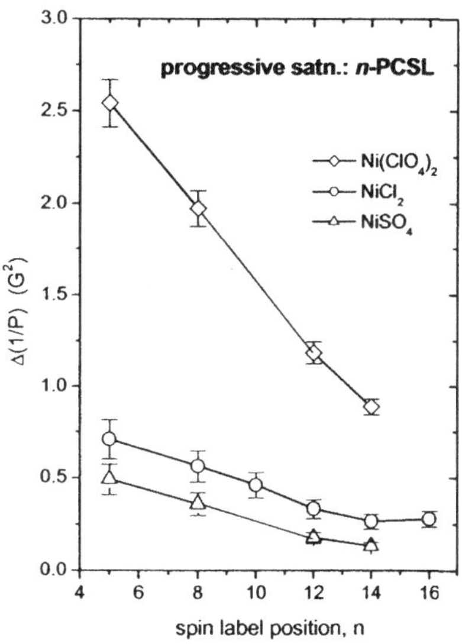

Figure 16. Dependence of the relaxation enhancement, $\Delta(1/P)$, in progressive saturation EPR (see Eq. 37) on spin label position, n, in the sn-2 chain of phosphatidylcholine for fluid dimyristoyl phosphatidylcholine membranes ($T = 39°C$) in the presence of 30 mM $Ni(ClO_4)_2$ (◇), $NiCl_2$ (○) or $NiSO_4$ (Δ) (Livshits et al., 2001).

Estimation of the possible dipolar contribution to T_1-relaxation enhancement by Ni^{2+} ions requires integration of $T_{1,dd}^{-1}$ (static) over the ion distribution in the membrane (Livshits et al., 2001). Relative to Heisenberg exchange, it is estimated that $\left\langle T_{1,dd}^{-1}(\text{static})\right\rangle / T_{1,HE}^{-1} \approx 5-8\times10^{8}$ cm² s⁻¹/D_T

and spin exchange therefore dominates for translational diffusion coefficients of the Ni^{2+} ion pairs with values $D_T > 5\text{-}8\times10^{-8}$ cm^2 s^{-1}, depending on the spin label position. Comparison with linebroadenings (that are equal to the spin-lattice relaxation enhancements, for Heisenberg exchange – see section 8.2), and examination of temperature dependences, confirms that the spin-label relaxation enhancements for Ni^{2+} salts are dominated by spin exchange with paramagnetic ions dissolved in the fluid membranes. The relaxation enhancements in Fig. 16 therefore rather directly reflect the penetration profile of the Ni^{2+} ions into the membranes. The strong dependence on the anionic counterion arises because partitioning into the membrane takes place as electroneutral ion pairs which minimises the Born energy penalty for burying a charge in a low-dielectric medium.

For ions with longer spin-lattice relaxation times, Cu^{2+} and Mn^{2+}, dipolar relaxation also contributes in addition to spin exchange (Livshits et al., 2001). The net result is a greater relaxation enhancement, but this is less straightforward to interpret in terms of a simple penetration profile because of the different relaxation mechanisms. For Cu^{2+} salts, additional relaxation enhancement arises from dipolar interactions with ions within the membrane. This is analysed in terms of the enhancement in T_2-relaxation, which is bigger than that in T_1-relaxation and therefore is diagnostic of a dipolar contribution (see sections 8.2 and 8.3). Again integrating the static dipolar contribution over the Cu^{2+} ion distribution in the membrane, it is estimated that $T_{2,dd}^{-1}(\text{dynamic})/\ T_{1,dd}^{-1}(\text{static}) \approx 0.3\text{-}2\times10^{-6}$ cm^2 s^{-1}/D_T, depending on spin label position (Livshits et al., 2001). Thus, the dynamic dipolar mechanism dominates over the static contribution for translational diffusion coefficients of the Cu^{2+} ion pairs with values $D_T > 2\times10^{-6}$ cm^2 s^{-1}. Most probably both dipolar mechanisms make appreciable contributions to the relaxation induced by intramembranous Cu^{2+} ions.

Only for Mn^{2+} ions do dynamic dipolar interactions make an appreciable contribution to T_1-relaxation enhancements. From Eqs. 53, 54 and 65, the enhancement, relative to that from Heisenberg exchange, is: $T_{1,dd}^{-1}(\text{dynamic})/T_{1,HE}^{-1} \approx 2.0\times10^{-12}\ \text{cm}^4\ \text{s}^{-2}/D_T^2$ (Livshits et al., 2001). Therefore spin exchange dominates only for diffusion coefficients of the Mn^{2+} ion pair with values $D_T > 2\times10^{-6}$ cm^2 s^{-1}. In the case of Mn^{2+} ions, however, static dipolar enhancements are contributed not only by ions in the membrane, but also by those in the aqueous phase.

These results have considerable implications for the use of paramagnetic ions as relaxants for determining residue accessibilities in site-directed spin-labelling. Ni^{2+} ions have the advantage of a purely contact spin-exchange mechanism for relaxation enhancement. Cu^{2+} ions potentially can give larger enhancements, but some selectivity may be lost as a result of the longer-range dipolar contributions. Mn^{2+} ions may be exploited to obtain distance information from the additional contribution from ions in the aqueous phase. This may be an alternative to using gel-phase membranes (as in the previous section), particularly with the less membrane-soluble sulphate salt. The results with other anions illustrate how membrane penetration of nickel salts may be enhanced. In applications of site-directed spin labelling, this is frequently achieved by using electroneutral chelate complexes.

10. OUTLOOK

The two sections, 6 and 9, on applications have illustrated something of the breadth of possibilities available for ST-EPR methodologies, in both their traditional and nonstandard forms. For rotational diffusion measurements, an imminent development will be the implementation of ST-EPR on high-field spectrometers. The ability to resolve the three canonical g_{xx}, g_{yy} and g_{zz} spectral regions at high-field will give sensitivity to both axial and off-axis rotations (see e.g., Kurad et al., 2001; Marsh et al., 2002). This enhanced orientational selection should help immediately to resolve some of the ambiguities associated with slow anisotropic rotational diffusion that were mentioned in Section 6.4.

The nonstandard forms of ST-EPR, particularly first-harmonic out-of-phase measurements, are still comparatively in their infancy. On the other hand, determination of relaxation enhancements is now a standard part of site-directed spin-labelling methodology, although it relies almost exclusively on progressive saturation experiments. Quite apart from a more unique sensitivity to T_1-relaxation, the rapid-passage experiment has the advantage relative to conventional progressive saturation studies in consisting of a single measurement. It thus can be more readily automated, and is therefore anticipated as a possible method for high-throughput screening of spin-labelled cysteine mutants. Such developments seem to show promise for the application of spin-labelling to proteomics. In short, this aspect of the work of Jim Hyde has a very bright future.

ACKNOWLEDGEMENTS

Our studies on V-ATPases and the 16-kDa *Nephrops* protein are supported by the European Union (Contract No. QLGI-CT 2000-01801). Work is also supported by joint travel grants between Germany and Hungary (DAAD-MÖB) and between Germany and the Russian Federation (Deutsche Forschungsgemeinschaft-Russian Academy of Sciences). DM and TP are members of the European COST D22 Action.

APPENDIX:

Calibrations for T_1-relaxation enhancements in nonlinear CW EPR

This Appendix collects together fitting parameters for calibration of the T_1-dependence of nonlinear CW EPR spectra. Both in-phase (V_1) and out-of-phase (V_1') first-harmonic nonlinear spectra are considered and isotropic rotational dynamics of the spin label is taken into account. Rotational frequencies, f_R (or correlation times, τ_R), must be obtained from the linear EPR spectra at subsaturating microwave powers – either by spectral simulation or from motional narrowing theory.

Table A1: Calibration of the fitting parameters, a and b, in Eq. A.2 for progressive saturation V_1-EPR with spin label rotational frequencies in the range 10^7 s^{-1} $< f_R <$ 10^9 s^{-1} which give rise to extreme motional broadening. Results are given for different values of the intrinsic linewidth, ΔH_0 ($=1/\gamma_e T_2^o$), and the range of T_1-values for which the fitting was performed is indicated. Zeeman modulation frequency: $\omega_m/2\pi$ = 100 kHz (Livshits et al., 1998b).

f_R (s^{-1})	ΔH_o (G)	T_1 (μs)	a (G²)	bx10^{-6} (G².s)
1.0x10^7	0.5 – 1.5	(0.2 – 6.6)	0.032 – 0.044	0.083 – 0.145
3.16x10^7	0.5 – 1.5	(0.2 – 6.6)	0.022 – 0.029	0.137 – 0.185
1.0x10^8	0.5 – 1.0	(0.2 – 5.1)	0.024 – 0.025	0.134 – 0.162
3.16x10^8	0.5 – 1.5	(0.2 – 6.6)	0.022 – 0.023	0.100 – 0.129
1.0x10^9	0.5 – 1.0	(0.2 – 5.1)	0.012 – 0.014	0.063 – 0.091

Table A2: Calibration of the fitting parameters, a_1 and b_1, in Eq. A.3 for progressive saturation V_1-EPR with spin label rotational frequencies in the ranges 10^5 s^{-1} $< f_R <$ 10^7 s^{-1} and 10^9 s^{-1} $< f_R <$ 3.16x10^{10} s^{-1} for which rotational broadening is not overwhelming. Values are given for different intrinsic linewidths, ΔH_0 ($=1/\gamma_e T_2^o$), and are fitted over the interval T_1 = 0.2-5.1 μs. Zeeman modulation frequency: $\omega_m/2\pi$ = 100 kHz (Livshits et al., 1998b).

f_R (s^{-1})	ΔH_o (G)	a_1x10^{-14} (s^{-2})	b_1
1.0x10^5	1.0 – 3.0	0.01 – 0.06	0.97 – 0.93
1.0x10^6	1.0 – 2.8	0.037 – 0.11	1.07 – 1.00
3.16x10^6	1.0 – 3.0	0.10 – 0.16	1.34 – 1.10
6.0x10^6	1.0 – 3.0	0.115 – 0.164	1.67 – 1.22
1.0x10^7	1.0 – 3.0	0.126 – 0.172	2.03 – 1.33
1.0x10^7	0.5 – 1.0	0.090 – 0.126	2.96 – 2.03
1.0x10^9	0.5 – 1.0	0.037 – 0.043	2.20 – 1.60
3.16x10^9	0.5 – 1.0	0.016 – 0.025	1.47 – 1.21
1.0x10^{10}	0.5 – 1.0	0.011 – 0.019	1.17 – 1.05
3.16x10^{10}	0.5 – 1.0	0.009 – 0.018	1.03 – 1.00

The first set of calibrations (Tables A1 and A2) is for progressive saturation experiments where the H_1-dependence of the in-phase V_1-spectral intensity, S, is given by:

$$S(H_1) = \iint V_1(H).d^2H = P_1H_1 / \sqrt{1 + PH_1^2} \tag{A.1}$$

where P_1 is a scaling parameter and P is the fitting parameter that depends on T_1. In the extreme motional broadening region, for rotational frequencies $10^7\ s^{-1} < f_R < 10^9\ s^{-1}$, the parameter P is independent of intrinsic linewidth. The dependence on T_1 is then given by (Livshits et al., 1998b):

$$1/P = a + b/T_1 \tag{A.2}$$

where the calibration parameters a and b are given in Table A1. For rotational frequencies in the fast (motional narrowing) and slow regimes, $f_R > 10^8\ s^{-1}$ and $f_R < 10^7\ s^{-1}$, $P = \gamma_e^2(T_1T_2)^{eff}$ is a saturation parameter that depends directly on T_2. In these regimes, the effective T_1T_2 relaxation time product is given by (Livshits et al., 1998b):

$$1/(T_1T_2)^{eff} = a_1 + b_1/(T_1T_2^o) \tag{A.3}$$

where the calibration parameters a_1 and b_1 are given in Table A.2.

Table A3: Calibration parameters ($\rho^o{}_1'$, a_1', b_1' and m) for the T_1-dependence of the first-harmonic absorption out-of-phase/in-phase ratio in terms of Eq. A.4 for both the double integral intensities and amplitudes.[a] Parameters are given for rigid-limit or extreme-narrowing spectra, with different values of the modulation frequency, ω_m, microwave field intensity, H_1, and intrinsic linewidth ΔH_0 ($=1/\gamma_eT_2^o$) (Livshits et al., 1998a).

$\omega_m/2\pi$ (kHz)	H_1 (G)	ΔH_o (G)	$\rho^o{}_1'$	a_1'	b_1'	m
integrals:						
100	0.5	2.0	0.0067	0.1562	0.754	2.0
100	0.4	2.0	0.0041	0.1265	0.680	2.0
100	0.3	2.0	0.0019	0.0894	0.590	2.0
25	0.5	2.0	0.0026	0.0347	0.0967	1.3
25	0.4	2.0	0.0035	0.0315	0.0897	1.3
25	0.3	2.0	0.0045	0.0263	0.0760	1.3
amplitudes:						
100	0.5	1.0	-0.0160	0.553	0.538	1.6
100	0.5	1.5	-0.0170	0.496	0.507	1.6
100	0.5	2.0	-0.0177	0.489	0.483	1.6
100	0.5	2.5	-0.0179	0.410	0.464	1.6
25	0.5	1.0	-0.0037	0.1055	0.0787	1.3
25	0.5	1.5	-0.0064	0.1014	0.0753	1.3
25	0.5	2.0	-0.0086	0.0975	0.0720	1.3
25	0.5	2.5	-0.0100	0.0938	0.0690	1.3

[a]Parameters a_1' and b_1' are given for T_1 in μsecs.

Table A4: Calibration parameters ($\rho_1^{0\prime}$, a_1', b_1' and $m = 2$) for the T_1-dependence of the first-harmonic absorption out-of-phase/in-phase intensity ratio (V_1'/V_1) in terms of Eq. A.4, for a Zeeman modulation frequency of 100 kHz, and rotational frequencies in the range 10^4-$3.16\text{x}10^{10}$ s^{-1} (Livshits and Marsh, 2000).[a]

f_R (s^{-1})	$\rho_1^{0\prime}$	a_1'	b_1'
10^4	0.0134 – 0.0037	0.202 – 0.133	0.91 – 0.69
10^5	0.0146 – 0.0044	0.180 – 0.123	0.81 – 0.64
$3.16\text{x}10^5$	0.0147 – 0.0047	0.162 – 0.114	0.78 – 0.61
10^6	0.0129 – 0.0042	0.142 – 0.104	0.78 – 0.62
$3.16\text{x}10^6$	0.0096 – 0.0031	0.115 – 0.094	0.63 – 0.57
10^7	0.0053 – 0.0022	0.113 – 0.093	0.56 – 0.55
$3.16\text{x}10^7$	0.0037 – 0.0016	0.116 – 0.094	0.57 – 0.55
10^8	0.0040 – 0.0017	0.120 – 0.097	0.60 – 0.57
$3.16\text{x}10^8$	0.0053 – 0.0022	0.134 – 0.105	0.68 – 0.60
10^9	0.0076 – 0.0026	0.164 – 0.118	0.80 – 0.65
$3.16\text{x}10^9$	0.0102 – 0.0031	0.190 – 0.128	0.88 – 0.68
10^{10}	0.0120 – 0.0034	0.200 – 0.133	0.91 – 0.69
$3.16\text{x}10^{10}$	0.0127 – 0.0035	0.204 – 0.134	0.92 – 0.70

[a]Results are given for different values of the intrinsic linewidth, ΔH_0 ($=1/\gamma_e T_2^o$), in the range 1-3 G, over the range T_1 = 0.2-5.1 μs for which fitting was performed. Microwave field amplitude, H_1, is 0.5 G. Values of a_1' and b_1' are given for T_1 in μs.

Table A5: Calibration parameters ($\rho_1^{0\prime}$, a_1', b_1' and $m = 1.3$) for the T_1-dependence of the first-harmonic absorption out-of-phase/in-phase intensity ratio (V_1'/V_1) in terms of Eq. A.4, for a Zeeman modulation frequency of 25 kHz, and rotational frequencies in the range 10^4-$3.16\text{x}10^{10}$ s^{-1} (Livshits and Marsh, 2000).[a]

f_R (s^{-1})	$\rho_1^{0\prime}$	a_1'	b_1'
$1\text{x}10^4$	0.0007 – -0.0025	0.0361 – 0.0303	0.1051 – 0.0724
$1\text{x}10^5$	0.0010 – -0.0023	0.0344 – 0.0291	0.1041 – 0.0701
$3.16\text{x}10^5$	0.0013 – -0.0021	0.0310 – 0.0273	0.0928 – 0.0632
$1\text{x}10^6$	0.0008 – -0.0023	0.0273 – 0.0254	0.0635 – 0.050
$3.16\text{x}10^6$	-0.0013 – -0.0030	0.0278 – 0.0254	0.0486 – 0.0430
$1\text{x}10^7$	-0.0030 – -0.0034	0.0300 – 0.0260	0.0490 – 0.0429
$3.16\text{x}10^7$	-0.0035 – -0.0037	0.0309 – 0.0263	0.0510 – 0.0438
$1\text{x}10^8$	-0.0032 – -0.0036	0.0305 – 0.0264	0.0542 – 0.0460
$3.16\text{x}10^8$	-0.0023 – -0.0033	0.0309 – 0.0269	0.0664 – 0.0529
$1\text{x}10^9$	-0.0012 – -0.0030	0.0334 – 0.0285	0.0868 – 0.0633
$3.16\text{x}10^9$	-0.0002 – -0.0028	0.0355 – 0.0294	0.0990 – 0.0673
$1\text{x}10^{10}$	-0.0003 – -0.0027	0.0363 – 0.0302	0.1033 – 0.0718
$3.16\text{x}10^{10}$	-0.0005 – -0.0026	0.0366 – 0.0305	0.1047 – 0.0735

[a]Results are given for different values of the intrinsic linewidth, ΔH_0 ($=1/\gamma_e T_2^o$), in the range 1-3 G, over the range T_1 = 0.2-5.1 μs for which fitting was performed. Microwave field amplitude, H_1, is 0.5 G. Values of a_1' and b_1' are given for T_1 in μs.

The second set of calibrations is given for the out-of-phase/in-phase ratios (V_1'/V_1) of the first-harmonic integrated spectral intensities (Tables A3-A5) and low-field spectral amplitudes (Tables A3, A6, A7). For all calibrations, the (V_1'/V_1) ratio depends on T_1 according to the following semi-empirical expression (Livshits et al., 1998a; Livshits and Marsh, 2000):

$$\rho_1' = \rho_1^{o}{}' + \frac{a_1' T_1^m}{1 + b_1' T_1^m} \quad \text{(A.4)}$$

where the parameters a_1' and b_1' depend quite strongly on H_1 and the Zeeman modulation frequency, ω_m, but less strongly on T_2 and the spin label rotational rate, f_R. The exponent m in Eq. A.4 depends on the modulation frequency and also on whether signal amplitudes or integrated intensities are measured. Table A3 gives calibrations established for the "no-motion" situation, which applies to both quasi-rigid limit and extreme motional narrowing spectra (Livshits et al., 1998a). Tables A4-A7 give corresponding calibrations that specifically take account of molecular motion (Livshits and Marsh, 2000). For the first-harmonic out-of-phase method, calibrations are given not only for the standard modulation frequency $\omega_m/2\pi$ = 100 kHz, but also for a modulation frequency of 25 kHz. Calibrations for $\omega_m/2\pi$ = 25 kHz show enhanced sensitivity to longer T_1 relaxation times.

Table A6: Calibration parameters [$\rho_1^{0}{}'(M_I=+1)$, $a_1'(M_I=+1)$, $b_1'(M_I=+1)$ and m = 1.6] for the T_1-dependence of the out-of-phase/in-phase ratio of the first-harmonic M_I=+1 amplitudes according to Eq. A.4, for a Zeeman modulation frequency of 100 kHz, and rotational frequencies in the range 10^4-3.16×10^{10} s^{-1} (Livshits and Marsh, 2000).[a]

f_R (s^{-1})	$\rho_1^{0}{}'(M_I = +1)$	$a_1'(M_I = +1)$	$b_1'(M_I = +1)$
1×10^4	-0.0060 – -0.0116	0.352 – 0.260	0.629 – 0.560
1×10^5	0.0007 – -0.0065	0.337 – 0.250	0.623 – 0.555
3.16×10^5	0.0007 – -0.0080	0.269 – 0.216	0.544 – 0.487
1×10^6	-0.0001 – -0.0076	0.219 – 0.190	0.555 – 0.485
3.16×10^6	-0.0060 – -0.0095	0.175 – 0.171	0.473 – 0.452
1×10^7	-0.0120 – -0.0110	0.174 – 0.169	0.436 – 0.437
3.16×10^7	-0.0134 – -0.0108	0.188 – 0.177	0.461 – 0.451
1×10^8	-0.0126 – -0.0120	0.251 – 0.218	0.528 – 0.498
3.16×10^8	-0.0173 – -0.0156	0.370 – 0.285	0.494 – 0.481
1×10^9	-0.0169 – -0.0171	0.440 – 0.330	0.500 – 0.480
3.16×10^9	-0.0151 – -0.175	0.464 – 0.352	0.507 – 0.483
1×10^{10}	-0.0141 – -0.0176	0.470 – 0.360	0.508 – 0.485
3.16×10^{10}	-0.0139 – -0.0176	0.470 – 0.361	0.508 – 0.486

[a]Results are given for different values of the intrinsic linewidth, ΔH_0 (=$1/\gamma_e T_2^o$), in the range 1-3 G, a peak-to-peak inhomogeneous linewidth of 1.4 G, over the range T_1 = 0.2-5.1 μs for which fitting was performed. Microwave field amplitude, H_1, is 0.5 G. Values of $a_1'(M_I = +1)$ and $b_1'(M_I = +1)$ are given for T_1 in μs.

A corresponding treatment for anisotropic rotational motion is given by Livshits et al. (2003). In this case, the angular amplitude of motion (or orientational order parameter) must be obtained from spectral simulations, in addition to the rotational frequency/correlation time.

Table A7: Calibration parameters [$\rho_1^{0\prime}(M_I=+1)$, $a_1'(M_I=+1)$, $b_1'(M_I=+1)$ and m = 1.3] for the T_1-dependence of the out-of-phase/in-phase ratio of the first-harmonic $M_I=+1$ amplitudes according to Eq. A.4, for a Zeeman modulation frequency of 25 kHz, and rotational frequencies in the range 10^4-3.16×10^{10} s^{-1} (Livshits and Marsh, 2000).[a]

f_R (s^{-1})	$\rho_1^{0\prime}(M_I = +1)$	$a_1'(M_I = +1)$	$b_1'(M_I = +1)$
1×10^4	-0.0027 – -0.0055	0.0677 – 0.0568	0.090 – 0.073
1×10^5	-0.0017 – -0.0053	0.0620 – 0.0540	0.087 – 0.063
3.16×10^5	-0.0007 – -0.0050	0.0530 – 0.0495	0.072 – 0.051
1×10^6	-0.0016 – -0.0057	0.0440 – 0.0458	0.040 – 0.036
3.16×10^6	-0.0059 – -0.0074	0.0439 – 0.0456	0.026 – 0.028
1×10^7	-0.0084 – -0.0082	0.0471 – 0.0465	0.027 – 0.029
3.16×10^7	-0.0085 – -0.0075	0.0491 – 0.0473	0.033 – 0.033
1×10^8	-0.0074 – -0.0077	0.0586 – 0.0537	0.056 – 0.050
3.16×10^8	-0.0092 – -0.0090	0.0846 – 0.0690	0.075 – 0.066
1×10^9	-0.0081 – -0.0093	0.0948 – 0.0780	0.079 – 0.071
3.16×10^9	-0.0071 – -0.0093	0.0972 – 0.0814	0.079 – 0.073
1×10^{10}	-0.0066 – -0.0092	0.0978 – 0.0826	0.079 – 0.074
3.16×10^{10}	-0.0065 – -0.0092	0.0978 – 0.0829	0.078 – 0.074

[a]Results are given for different values of the intrinsic linewidth, ΔH_0 ($=1/\gamma_e T_2^o$), in the range 1-3 G, a peak-to-peak inhomogeneous linewidth of 1.4 G, over the range T_1 = 0.2-5.1 μs for which fitting was performed. Microwave field amplitude, H_1, is 0.5 G. Values of $a_1'(M_I = +1)$ and $b_1'(M_I = +1)$ are given for T_1 in μs.

REFERENCES

Abragam, A. (1961). *The Principles of Nuclear Magnetism.* Oxford: Oxford University Press, 1961.

Baroin A., Thomas D.D., Osborne B., Devaux P.F. (1977) Saturation transfer electron paramagnetic resonance on membrane-bound proteins. I. Rotational diffusion of rhodopsin in the visual receptor membrane. *Biochem Biophys Res Commun* **78**, 442-447.

Bloembergen N. (1949). On the interaction of nuclear spins in a crystalline lattice. *Physica* **15**, 386-426.

Cherry R.J., Godfrey R.E. (1981). Anisotropic rotation of bacteriorhodopsin in lipid membranes. *Biophys J.* **36**, 257-276.

Clegg R.M., Vaz W.L.C. (1985). "Translation diffusion of proteins and lipids in artificial lipid bilayer membranes. A comparison of experiment with theory." In *Progress in Protein-Lipid Interactions.*, Watts A., de Pont J.J.H.H.M., eds. Amsterdam: Elsevier, , pp. 173-229.

Corless J.M., McCaslin D.R., Scott B.L. (1982). Two-dimensional rhodopsin crystals from disk membranes of frog retinal rod outer segments. *Proc Natl Acad Sci USA* **79**, 1116-1120.

Deatherage J.F., Henderson R., Capaldi R.A. (1982) Relationship between membrane and cytoplasmic domains in cytochrome c oxidase by electron microscopy in media of different density. *J Mol Biol* **158**, 501-514.

Downer N.W. (1985) Cross-linking of dark adapted frog photoreceptor disk membranes - evidence for monomeric rhodopsin. *Biophys J* **47**, 285-293.

Eastman M.P., Kooser R.G., Das M.R., Freed J.H. (1969) Studies of Heisenberg spin exchange in ESR spectra. I. Linewidth and saturation effects. *J Chem Phys* **51**, 2690-2709.

Esmann M., Hideg K., Marsh D. (1994) Influence of poly(ethylene glycol) and aqueous viscosity on the rotational diffusion of membranous Na, K-ATPase. *Biochemistry* **33**, 3693-3697.

Esmann M., Marsh D. (1992) Local translational diffusion rates of membranous Na^+,K^+-ATPase measured by saturation transfer ESR spectroscopy. *Proc Natl Acad Sci USA* **89**, 7606-7609.

Evans C.A. (1981) Use of integral of saturation transfer electron paramagnetic spectra to determine molecular rotational correlation times. Slowly tumbling spin labels in the presence of rapidly tumbling spin labels. *J Magn Reson* **44**, 109-116.

Fajer P., Knowles P.F., Marsh D. (1989) Rotational motion of yeast cytochrome oxidase in phosphatidylcholine complexes studied by saturation-transfer electron spin resonance. *Biochemistry* **28**, 5634-5643.

Fajer P., Marsh D. (1983a) Analysis of dispersion mode saturation-transfer ESR spectra. Application to model membranes. *J Magn Reson* **55**, 205-215.

Fajer P., Marsh D. (1983b) Sensitivity of saturation transfer ESR spectra to anisotropic rotation. Application to membrane systems. *J Magn Reson* **51**, 446-459.

Fajer P., Thomas D.D., Feix J.B., Hyde J.S. (1986) Measurement of rotational molecular motion by time-resolved saturation transfer electron paramagnetic resonance. *Biophys J* **50**, 1195-1202.

Halbach K. (1954) Über eine neue Methode zur Messung von Relaxationszeiten und über den Spin von Cr^{53}. *Helv Phys Acta* **27**, 259-282.

Hidalgo C., Thomas D.D., Ikemoto N. (1978) Effect of the lipid environment on protein motion and enzymatic activity of the sarcoplasmic reticulum calcium ATPase. *J Biol Chem* **253**, 6879-6887.

Hoffmann W., Pink D.A., Restall C., Chapman D. (1981) Intrinsic molecules in fluid phospholipid bilayers. Fluorescence probe studies. *Eur J Biochem* **114**, 585-589.

Holzenburg A., Jones P.C., Franklin T., Páli T., Heimburg T., Marsh D., Findlay J.B.C., Finbow M.E. (1993) Evidence for a common structure for a class of membrane channels. *Eur J Biochem* **213**, 21-30.

Horváth L.I., Brophy P.J., Marsh D. (1988) Influence of lipid headgroup on the specificity and exchange dynamics in lipid-protein interactions. A spin label study of myelin proteolipid apoprotein-phospholipid complexes. *Biochemistry* **27**, 5296-5304.

Horváth L.I., Brophy P.J., Marsh D. (1993) Exchange rates at the lipid-protein interface of the myelin proteolipid protein determined by saturation transfer electron spin resonance and continuous wave saturation studies. *Biophys J* **64**, 622-631.

Horváth L.I., Dux L., Hankovszky H.O., Hideg K., Marsh D. (1990) Saturation transfer electron spin resonance of Ca^{2+}-ATPase covalently spin-labeled with β-substituted vinyl ketone- and maleimide-nitroxide derivatives. Effects of segmental motion and labeling levels. *Biophys J* **58**, 231-241.

Horváth L.I., Marsh D. (1988) Improved numerical evaluation of saturation transfer electron spin resonance spectra. *J Magn Reson* **80**, 314-317.

Horváth L.I., Marsh D. (1983) Analysis of multicomponent saturation transfer ESR spectra using the integral method: application to membrane systems. *J Magn Reson* **54**, 363-373.

Hubbell W.L., Altenbach C. (1994) "Site-directed spin-labeling of membrane proteins." In *Membrane protein structure: experimental approaches.*, White S.H., ed. New York: Oxford University Press, pp. 224-248.

Hyde J.S. (1978) "Saturation-transfer spectroscopy." In *Methods in Enzymology*, Hirs C.H.W., Timasheff S.N., eds. New York: Academic Press, pp. 480-511.

Hyde J.S., Dalton L. (1972) Very slowly tumbling spin labels: adiabatic rapid passage. *Chem Phys Lett* **16**:568-572.

Hyde J.S., Dalton L.R. (1979) "Saturation-transfer spectroscopy." In *Spin-Labeling II. Theory and Applications*, Berliner L.J., ed. New York: Academic Press, pp. 71-113.

Hyde J.S., Eriksson L.E.G., Ehrenberg A. (1970) EPR relaxation of slowly moving flavin radicals: "Anomalous" saturation. *Biochim Biophys Acta* **222**, 688-692.

Hyde J.S., Subczynski W.K. (1989) "Spin-label oximetry." In *Spin Labeling. Theory and Applications*, Berliner L.J., Reuben J., eds. New York and London: Plenum Press, pp. 399-425.

Hyde J.S., Thomas D.D. (1973) New EPR methods for the study of very slow motion: application to spin-labeled hemoglobin. *Ann N Y Acad Sci* **222**, 680-692.

Jesaitis A.J., Yguerabide J. (1986) The lateral mobility of the (Na^+,K^+)-dependent ATPase in Madin-Darby canine kidney cells. *J Cell Biol* **102**, 1256-1263.

Johnson M.E., Hyde J.S. (1981) 35-GHz (Q-band) saturation transfer electron paramagnetic resonance studies of rotational diffusion. *Biochemistry* **20**, 2875-2880.

Kirino Y., Ohkuma T., Shimizu H. (1978) Saturation transfer electron spin resonance study on the rotational diffusion of calcium- and magnesium-dependent adenosine triphosphatase in sarcoplasmic reticulum membranes. *J Biochem (Tokyo)* **84**, 111-115.

Knowles P.F., Watts A., Marsh D. (1979) Spin label studies of lipid immobilization in dimyristoylphosphatidylcholine-substituted cytochrome oxidase. *Biochemistry*, **18**, 4480-4487.

Kurad D., Jeschke G., Marsh D. (2001) Spin-label HF-EPR of lipid ordering in cholesterol-containing membranes. *Appl Magn Reson* **21,** 469-481.

Kusumi A., Hyde J.S. (1982) Spin-label saturation-transfer electron spin resonance detection of transient association of rhodopsin in reconstituted membranes. *Biochemistry* **21**, 5978-5983.

Kusumi A., Ohnishi S., Ito T., Yoshizawa T. (1978) Rotational motion of rhodopsin in the visual receptor membrane as studied by saturation-transfer spectroscopy. *Biochim Biophys Acta* **507**, 539-543.

Langen R., Isas J.M., Luecke H., Haigler H.T., Hubbell W.L. (1998) Membrane-mediated assembly of annexins studied by site-directed spin labeling. *J Biol Chem* **273**, 22453-22457.

Leigh Jr. J.S. (1970) ESR rigid-lattice line shape in a system of two interacting spins. *J Chem Phys* **52**, 2608-2612.

Livshits V.A. (1974) Slow anisotropic tumbling in ESR spectra of nitroxyl radicals. *J Magn Reson* **24**, 307-313.

Livshits V.A., Dzikovski B.G., Marsh D.(2001) Mechanism of relaxation enhancement of spin labels in membranes by paramagnetic ion salts: dependence on *3d* and *4f* ions and on the anions. *J Magn Reson* **148**, 221-237.

Livshits V.A., Dzikovski B.G., Marsh D. (2003) Anisotropic motion effects in CW non-linear EPR spectra. Spin relaxation enhancement of lipid spin labels. *J Magn Reson*, in press.

Livshits V.A., Marsh D. (2000) Spin relaxation measurements using first-harmonic out-of-phase absorption EPR signals: rotational motion effects. *J Magn Reson* **145**, 84-94.
Livshits V.A., Páli T., Marsh D. (1998a) Spin relaxation measurements using first-harmonic out-of-phase absorption EPR signals. *J Magn Reson* **134**, 113-123.
Livshits V.A., Páli T., Marsh D. (1998b) Relaxation time determinations by progressive saturation EPR: Effects of molecular motion and Zeeman modulation for spin labels. *J Magn Reson* **133**, 79-91.
Marsh D. (1980) Molecular motion in phospholipid bilayers in the gel phase: long axis rotation. *Biochemistry* **19**, 1632-1637.
Marsh D. (1990) Sensitivity analysis of magnetic resonance spectra from unoriented samples. *J Magn Reson* **87**, 357-362.
Marsh D. (1992a) Influence of nuclear relaxation on the measurement of exchange frequencies in CW saturation EPR studies. *J Magn Reson* **99**, 332-337.
Marsh D. (1992b) Exchange and dipolar spin-spin interactions and rotational diffusion in saturation transfer EPR spectroscopy. *Appl Magn Reson* **3**, 53-65.
Marsh D. (1995) Sensitivity of the continuous wave saturation of conventional EPR spectra to slow rotational motion. *Spectrochim Acta* **51A**, 111-114.
Marsh D., Horváth L.I. (1989) "Spin-label studies of the structure and dynamics of lipids and proteins in membranes." In *Advanced EPR. Applications in Biology and Biochemistry*, Hoff A.J., ed. Amsterdam: Elsevier, pp. 707-752.
Marsh D., Horváth L.I. (1992a) A simple analytical treatment of the sensitivity of saturation transfer EPR spectra to slow rotational diffusion. *J Magn Reson* **99**, 323-331.
Marsh D., Horváth L.I. (1992b) Influence of Heisenberg spin exchange on conventional and phase-quadrature EPR lineshapes and intensities under saturation. *J Magn Reson* **97**, 13-26.
Marsh D., Horváth L.I. (1998) Structure, dynamics and composition of the lipid-protein interface. Perspectives from spin-labelling. *Biochim Biophys Acta* **1376**, 267-296.
Marsh D., Kurad D., Livshits V.A. (2002) High-field electron spin resonance of spin labels in membranes. *Chem Phys Lipids* **116**, 93-114.
Marsh D., Livshits V.A., Páli T. (1997) Non-linear, continuous-wave EPR spectroscopy and spin-lattice relaxation: spin-label EPR methods for structure and dynamics. *J Chem Soc Perkin Trans* **2**, 2545-2548.
Marsh D., Páli T., Horváth L.I. (1998) "Progressive saturation and saturation transfer EPR for measuring exchange processes and proximity relations in membranes." In *Spin Labeling. The Next Millenium.*, Berliner L.J., ed. New York: Plenum Press, pp. 23-82.
Maunsbach A.B., Skriver E., Soderholm M., Hebert H. (1989) "Three-dimensional structure and topography of membrane bound Na,K-ATPase." In *The Na,K-Pump. Part A: Molecular Aspects. Progress in Clinical and Biological Research.*, Skou J.C., Norby J.G., Maunsbach A.B., Esmann M., eds. New York: Alan Liss, pp. 39-56.
McLaughlin A., Gratwohl G., McLaughlin S. (1978) The adsorption of divalent cations to phosphatidylcholine bilayer membranes. *Biochim Biophys Acta* **513**, 338-357.
Molin Y.N., Salikov K.M., Zamaraev K.I. (1980) *Spin Exchange. Principles and Applications in Chemistry and Biology*. Berlin: Springer Verlag.
Páli T., Bartucci R., Horváth L.I., Marsh D. (1992) Distance measurements using paramagnetic ion-induced relaxation in the saturation transfer electron spin resonance of spin- labeled biomolecules. Application to phospholipid bilayers and interdigitated gel phases. *Biophys J* **61**, 1595-1602.
Páli T., Finbow M.E., Marsh D. (1999) Membrane assembly of the 16-kDa proteolipid channel from *Nephrops norvegicus* studied by relaxation enhancements in spin-label ESR. *Biochemistry* **38**, 14311-14319.

Páli T., Horváth L.I., Marsh D. (1993) Continuous-wave saturation of two-component, inhomogeneously broadened, anisotropic EPR spectra. *J Magn Reson* A**101**, 215-219.

Páli T., Livshits V.A., Marsh D. (1996) Dependence of saturation-transfer EPR intensities on spin-lattice relaxation. *J Magn Reson B* **113**, 151-159.

Portis A.M. (1955) Rapid passage in electron spin resonance. *Phys Rev* **100**, 1219-1224.

Robinson B.H., Dalton L.R. (1980) Anisotropic rotational diffusion studied by passage saturation transfer electron paramagnetic resonance. *J Chem Phys* **72**, 1312-1324.

Rousselet A., Devaux P.F. (1977) Saturation transfer electron paramagnetic resonance on membrane bound proteins. II. Absence of rotational diffusion of the cholinergic receptor protein in *Torpedo marmorata* membrane fragments. *Biochem Biophys Res Commun* **78**, 448-454.

Ryba N.J.P., Marsh D. (1992) Protein rotational diffusion and lipid/protein interactions in recombinants of bovine rhodopsin with saturated diacylphosphatidylcholines of different chain lengths studied by conventional and saturation transfer electron spin resonance. *Biochemistry* **31**, 7511-7518.

Saffman P.G., Delbrück M. (1976) Brownian motion in biological membranes. *Proc Natl Acad Sci USA* **72**, 3111-3113.

Slichter, C. P. (1978) *Principles of Magnetic Resonance*. Berlin-Heidelberg-New York: Springer-Verlag,.

Snel M.M.E., Marsh D. (1994) Membrane location of apocytochrome *c* and cytochrome *c* determined from lipid-protein spin exchange interactions by continuous wave saturation electron spin resonance. *Biophys J* **67**, 737-745.

Solomon I. (1955) Relaxation processes in a system of 2 spins. *Phys Rev* **99**, 559-565.

Squier T.C., Thomas D.D. (1986) Methodology for increased sensitivity and precision in saturation transfer electron paramagnetic resonance studies of molecular dynamics. *Biophys J* **49**, 921-935.

Tardieu A. (1972) Thesis/Dissertation. Université de Paris-Sud..

Thomas D.D., Dalton L.R., Hyde J.S. (1976) Rotational diffusion studied by passage saturation transfer electron paramagnetic resonance. *J Chem Phys* **65**, 3006-3024.

Thomas D.D., McConnell H.M. (1974) Calculation of paramagnetic resonance spectra sensitive to very slow rotational motion. *Chem Phys Lett* **25**, 470-475.

Thomas D.D., Seidel J.C., Hyde J.S., Gergely J. (1975) Motion of subfragment-1 in myosin and its supramolecular complexes: saturation transfer electron paramagnetic resonance. *Proc Natl Acad Sci USA* **72**, 1729-1733.

Weger M. (1960) Passage effects in paramagnetic resonance experiments. *Bell Syst Tech J* **39**, 1013-1112.

Yin J.J., Hyde J.S. (1987) Application of rate equations to ELDOR and saturation recovery experiments on N-14-N-15 spin label pairs. *J Magn Reson* **74**, 82-93.

Chapter 12

Saturation Transfer EPR:

Rotational Dynamics of Membrane Proteins

Albert H. Beth and Eric J. Hustedt

Department of Molecular Physiology and Biophysics, Vanderbilt University, Nashville, TN 37232, USA

Abstract: Saturation transfer EPR (ST-EPR) has been utilized to characterize the very slow rotational diffusion of a wide range of membrane proteins over the past three decades. Development of computational algorithms for analyzing the ST-EPR data from membrane proteins undergoing uniaxial rotational diffusion and inclusion of these algorithms into global nonlinear least squares fitting routines has permitted detailed analyses of experimental data. These advances, coupled with access to multifrequency measurements, has established ST-EPR as an important complement to transient optical anisotropy for unraveling the complex rotational dynamics of intrinsic membrane proteins in cell membranes.

1. INTRODUCTION

The membrane that surrounds all eukaryotic cells and their subcellular organelles is composed of lipids and proteins in a bilayer configuration, an arrangement that has been qualitatively described for the past three decades by the fluid mosaic model (Singer, 1972). These remarkable, dynamic structures play essential roles in regulating the substances and signals that enter and leave the cell and its organelles in order to support and regulate the complex physiological processes that take place within. There are vast amounts of data that define the composition of cell membranes including the heterogeneity of lipids that assemble to form the permeability barrier and the relative amounts of lipid and protein that are present. Spectroscopic methods including EPR, NMR, and fluorescence have provided important insights into the lateral organization and the dynamics of the lipid components of membranes. EPR, in particular, has provided unique characterization of the structure and dynamics of the lipids that solvate the periphery of intrinsic membrane proteins, the so-called boundary lipids (e.g. Griffith and Jost, 1976; Ge and Freed, 1993; Borbat et al., 2001). Most recently, these basic techniques and in particular EPR have been utilized to investigate the lateral organization of lipids in cell membranes including the sizes and stabilities of microdomains (e.g. Kawasaki et al., 2001). These investigations have taken on added importance with the discovery of specialized structures including lipid "rafts" that are hypothesized to play important roles in modulating protein-membrane interactions and in cellular signaling pathways (e.g. Zacharias et al., 2002; Anderson and Jacobson, 2002).

Less is known about the organization and dynamics of intrinsic membrane proteins despite the central role that they play in regulating cellular functions. Studies of membrane protein global dynamics are particularly challenging due in large part to the anisotropic, viscous nature of the bilayer, the number of different proteins that are present, and the very high local concentrations of total protein. The latter property merits serious attention. While there is variability in the percentage by weight of lipid and protein in different specialized membranes (e.g. myelin versus inner mitochondrial), the plasma membranes of cells and many subcellular organelle membranes are often composed of roughly equal percentages of lipid and protein by weight. Using average values for molecular weights and sizes of lipids and intrinsic membrane proteins, on the order of 30% of the outer membrane surface area can be occupied by protein and the effective concentration of protein within the volume of the bilayer can exceed 100 mg/mL! Given this extremely crowded two-dimensional environment, there are obvious questions regarding the extent of weak interactions between membrane proteins that would not be observed following extraction with non-denaturing detergents and how such interactions might modulate the functions of membrane proteins. Currently, there is unprecedented interest in determining the static structures of membrane proteins and in understanding their functional dynamics. As the structural database of individual membrane proteins continues to expand, studies designed to determine the structural consequences of interactions between proteins in the crowded environment of the cell membrane will undoubtedly take on increased importance.

A number of different experimental approaches have been utilized to examine the oligomeric state and higher order lateral organization of membrane proteins *in situ* including chemical cross-linking, visualization by freeze-fracture electron microscopy, visualization by fluorescence microscopy, and target size analysis by neutron inactivation to cite a few. Each of these approaches has provided valuable information but each is limited in its ability to provide information on dynamic aspects of protein-

protein interactions under native conditions. Fluorescence resonance energy transfer (FRET; Stryer, 1978) between labeled subunits has been utilized extensively to provide direct information on the stable assembly and dynamic proximity of membrane proteins (e.g. Blackman et al., 1998) including changes in assembly associated with cellular signaling events in living cells (e.g. Martin-Fernandez et al., 2002).

Studies of the translational or rotational dynamics of membrane proteins have been carried out under a wide range of experimental conditions including in living cells and these studies have provided important insights into static and dynamic aspects of their assembly and interactions. Most studies of translational diffusion have utilized fluorescence recovery after photobleaching (FRAP; Frye and Edidin, 1970; Webb, 1981) due in large part to the excellent signal-to-noise ratio and dynamic range that this methodology provides. More recently, single particle tracking experiments have provided new insights into the constraints on the lateral motion of membrane proteins (e.g. Cherry et al., 1998). While these optical techniques provide outstanding capabilities for investigating the molecular basis for constraints to the long range lateral movement of membrane proteins on the cell surface (e.g. Koppel et al., 1981), the translational diffusion coefficient (D_t) is weakly dependent on the effective size of the diffusing species (Saffman and Delbrück, 1975) and hence, to oligomeric state or to other localized, transient interactions. On the other hand, the rotational diffusion coefficient, (D_r), of an integral membrane protein undergoing global uniaxial rotational diffusion (URD) is inversely proportional to the square of the cross sectional radius of the protein complex (Saffman and Delbrück, 1975; Jähnig, 1986). Thus, measurements of rotational diffusion should be very sensitive to the presence of homo and hetero oligomeric protein complexes in the membrane.

Given the predicted high sensitivity of the correlation time for URD, which can be defined as $\tau_r = 1/6D_r$, on the effective sizes of membrane proteins and their complexes in membranes, it is not surprising that considerable effort during the past three decades has been devoted to developing spectroscopic methods that can measure their rotational dynamics in systems ranging from reconstituted proteoliposomes to intact, living cells (Edidin, 1974). These efforts have led to the development of two complementary approaches that have been successfully applied to measuring the very slow rotational dynamics, on the timescale of μsec to msec, that are characteristic of many integral membrane proteins in their natural membranes or in proteoliposomes.

The first approach is a collection of related optical methods that measure the transient optical anisotropy (TOA) or dichroism, following a short excitation pulse of light, of long lived triplet probes coupled to a target protein (Cone, 1972; Naqvi et al., 1973; Cherry et al., 1976; Cherry and Schneider, 1976). TOA methods and their applications to measure the rotational diffusion of membrane proteins have been reviewed previously (e.g. Cherry, 1978; 1979; 1981; Thomas et al., 1985). Analytical expressions have been derived for the amplitudes and rates of anisotropy decays for a variety of diffusion models including URD (Cherry, 1979), constrained URD (Wahl, 1975, Szabo, 1984), wobble in a cone (Szabo, 1984), and generalized anisotropic diffusion (Weber, 1971; Chuang and Eisenthal, 1972; Belford et al., 1972; Ehrenberg and Rigler, 1972). These various models provide a sound theoretical base upon which experimental data can be analyzed in a model dependent fashion to determine the values of D_r of membrane proteins as well as the number of different rotational species that are present. Unique interpretation of TOA data from anisotropic systems requires determination of the orientation of the molecular probe reference frame relative to the molecular reference frame and the uniqueness of this orientation. Determination of the orientations of the absorption and emission dipoles of a chromophore relative to the molecular reference frame can be quite challenging (e.g. Blackman et al., 1996) in comparison with the relative ease with which the orientation of a spin label can be determined by EPR (e.g. Hustedt and Beth, 1996).

Typically, TOA data are analyzed by fitting to a multi-exponential decay curve. The biggest challenge in interpreting the anisotropy decays from membrane proteins is the assignment of the multiple decay components to distinct molecular species and to specific dynamic processes. Even for simple unconstrained URD, two decay components are predicted for each distinct oligomeric species that is present (Nigg and Cherry, 1980). Experimentally, one often observes many decay components and it is possible to fit the complex decays to a variety of different multi-component models with similar statistical agreement between experiment and theory. Matayoshi and Jovin (1991) and Blackman et al., (1996; 2001) have addressed some of the complexities that are involved in uniquely interpreting TOA data from membrane proteins in cell membranes.

The second approach that has been utilized to characterize the rotational diffusion of intrinsic membrane proteins is an EPR method, developed by Hyde and Dalton (1972) which measures the effects of rotational diffusion on spectral lineshapes from nitroxide spin labeled proteins under conditions of continuous, partially saturating microwave excitation. The EPR method, which derives its sensitivity to μsec motions from the "transfer" of microwave "saturation" between different orientational resonance positions of the spectrum due to rotational diffusion, has been universally referred to as saturation transfer EPR (ST-EPR; Thomas et al., 1976). A number of reviews of ST-EPR and its applications to studies of very slow rotational dynamics of spin labeled proteins, including membrane proteins, have appeared in the literature (e.g. Hyde and Thomas, 1974; Hyde, 1978; Hyde and Dalton, 1979; Hyde and Thomas, 1980; Thomas, 1985; Thomas et al., 1985; Robinson et al., 1985; Hemminga and de Jager, 1989; Beth and Robinson, 1989). These reviews provide an introduction to ST-EPR, discussions of the sensitivity of ST-EPR to rotational motions in the μsec to msec time range, and extensive discussions of applying the method to systems undergoing anisotropic rotational diffusion. The interested reader is also referred to a small collection of seminal papers that describe the discovery, early applications, and the theory underlying the method (Hyde et al., 1970; Hyde and Dalton, 1972; Thomas and McConnell, 1974; Robinson et al., 1974; Thomas et al., 1976). These papers, and the comprehensive review by Hyde and Dalton (1979), will provide readers with a thorough introduction to the basic principles of ST-EPR.

The remainder of this chapter will discuss the capabilities of modern multi-frequency ST-EPR for determining the very slow global rotational dynamics of membrane proteins and in particular, recent advances that have permitted rigorous testing of diffusion models and determination of correlation times from global nonlinear least squares fitting of experimental data sets (Hustedt et al., 1993; Hustedt and Beth, 1995; 2001). While the major emphasis is on rotational dynamics of integral membrane proteins, it should be noted that the methods and computational approaches that have been developed can easily be adapted to investigations of the very slow rotational dynamics of other spin labeled macromolecules. Also, though this chapter only addresses the capabilities of continuous wave (CW) ST-EPR, there are other nonlinear saturation transfer EPR approaches, such as continuous wave and time-domain electron-electron double resonance (ELDOR; Hyde and Feix, 1989; Borbat et al., 2001), that continue to be explored for their potential to provide additional insights into the rotational motions of spin labeled proteins. In addition, multiquantum EPR (MQ-EPR), a technique developed by Hyde and coworkers (Sczaniecki et al., 1990; Mchaourab and Hyde, 1993), is potentially useful for obtaining very slow rotational motion.

2. METHODS FOR ANALYSIS OF ST-EPR DATA

Linear EPR lineshapes are sensitive to motions which effectively average elements of the nitroxide *A*- and *g*-tensors, and thus the resonant field position, within a timescale determined by the spread in resonant positions of the manifolds that make up the rigid limit spectrum. ST-EPR derives its sensitivity to motion from a complex competition between the rotational dynamics, Zeeman modulation, T_{1e}, and T_{1n}. This competition is in response to the partial saturation effects of a nonlinear microwave field. ST-EPR lineshapes also depend on the relative orientation of the spin label reference frame relative to the diffusion tensor, on the microwave frequency and amplitude, and to a lesser extent, on the principal values of the *A*- and *g*-tensors. Fortunately, many of these parameters can be selected (e.g. microwave and modulation frequencies and amplitudes) or measured (e.g. T_{1e} and T_{1n} using saturation recovery EPR (Percival and Hyde, 1976; Robinson et al., 1994); *A*- and *g*-tensor elements from linear EPR measurement, (e.g. Hustedt et al., 1993)). With this information, it is possible to carry out detailed simulations of experimental data in order to determine the unique elements of the diffusion tensor. Even in the absence of precise values for T_{1e} and T_{1n}, it is possible to analyze experimental ST-EPR data using estimated values for these parameters (e.g. Robinson et al., 1994) and obtain very good agreement between theory and experiment.

The paucity of computational algorithms for analyzing experimental ST-EPR data has undoubtedly limited its use for measuring the global URD of membrane proteins as compared with TOA. URD, and other anisotropic rotational diffusion processes, present special challenges for the analysis of ST-EPR data since the orientation of the spin label reference frame relative to the molecular reference frame and the uniqueness of this orientation both have profound spectral effects (Robinson and Dalton, 1980; 1981; Robinson et al., 1985; Beth and Robinson, 1989). Explicit consideration of this additional complexity, compared to simple isotropic rotational diffusion, and the inclusion of nonlinear microwave and modulation field effects, results in a computationally demanding problem for calculation of ST-EPR lineshapes. Therefore, the early ST-EPR literature on membrane protein rotational dynamics is dominated by estimation of effective correlation times via the comparison of motion-sensitive ratio parameters measured from experimental spectra with calibration curves of the same parameters obtained from an isotropic model system such as spin labeled hemoglobin in glycerol/water solutions (Thomas et al., 1976). While this model system approach has permitted very useful assessments of rotational dynamics, including changes in rotational dynamics of membrane proteins in different functional states (e.g. Beth et al., 1986; Bigelow and Thomas, 1987; Squier et al., 1988; Squier and Thomas, 1988; Cobb et al., 1990; Schwarz et al., 1990; Lewis and Thomas, 1991; Mahaney and Thomas, 1991; Esmann et al., 1992; Qiu et al., 1992; Mahaney and Grisham, 1992; Rousseau et al., 1993; Cornea and Thomas, 1994; Middleton et al., 1995; Negash et al., 1996; Wolkers et al., 1997; Schwarz et al., 1999; Marsh and Henderson, 2001), it has not permitted the determination of true rotational correlation times nor the rigorous testing of rotational diffusion models (Robinson and Dalton,

1980; Beth and Robinson, 1989). Both of these provide important insights into the size(s), and hence the oligomeric state(s) of membrane proteins and the nature and extent of their interactions with other proteins (e.g. Nigg and Cherry, 1980; Cherry and Godfrey, 1981; Hustedt and Beth, 1995; 2001).

The importance of developing computational approaches for analyzing experimental ST-EPR data was appreciated at the time of its discovery as discussed in several early papers (e.g. Thomas and McConnell, 1974; Robinson et al., 1974; Coffey et al., 1976; Thomas et al., 1976; Perkins et al., 1976). However, the limitations of computers at that time necessitated that a number of approximations be made, including axially symmetric magnetic tensors and simple isotropic rotational diffusion, in order to speed up the simulations. These approximations precluded the use of these early algorithms for rigorously analyzing the ST-EPR data from integral membrane proteins undergoing anisotropic rotational diffusion by optimization of the agreement between experiment and theory. However, this early work established two independent ways of treating the effects of rotational diffusion on ST-EPR spectra. The first, based on the work of McCalley et al. (1972), treats molecular motion in terms of the transition rate matrix approach (Thomas and McConnell, 1974; Thomas et al., 1976). The second is based on the orthogonal eigenfunction expansion approach (Robinson et al., 1974). Both of these basic approaches have been widely employed in the evolution of computational algorithms for simulation of ST-EPR spectra (Robinson and Dalton, 1980; 1981; Howard et al., 1993; Hustedt and Beth, 1995; 2001).

The phenomenal increase in computation speed and memory of modern computers, compared to what was available at the time of the development of ST-EPR, makes it possible to include full A- and g-tensor anisotropy, relevant relaxation rates, nonlinear microwave and modulation effects, anisotropic rotational diffusion, and spin label orientation in ST-EPR calculations while maintaining acceptable computation times even on inexpensive desktop personal computers. Computational algorithms, based upon the transition rate matrix approach, have been developed for unconstrained URD (Hustedt and Beth, 1995) and for constrained URD (Hustedt and Beth, 2001). These algorithms can be utilized to analyze single experiments or to globally analyze multiple ST-EPR spectra using automated nonlinear least squares methods to optimize the agreement between experiment and theory (Hustedt et al., 1993; Hustedt and Beth, 1995; 2001). These methods have been employed to globally analyze multiple ST-EPR spectra from the anion exchange protein (AE1) in human erythrocytes (Blackman et al., 2001) and from the EGF receptor in A-431 membrane vesicles (Stein et al., 2002) that were recorded at different microwave or different Zeeman modulation frequencies. These analyses have permitted testing of the adequacy of the URD model for rotational diffusion of membrane proteins, determination of the best fit values for the correlation time(s) for URD, constraints to the amplitude of URD, and the number of different motional species that are present. These data analysis methods, in combination with the computational algorithms based upon the orthogonal eigenfunction expansion approach that have been developed by Robinson and Dalton (1980; 1981), provide users with a fairly comprehensive set of tools that can be utilized to analyze ST-EPR data from intrinsic membrane proteins at a quantitative level. Using these methods, ST-EPR can provide detailed insights into the assembly and interactions of membrane proteins thereby providing an important complement to TOA methods for understanding the complex interactions and dynamics of membrane proteins.

3. OVERVIEW OF THEORY FOR CALCULATION OF ST-EPR SPECTRA

3.1 The Uniaxial Rotational Diffusion Model

The anisotropic arrangement of the lipid bilayer places severe constraints on the global rotational diffusion of integral membrane proteins since rotations about axes that are orthogonal to the membrane normal require the cooperative displacement of lipids. Therefore, most studies of rotational diffusion by TOA and by ST-EPR have utilized the URD model as a starting point for analysis of experimental data. Though there is not an extensive literature that evaluates the adequacy of the URD model for a wide range of intrinsic membrane proteins, the spectroscopic data from studies of global rotational diffusion of bacteriorhodopsin reconstituted in proteoliposomes (Cherry and Godfrey, 1981) and AE1 in erythrocyte membranes (Hustedt and Beth, 1995) can be accurately described using this model. The study by Cherry and Godfrey (1981), in particular, showed that the TOA data from bacteriorhodopsin reconstituted in liposomes at low protein/lipid ratios could be fit quantitatively with the URD model using the known orientation of the intrinsic chromophore relative to the URD axis. Moreover, the correlation time for global URD (20 μsec) that was determined was consistent with the intramembrane dimensions of the protein and the effective viscosity of the bilayer and in reasonable agreement with the theoretical predictions of Saffman and Delbrück (1975).

ST-EPR spectra provide the greatest sensitivity to the correlation time for rotational diffusion at those positions of the spectrum where the change in resonance condition (H_{res}) with a change in the orientation of the spin label with respect to the external magnetic field due to molecular reorientations is large (i.e. where $|\partial H_{res}/\partial \Omega|$ is maximum. Here, the variable Ω represents the polar angles between the spin label principal axes (x, y, and z) and the external field (Z); Thomas et al., 1976; Fajer and Marsh, 1983; Beth and Robinson, 1989). The plots in Figure 1 show the regions of the second harmonic out-of-phase absorption ST-EPR signal, denoted V_2', where sensitivity to rotational motions about the three spin label principal axes is maximized at microwave frequencies of 9.8, 34, and 94 GHz. These plots, which complement those published previously by Fajer and Marsh (1983) for X-band, provide qualitative insight into how URD would be manifested in the observed spectrum and why the orientation of the spin label reference frame relative to the URD axis, as defined in Figure 2, would be such an important determinant of lineshape. For example, if the nitroxide z-axis is aligned with the URD axis, rotational motion will only lead to interconversion of the nitroxide x- and y-axes with no motional averaging of the z-axis into the minor plane while alignment of the nitroxide x-axis with the URD axis leads to motional averaging of the nitroxide y- and z-axes.

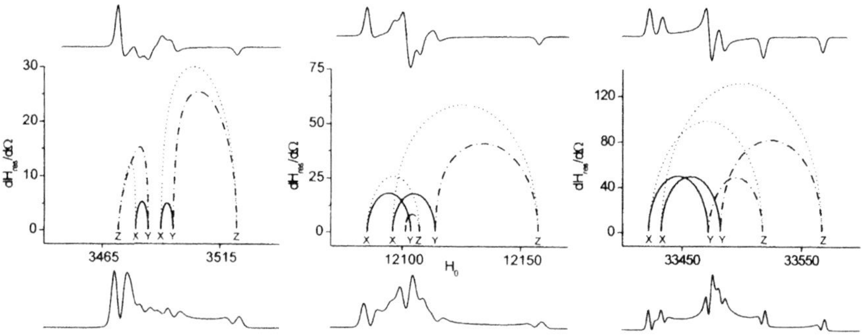

Figure 1. Top: Linear EPR signals at X- (*left panel*), Q- (*center panel*), and W-band (*right panel*) respectively calculated for a [^{15}N,^{2}H]-nitroxide spin label. *Middle*: Plots of $dH_{res}/d\Omega$ (Gauss/radian). *Bottom*: V_2' ST-EPR spectra calculated assuming $\tau = 1/6D = 10$ μsec; $\theta = 38°$; and $\phi = 0°$.

Beth and Robinson (1989) provided an extensive discussion of the sensitivity of ST-EPR signals to a variety of rotational diffusion models including URD at 3.0, 9.8, and 22 GHz. A generalization of these discussions is that the higher microwave frequencies (i.e. X- and K-bands) provide increased sensitivity to the details of anisotropic motion relative to the nearly axial character of spectra recorded at S-band, and below. V_2' signals at Q-band show excellent sensitivity to the correlation time for URD at all labeling geometries (Hustedt and Beth, 2001; Figure 4, below). As originally pointed out by Johnson and Hyde (1981), Q-band offers some unique capabilities for detailed studies of rotational diffusion due to the comparable anisotropy arising from the *g*- and *A*-tensor interactions. Commercial EPR spectrometers that operate at Q-band and that provide excellent signal-to-noise ratios with aqueous samples have permitted the use of this higher microwave frequency for ST-EPR studies on membrane proteins even at their natural levels in the cell membrane (Blackman et al., 2001). Though ST-EPR studies at W-band have not been explored experimentally to date, the plots of $|\partial H_{res}/\partial\Omega|$ at W-band in Figure 1 and calculated spectra (Hustedt and Beth, 2001; Figure 5, below) suggest that this frequency could provide outstanding sensitivity to the correlation time for URD regardless of the geometry of the spin label relative to the URD axis. Access to spectrometers that operate at Q- and W-band is increasing and this should facilitate the use of these higher microwave frequencies as a routine complement to X-band for detailed characterization of the dynamics of membrane proteins or to any systems that exhibit complex anisotropic rotational motions.

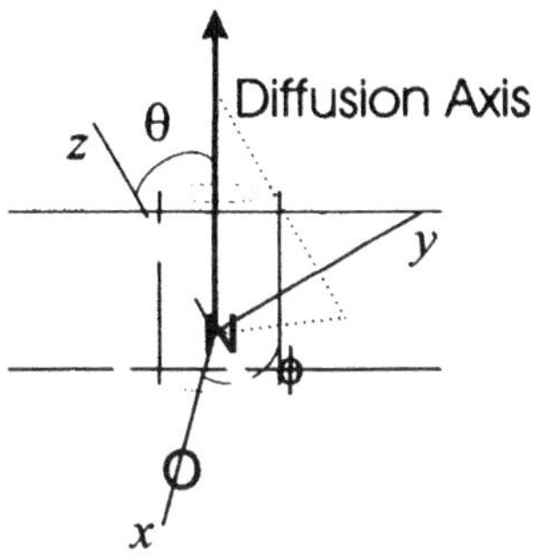

Figure 2. Orientation of the nitroxide with respect to the URD axis.

3.2 Transition Rate Matrix

ST-EPR utilizes nonlinear microwave power and Zeeman modulation amplitude to maximize sensitivity to slow rotational diffusion. Therefore, calculation of ST-EPR spectra requires treatment of not only the orientation dependent spin Hamiltonian and the stochastic reorientation processes, but also, the nonlinear microwave power and Zeeman modulation amplitude. Two approaches have been employed to simulate the effects of rotational motion on ST-EPR spectra. In the first approach, the spin dynamics are treated in terms of Bloch equations for the absorption, dispersion, and *z*-

components of the magnetization and the orientation variables are quantized to a grid of N sites.

$$\dot{\vec{M}}_n = \vec{M}_n \times \left\{ \gamma_e h_1 \hat{i} + \left[H(\Omega(t)) + \gamma_e h_m \cos \omega_m t \right] \hat{k} \right\} - \Gamma_R \left(\vec{M}_n - \vec{M}_n^{eq} \right) - \Gamma_\Omega \vec{M}_n \tag{1}$$

where γ_e is the electron gyromagnetic ratio; $H(\Omega(t))$ is the orientation dependent spin Hamiltonian; h_1 is the microwave amplitude; h_m is the Zeeman modulation amplitude; ω_m is the Zeeman modulation frequency; and Γ_R is the phenomenological spin relaxation operator. Rotational dynamics are modeled as hopping between nearest neighbor sites on the angular grid. The transition rate matrix, Γ_Ω, which determines the rates at which transitions between neighboring orientations occurs, is constructed based on the equilibrium orientation distribution. The application of the transition rate matrix approach is limited, for practical purposes, to models of rotational dynamics that are inherently one-dimensional such as isotropic rotational diffusion under the assumption that the nitroxide A- and g-tensors are axial (Thomas and McConnell, 1974; Thomas et al., 1976; Howard et al., 1993) or URD with nonaxial A- and g-tensors (Hustedt and Beth, 1995).

ST-EPR spectra are typically detected as that component of the absorption that is out-of-phase at the second harmonic of the Zeeman modulation frequency, denoted V_2', or that component of the dispersion that is out-of-phase at the first harmonic of the Zeeman modulation frequency, denoted U_1' (Hyde and Dalton, 1972; Thomas et al., 1976). Experimentally, both of these ST-EPR signals depend on the Zeeman modulation amplitude (Robinson, 1983) and on the modulation frequency (Hyde and Dalton, 1972; Beth and Robinson, 1989). In order to simulate these effects, Zeeman modulation must be explicitly included in the Bloch equations. The components of the magnetization are written as a Fourier series at harmonics of the Zeeman modulation frequency. The steady-state solution to the Bloch equations gives the Fourier coefficients which are the observable EPR signals. The Fourier coefficients at the kth harmonic depend on those at the $(k\text{-}1)^{th}$ harmonic and are also coupled to those at the $(k+1)^{th}$ harmonic. The back coupling of terms at one harmonic to those at the next lowest harmonic determine the nonlinear effects of Zeeman modulation.

In practice, the infinite series of modulation harmonics can be truncated at k_{max}, where k_{max} is defined operationally as the number of harmonics required for the calculated signal to converge to a constant lineshape as additional harmonics are included. Alternatively, as an approximation, all back coupling terms can be neglected. The simulated lineshape is then independent of Zeeman modulation amplitude and the Fourier coefficients at each harmonic are determined progressively. Computational algorithms using both of these approaches have been developed for unconstrained URD as described in detail in Hustedt and Beth (1995).

3.3 Orthogonal Eigenfunction Expansion

An alternative approach to the simulation of ST-EPR spectra has been developed by Robinson and Dalton (1980; 1981). Starting with the Stochastic Liouville Equation (SLE) one writes

$$\dot{\chi} = i\left[H(\Omega(t)) + \gamma_e h_1 + \gamma_e h_m \cos\omega_m t, \chi\right] - \Gamma_R \chi - \Gamma_\Omega \chi - i\left[H(\Omega(t)) + \gamma_e h_1 + \gamma_e h_m \cos\omega_m t, \sigma^0\right] \quad (2)$$

Individual elements of the reduced density matrix, $\chi = \sigma - \sigma^0$, are written in terms of the spherical harmonics, Y_{lm}, and as a Fourier series at harmonics of the Zeeman modulation frequency.

$$\chi = \sum_{k,l,m} \chi(k,l,m) Y_{lm} e^{ik\omega_m t} \quad (3)$$

The simulation of ST-EPR spectra using the eigenfunction expansion method is complicated by the fact that there are two indices in the expansion of χ which range to infinity. These are the index k over the harmonics of ω_m and the index l over the spherical harmonics. Treatment of the expansion in k is the same as discussed for the transition rate matrix approach above. The expansion over l is more problematic since the eigenfunction expansion is inherently numerically inefficient in the slow motion regime applicable to ST-EPR. The eigenfunction expansion approach converges to a constant lineshape very efficiently at low microwave frequencies (i.e. X-band, S-band, L-band). However, convergence requires the inclusion of much higher values of l and m as the microwave frequency is increased. Various extrapolation schemes have been implemented to determine, point-by-point, the number of l and m indices that are required for convergence. These extrapolations greatly speed up calculations since the calculated signals converge to a constant lineshape much more rapidly at field position where $|\partial H_{res}/\partial\Omega|$ approaches 0 (i.e. at the turning points) than they do at regions where $|\partial H_{res}/\partial\Omega|$ is large. In practice, the existing eigenfunction algorithms are very tractable at X-band microwave frequencies, and below. However, they can be problematic at Q-band and very difficult at W-band, particularly at correlation times near 1 msec.

Both the transition rate matrix approach (in the limit that the number of angles in the grid is infinite) and the eigenfunction expansion approach (in the limit that all l up to $l = \infty$ are included) will, in principle, yield identical results. In applying these approaches to the calculation of ST-EPR spectra, these limits are not met and other approximations have been made. In the transition rate matrix approach three significant approximations have been employed: 1) the use of reflective rather than cyclic boundary conditions to model the rotational dynamics; 2) the neglect of nitrogen nuclear spin-lattice relaxation time, T_{1n}; and 3) the use of an approximate treatment of the pseudosecular terms in the spin Hamitonian. Algorithms have been developed using the transition rate matrix approach to calculate ST-EPR spectra for a nitroxide undergoing URD which either include (Algorithm I) or neglect (Algorithm II) Zeeman overmodulation effects. A direct comparison between the eigenfunction expansion method developed by Robinson and coworkers and Algorithm II has shown that these three approximations do not significantly alter the calculated ST-EPR lineshape at correlation times greater than 1 μsec. Comparison of Algorithms I and II has shown that either algorithm can be used to obtain meaningful parameters describing the rotational dynamics (Hustedt and Beth, 1995).

3.4 Constrained URD

Linear EPR and ST-EPR are sensitive both to the rate of rotational dynamics and also to the RMS angular amplitude of motion when rotational dynamics are restricted. Just as the sensitivities of the two techniques to the rate of motion are different, so also are the sensitivities to the amplitude of rotational dynamics. The first work to directly address the sensitivity of ST-EPR spectra to the RMS amplitude of rotation for a restricted rotational diffusion model was done by Thomas and coworkers (Howard et al., 1993). In this work, an algorithm was developed, based upon the transition rate matrix approach, to treat the case of motion in a cone under the assumption of axially symmetric magnetic tensors. The calculations of Howard et al. (1993), carried out at X-band using 5 Gauss modulation amplitude and a modulation frequency of 50 kHz, established sensitivity limits of ST-EPR under these conditions for detection of restricted amplitude motion for the model considered. Three general conclusions can be drawn from this work. First, ST-EPR has limited sensitivity to the angular amplitude of motion. Under all the conditions tested by Howard et al. (1993), maximal effects were obtained for rotational motion that was constrained to an angular amplitude of 90°, or less. Second, the sensitivity of ST-EPR to the amplitude of angular motion decreases as the rate of motion decreases. Third, the effects on the ST-EPR lineshape of reducing the rate or the angular amplitude of diffusion are similar. Nonetheless, there are unique spectral features for each combination of rate and amplitude.

Methods previously developed to simulate ST-EPR spectra for unrestricted URD based on the transition rate matrix approach (Hustedt and Beth, 1995) have been adapted to treat URD within a square-well restriction

(Hustedt and Beth, 2001). URD in a square-well of width Δ is modeled as random jumps between equally spaced sites on an angular grid between $\phi_0 - \Delta/2$ and $\phi_0 + \Delta/2$. The final simulation is obtained as the sum for a number of different values of ϕ_0. Simulations have been performed for different values of Δ, τ, and θ at 9.8, 34, and 94 GHz (Hustedt and Beth, 2001). The results of these simulations for restricted amplitude URD, which are discussed in section 5.2, below, mirror the conclusions that were drawn from the diffusion in a cone model by Thomas and coworkers (Howard et al., 1993).

4. NONLINEAR LEAST SQUARES METHODS OF DATA ANALYSIS

With the enormous increase in the speed of computers since ST-EPR was first developed in the early 1970's, along with the accompanying decrease in their cost, it became practical in the past decade to adapt routines capable of simulating ST-EPR spectra into nonlinear least squares data analysis programs. Initially, the eigenfunction expansion algorithm developed by Robinson, Dalton, and coworkers (Dalton et al., 1976; Robinson and Dalton, 1980; 1981) was incorporated into the nonlinear least squares global analysis package developed by Beechem and colleagues for analyzing fluorescence data (Beechem, 1992). Using this approach, the ST-EPR spectra of labeled bovine serum albumin in glycerol/water mixtures were successfully analyzed according to an isotropic rotational diffusion model (Hustedt et al., 1993). Subsequently, transition rate matrix algorithms were developed and incorporated into this global analysis package (Hustedt and Beth, 1995; 2001). These algorithms have been employed in fitting the ST-EPR data obtained from spin labeled AE1 of the human erythrocyte (Hustedt and Beth, 1995; Blackman et al., 2001), and spin labeled EGF bound to the EGF receptor in A-431 cell membranes (Stein et al., 2002).

Two important features of the global analysis approach are that it allows the simultaneous analysis of data obtained at multiple modulation and microwave frequencies according to the same rotational diffusion model and it also allows determination of the uncertainties in fitting parameters that are recovered from the analyses. As it has been applied to data from ST-EPR experiments, the global analysis approach uses the Marquardt-Levenberg algorithm to perform the nonlinear least squares analysis. The integrated amplitude of a V_2' ST-EPR signal is a function of many factors including both the Zeeman modulation frequency and the rotational dynamics of the spin label (Dalton et al., 1976; Hyde and Dalton, 1979; Beth and Robinson, 1989). As a result, components with different rotational mobility, if present in the sample, will make a varying degree of contribution to the total V_2' ST-

EPR spectra collected at different modulation frequencies. Hustedt and Beth (1995) simultaneously analyzed spectra from spin labeled AE1 in erythrocyte membranes that were obtained at 100, 50, and 25 kHz Zeeman modulation frequencies. The simultaneous analysis of data at these multiple modulation frequencies strengthened the argument that nearly all copies of AE1 were undergoing URD at a rate that was consistent with an AE1 dimer. The influence of the *g*-tensor anisotropy scales directly with the microwave frequency. As a result, the separation between the *x*, *y*, and *z* turning points of the ST-EPR spectrum shifts dramatically with ω_0. Thus, the global analysis of data obtained at different ω_0 should aid in defining the relationship between the diffusion axis and the spin label for anisotropic rotational diffusion models such as URD. This is an important avenue for future applications of ST-EPR.

5. MODEL CALCULATIONS OF ST-EPR SPECTRA USING THE TRANSITION RATE MATRIX APPROACH

5.1 Sensitivity of ST-EPR Signals to τ_r, Label Geometry, and Microwave Frequency.

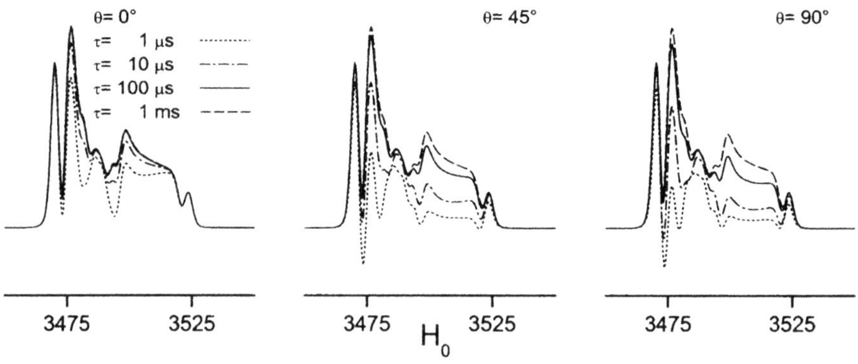

Figure 3. Calculated X-band V_2' ST-EPR spectra for a $[^{15}N,^2H]$-nitroxide undergoing URD as a function of $\tau = 1/6D = 1$ μsec (*dotted line*), $\tau = 10$ μsec (*dashed dotted line*), $\tau = 100$ μsec (*solid line*), and $\tau = 1$ msec (*dashed line*) for three different labeling geometries $\theta = 0°$ (*left panel*), $\theta = 45°$ (*middle panel*), and $\theta = 90°$ (*right panel*); $\phi = 0°$.

The computational algorithms that have been developed for unconstrained URD (Hustedt and Beth, 1995), described in section 3.2 of this chapter, can be utilized to predict the sensitivity of ST-EPR lineshapes to the correlation time for URD for a wide range of experimental parameters (microwave frequency and amplitude, modulation frequency and amplitude, harmonic and phase of the modulation, etc.) and for any selected geometry between the spin label reference frame and the URD axis. The calculated V_2' spectra in Figure 3 show the dramatic changes in lineshapes that occur as a function of correlation time at 9.8 GHz when the labeling geometry is θ= 0°, 45°, or 90°. These calculations, which show excellent agreement with previous calculations that were carried out using the eigenfunction expansion approach (Beth et al., 1983; Beth and Robinson, 1989), demonstrate that sensitivity to the correlation time for URD is minimal when the spin label *z*-axis is aligned with the diffusion axis and is maximum when the spin label *z*-axis is orthogonal to the diffusion axis. A major point to be made from these calculations is that the spectral shapes at, or near, major tilt angles of 0 or 90° deviate significantly from the lineshapes for isotopic rotational diffusion (i.e. there is no single correlation time for isotropic motion that will reproduce the spectra shown in Figure 3, left, and Figure 3, right, in the correlation time range from 1 to 100 μsec). At intermediate labeling geometries and at correlation times near 1 msec, however, the differences between URD and isotropic motion can be remarkably subtle (Hustedt and Beth, 1995).

The calculations shown in Figure 3 were carried out using *A*- and *g*-tensor values and line widths that are appropriate for high resolution [^{15}N,^{2}H] spin labels. However, the same sensitivities are predicted for conventional [^{14}N,^{1}H] spin labels (see Beth and Robinson, 1989, for an extensive comparison of lineshapes from ^{14}N- and ^{15}N-nitroxide spin labels). When carrying out spin labeling experiments on membrane proteins, it is generally not straightforward to predict *a priori* what the label geometry will be nor to alter it in those unfortunate cases where spectral sensitivity at X-band or lower microwave frequencies is minimized due to alignment, or near alignment, of the spin and diffusion tensors. When this situation is encountered, the higher microwave frequencies provide an increase in the minor element anisotropy due to the non-axial characteristics of the *g*-tensor and hence, increased sensitivity to motions that modulate the interconversion of the spin label *x*- and *y*-axes (see Beth and Robinson, 1989, for additional discussion of this point for ST-EPR studies at 3.0, 9.8, and 22 GHz).

Figure 4 demonstrates the high sensitivity of V_2' signals at Q-band to the correlation time for URD and the spin label geometry. In particular, Figure 4, left, illustrates the dramatic increase in sensitivity to motional averaging of the nitroxide *x*- and *y*-axes by URD compared with the modest sensitivity at X-band (Figure 3, left). At the intermediate labeling geometry of 45° (Figure 4, middle), monotonic lineshape changes are seen throughout the spectrum as the correlation time increases in the range from 1 μsec to 1 msec as reported previously for isotropic rotational diffusion (Johnson and Hyde, 1981). Alignment of the nitroxide *x*-axis with the URD axis (Figure 4, right) results in motional averaging of the *y*- and *z*-axes and large changes in lineshape in the high field end of the V_2' spectrum with almost no changes in the low field region. These results are entirely consistent with the predicted regions of spectral sensitivity in Figure 1, middle. These model calculations show that it is possible to get a qualitative sense of spin label geometry from simple inspection of the symmetry of the V_2' spectrum when using ^{15}N-spin labels at Q-band. Even though the situation is more complex with ^{14}N-spin labels due to the severe overlap of turning points from the three nuclear manifolds in the central region of the spectrum, Q-band still provides improved sensitivity to the anisotropy of motion and spin label geometry relative to X-band (unpublished calculations). Early work by Johnson and coworkers (Johnson et al., 1982a; 1982b) demonstrated the advantages of Q-band ST-EPR to observe the effects of anisotropic motion.

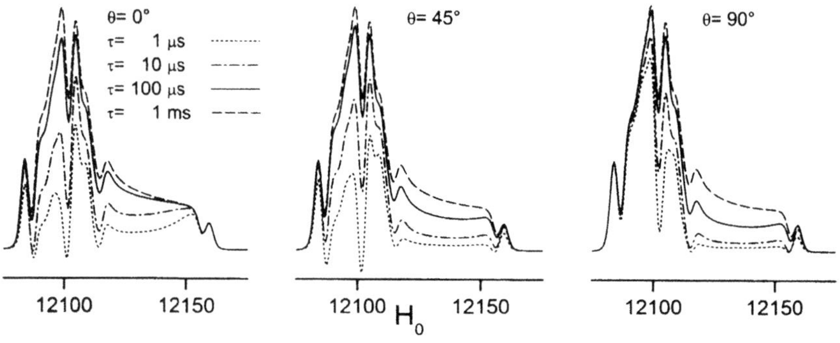

Figure 4. Calculated Q-band V_2' ST-EPR spectra for a [^{15}N,^{2}H]-nitroxide undergoing URD as a function of τ= 1 μsec (*dotted line*), τ= 10 μsec (*dashed dotted line*), τ= 100 μsec (*solid line*), and τ= 1 msec (*dashed line*) for θ= 0° (*left panel*), θ= 45° (*middle panel*), and θ= 90° (*right panel*); ϕ= 0°.

EPR spectrometers are now commercially available that provide excellent signal-to-noise ratio, stability, and reproducibility for carrying out ST-EPR experiments on aqueous samples at Q-band (e.g. Blackman et al., 2001). Given the significant improvements in sensitivity to details of anisotropic motion and spin label geometry, this microwave frequency should become a routine complement to X-band for detailed studies of rotational diffusion of any system undergoing anisotropic motion including the URD of intrinsic membrane proteins.

W-band has recently emerged as an important complement to X- and Q-band for making EPR measurements on spin labels in aqueous solutions and in membrane systems (e.g. Smirnov et al., 1995; Hustedt et al., 1997; Gaffney and Marsh, 1998; Smirnov et al., 1998; Rohrer et al., 2001; Mangels et al., 2001). The higher magnetic field at W-band leads to complete resolution of the six turning points of a ^{15}N-spin label and the *g*-tensor anisotropy is larger than the *A*-tensor anisotropy (see Figure 1, right). The calculated V_2' spectra in Figure 5 show that like Q-band, W-band provides impressive sensitivity both to the correlation time for URD and the spin label geometry relative to the URD axis. To date, ST-EPR studies at W-band have not been reported in the literature. Therefore, there are no experimental data available to compare with these calculated lineshapes. It should be noted that calculations at W-band are more demanding than at the lower microwave frequencies both because of the need to calculate spectra over a wider range of magnetic fields and to couple over a finer orientational grid in order to achieve convergence. Efforts to compare spectra calculated with the transition rate matrix approach with those calculated using the eigenfunction expansion approach have been problematic due to the extreme number of eigenstates that have to be included in the calculations to approach convergence and the resulting very long computation times required.

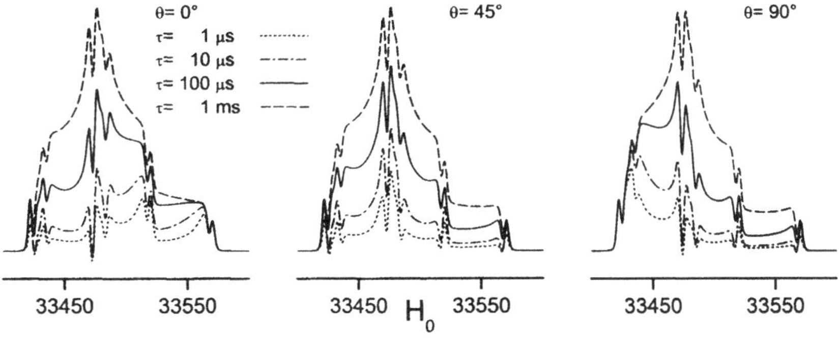

Figure 5. Calculated W-band V_2' ST-EPR spectra for a [^{15}N,^{2}H]-nitroxide undergoing URD as a function of τ= 1 μsec (*dotted line*), τ= 10 μsec (*dashed dotted line*), τ= 100 μsec (solid line), and τ= 1 msec (*dashed line*) for θ= 0° (*left panel*), θ= 45° (*middle panel*), and θ= 90° (*right panel*); ϕ= 0°.

The model calculations in Figure 5 indicate that ST-EPR studies at W-band could potentially provide a very useful avenue for investigation of membrane protein dynamics. However, it will be important to carefully choose systems that are amenable to investigation at this higher frequency due to sample volume limitations for aqueous samples using current resonator designs. For those systems that can be concentrated into a small volume (e.g. 80 nL) and still maintain sample integrity, studies carried out at W-band are predicted to yield very detailed information on rotational dynamics throughout the very slow motional regime.

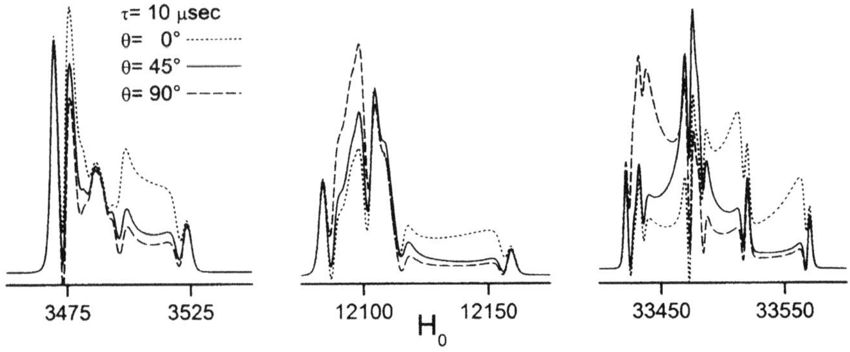

Figure 6. Calculated X-band (*left panel*), Q-band (*middle panel*), and W-band (*right panel*) V_2' ST-EPR spectra for a [^{15}N,^{2}H]-nitroxide undergoing URD with τ= 10 μsec as a function of labeling geometry. θ= 0° (*dotted line*), θ= 45° (*solid line*), and θ= 90° (*dashed line*); ϕ= 0°.

The calculated spectra in Figure 6 further illustrate the range of ST-EPR lineshapes that are predicted at X-, Q-, and W-bands for three different orientations of the spin label reference frame relative to the URD axis. These spectra were all calculated for a URD model using a correlation time of 10 μsec. At X-band, the overall lineshapes are remarkably similar at tilt angles of 90 and 45° (Figure 6, left). It is not until the angle approaches 0° that the spectrum begins to deviate significantly from a fairly isotropic appearance. The net effect of this lack of sensitivity is that different models will fit the experimental data with rather similar values of χ^2. At Q-band, there is a steady change in overall lineshape from 0 to 90° as shown in Figure 6, middle. This sensitivity to orientation of the spin label is even more dramatic at W-band as shown in Figure 6, right. The spectra at 0 and at 90° at these higher frequencies are diagnostic of the label geometry and they are absolutely distinct from any isotropic motion lineshape. Since 45° is near the magic angle, the spectra at this angle appear similar to those for isotropic motion at a longer correlation time as predicted (Robinson and Dalton, 1980; Beth and Robinson, 1989). However, the lineshapes deviate from an isotropic appearance at the magic angle more quickly as the observer frequency is increased. These data suggest that it should be possible to recover much more accurate values for the correlation time for URD and the angle between the spin label and the URD axis at the higher microwave frequencies using nonlinear least squares optimization between experiment and theory. Moreover, it should be possible to obtain better discrimination between different diffusion models at the higher frequencies except very near the magic angle.

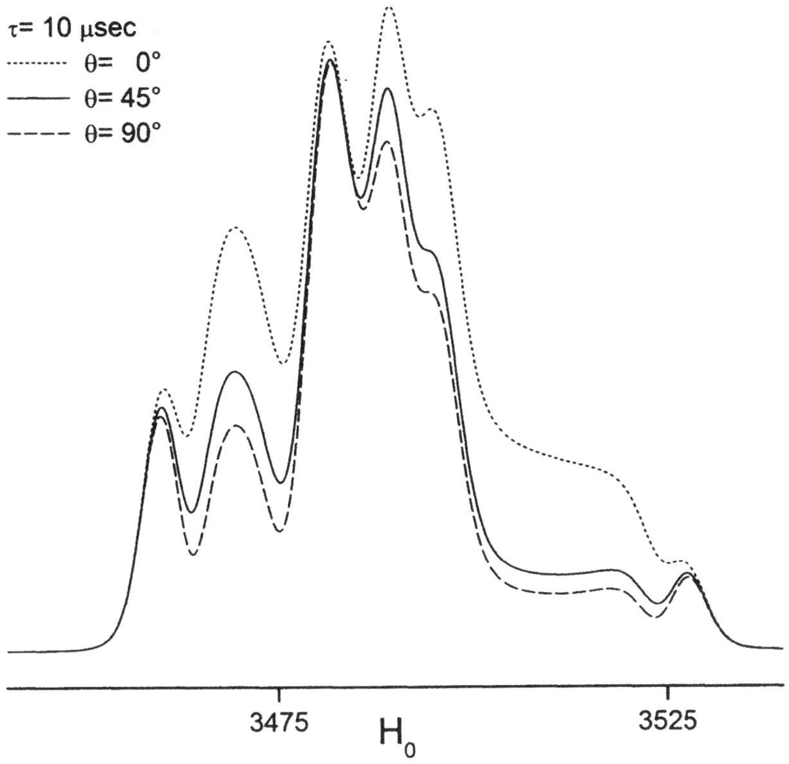

Figure 7. Calculated X-band V_2' ST-EPR spectra of a [^{14}N,H]-nitroxide undergoing URD with τ= 10 μsec as a function of labeling geometry. θ= 0° (*dotted line*), θ= 45° (*solid line*), θ= 90° (*dashed line*); ϕ= 0°.

Most of the ST-EPR studies that have been carried out to-date have utilized normal isotope [^{14}N,^{1}H]-spin labels and X-band. The improvements in spectral resolution and signal-to-noise ratio that are provided by the use of [^{15}N,^{2}H]-spin labels have been reviewed previously (Beth and Robinson, 1989) and will not be repeated in this chapter. However, it is instructive to make one direct comparison. The V_2' lineshapes in Figure 7 were calculated using tensor values and line widths that are appropriate for [^{14}N,^{1}H]-spin labels and the same motional model as in Figure 6. The major spectral change that is observed in going from 0° to 90° tilt of the spin label relative to the URD axis is apparent faster motion throughout the spectrum. Each of these calculated spectra has a remarkably isotropic appearance and each can be fit fairly convincingly using an isotropic rotational diffusion model and reasonable parameters for line width and a correlation time that is longer than the true correlation time for URD. This loss of discrimination between diffusion models represents a limitation for most studies of membrane protein dynamics. Going to higher microwave frequencies will improve the discrimination between different diffusion models. However, it will be more straightforward to analyze the data from [^{15}N,^{2}H]-spin labels and to obtain more precise values for all parameters recovered from the nonlinear least squares fitting of experimental data. In most applications, the increase in signal-to-noise ratio, the increased resolution of spectral features, the better resolution of spectral turning points, and the decreased computation times will offset the modest expense of the isotopically substituted probes.

5.2 Effects of Constrained URD on ST-EPR Lineshapes

Many intrinsic membrane proteins interact via their cytoplasmic domains with a variety of intracellular proteins including those that form the cytoskeleton. This type of interaction can restrict the global rotation of the transmembrane domain of an intrinsic membrane protein to an extent that will be determined by the flexibility of the linker region of the cytoplasmic domain (Nigg and Cherry, 1980; Matayoshi and Jovin, 1991; Blackman et al., 2001). When studies of rotational diffusion are carried out on membrane proteins in intact cell membranes, it is necessary to determine if there is a restriction in the amplitude of global URD in order to carry out a reliable analysis of experimental data.

Square-well restriction (Wahl, 1975) and harmonic-well restriction (Szabo, 1984) of URD has been considered for TOA. A restriction in the amplitude of URD results in an increase in r_∞ and a decrease in the apparent amplitudes of anisotropy decays with minimal effects on the rates of the decays (see Blackman et al., 2001). For most fluorophores and most label geometries, even weak restrictions (e.g. $\Delta \leq 180^\circ$ where Δ is the full width of a square-well restriction) produce measurable changes in $r_\infty$and there is excellent sensitivity to Δ in the range from 0 to approximately 100°.

Recent work has considered the effects of restricted amplitude URD on ST-EPR lineshapes (Hustedt and Beth, 2001). Figure 8 shows the dependence of the V_2' signal on the full width (Δ) of a square-well restriction at X- Q-, and W-band microwave frequencies for a URD model with $\tau = 10$ μsec and $\theta = 90^\circ$. Three general conclusions are supported by these calculations. First, at all three microwave frequencies, there is excellent sensitivity to Δ between 0 and 30°, limited sensitivity between 30 and 90°, and almost no sensitivity between 90 and 360°. Sensitivity to Δ depends weakly on the spin label geometry relative to the URD axis. However, it is highly dependent on the correlation time for URD (Hustedt and Beth, 2001). At a correlation time of 1 μsec, the V_2' signal shows measurable changes out to at least 90°. At a correlation time of 100 μsec, almost no changes in lineshape occur past 30°.

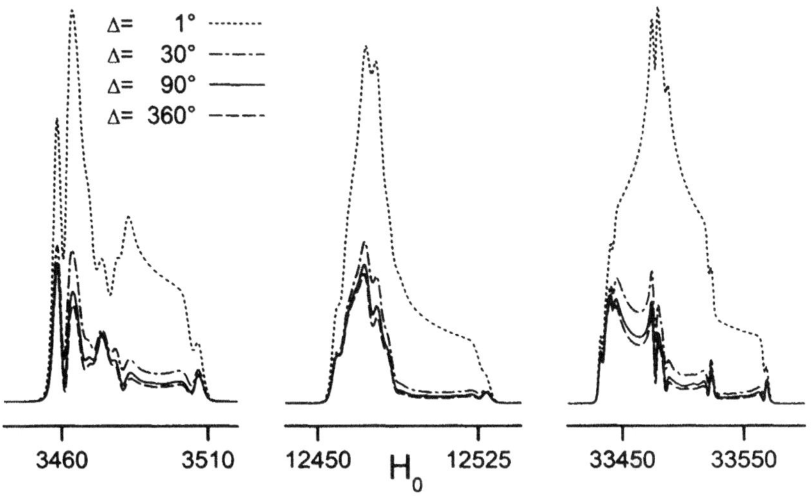

Figure 8. Calculated X-band (*left panel*), Q-band (*middle panel*), and W-band (*right panel*) V_2' ST-EPR spectra for a [^{15}N,^2H]-nitroxide undergoing restricted URD with τ= 10 μsec; θ= 90°; ϕ= 0°; and Δ= 1°(*dotted line*), Δ= 30° (*dashed dotted line*), Δ= 90° (*solid line*), and Δ= 360° (*dashed line*).

Second, Q- and W-band are only marginally more sensitive to large values of Δ than X-band. While this result might be unexpected, it should be noted that there has not been a systematic evaluation of the effects of various parameters, including modulation amplitude, on the sensitivity of these higher frequency measurements to Δ. The sensitivity of ST-EPR to rotational dynamics is determined, in part, by the amplitude of the Zeeman modulation. For most biological samples, large Zeeman modulation amplitudes (typically 5 Gauss) are required to collect V_2' spectra with adequate signal-to-noise ratios. The use of higher modulation amplitudes leads to spectral broadening and to loss of resolution of spectral features without additional useful increases in the signal-to-noise ratio or increased sensitivity to rotational dynamics at X-band. However, the increased spectral width at the higher frequencies means that fewer orientations are sampled during a modulation cycle. It may prove useful to evaluate the effects of modulation amplitude on the sensitivity of these higher frequency measurements to restricted amplitude motion in the future.

Third, at all three microwave frequencies, the V_2' signal changes dramatically for even small values of Δ. This is good news if one is interested in measuring a strong restriction on URD. However, it is troubling news with regards to the potential for unwanted contributions of local probe motion on V_2' lineshapes. One limitation of ST-EPR is that the signals are dominated by the fastest motional processes that are present. The calculations in Figure 8 and those in previous work (Howard et al., 1993; Hustedt and Beth, 2001) indicate that even a 10° local mode of motion at a correlation time of 1 μsec, or less, will give the appearance of significantly faster overall motion thereby compromising determination of the true global rotational diffusion of the system. Very little work has been directed at designing spin labels that are rigidly coupled to target proteins, particularly since the advent of site directed spin labeling approaches. This is an area that should be emphasized as methods for quantitative analysis of ST-EPR data evolve.

5.3 Opportunities for Further Advances in ST-EPR

ST-EPR lineshapes are dependent on the amplitudes of the microwave observer field and the Zeeman modulation field at the sample. Also, since they are detected 90° out-of-phase with respect to the field modulation and the in-phase signals are much larger in amplitude, they are very dependent on the phase of the modulation at the sample. Most X-band ST-EPR studies to-date have been carried out in conventional TE_{102} or TM_{110} cavities with sample volumes ranging from 10-50 μL (in capillaries) to >200 μL. (in agueous flat çells). The sample geometry normally employed gives rise to a distribution of microwave and modulation fields and to phase shifts over the active dimensions of the sample (see Fajer and Marsh, 1982; Hemminga and de Jager, 1989). Figure 9 shows the experimentally measured values for each of these parameters over a 3.2 cm long aqueous sample positioned vertically in a commercial Bruker TM_{110} resonator. Also shown are the calculated V_2' signals from a point sample positioned at the center of the cavity (defined operationally as the position where h_1 and h_m are maximum) and the calculated composite signal that explicitly accounts for each of these distributions. These calculations show that there can be discernable distortions of the experimental lineshape even with a moderate sample size. For samples that are amenable, such as membrane proteins reconstituted in liposomes at high effective concentrations, loop-gap resonators may provide better characteristics due to the smaller sample size. Though extensive studies have not been carried out in commercial TE_{011} resonators at Q-band, preliminary work has indicated that the distribution of modulation amplitudes at this higher frequency is appreciable, perhaps worse than at X-band, even with the small sample volumes normally employed.

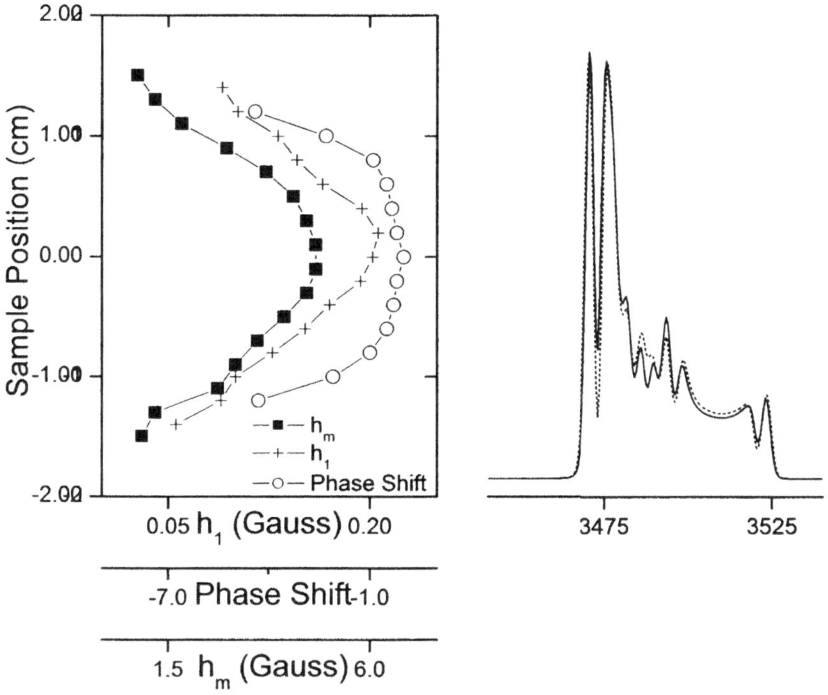

Figure 9. Left: Variation of h_1 (*plus signs)*, h_m (*solid squares*), and modulation phase (*open circles)* as a function of sample position in a TM_{110} cavity. These values were measured using a point sample of PADS in a 50 μL capillary. *Right:* Calculated V_2' ST-EPR spectra for a point sample at the center position (*solid line*) and a composite signal calculated for a 3.2 cm long sample (*dotted line*); τ= 10 μsec; θ= 38°; ϕ= 0°.

In theory, it would be straightforward to include each of these distributions in the nonlinear least squares fitting of experimental data. However, this added complexity results in a very demanding computational problem given the number of additional lineshapes that have to be calculated at each iteration. Recent work by Hyde and coworkers (Mett et al., 2001; Anderson et al., 2002) has shown that alternative cavity structures can be designed that provide a homogeneous h_1 field over an extended sample volume. These structures may prove extremely useful for future ST-EPR applications and for other EPR investigations that require nonlinear h_1 fields.

Loop-gap resonators also have very favorable h_1 field characteristics for ST-EPR (reviewed in Hyde and Froncisz, 1989). It remains an important goal to design modulation coils that will improve h_m field and phase homogeneity over the dimensions of most samples that are studied using ST-EPR. Until further advances are made with regards to homogeneity of each of these parameters, there will be some systematic error in determining correlation times from nonlinear least squares fitting of experimental data and some loss of discrimination in comparing the accuracy of different rotational diffusion models.

Previous work has shown that global analysis of V_2' ST-EPR data sets obtained at different microwave and modulation frequencies enables determination of rotational diffusion coefficients for membrane proteins, testing of the adequacy of different rotational diffusion models, determination of the number of different overlapping motional species that are present, and assessment of the confidence levels for each recovered parameter (e.g. Hustedt and Beth, 1995; 2001; Blackman et al., 2001; Stein et al., 2002). One of the strengths of global analysis is the ability to simultaneously optimize the fit of a large number of different experiments (see Beechem, 1992; Hustedt et al., 1993). Previous work has shown that several EPR signals, both in- and out-of-phase and at the first and second harmonics of the modulation as well as dispersion signals, are sensitive to the correlation time for rotational diffusion (e.g Perkins et al., 1976). The V_2' signal has been utilized extensively in past ST-EPR studies due to its sensitivity to rotational dynamics in the μsec to msec correlation time range, its favorable signal-to-noise ratio relative to the U_1' signal when using a conventional cavity, and the capability of most commercial spectrometers to record this signal. Recent advances including time-locked subsampling (TLSS; Hyde et al., 1998) have enabled the simultaneous recording of all eight unique signals at the first and second harmonics of the modulation with preservation of the relative amplitudes of the individual signals. Given that the computational approaches that have been developed also calculate these same signals (Robinson and Dalton, 1980; Hustedt and Beth, 1995; 2001), it seems logical to explore what additional information can be gained from global analysis of all eight signals and their relative amplitudes rather than discarding all of the signals except V_2'. Recent advances in resonator design and in the development of low phase-noise microwave oscillators permit the routine detection of dispersion signals with excellent signal-to-noise ratios as described in detail in the chapter by Hyde in this volume (ch. 13).

Digital detection also provides a convenient way to overcome problems of proper selection of the modulation phase for detecting true out-of-phase signals (U_1', V_2', etc.) Automated methods have been developed that permit post-processing of the data for accurate phasing of the signals once they have been recorded digitally (Auteri et al., 1988). This advance circumvents the need to carry out time-consuming self-nulling of the signals (see Thomas et al., 1976) prior to phase sensitive detection.

There are a number of intrinsic membrane proteins that contain only a single membrane spanning α-helix. These include proteins like the glycophorins in erythrocytes and the growth factor receptors including the ErB family. The rotational motions of these proteins have been difficult to measure due to the fact that they exhibit correlation times in the sub-microsecond time window. Rotational motions in the 100 nsec to 1 μsec time window are difficult to measure using conventional time-resolved optical approaches due to the nsec fluorescent lifetimes of most fluorophores and the difficulties of PMT gating and rejection of photons from the tail of the laser pulse in phosphorescence experiments (e.g. Herman et al., 1992). Linear EPR provides very limited sensitivity to motion in this time window (McCalley et al., 1972). Using standard 50 kHz field modulation, the V_2' ST-EPR signal also shows very limited sensitivity to the correlation time for isotropic motion in this range (Thomas et al., 1976) and URD with modulation frequencies up to 100 kHz shows low sensitivity (Stein et al., 2002). While the U_1' ST-EPR signal shows good sensitivity to motion in this range (Hyde and Dalton, 1972), it has a greatly reduced amplitude relative to that observed at correlation times longer than 1 μsec. This reduction in amplitude, coupled with the difficulty in obtaining U_1' ST-EPR signals using conventional cavities at X-band, has precluded its use in most studies of membrane protein dynamics. Even though ST-EPR can provide a signal-to-noise ratio that is comparable to linear EPR, most applications of the technique in cells and in intact membrane preparations require long data acquisition times to obtain high quality data due to the low concentration of protein in a loosely packed suspension of cells or membranes (even though the concentration within the bilayer is high, the concentration per unit volume of cells or intact membranes is low due to the large intracellular and extracellular aqueous compartments that contain no protein).

The calculated spectra in Figure 10 suggest that it might be possible to increase the sensitivity of the V_2' signal to motion in this time window by utilizing higher modulation frequencies. Currently available commercial spectrometers do not provide the capability of recording V_2' signals at frequencies higher than 100 kHz. Undoubtedly, this has limited experimental testing of the potential advantages of higher modulation frequencies in previous applications of the technique.

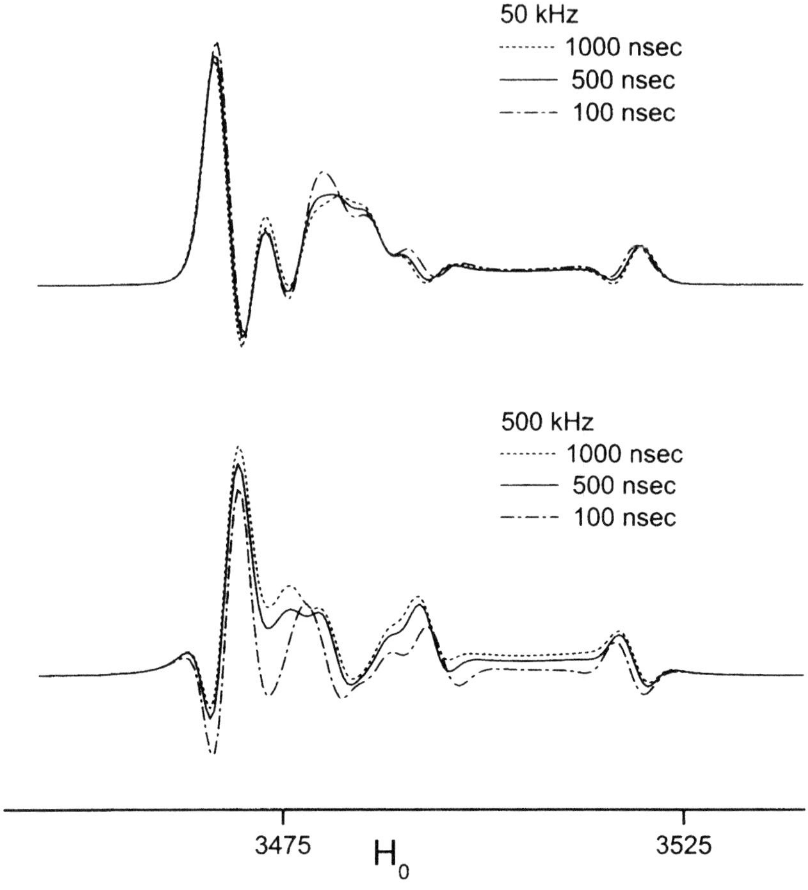

Figure 10. Calculated V_2' X-band ST-EPR spectra for a [$^{15}N,^2H$]-nitroxide undergoing URD with $\theta = 90°$, $\phi = 0°$, and $\tau = 1000$ nsec (*dotted line*), $\tau = 500$ nsec (*solid line*), and $\tau = 100$ nsec (*dashed dotted line*) for modulation frequencies of $\upsilon_m = 50$ kHz (*upper panel*) and $\upsilon_m = 500$ kHz (*lower panel*).

6. APPLICATIONS OF ST-EPR TO MEMBRANE PROTEINS

6.1 Applications Based upon Isotropic Model System Analyses

ST-EPR has been utilized to characterize the rotational dynamics of a wide range of intrinsic and peripheral membrane proteins including: Ca-ATPase in sarcoplasmic reticulum, rhodopsin, mitochondrial electron transport enzymes, cytochrome P-450, acetyl choline receptor, and a number of erythrocyte membrane proteins including band 3 (AE1), ankyrin, spectrin, hemoglobin, and GAPDH (see Thomas, 1985, for a comprehensive review of these early studies). Since that time, studies have been reported on a variety of membrane proteins including: erythrocyte anion exchange protein (AE1; Beth et al., 1986; Anjaneyulu et al., 1988; 1989; Cobb et al., 1990); cytochrome c reductase (Gwak et al., 1986); Ca-ATPase (e.g. Bigelow and Thomas, 1987; Squier et al., 1988; Squier et al., 1988; Lewis and Thomas, 1991; Mahaney and Thomas, 1991; Cornea and Thomas, 1994; Negash et al., 1996); Na,K-ATPase (Mahaney et al., 1990; Mahaney and Grisham, 1992; Esmann et al., 1992); cytochrome P-450's (Schwarz et al., 1990; 1999); cytochrome c oxidase (Qiu et al., 1992); EGF receptor (Rousseau et al., 1993); H/K-ATPase (Middleton et al., 1995); spectrin (Fung et al., 1996); bacteriophage M13 coat protein (Wolkers et al., 1997); and GalP (Marsh and Henderson, 2001). Effective rotational correlation times for the proteins examined in these studies have been determined by comparison of motion sensitive ratio parameters with those obtained from isotropic model systems such as spin labeled hemoglobin (Thomas et al., 1976) or spin labeled GAPDH (Beth et al., 1981) or by fitting experimental ST-EPR spectra based on an isotropic rotational diffusion model (Rousseau et al., 1993). Collectively, these studies have demonstrated the very slow rotational diffusion of membrane proteins that has been predicted by hydrodynamic theory (Saffman and Delbruck, 1975; Jahnig, 1986) and they have provided important qualitative insights into changes in rotational dynamics that are related to biological function.

6.2 Applications of Computational Approaches to the Analysis of ST-EPR Data Obtained From Membrane Proteins

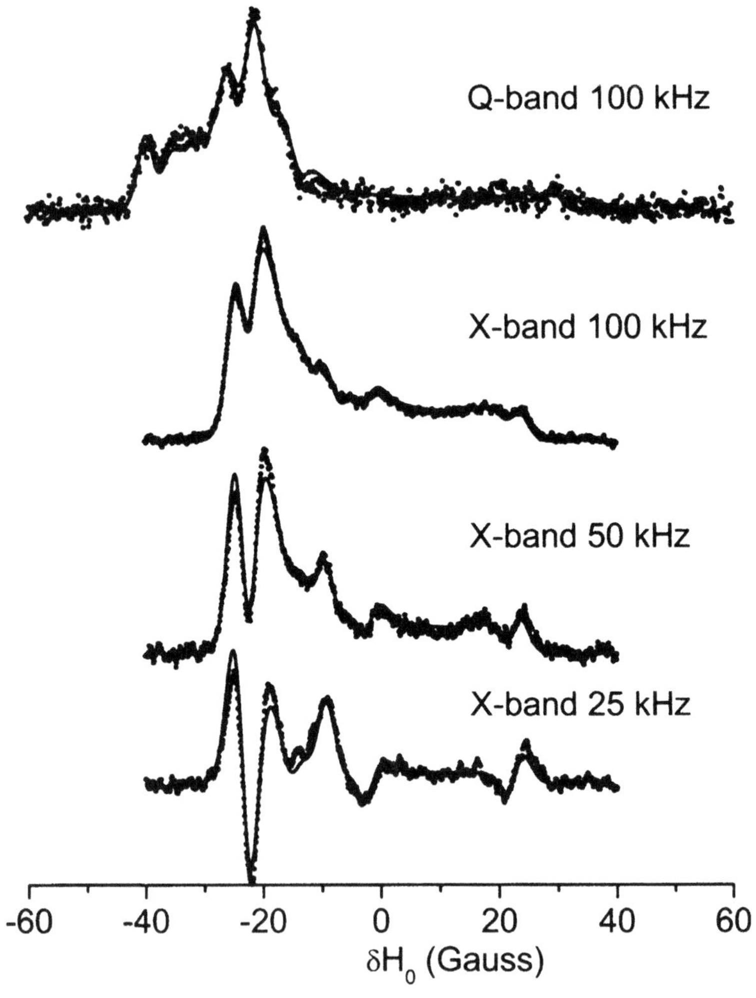

Figure 11. Experimental V_2' ST-EPR spectra (*dots*) obtained from spin labeled AE1 in ghost membranes at Q-band and υ_m= 100 kHz (*upper panel*), X-band and υ_m= 100 kHz (*second panel*), X-band and υ_m= 50 kHz (*third panel*), X-band and υ_m= 25 kHz (*bottom panel*). The overlaid best fit calculations (*solid lines*) were obtained for τ= 13 μsec, θ= 39°, and ϕ= 68° (see Blackman et al. 2001).

Three recent ST-EPR studies of membrane protein rotational dynamics have taken advantage of the currently available computational algorithms to analyze experimental data. In the first study, Hustedt and Beth (1995) utilized the global analysis approach (Hustedt et al., 1993) to determine the simultaneous best-fit of the V_2' ST-EPR signals obtained at modulation frequencies of 25, 50, and 100 kHz from spin labeled AE1 (band 3) in human erythrocytes. The analyses indicated that the data at all three modulation frequencies could be simultaneously fit with a URD model with a single rotational correlation time of 13 μsec and a tilt of the spin label relative to the URD axis of 39°. Attempts to improve the statistics (χ^2) of the fits by utilizing more than one URD population were unsuccessful. At the time, these results were controversial since previous TOA studies had suggested that multiple rotational sub-populations of AE1 existed in the erythrocyte membrane and that as much as 75% of AE1 was highly clustered and characterized by rotational correlation times in the msec time range or longer (e.g. Corbett et al., 1994; see also Matayoshi and Jovin, 1991; Blackman et al., 1998; Blackman et al., 2001). In subsequent work, Hustedt and Beth (1996) determined the orientation of the spin label employed in these studies relative to the membrane normal axis by orienting erythrocytes by flow in an EPR flat cell. These studies showed that the *z*-axis of the nitroxide was tilted from the membrane normal axis by 38°. Since URD of intrinsic membrane proteins is predicted to occur about an axis that is parallel with the membrane normal, this result reinforced the conclusions obtained from analysis of the ST-EPR data (Hustedt and Beth, 1995). More recent work has shown that global analysis of X-band data obtained at modulation frequencies of 25, 50, and 100 kHz along with Q-band data obtained at a modulation frequency of 100 kHz is best fit by a single unconstrained URD model with a correlation time of 13 μsec and a major tilt angle of 39° (Figure 11) thereby adding additional support to the previous data analysis carried out at X-band alone.

In the second study, the data from spin labeled AE1 was reevaluated using a constrained URD model (Blackman et al., 2001). Previous TOA studies had suggested that a subpopulation of AE1 might exhibit restricted amplitude rotational diffusion due to interaction of its cytoplasmic domain with the membrane skeleton (e.g. Nigg and Cherry, 1980). In these studies, a major change in anisotropy decay (e.g. a significant reduction in r_∞) was observed following proteolytic cleavage of the link between the transmembane and cytoplasmic domains of AE1, a result that was consistent with a release of a constraint on the amplitude of URD of that subpopulation of AE1 that was interacting with the membrane skeleton (see also Matayoshi and Jovin, 1991; Blackman et al., 1996; 2001). Blackman et al (2001) determined that: 1) 28 % of the copies of AE1 interact with the membrane skeleton in intact erythrocyte membranes; 2) there was no detectable change in the V_2' ST-EPR signals recorded at X- and Q-band microwave frequencies following cleavage of the cytoplasmic domain with trypsin; and 3) these ST-EPR results are entirely consistent with a weak constraint of 71° that was determined from the analysis of the TOA data.

In the third study, Stein et al (2002) employed X-band ST-EPR to characterize the rotational diffusion of spin labeled EGF (epidermal growth factor) bound to its receptor in A-431 membrane vesicles. Spectra recorded at modulation frequencies of 60, 80, and 100 kHz were globally analyzed using a one or a two component unconstrained URD model. These analyses indicated that the experimental data was best fit with a two-component model characterized by correlation times of 0.12 and 0.78 μsec and with different orientations of the spin label relative to the URD axis. The correlation times recovered from the analyses were reasonably consistent with those predicted for monomeric and dimeric EGF receptors and they provided a possible explanation for the known existence of high and low affinity receptors in the membrane.

These three studies demonstrate that the computational algorithms described in this chapter can be employed to analyze experimental ST-EPR data obtained from membrane proteins. These analyses have shown that it is possible to extract rotational correlation times from the data and that the adequacy of rotational diffusion models can be evaluated. With these computational tools, ST-EPR provides an important complement to TOA for detailed studies of the rotational diffusion of membrane proteins. These same computational tools should also be useful for analyzing the experimental data from other systems that exhibit URD.

6.3 Future Prospects

The past decade has seen an amazing emphasis on determining the static structures of intrinsic membrane proteins using a wide range of experimental techniques. This emphasis will undoubtedly continue in the next decade given the central role that membrane proteins play in regulating cellular functions under normal physiological conditions and the number of membrane proteins that are candidates as specific drug targets in a wide range of pathological settings. As the data base of atomic resolution static structures of membrane proteins expands, studies of their functional dynamics and their interactions within cell membranes will take on added importance. The advances in site-directed spin labeling of membrane proteins that have taken place in the past decade (for recent reviews see: Hubbell et al., 1996; Feix and Klug, 1998; Hustedt and Beth, 1999; Hubbell et al., 2000; Columbus and Hubbell, 2002) places ST-EPR in a unique position to contribute to characterizing the dynamics of structural transitions of membrane proteins that are related to their functions and to defining important protein-protein interactions. The μsec to msec time range not only corresponds to the correlation time range for global rotational diffusion of membrane proteins but it also corresponds to the time range for functionally important structural transitions of a large number of membrane proteins including many that facilitate the transport or exchange of substrates across the bilayer and those that activate intracellular signal transduction pathways.

Given its sensitivity to rotational dynamics in the very slow motional range, the excellent signal-to-noise ratio, the existence of facile methods for site-directed spin labeling, and the availability of computational methods for analysis of experimental data using a variety of different diffusion models, ST-EPR is well positioned to contribute in a substantial way to understanding the functional dynamics of membrane proteins in the next decade and beyond. Little work has been done to date that takes advantage of site-directed spin labeling for optimal positioning of probes to characterize functional dynamics of membrane proteins or even of soluble proteins. There are many opportunities for elegant work in this area.

There are also opportunities for additional improvements in hardware including the development of resonators that have uniform h_l and h_m fields over the active dimensions of the sample and access to a wider range of modulation frequencies in the 100 to 500 kHz range for characterizing the rotational motions of membrane proteins including cell surface receptors that have only a single transmembrane α-helix. In the area of data analysis, there are many opportunities for explicit inclusion of relaxation rates in computational algorithms and for refining existing global analysis methods for simultaneously analyzing ST-EPR data sets collected at different microwave and modulation frequencies. In the area of spin label design and synthesis, it would be very advantageous to develop probes that would attach rigidly to elements of defined secondary structure such as α-helices and β-sheets in order to overcome problems with local probe motion contributing to the observed ST-EPR signals.

ACKNOWLEDGEMENTS

This work supported by NIH R37 HL34737 (AHB) and NIH RO1 GM 60538 (EJH). The authors wish to thank Hassane Mchaourab for helpful discussions and for critical reading of this chapter.

7. REFERENCES

Anderson, J.R., Mett, R.R., and Hyde, J.S. (2002). Cavities with axially uniform fields for use in electron paramagnetic resonance. II. Free space generalization. *Rev. Sci. Instrum.*, **73**, 3027-3037.

Anderson, R.G.W., and Jacobson, K. (2002). A role for lipid shells in targeting proteins to caveoli, rafts, and other lipid domains. *Science*, **296**, 1821-1825.

Anjaneyulu, P.S.R., Beth, A.H., Sweetman, B.J., Faulkner, L.A., and Staros, J.V. (1988). Bis(Sulfo-N-succinimidyl)-[$^{15}N,^{2}H_{16}$]-doxyl-2-spiro-4'-pimelate, a stable isotope-substituted, membrane-impermeant bifunctional spin label for studies of the dynamics of membrane proteins: Application to the anion exchange channel in intact human erythrocytes. *Biochemistry* **27**, 6844-6851.

Anjaneyulu, P.S.R., Beth, A.H., Cobb, C.E., Juliao, S.F., Sweetman, B.J., and Staros, J.V. (1989). Bis(sulfo-N-succinimidyl) doxyl-2-spiro-5'-azelate: Synthesis, characterization, and reaction with the anion-exchange channel in intact human erythrocytes. *Biochemistry* **28**, 6583-6590.

Auteri, F.P., Beth, A.H., and Robinson, B.H. (1988). Finding the proper phase in CW-EPR spectroscopy: An automated method. *J. Magn. Reson.* **80**, 493-501.

Beechem, J.M. (1992). Global analysis of biochemical and biophysical data. *Methods Enzymol.* **210**, 37-54.

Belford, G.G., Belford, R.L., and Weber, G. (1972). Dynamics of fluorescence polarization in macromolecules. *Proc. Natl. Acad. Sci. USA*, **69**, 1392-1393.

Beth, A.H., Venkataramu, S.D., Balasubramanian, K., Dalton, L.R., Robinson, B.H., Pearson, D.E., Park, C.R. and Park, J.H. (1981). ^{15}N-and ^{2}H-substituted maleimide spin labels: Improved sensitivity and resolution for biological EPR studies. *Proc. Natl. Acad. Sci.* **78**, 967-971.

Beth, A.H., Balasubramanian, K., Robinson, B.H., Dalton, L.R., Venkataramu, S.D., and Park, J.H. (1983). Sensitivity of V_2' saturation transfer electron paramagnetic resonance signals to anisotropic rotational diffusion with [15N]nitroxide spin-labels. Effects of noncoincident magnetic and diffusion tensor principal axes. *J. Phys. Chem.* **87**, 359-367.

Beth, A.H., Conturo, T.E., Venkataramu, S.D., and Staros, J.V. (1986). Dynamics and interactions of the anion channel in intact human erythrocytes: An electron paramagnetic resonance spectroscopic study employing a new membrane-impermeant bifunctional spin label. *Biochemistry* **25**, 3824-3832.

Beth, A.H., and Robinson, B.H. (1989). Nitrogen-15 and deuterium substituted spin labels for studies of very slow rotational motion. *In: Biological Magnetic Resonance, Vol. 8, Spin Labeling - Theory and Applications*, L.J. Berliner and J. Reuben, eds., Plenum Press, New York, pp. 179-253.

Bigelow, D.J., and Thomas, D.D. (1987). Rotational dynamics of lipid and the Ca-ATPase in sarcoplasmic reticulum. The molecular basis of activation by diethyl ether. *J. Biol. Chem.* **262**, 13449-13456.

Blackman, S.M., Cobb, C.E., Beth, A.H., and Piston, D.W. (1996). The orientation of eosin-5-maleimide on human erythrocyte band 3 measured by fluorescence polarization microscopy. *Biophys. J.* **71**, 194-208.

Blackman, S.M., Piston, D.W., and Beth, A.H. (1998). Oligomeric state of human erythrocyte band 3 measured by fluorescence resonance energy homotransfer. *Biophys. J.* **75**, 1117-1130.

Blackman, S.M., Hustedt, E.J., Cobb, C.E., and Beth, A.H. (2001). Flexibility of the cytoplasmic domain of the anion exchange protein, band 3, in human erythrocytes. *Biophys. J.* **81**, 3363-3376.

Borbat, P.P., Costa-Filho, A.J., Earle, K.A., Moscicki, J.K., and Freed, J.H. (2001). Electron spin resonance in studies of membranes and proteins. *Science* **291**, 266-269.

Cherry, R.J., Cogoli, A., Oppliger, M., Schneider, G., and Semenza, G. (1976). A spectroscopic technique for measuring slow rotational diffusion of macromolecules. 1. Preparation and properties of a triplet probe. *Biochemistry* **15**, 3653-3656.

Cherry, R.J., and Schneider, G. (1976). A spectroscopic technique for measuring slow rotational diffusion of macromolecules. 2. Determination of rotational correlation times of proteins in solution. *Biochemistry* **15**, 3657-3661.

Cherry, R.J. (1978). Measurement of protein rotational diffusion in membranes by flash photolysis. *Methods Enzymol.* **54**, 47-61.

Cherry, R.J. (1979). Rotational and lateral diffusion of membrane proteins. *Biochim. Biophys. Acta* **559**, 289-327.

Cherry, R.J., Smith, P.R., Morrison, I.E., and Fernandez, N. (1998). Mobility of cell surface receptors: a re-evaluation. *FEBS Lett.* **430**, 88-91.

Cherry, R.J., and Godfrey, R.E. (1981). Anisotropic rotation of bacteriorhodopsin in lipid membranes. Comparison of theory with experiment. *Biophys. J.* **36**, 257-276.

Chuang, T.L., and Eisenthal, K.B. (1972). Theory of fluorescence depolarization by anisotropic rotational diffusion. *J. Chem. Phys.* **57**, 5094-5097.

Cobb, C.E., Juliao, S., Balasubramanian, K., Staros, J.V., and Beth, A.H. (1990). Effects of diethyl ether on membrane lipid ordering and on rotational dynamics of the anion exchange protein in intact human erythrocytes: Correlations with anion exchange function. *Biochemistry* **29**, 10799-10806.

Coffey, P., Robinson, B.H., and Dalton, L.R. (1976). Rapid computer simulation of ESR spectra. II. Saturation transfer spectroscopy of axially symmetric ^{14}N-nitroxide spin labels. *Mol. Phys.* **31**, 1703-1715.

Columbus, L., and Hubbell, W.L. (2002). A new spin on protein dynamics. *TIBS* **27**, 288-295.

Cone, R.A. (1972). Rotational diffusion of rhodopsin in the visual receptor membrane. *Nature New Biol.* **236**, 39-43.

Corbett, J.D., Agre, P., Palek, J., and Golan, D.E. (1994). Differential control of band 3 lateral and rotational mobility in intact red cells. *J. Clin. Invest.* **94**, 683-688.

Cornea, R.L., and Thomas, D.D. (1994). Effects of membrane thickness on the molecular dynamics and enzymatic activity of reconstituted Ca-ATPase. *Biochemistry* **33**, 2912-2920.

Dalton, L.R., B.H. Robinson, L.A. Dalton, and P. Coffey. (1976). Saturation transfer spectroscopy, in *Advances in Magnetic Resonance*, Volume 8, J.S. Waugh, ed.. Academic Press. New York, pp. 149-249.

Edidin, M. (1974). Rotational and translational diffusion in membranes. *Ann. Rev. Biophys. Bioeng.* **3**, 179-201.

Ehrenberg, M., and Rigler, R. (1972). Polarized fluorescence and rotational Brownian motion. *Chem. Phys. Lett.* **14**, 539-544.

Essman, M., Hideg, K., and Marsh, D. (1992). Conventional and saturation transfer EPR spectroscopy of Na+/K(+)-ATPase modified with different maleimide-nitroxide derivatives. *Biochim. Biophys. Acta* **1159**, 51-59.

Fajer, P., and Marsh, D. (1982). Microwave and modulation field inhomogeneities and the effect of cavity Q in saturation transfer ESR spectra. Dependence on sample size. *J. Magn. Reson.* **49**, 212-224.

Fajer, P., and Marsh, D. (1983). Sensitivity of saturation transfer ESR spectra to anisotropic rotation. Application to membrane systems. *J. Magn. Reson.* **51**, 446-459.

Feix, J.B., and Klug, C.S. (1998). Site-directed spin labeling of membrane proteins and peptide-membrane interactions. *In: Biological Magnetic Resonance, Volume 14: Spin Labeling: The Next Millennium*, L.J. Berliner, Ed., Plenum Press, New York, 251-281.

Frye, L.D., and Edidin, M. (1970). The rapid intermixing of cell surface antigens after formation of mouse-human heterokaryons. *J. Cell Sci.* **7**, 319-335.

Fung, L.W., Kalaw, B.O., Hatfield, R.M., and Dias, M.N. (1996). Erythrocyte spectrin maintains its segmental motions on oxidation: a spin-label EPR study. *Biophys. J.* **70**, 841-851.

Gaffney, B.J., and Marsh, D. (1998). High-frequency, spin-label EPR of nonaxial lipid ordering and motion in cholesterol-containing membranes. *Proc. Natl. Acad. Sci. USA*, **95**, 12940-12943.

Ge, M., and Freed, J.H. (1993). An electron spin resonance study of interactions between gramicidin A' and phosphatidylcholine bilayers. *Biophys. J.* **65**, 2106-2123.

Griffith, O.H., and Jost, P.C. (1976). Lipid spin labels in biological membranes. *In: Spin Labeling: Theory and Applications*, L.J. Berliner, ed., Academic Press, New York, pp. 453-523.

Gwak, S.H., Yu, L., and Yu, C.A. (1986). Spin-label electron paramagnetic resonance and differential scanning calorimetry studies of the interaction between mitochondrial succinate-ubiquinone and ubiquinol-cythchrome c reductases. *Biochemistry* **25**, 7675-7682.

Hemminga, M.A., and de Jager, P.A. (1989). Saturation transfer spectroscopy of spin labels. Techniques and interpretation of spectra. *In: Biological Magnetic Resonance, Vol. 8. Spin Labeling Theory and Applications.* L.J. Berliner and J. Reuben, eds., Plenum Press, New York, pp. 131-178.

Herman, J.R., Londo, T.R., Rahman, N.A., and Barisas, B.G. (1992). Normally on photomultiplier gating circuit with reduced post-gate artifacts for use in transient luminescence measurements. *Rev. Sci. Instrum.* **63**, 5454-5458.

Howard, E.C., Lindahl, K.M., Polnaszek, C.F., and Thomas, D.D. (1993) Simulation of saturation transfer electron paramagnetic resonance spectra for rotational motion with restricted angular amplitude. *Biophys. J.* **64**, 581-593.

Hubbell, W.L., Mchaourab, H.S., Altenbach, C., and Lietzow, M.A. (1996). Watching proteins move using site-directed spin labeling. *Structure* **4**, 779-783.

Hubbell, W.L., Cafiso, D.S., and Altenbach, C., (2000). Identifying conformational changes with site-directed spin labeling. *Nat. Struct. Biol.* **7**, 735-739.

Hustedt, E.J., Cobb, C.E., Beth, A.H., and Beechem, J.M. (1993) Measurement of rotational dynamics by the simultaneous non-linear analysis of optical and EPR data. *Biophys. J.* **64**, 614-621.

Hustedt, E.J., and Beth, A.H. (1995). Analysis of saturation transfer electron paramagnetic resonance spectra of a spin-labeled integral membrane protein, band 3, in terms of the uniaxial rotational diffusion model. *Biophys. J.* **69**, 1409-1423.

Hustedt, E.J., and Beth, A.H. (1996). Determination of the orientation of a band 3 affinity spin-label relative to the membrane normal axis of the human erythrocyte. *Biochemistry*, **35**, 6944-6954.

Hustedt, E.J., Smirnov, A.I., Laub, C.F., Cobb, C.E., and Beth, A.H. (1997). Molecular distances from dipolar coupled spin-labels: the global analysis of multifrequency continuous wave electron paramagnetic resonance data. *Biophys. J.* **74**, 1861-1877.

Hustedt, E.J., and Beth, A.H. (1999). Nitroxide spin-spin interactions: Applications to protein structure and dynamics. *Annu. Rev. Biophys. Biomol. Struct.* **29**, 129-153.

Hustedt, E.J. and Beth, A.H. (2001). Simulation of saturation transfer electron paramagnetic resonance spectra for a restricted uniaxial rotational diffusion model. *Biophys. J.* **81**, 3156-3165.

Hyde, J.S., Eriksson, L.E.G., and Ehrenberg, A. (1970). EPR relaxation of slowly moving flavin radicals: "Anomalous" saturation. *Biochim. Biophys. Acta* **222**, 688-692.

Hyde, J.S., and Dalton, L.R. (1972). Very slow tumbling spin labels: Adiabatic rapid passage. *Chem. Phys. Lett.* **16**, 568-572.

Hyde, J.S., and Thomas, D.D. (1974). New EPR methods for the study of very slow motion: Application to spin-labeled hemoglobin. *Ann. N. Y. Acad. Sci.* **222**, 680-692.

Hyde, J.S. (1978). Saturation-transfer spectroscopy. *In: Methods in Enzymology*, C.H.W. Hirs and S.N. Timasheff, eds., Academic Press, New York, **49G**, pp. 480-511.

Hyde, J.S., and Dalton, L.R. (1979). Saturation-transfer spectroscopy. *In: Spin Labeling II: Theory and Applications*, L.J. Berliner, ed., Academic Press, New York, pp. 1-70.

Hyde, J.S., and Thomas, D.D. (1980). Saturation-transfer spectroscopy. *Ann. Rev. Phys. Chem.* **31**, 293-317.

Hyde, J.S., and Froncisz, W. (1989). Loop gap resonators. *In: Advanced EPR: Applications in Biology and Biochemistry*, A.J. Hoff, ed., Elsevier, Amsterdam, pp. 227-306.

Hyde, J.S., and Feix, J.B. (1989). Electron-electron double resonance. *In: Biological Magnetic Resonance: Volume 8: Spin Labeling*, L.J. Berliner and J. Reuben, eds., Plenum Press, New York, pp. 305-337.

Hyde, J.S., Mchaourab, H.S., Camenisch, T.G., Ratke, J.J., Cox, R.W., and Froncisz, W. (1998). Electron paramagnetic resonance detection by time-locked subsampling. *Rev. Sci. Instrum.* **69**, 2622-2628.

Jahnig, F. (1986). The shape of a membrane protein derived from rotational diffusion. *Eur. Biophys. J.* **14**, 63-64

Johnson, M.E., and Hyde, J.S. (1981). 35-GHz (Q-band) saturation transfer electron paramagnetic resonance studies of rotational diffusion. *Biochemistry* **20**, 2875-2880.

Johnson, M.E., Lee, L., and Fung, L.W.-M. (1982a). Models for slow anisotropic rotational diffusion in saturation transfer electron paramagnetic resonance at 9 and 35 GHz. *Biochemistry* **21**, 4459-4467.

Johnson, M.E., Thiyagarajan, P., Bates, B., and Currie, B. (1982b). A comparison of resolution-enhancement methods in saturation transfer EPR. 15N isotopically substituted spin labels and 35 GHz high-frequency operation. *Biophys. J.* **37**, 553-557.

Kawasaki, K., Yin, J.-J., Subczynski, W.K., Hyde, J.S., and Kusumi, A. (2001). Pulse EPR detection of lipid exchange between lipid-rich raft and bulk domains in the membrane: Methodology development and its application to studies of influenza viral membrane. *Biophys. J.* **80**, 738-748.

Koppel, D.E., Sheetz, M.P., and Schindler, M. (1981). Matrix control of protein diffusion in biological membranes. *Proc. Natl. Acad. Sci. USA* **78**, 3576-3580.

Lewis, S.M., and Thomas, D.D. (1991). Microsecond rotational dynamics of spin-labeled Ca-ATPase during enzymatic cycling initiated by photolysis of caged ATP. *Biochemistry* **30**, 8331-8339.

Mahaney, J.E., Girard, J.P., and Grisham, C.M. (1990). Saturation transfer EPR measurements of the rotational diffusion of a strongly immobilized ouabain spin label on renal Na,K-ATPase. *FEBS Lett.* **260**, 160-164.

Mahaney, J.E., and Thomas, D.D. (1991). Effects of melittin on molecular dynamics and Ca-ATPase activity in sarcoplasmic reticulum membranes: electron paramagnetic resonance. *Biochemistry* **30**, 7171-7180.

Mahaney, J.E., and Grisham, C.M. (1992). Effects of ouabain on the rotational dynamics of renal Na,K-ATPase studied by saturation-transfer EPR. *Biochemistry* **31**, 2025-2034.

Mangels, M.L., Harper, A.C., Smirnov, A.I., Howard, K.P., and Lorigan, G.A. (2001). Investigating magnetically aligned phospholipids bilayers with EPR spectroscopy at 94 GHz. *J. Mang. Reson.* **151**, 253-259.

Marsh, D., and Henderson, P.J. (2001). Specific spin labeling of the sugar-H(+) symporter, GalP, in cell membranes of Escherichia coli: site mobility and overall rotational diffusion of the protein. *Biochim. Biophys. Acta* **1510**, 464-473.

Martin-Fernandez, M., Clarke, D.T., Tobin, M.J., Jones, S.V., and Jones, G.R. (2002). Preformed oligomeric epidermal growth factor receptors undergo an ectodomain structure change during signaling. *Biophys. J.* **82**, 2415-2427.

Matayoshi, E.D., and Jovin, T.M. (1991). Rotational diffusion of band 3 in erythrocyte membranes. 1. Comparison of ghosts and intact cells. *Biochemistry* **30**, 3527-3538.

Mchaourab, H.S., and Hyde, J.S. (1993). Continuous wave multiquantum electron paramagnetic resonance spectroscopy. III. Theory of intermodulation sidebands. *J. Chem. Phys.* **98**, 1786-1796.

McCalley, R.C., Shimshick, E.J., and McConnell, H.M. (1972). The effect of slow rotational motion on paramagnetic resonance spectra. *Chem. Phys. Lett.* **13**, 115-119.

Mett, R.R., Froncisz, W., and Hyde, J.S. (2001). Axially uniform resonant cavity modes for potential use in electron paramagnetic resonance spectroscopy. *Rev. Sci. Instrum.* **72**, 4188-4200.

Middleton, D.A., Reid, D.G., and Watts, A. (1995). The conformations of a functional spin-labeled derivative of gastric H/K-ATPase investigated by EPR spectroscopy. *Biochemistry* **34**, 7420-7429.

Naqvi, K.R., Gonzalez-Rodriguez, J., Cherry, R.J., and Chapman, D. (1973). Spectroscopic technique for studying protein rotation in membranes. *Nature New Biol.* **245**, 249-251.

Negash, S., Chen, L.T., Bigelow, D.J., and Squier, T.C. (1996). Phosphorylation of phospholamban by cAMP-dependent protein kinase enhances interactions between Ca-ATPase polypeptide chains in cardiac sarcoplasmic reticulum membranes. *Biochemistry* **35**, 11247-11259.

Nigg, E.A., and Cherry, R.J. (1980). Anchorage of a band 3 population at the erythrocyte cytoplasmic membrane surface: Protein rotational diffusion measurements. *Proc. Natl. Acad. Sci. USA* **77**, 4702-4706.

Percival, P.W., and Hyde, J.S. (1976). Saturation recovery measurements of the spin-lattice relaxation times of some nitroxides in solution. *J. Magn. Reson.* **23**, 249-257.

Perkins, R.C., Lionel, T., Robinson, B.H., Dalton, L.A., and Dalton, L.R. (1976). Saturation transfer spectroscopy: Signals sensitive to very slow molecular reorientation. *Chem. Phys.* **16**, 393-404.

Qiu, Z.H., Yu, L., and Yu, C.A. (1992). Spin-label electron paramagnetic resonance and differential scanning calorimetry studies of the interaction between mitochondrial cytochrome c oxidase and adenosine triphosphate synthase complex. *Biochemistry* **31**, 3297-3302.

Robinson, B.H., Dalton, L.R., Dalton, L.A., and Kwiram, A.L. (1974). Fast computer calculation of ESR and nonlinear spin response spectra from the fast motion to the rigid lattice limits. *Chem. Phys. Lett.* **29**, 56-64.

Robinson, B.H., and Dalton, L.R. (1980). Anisotropic rotational diffusion studied by passage saturation transfer electron paramagnetic resonance. *J. Chem. Phys.* **72**, 1312-1324.

Robinson, B.H., and Dalton, L.R. (1981). Approximate methods for the fast computation of EPR and ST-EPR spectra. V. Application of the perturbation approach to the problem of anisotropic motion. *Chem. Phys.* **54**, 253-259.

Robinson, B.H. (1983). Effects of overmodulation on saturation transfer EPR signals. *J. Chem. Phys.* **78**, 2268-2273.

Robinson, B.H., Thomann, H. Beth, A.H., Fajer, P. and Dalton, L.R. (1985). EPR and advanced EPR studies of biological systems, L.R. Dalton, ed., CRC Press, Boca Raton, Florida, pp. 1-314.

Robinson, B.H., Haas, D.A., and Mailer, C. (1994). Molecular dynamics in liquids: Spin-lattice relaxation of nitroxide spin labels. *Science* **263**, 490-493.

Rohrer, M., Prisner, T.F., Brugmann, O., Kass, H., and Spoerner, M. (2001). Structure of the metal-water complex in Ras x GDP studied by high-field EPR spectroscopy and 31P NMR spectroscopy. *Biochemistry* **40**, 1884-1889.

Rousseau, D.L., Guyer, C.A., Beth, A.H., Papayannopoulos, I.A., Wang, B., Wu, R., Mroczkowski, B., and Staros, J.V. (1993). Preparation and characterization of a bifunctionally spin-labeled mutant of murine epidermal growth factor for saturation-transfer electron paramagnetic resonance studies of the growth factor/receptor complex. *Biochemistry* **32**, 7893-7903.

Saffman, P.G., and Delbruck, M. (1975). Brownian motion in biological membranes. *Proc. Natl. Acad. Sci. USA* **72**, 3111-3113.

Schwarz, D., Pirrwitz, J., Meyer, H.W., Coon, M.J., and Ruckpaul, K. (1990). Membrane topology of microsomal cytochrome P-450: Saturation transfer EPR and freeze-fracture electron microscopy studies. *Biochem. Biophys. Res. Comm.* **171**, 175-181.

Schwarz, D., Chernogolov, A., and Kisselev, P. (1999). Complex formation in vesicle-reconstituted mitochondrial cytochrome P450 systems (CYP11A1 and CYP11B1) as evidenced by rotational diffusion experiments using EPR and ST-EPR. *Biochemistry* **38**, 9456-9464.

Sczaniecki, P.B., Hyde, J.S., and Froncisz, W. (1990). Continuous wave multiquantum electron paramagnetic resonance spectroscopy. *J. Chem. Phys.* **93**, 3891-3898.

Singer, S.J. (1972). A fluid lipid-globular protein mosaic model of membrane structure. *Ann. N.Y. Acad. Sci.* **195**, 16-23.

Squier, T.C., and Thomas, D.D. (1988). Relationship between protein rotational dynamics and phosphoenzyme decomposition in the sarcoplasmic reticulum Ca-ATPase. *J. Biol. Chem.*, **263**, 9171-9177.

Squier, T.C., Hughes, S.E., and Thomas, D.D. (1988). Rotational dynamics and protein-protein interactions in the Ca-ATPase mechanism. *J. Biol. Chem.* **263**, 9162-9170.

Smirnov, A.I., Belford, R.L., and Clarkson, R.B. (1998). Comparative spin label spectra at X-band and W-band. *In: Spin Labeling: The Next Millenium*, L.J. Berliner, ed., Plenum Press, New York, pp. 83-107.

Smirnov, A.I., Smirnova, T.I., and Morse, P.D. (1995). Very high frequency electron paramagnetic resonance of 2,2,6,6-tetramethyl-1-piperidinyloxy in 1,2-dipalmitoyl-sn-glycero-3-phosphaticylcholine liposomes: Partitioning and molecular dynamics. *Biophys. J.* **68**, 2350-2360.

Stein, R.A., Hustedt, E.J., Staros, J.V., and Beth, A.H. (2002). Rotational dynamics of the epidermal growth factor receptor. *Biochemistry* **41**, 1957-1964.

Stryer, L. (1978). Fluorescence energy transfer as a spectroscopic ruler. *Ann. Rev. Biochem.* **47**, 819-846.

Szabo, A. (1984). Theory of fluorescence depolarization in macromolecules and membranes. *J. Chem. Phys.* **81**, 150-167.

Thomas, D.D., and McConnell, H.M. (1974). Calculation of paramagnetic resonance spectra sensitive to very slow rotational motion. *Chem. Phys. Lett.* **25**, 470-475.

Thomas, D.D., Dalton, L.R., and Hyde, J.S. (1976). *J. Chem. Phys.* **65**, 3006-3024.

Thomas, D.D., Eads, T.M., Barnett, V.A., Lindahl, K.M., Momont, D.A., and Squier, T.C. (1985). Saturation transfer EPR and triplet anisotropy: Complementary techniques for the study of microsecond rotational dynamics. *In: Molecular Biological Systems*, P. Bayley and R. Dale, eds., Academic Press, New York, pp. 239-257.

Thomas, D.D. (1985). Saturation transfer EPR studies of microsecond rotational motions in biological membranes. *In: The Enzymes of Biological Systems*, A.N. Martonosi, ed., Plenum, New York, pp. 287-312.

Wahl, Ph. (1975) Fluorescence anisotropy of chromophores rotating between two reflecting barriers. *Chem. Phys.* **7**, 210-219.

Webb, W.W. (1981). Luminescence measurements of macromolecular mobility. *Ann. N. Y. Acad. Sci.* **366**, 300-314.

Weber, G. (1971). Theory of fluorescence depolarization by anisotropic Brownian rotations. Discontinuous distribution approach. *J. Chem. Phys.* **55**, 2399-2411.

Wolkers, W.F., Spruijt, R.B., Kaan, A., Konings, R.N., and Hemminga, M.A. (1997). Conventional and saturation-transfer EPR of spin-labeled mutant bacteriophage M13 coat protein in phospholipids bilayers. *Biochim. Biophs. Acta* **1327**, 5-16.

Zacharias, D.A., Violin, J.D., Newton, A.C., and Tsien, R.Y. (2002). Partitioning of lipid-modified monomeric GFP's into membrane microdomains of live cells. *Science* **296**, 913-916.

Chapter 13

Trends in EPR Technology

James S. Hyde
Department of Biophysics, Medical College of Wisconsin, 8701 Watertown Plank Road, Milwaukee, WI 53226

Abstract: A personal view of trends in EPR technology is presented. It is unlikely that the fundamental structure of the field will change, but it will be strongly influenced by the rapid increase in computer power, digital storage, and signal processing capability. In the author's laboratory current themes are resonator enhancement by electromagnetic field finite element modeling, analysis of noise, and digital detection and acquisition of data at multiple microwave frequencies. Some trends foreseen are (1) optimization of resonators for ultra-small samples; (2) step-recovery pulse EPR in which the initial conditions may be established by a step in some experimental condition such as light level or nuclear frequency irradiation; (3) blurring of the distinction between pulse and CW EPR as temporal changes in the resonant condition of a "CW" measurement are changed in times of the order of spin relaxation times; and (4) increased use of ELDOR.

1. INTRODUCTION

The EPR field is composed of a large number of application areas, each with a small number of active participants. The systems manufacturers produce flexible general purpose spectrometers with numerous accessories in order to serve this fragmented market. Although EPR spectroscopy is a fundamental measurement tool that will remain active indefinitely, and although there will continue to be inventions, discoveries and new applications, it is unlikely that the fundamental structure of the field will change. EPR will continue to be used for research in physics, chemistry and biology to examine samples in the liquid, solid and gas phases over a range of temperature and other conditions.

From its earliest days, the underlying technology was based on military developments in radar. In recent years, computers, cellular telephone technology and advances in digital devices have become increasingly important in contributing to the technological foundation of EPR spectrometers.

This reality – that our field is relatively small and dependent to a considerable degree on technology from larger scale development activities – places constraints on future progress of the field. The fundamental technological event of our times is rapid increase in computing power and the performance of associated digital and mass storage devices. Our future, from a technological perspective, will be based on this fact. Advances in EPR digital detection, in data capture and storage as well as in use of advanced signal processing methods are discussed in the chapter on digital detection (see Ch. 7).

Even though EPR instrumentation is strongly dependent on technology developed in other fields, there are areas where the special constraints of EPR have led to significant technological advances. Some of these are discussed below.

2. RESONATORS

Basic contributions to EPR technology that have been developed from within the EPR discipline include resonator development. The requirements of sample access, variable microwave coupling, resonators free from impurities, wall penetrability by high frequency field modulation, temperature control, etc. place constraints on resonator design that we ourselves must face – there is little in the way of technology to borrow from other fields. It is appropriate for the EPR instrumental futurist to predict advances in those specific areas where we control the technology and an attempt is made here.

The Varian multipurpose cavity oscillating in the rectangular TE_{102} mode served as the default EPR resonator for many years. See Rempel et al. (1964) and Hyde (1995) for details on the design. This structure was designed to enable a number of specialized EPR experiments as follows: i) light irradiation, ii) dewar insert for flowing temperature controlled gas, iii) dewar insert for liquid nitrogen, iv) so-called "flat cells" for aqueous samples, v) flat cells for tissue samples, vi) a mixing chamber geometry for stopped and continuous flow EPR, and vii) an electrochemical cell. A collet system was developed to support the dewars, flat cells, and sample tubes of various sizes. Two cavity bodies were bolted together to form the dual sample cavity (Hyde, 1965a) oscillating in the TE_{104} mode in order to examine a reference sample and a sample of interest simultaneously. A top coupled section was added (Piette et al., 1962) to form a TE_{103} cavity that permitted simultaneous optical absorption and EPR experiments on a sample. The demountable sidewall construction permitted disassembly and cleaning. Dr. Robert Rempel was the leader of this initiative with substantial contributions by Dr. Lawrence Piette, both of whom were Varian scientists. This outpouring of technology provided the basis for transfer of EPR spectroscopy from its original base in physics into chemistry and biology.

A number of special purpose X-band cavities were developed commercially. A list of some of these includes: cylindrical TE_{011} "wirewound" cavity (Hyde et al., 1965b, 1966) for enhanced sensitivity in certain classes of samples, ENDOR cavities, cavities for use at cryogenic temperatures, $H_1 \| H_0$ cavity, ELDOR cavities and the TM_{110} cavity for aqueous samples (Hyde, 1975).

This rich array of X-band resonators is the primary reason that this microwave frequency remains dominant in EPR spectroscopy. Nothing comparable exists at either higher or lower microwave frequencies. This is a significant opportunity for future technological development. Certainly all of these X-band capabilities could be replicated at Q-band (35 GHz) and S-band (ca 3 GHz), which would greatly strengthen the concept of "multifrequency EPR."

In 1965, the author introduced a system of classification for EPR samples from the perspective of sensitivity and optimum microwave configuration: namely, eight classes of samples depending on yes or no answers to three questions: i) Does it saturate? ii) Is it limited in size or availability? iii) Does it exhibit substantial dielectric loss (Varian Associates, 1965)? Of course, real samples may not correspond to one of these classes, but would lie in some intermediate category. Nevertheless, this classification system has proven helpful in thinking about resonator design in EPR spectroscopy. To a certain degree, it lies at the heart of the development of loop-gap resonators (LGR) for use in EPR spectroscopy, largely by W. Froncisz and the author (See Froncisz and Hyde, 1982; Hyde and Froncisz, 1989; and ch. 2 in the present volume). This class of resonators fundamentally improves sensitivity for those four classes of samples that correspond to a YES answer concerning question ii): Is the sample limited in size or availability? The benefits with respect to sensitivity can be quite substantial. The X-band LGR with 1 mm diameter sample access hole is one of the enabling technologies for the development of site directed spin labelling. The LGR is also a central technology for *in vivo* small animal imaging and spectroscopy (see ch 9 and ch. 11 in volume 23.)

Simulation of electromagnetic fields in microwave cavity resonators using finite-element computer-driven solutions of Maxwell's equations holds great promise for future development of EPR resonators. The author is aware of six papers in the literature where finite element modeling of electromagnetic fields was employed in a context of microwave resonators for EPR use. The earliest papers were from A. Schweiger's group in Zurich (Pfenninger et al., 1988; Forrer et al., 1996). In Pfenninger et al. (1988) the program known as MAFIA (Solution of **MA**xwell's equation by the **F**inite **I**ntegration **A**lgorithm) was used as an aid to development of bridged loop gap resonators. The work of this group in this area was updated in Forrer et al., 1996. Also in that year, the MCW group used the Hewlett-Packard version of High Frequency Structure Simulator, HFSS, to develop a bimodal loop gap resonator (Piasecki et al., 1996).

Recently, the author and his colleagues wrote a series of three papers on a new class of microwave resonators that we chose to call uniform field (UF) cavities (Mett et al., 2001; Anderson et al., 2002; Hyde et al., 2002). Transverse magnetic field (TM) modes exist, that exhibit uniformity of the fields along the axis that is perpendicular to the transverse field. This is because Maxwell's equations allow magnetic fields that are tangential to conducting end walls to exist. The familiar TM_{110} cylindrical cavity widely used in EPR for aqueous samples is an example of such a structure. However, previous to these three papers there were no known transverse electric field (TE) modes that exhibited this property of uniformity of the fields along the perpendicular axis. This is because components of the microwave electric field tangential to conducting end walls cannot exist. The UF modes described in these three papers consisted of a central section where the fields were precisely uniform, with end sections designed to satisfy the boundary conditions at the end walls.

These three papers made extensive use of high frequency structure simulation (HFSS) finite element software. UF resonators were, in fact, discovered using HFSS. We were working on an unrelated problem and noticed some unusual microwave field patterns that we did not understand. When the structures were built, they worked exactly as designed. Although they were largely theoretical, experimental "proof-of-principle" data were presented. Design of microwave coupling structures and sample entrance structures (so-called "stacks") for these prototypical resonators was also carried out using HFSS.

The HP HFSS software, which is no longer available, operated in the so-called driven mode. We found it very difficult to use for microwave cavity development because every change in the structure required a tedious hunt for the slightly shifted resonant frequency. When Hewlett-Packard dropped out of this business, we shifted to Ansoft HFSS (Ansoft Corp., Pittsburgh, PA). This software incorporates both the driven mode and the 3D modal frequency or "eigenmode" solution method (Brauer, 1997). This method is also available in MAFIA. We find that this method is extremely useful for EPR resonator development (Mett et al., 2001; Anderson et al., 2002; Hyde et al., 2002). Since resonators are characterized by a single resonant mode of interest, use of the eigenmode method permits focus only on that mode; a frequency sweep is not needed. Changes to the structure can be made and the resulting effect seen on the fields, resonance frequency and Q-value without having to track the changes through a frequency scan. Second, using the eigenmode method, we can focus on the design of the resonator without introducing a coupling structure. Mechanical drawing time and computation time are each reduced by one to two orders of magnitude due to the increased symmetry of the structure. Introduction of a coupling structure as well as sample access stacks is straightforward in the final stages of a design. We have also used the software to calculate the effects of sample support structures, dewar inserts, and the effect of the dielectric properties of the sample on the electromagnetic field distribution. The resonator efficiency parameter Λ (Hyde and Froncisz, 1989) can be calculated, which permits an estimate of EPR sensitivity. Currently we run this code on a Compaq W8000 workstation with dual Xeon 1.7 GHz processors with 2 GB of RAM.

In another cavity development initiative, we gave ourselves the goal of finding a way to examine a normal X-band sample in a 3 mm i.d., 4 mm o.d. quartz sample tube at Q-band. The problem was to determine whether the microwave sample access stacks would be beyond cutoff. As the sample diameter increases, the stack, viewed as a cylindrical waveguide propagating in either the cylindrical TE_{11} or cylindrical TE_{01} modes will first become evanescent and eventually radiate power into space. It was found that practical structures for this purpose can be constructed providing that the stack is sufficiently long.

Ansoft offers another software package known as Maxwell 3D that is well configured for computation of the 100 kHz field modulation patterns and eddy currents in surrounding metallic structures. Four approaches to the design of field modulation assemblies for use in EPR can be identified from the existing literature: i) Use of electrically thin walls – less than one skin depth at 100 kHz but many skin depths at the microwave frequency. ii) The wirewound technique introduced by the author for use with the cylindrical TE_{011} mode (Hyde, 1965, 1966, and Fig. 1). All microwave currents must be substantially parallel to the wires. iii) Partially cut-through slots for use with loop gap resonators (Fig. 2). iv) Direct insertion into the resonator of wires or rods that carry field modulation current. In addition to modeling of fields produced by modulation coils, Maxwell 3D is well suited for modeling of fields in a context of ENDOR and RF coils. It is apparent that these various approaches can also affect the microwave fields, and that the final design will require iteration between both software packages.

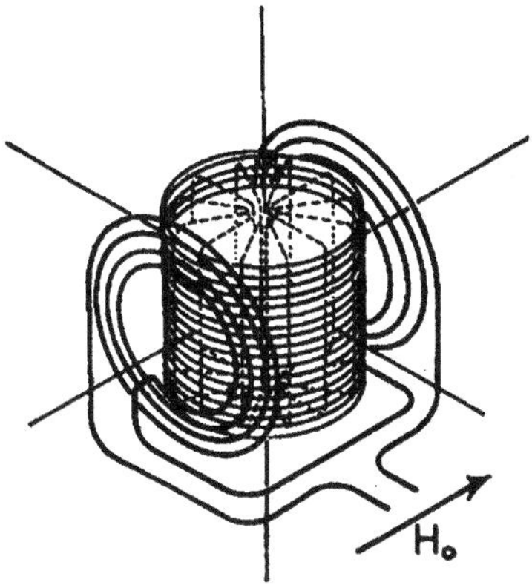

Figure 1. Schematic diagram of wire-wound TE_{011} cavity and field modulation coils.

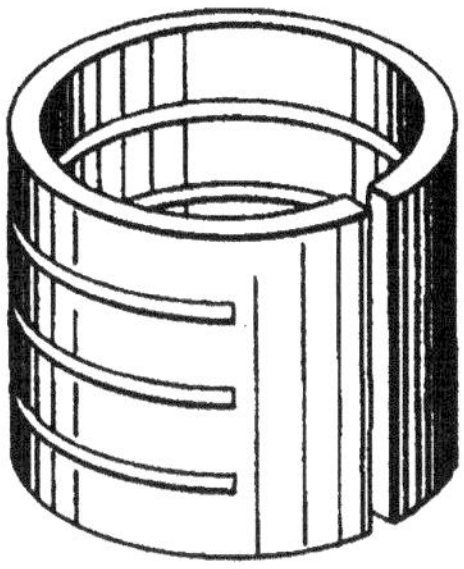

Figure 2. Outline drawing of the one-loop—one-gap resonator used for multifrequency EPR in the frequency range of 0.5 to 8 GHz. Hyde and Froncisz (1989) contains additional detail. The horizontal slots permit penetration of the 100 kHz magnetic field modulation.

There are several consequences of these advances in simulation of electromagnetic microwave and modulation fields. i) Design efficiency is greatly improved, which facilitates development of resonators that are optimized for specific applications. This, in turn, leads to an increase in sensitivity for the particular application. Previously, a resonator that was suitable for a broad class of applications was generally employed, which often resulted in a compromise with respect to sensitivity. ii) Cavity coupling structures, sample access stacks, dewar inserts and sample geometries can now be designed on a rational basis rather than the highly intuitive basis of the past. iii) A proliferation of resonators for use at frequencies above and below X-band can be expected that are modeled after the rich array of X-band designs.

It is concluded from this experience that the software and computing capability has reached the point where optimization of design is better carried out by computer than by experiment. Lower symmetry structures with a large number of design parameters can be investigated by computer modeling that would be impractical to investigate by construction of a series of actual cavities that could be experimentally probed to explore this parameter space.

The conclusion of this section is that we are truly on the verge of a paradigm shift in EPR resonator design that will enable optimization of the structure for any set of sample constraints. The new modeling tools fundamentally increase efficiency in resonator design. One can speculate that commercial design of resonators optimized for specific problems and sold in small numbers may become economically feasible. And there may well be other novel structures waiting to be discovered, in analogy to the discovery of UF modes.

3. NOISE

In the previous section, we discussed optimization of microwave resonator design to maximize the EPR signal. Now we discuss minimizing noise. Together, the goal is to identify future opportunities to improve the signal-to-noise ratio (SNR).

In continuous wave EPR, noise originating in the microwave oscillator and noise originating in the detection system as characterized by the overall receiver noise figure are of primary concern. At high microwave powers incident on the EPR resonator, noise from the oscillator dominates, while at low incident powers, receiver noise dominates. We have found that plots of spectrometer noise versus incident power give considerable insight into spectrometer performance. Six such plots in log-log format are shown in Figs. 3-5. They have been redrawn for display consistency from figures in (Hyde et al., 1982; Froncisz et al., 1986; Hyde et al., 1991; Lesniewski and Hyde, 1990; Pfenninger et al., 1995). Figures 3 and 5 are plots of noise voltage and Fig. 4 is of noise power. Note the differences in scales of the ordinate and abscissa comparing Figs. 3 and 5 with Fig. 4. At low incident powers, noise is independent of power, while at high incident power, the noise voltage varies linearly with incident power. The intercept of these two dependencies results in a "breakpoint" where the noise from each source is the same.

Figure 3a was taken from (Hyde et al., 1982), which was concerned with phase noise 100 kHz from the carrier frequency of the oscillator when tuned to the dispersion. The overall receiver noise figure that limits spectrometer performance at low powers in Fig. 3a is estimated to be 11 dB. Comparison was made between a loop gap resonator with an unloaded Q-value of 650 and a standard Varian multipurpose rectangular TE_{102} cavity with an unloaded Q-value of 7900. Demodulation of phase noise depends on the resonator Q. One expects the ratio of break-point powers to be given by the square of the ratio of Qs, i.e., 22 dB, compared with the measured value of 25 dB, which was considered to be fair agreement. Not only is phase noise less troublesome when using low Q resonators, but the energy density in the LGR employed in this study was 65 times higher than in the cavity. It follows that studies in the dispersion mode can be carried out with the loop gap resonator over a much greater range of energy densities at the sample without encountering source phase noise – about 43 dB.

Figure 3b is similar to 3a, but displays data taken from Froncisz et al. (1986), which was a comparison at Q-band (35 GHz) of phase noise when using a cylindrical TE_{011} cavity resonator versus a two-loop—one-gap resonator. Comparison of Figs. 3a and 3b suggests the presence of substantially greater receiver noise at Q-band than X-band. The breakpoints occur at higher incident powers even though one would expect about 10 dB more phase noise from the oscillator at Q-band than at X-band (Robins, 1982). This can only occur if the receiver noise figure is substantially higher at Q-band than at X-band.

In order to improve EPR performance at Q-band, a low phase noise Gunn diode oscillator was developed in the author's laboratory by R.A. Strangeway (Strangeway et al., 1995). A phase noise level in the range of -132 to -125 dBc/Hz at 100 kHz offset from the central frequency was achieved. This range can be compared with a measured value of -103 of the specific klystron in our Varian Q-band bridge (which was 4 dB worse than the manufacturer's specification for a new klystron) as well as published values from the literature that fall in the range of -70 to -115. The -115 value was also from a cavity-stabilized Gunn oscillator. See Lesniewski and Hyde (1990) for additional detail. The phase noise that was achieved in Strangeway's design remains, to the best of our knowledge, the lowest that has been reported at Q-band.

It is apparent from Figs. 3-5 that oscillator phase noise will become more evident if the overall receiver noise figure is reduced. Improvement in spectrometer performance over the widest range of conditions requires *simultaneous* reduction in receiver and oscillator noise. This concept was developed in detail in Hyde et al. (1991) and key results are displayed in Fig. 4. In that study, several changes were progressively made in the Q-band microwave bridge configuration, and noise was measured in each configuration as a function of incident microwave power while tuned to the dispersion using a TE_{011} cylindrical cavity resonator.

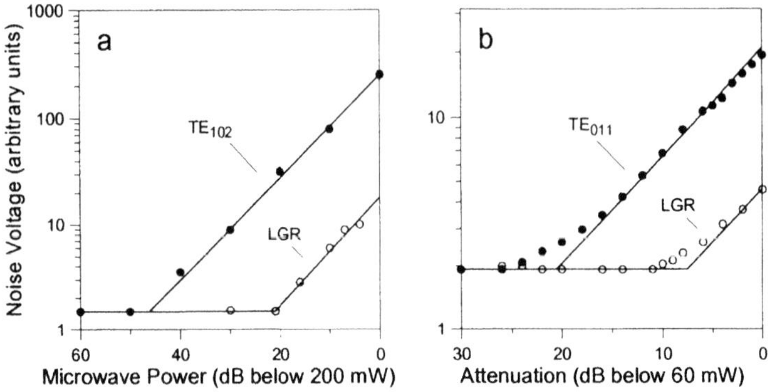

Figure 3. Comparison of LGR and cavity resonators. a) X-band (Hyde et al., 1982) and b) Q-band (Froncisz et al., 1986).

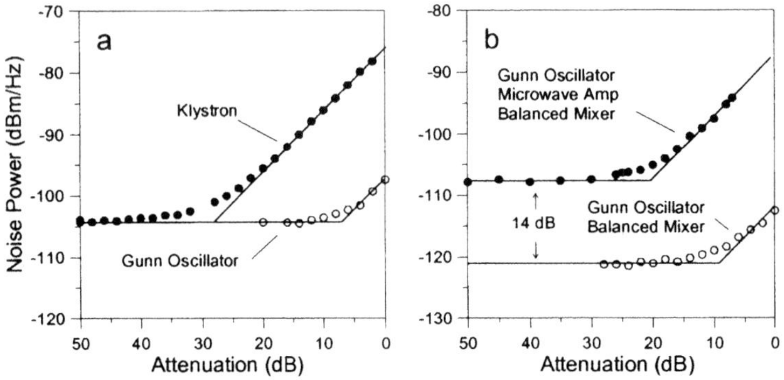

Figure 4. Noise plots at Q-band (Hyde et al., 1991). a) Low phase noise oscillator compared with a klystron in standard bridge; b) comparison with and without an LNA.

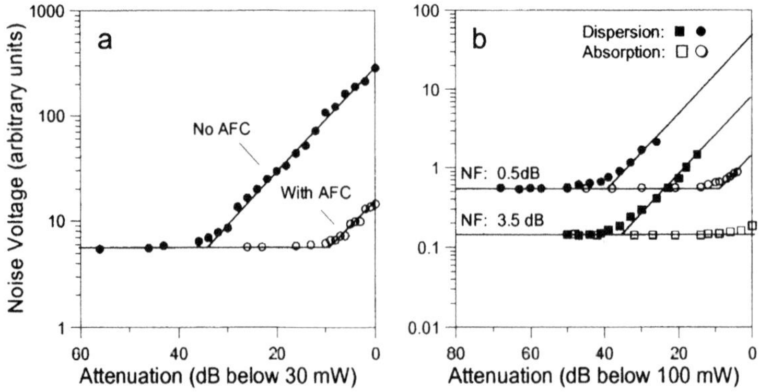

Figure 5. Noise plots; a) Reduction of phase noise using a broadband AFC (Lesniewski and Hyde, 1990); b) Comparison of cryogenic and room temperature LNAs (Pfenninger et al., 1995).

In Fig. 4a, phase noise levels at the microwave bridge output utilizing a klystron and the low phase noise Gunn diode oscillator are compared. Breakpoints differed by 22 dB, which was in fair agreement with the value of 26 dB that was carefully measured on a microwave test bench. The data for this figure were obtained using a 1N53D point contact detector diode. The Gunn oscillator experiment of Fig. 4a was repeated using a more modern balanced mixer detector, and the results are shown in the lower graph of Fig. 4b. (Note that the results of Fig. 4a and 4b are not directly comparable because the gain of the preamplifier following microwave detection was reduced in the experiment of Fig. 4b.) A low noise microwave amplifier (LNA) with a specified noise figure of 4.3 dB and a gain of 25.5 dB was then inserted into the microwave bridge. Results are shown in the upper graph of Fig. 4b.

Notably, at low power the noise increased by only 14 dB, although the gain increased by 25.5 dB. A value for the noise figure of the balanced mixer 100 kHz from the carrier is readily calculated: 25.5 – 14 + 4.3 = 15.8 dB. This is not a very good number relative to X-band standards, but it was a considerable improvement relative to the 1N53D. At high powers, the two graphs in Fig. 4b are displaced by 25.5 dB, the gain of the LNA, as expected. The breakpoints are separated by 13 dB, in good agreement with the low power separation of 14 dB, demonstrating consistency of the data.

Figure 5a is derived from Lesniewski and Hyde (1990). It shows data at 19 GHz using a microwave bridge with a varactor-tuned Gunn oscillator. An AFC circuit was developed where the feedback loop had high gain at 100 kHz from the carrier. This enabled reduction of phase noise by 23 dB. This is viewed as an interesting and relatively unexplored strategy for reduction of microwave phase noise.

Pfenninger et al. (1995) reports data obtained at X-band using a cryogenic microwave preamplifier cooled to the temperature of liquid nitrogen with a 0.5 dB noise figure corresponding to 35 K equivalent noise temperature. As expected, phase noise originating from the microwave oscillator becomes more evident as shown in Fig. 5b. Data in this figure were obtained using a loop gap resonator observing either the dispersion or the absorption. The cryogenic amplifier had 15 db higher gain and 3 dB lower noise figure than the room temperature LNA with which it was compared, resulting in about 12 dB separation of the noise levels, as observed. The separation of the breakpoints for dispersion and absorption was about 35 dB. This value is dependent on the resonator Q, reference arm power, amplifier gains, and other details of the bridge setup and is not necessarily the same for the two amplifiers. Excess oscillator noise was clearly observable in the absorption mode configuration when using the cryogenic amplifier, and it would have been more apparent if a high Q cavity resonator had been utilized. Detailed analysis presented in this paper as well as in Rinard et al. (1999) predicted further improvement if the microwave circuit elements in the bridge were to be cooled. This paper is cast not only in the context of CW EPR, but also pulse EPR.

From the perspective of "Future Trends," modern LNAs operating at room temperature seem adequate. Cooling of the amplifier combined with cooling of microwave bridge components may be desirable for selected experiments, but seems unlikely to become the general practice. However, improvement in microwave oscillator phase noise in CW EPR is feasible and desirable for two reasons: i) With an LNA in place, simultaneous detection of dispersion and absorption followed by a Hilbert transform of the dispersion and then addition of the two signals results in a highly desirable 3 dB improvement in SNR. Inspection of Figs. 3-5 indicates that phase noise remains excessive under practical conditions. Further reduction in oscillator phase noise is required in order to attain this improvement at higher levels of incident power. ii) As seen from the discussion concerning resonators, high Q cavities yield optimum performance for many classes of samples. In this case, reduced phase noise will be of value both for dispersion and absorption.

The Gunn diode itself is inherently noisy: the noise figure is on the order of 25 dB. Modern microwave transistors that could be used in an oscillator circuit have a noise figure on the order of a few dB. Continued improvements in transistor-based microwave oscillators have led to the recent development of low noise Yttrium-Iron-Garnet (YIG) oscillators. For example, standard manufactured YIG oscillators with a ± 1 GHz electronic tuning range, 14 dBm power output and a center frequency anywhere in the 3 - 10 GHz range have a typical phase noise of -125 dBc/Hz at 100 kHz offset (Korber et al., 2002). Further improvement in future years can be anticipated.

Microwave frequency translation is employed in an increasing number of modern EPR microwave bridges. Examples include multiquantum EPR (Hyde, 1998a), time-locked sampling (ch. 7 in this volume), and translation of complex irradiation patterns formed at X-band to Q-band and also W-band for sample irradiation and then back to X-band for detection as employed in current Bruker spectrometer systems. Frequency translation can introduce additional sources of noise. The appendix of Strangeway et al. (1995) discusses several different approaches to frequency translation. One is to mix the output of a microwave oscillator with the output of a frequency synthesizer. Noise aspects of frequency synthesizers in the context of EPR spectroscopy are discussed in ch 7 of this volume.

As noise originating in the EPR spectrometer decreases, it is possible that noise originating in the environment will become dominant. Environmental noise includes microphonics in the audio frequency range and below, line frequency noise, instabilities in connections to ground, radio frequency interference, and temperature instabilities. The author has encountered each of these. It seems likely that future EPR spectrometer installations will be in shielded rooms with independent connections to ground, much as in magnetic resonance imaging installations. Mass storage devices, for example RAID arrays, permit storage of raw data with no signal averaging or filtering. This will allow inspection of the data for spurious environmental noise. It will also permit development of optimum filters of environmental noise.

EPR spectra are often obtained as a function of one or more additional experimental parameters, for example, temperature or time. Almost always in modern spectrometers the information is available in digital format, permitting the use of a variety of digital filters. Ideally, such filtering should be performed in multidimensional parameter space making use of all available information. An example is provided in ch 6 of volume 23, by N. Hogg. The EPR literature in this area is sparse, but will undoubtedly increase in the years ahead. In the author's view this is a particularly promising opportunity for further improvement of the SNR in EPR spectroscopy.

4. MULTIFREQUENCY EPR

In 1963, Varian initiated a project to develop a commercial EPR bridge at Q-band (35GHz) headed by the author. The bridge circuit was the first from Varian to employ a circulator and reference arm. It was a modest commercial success – about 100 units were sold over the ensuing years. Analysis of the business opportunity presented by development of a bridge operating at a third microwave frequency, S-band, was not favorable.

After establishment of the National Biomedical EPR Center at the Medical College of Wisconsin, funds became available to develop an S-band bridge for use in collaborative projects at the center. The project was headed by Wojciech Froncisz, visiting from Jagiellonian University, Krakow, Poland. A commercially available tunable 2 to 4 GHz (octave bandwidth) microwave oscillator was used in the bridge. Although microwave cavities were initially used, the ungainly size of these structures led directly to the development of loop gap resonators.

In the ensuing years, octave bandwidth bridges were developed from 0.5 to 1, 1 to 2, and 4 to 8 GHz, resulting in continuous coverage from 0.5 to 8 GHz in coaxial transmission line configurations. An additional bridge was developed at 18 GHz, resulting in EPR capabilities at discrete frequencies in three narrow bands centered at 9.5, 18 and 35 GHz. This multifrequency capability was viewed as a national resource available to all EPR spectroscopists.

The 18 GHz bridge was designed to be suitable both for CW and saturation recovery (SR) EPR. An S-band module was developed for use with Froncisz's S-band bridge, resulting in multifrequency capabilities for saturation recovery at S-, X-, and K-band (see ch. 1 in this volume, by Eaton and Eaton).

Bruker has developed an approach that may well simplify the problem of achieving pulse multifrequency capabilities. The firm's W-band (94 GHz) bridge is based on an X-band bridge with advanced pulse capabilities. The output of the X-band bridge is mixed with the output of an 84 GHz oscillator to form a W-band microwave source, and the reflected signal from the W-band resonator is translated back to X-band for detection. The same strategy has recently been developed at Q-band, translating again from X-band. Tradeoffs made by this approach have not yet been analyzed in the literature.

Several initiatives were undertaken at MCW to arrive at multifrequency ENDOR and ELDOR capabilities in the S-band range. Newton and Hyde (1991) observed that the ENDOR enhancement effect increases inversely with the applied static magnetic field, which makes S-band a favorable microwave frequency for ENDOR of low moment nuclei such as ^{14}N. They demonstrated ^{14}N ENDOR of the P1 center in diamond. Pace et al. (1993) used S-band ENDOR to observe hyperfine interaction of ^{1}H and ^{14}N nuclei in trinitrophenylmethylnitroxide. Christidis et al. (1994) describe a probehead with both interchangeable LGRs and RF coils for multifrequency ENDOR in the range of 1 – 10 GHz. An ELDOR capability at S-band using two separate S-band bridges and a bimodal resonator was described by Piasecki et al. (1996).

Extensions of X-band multiquantum EPR to MQ ENDOR (Mchaourab et al., 1993) and electron-nuclear-electron triple resonance (Christidis et al., 1996) were reported. MQ ELDOR is described by Mchaourab et al. (1991). The center in Milwaukee has recently completed a Q-band bridge with conventional CW and ELDOR capability, MQ and MQ ELDOR capability and saturation recovery. Use of all five displays for ENDOR seems straightforward. The sample and resonator need not be changed when switching between modes of operation. Historically, each of these advanced capabilities was developed one at a time; it is now possible to combine them in a single microwave configuration. Translation to another microwave frequency following Bruker's approach seems feasible, although multiquantum EPR presents a special problem. The translation processes may generate unacceptable levels of intermodulation sidebands. A possible solution to this problem would be to translate the two closely spaced irradiating microwave frequencies independently and combine them after translation.

The continuous range of microwave frequencies that is available from 0.5 to 8 GHz presents an opportunity for future instrumental development. To date, the number of resonators available in this range in the author's laboratory is small – two in the S-band range and one in each of the other three octave bandwidths. Development of a multiplicity of resonators over this entire range seems impractical. What would suffice, as an example, is four resonators each with octave bandwidth resonator tunability in order to arrive at a true overall four-octave capability. Figures 6 and 7 are cartoons of possible resonator designs that indicate how this capability might be achieved. Figure 6 illustrates a re-entrant TE_{102} cavity. The re-entrant rods lie parallel to the electric field in regions of electric field maxima. They short out the field, reducing the resonant frequency. Rotation of the central section, as indicated in the figure, will dramatically shift the resonant frequency. Figure 7 illustrates a tunable bridged loop gap resonator (Pfenninger et al., 1988; Symons, 1995). As the plastic tuner is rotated, the capacitance of the resonator (gap in parallel with the two bridge capacitances in series) changes. Automated collection of spectra over an octave bandwidth at closely spaced microwave frequencies would appear to be possible, providing that the scientific rationale for the capability is sufficiently strong to merit the technological development effort

A general weakness in the multifrequency approach is that the wide range of resonator configurations presently available at X-band seems unlikely to be developed over the full range of frequencies of interest. However, it does seem feasible, as mentioned above, to develop a range of resonators at one or two additional frequencies, for example, S- and Q-band.

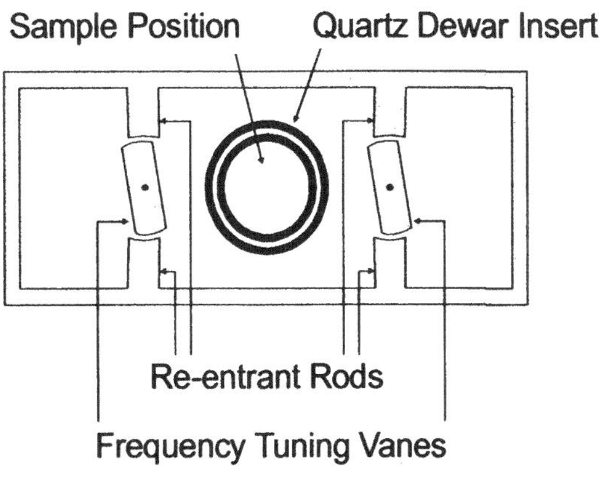

Figure 6. Variable frequency re-entrant rectangular TE_{102} cavity.

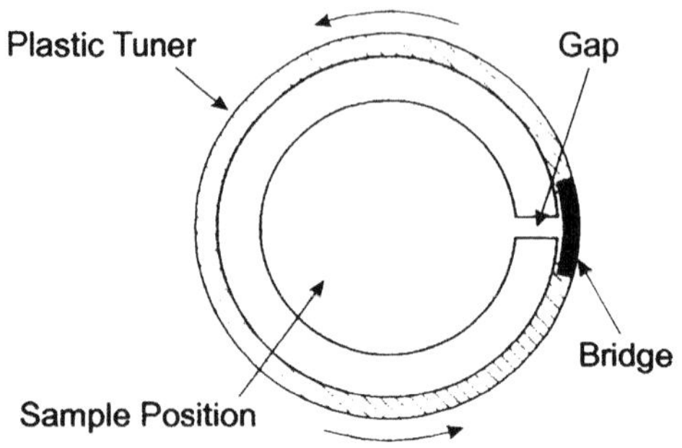

Figure 7. Variable frequency bridged loop gap resonator.

5. EPR FOR ROUTINE ANALYSIS

The use of EPR spectroscopy in routine commercial applications began as early as 1961, when Saraceno, Fanale and Coggeshall of Gulf Oil (Saraceno et al., 1961; Saraceno, 1963) developed techniques for using EPR to measure vanadium in crude oil. The vanadium poisoned catalysts used in refineries. Under contract with Gulf, Varian Associates developed a portable EPR spectrometer with various automatic calibration features that permitted its reliable operation by unskilled personnel in the environment of a refinery (Nelson and Baker, 1967).

Another effort occurred in 1971 when President Nixon requested and Congress agreed to the establishment of a Special Action Office of Drug Abuse, primarily for compulsory treatment and rehabilitation of addicted Vietnam veterans. Syva Corporation, a cooperative venture between Syntex and Varian, developed the Frat™ EPR spin immunoassay systems to monitor morphine in the urine of addicts. The EPR instrument was highly automated and approximately 50 spectrometers were built. Many were used in extremely difficult conditions in Vietnam, making several million examinations. See Hyde (1998b) for additional details.

In 1973, Hwang and Pusey developed an oil well logging process for determining paleotemperatures (Pusey, 1973; Hwang and Pusey, 1973). They measured the EPR free radical signal that occurs naturally in kerogen. The process was made commercially available by Robertson Research International, Ltd.

One of the most promising applications for routine EPR analysis is for radiation dosimetry using the tissue-equivalent alanine dosimeter. The National Institute of Standards and Technology (NIST) offers a number of services based on EPR analysis of the free radical content in alanine pellets exposed to high doses of ionizing radiation, including calibration services. Bruker offers an automated spectrometer, the *e-scan*, specially configured for this measurement. In 2002, a news release issued jointly by Kodak and Bruker describes a system for alanine dosimetry based on coating of microcrystalline alanine on a film, which is sold under the trademark *Biomax*. The system includes a bar code and alphanumeric identifier for sample tracking. The news release describes the system as an unmatched method for high-volume verification of radiation in processes ranging from the sterilization of medical devices to decontamination of food.

Another industrial application of EPR was introduced by Uchida and Ono in 1996 (Uchida and Ono, 1996). They employed a spin trapping strategy to measure the quantity of anti-oxidants in beer (Uchida et al., 1996; Uchida and Ono, 2000). This measurement correlated with shelf-life of the product. A patent was issued in 1998 (Ono and Uchida, 1998) for the process. Bruker has also configured the e-scan spectrometer for this application.

Modern approaches to EPR spectrometer design can be expected to result in enhanced performance and reduced cost for commercial applications of EPR, providing that the markets are sufficiently large. Using finite element modeling of microwave fields, resonator design and sample geometry can be optimized efficiently for each application. In addition, microwave circuits can be miniaturized and optimized by computer design. It is highly desirable to use the smallest magnet gap possible. In a patent entitled "Narrow Cavity Low Cost EPR Spectrometer" (Hyde, 1987), the author used a TE_{102} or TM_{110} cavity just 2 mm wide in order to reduce the mass of the magnet. Loop gap resonators provide ample opportunity to reduce magnet size in process-control or medical diagnosis applications. Modern PCs plus digital detection permit the equivalent of phase sensitive detection in the computer (ch. 7, this volume), manipulation of the gain and bandwidth of automatic frequency control circuits, and incorporation of quite sophisticated signal processing software.

Process control applications of EPR have been slow to develop, but the success that Bruker has been experiencing in recent years indicates progress. Newer computer-based engineering design tools, the introduction of digital detection, and modern computer-based signal analysis tools can be expected to provide the technological basis for further progress.

Development of medical diagnosis applications of EPR that are widely used has been a long-standing goal, but has remained elusive. It has become clear that success can only be achieved if there is a sound scientific rationale for EPR spectroscopy in comparison with other more familiar spectroscopic methods. Spin trapping in a context of free radical chemistry offers one opportunity, since it can be argued that EPR is generally a preferred spectroscopy when free radicals are involved.

6. DISCUSSION

From the author's own laboratory, several "mini-trends" that may flourish in the years ahead can be identified:

1) Optimization of resonators for ultra-small samples, i.e., submicroliter. In a context of loop gap resonators, electric discharge machining (EDM) permits cutting of very small gaps. In a poster (Camenisch et al., 2002), a Q-band resonator with a sample volume as small as 30 nl containing 1 picomole of protein was successfully used in a multiquantum experiment.

2) Step recovery pulse EPR. The concept is to generalize the step of incident microwave power used in saturation recovery to include other steps in experimental conditions including light level, temperature, pH value, stop flow mixing of reacting species, applied electric field, nuclear irradiation level in ENDOR, pressure, etc., and observe the time-dependent evolution to a new equilibrium condition. Observation of the transient response using either spin echo methods or CW methods is appropriate. The technological challenge lies in appropriate delivery of the step. Change in sample composition or temperature can result in changes in microwave resonance frequency or coupling, that require dynamic adjustment. Changes in the EPR field/frequency relationship during approach to a new equilibrium may require attention.

3) Blurring of the distinction between CW EPR and pulse EPR. High frequency field modulation is a form of pulse EPR, since it involves inversion of the magnetization during each sweep of the instantaneous modulation field through resonance. Multiquantum EPR is a temporal modulation of the saturation conditions that can be described equally well in the time domain or the frequency domain. MQ in the presence of high frequency field modulation has been demonstrated in unpublished work in the author's laboratory.

4) Increased use of ELDOR. Intense saturation transfer mechanisms are one of the distinguishing characteristics of EPR spectroscopy. The technological challenge in exploiting this opportunity lies largely in resonator development. Frequency sweep of the pumping microwaves is highly desirable, but technically challenging. These methods were pioneered by the author in the 1960s. The hypothesis can be made that modern technology creates opportunities for improved bimodal resonators

with frequency sweep of the pumping mode. ELDOR is a powerful method that remains underutilized because of technological complexities.

The material in this chapter has been largely derived from the author's experience and must be considered to be a personal view of technological trends in EPR. Several significant instrumental trends, including developments in pulse spin echo EPR, *in vivo* EPR and ultra-high frequency EPR, are not discussed. However, the themes of this article, viz, resonator enhancement by electromagnetic field finite element modeling, analysis of noise, digital detection and acquisition of data at multiple microwave frequencies, are relevant to pulse, *in vivo* and ultra-high frequency EPR.

The chapter by Freed and Peisach in this volume gives additional information on pulse EPR and high frequency EPR, and the chapters by Swartz (ch.10) Halpern (ch. 11), and Krishna (ch. 12) in volume 23 provides an overview of *in vivo* EPR. The recent monograph on pulse EPR by Schwieger and Jeschke (2001) provides a foundation for future progress in that field.

7. REFERENCES

Anderson, J.R., Mett, R.R. and Hyde, J.S. (2002). Cavities With Axially Uniform Fields For Use In Electron Paramagnetic Resonance. II. Free Space Generalization. *Rev. Sci. Instrum.* **73**, 3027-3037.

Brauer, J.R. (1997). *IEEE Trans. Microwave Theory Tech.*, **45**, 810.

Camenisch, T.G., Klug, C.R., Ratke, J.J., Hubbell, W.L. and Hyde, J.S. (2002). Multiquantum EPR of Arrestin K267C-MT SL at 35 GHz. *Proc.44th Rocky Mountain Conf. on Analytical Chemistry,* p. 43.

Christidis, T.C., Froncisz, W., Oles, T. and Hyde, J.S. (1994). Probehead with Interchangeable Loop-Gap Resonators and RF Coils for Multifrequency EPR/ENDOR. *Rev. Sci. Instrum.* **65**, 63-67.

Christidis, T.C., Mchaourab, H.S. and Hyde, J.S. (1996). Hyperfine Selectivity Using Multiquantum Electron-Nuclear-Electron Triple Resonance. *J. Chem. Phys.* **104**, 9644-9646.

Forrer, J., Pfenninger, S., Sierra, G. Jeschke, G. and Schweiger, A., Wagner, B. and Weiland, T. (1996). Probeheads And Instrumentation For Pulse EPR And ENDOR Spectroscopy With Chirped Radio Frequency Pulses And Magnetic Field Steps. *Appl. Magn. Reson.* **10**, 263-279.

Froncisz, W. and Hyde, J.S. (1982). The Loop-Gap Resonator: A New Microwave Lumped Circuit ESR Sample Structure. *J. Magn. Reson.* **47**, 515-521.

Froncisz, W., Oles, T. and Hyde, J.S. (1986) *Q*-Band Loop-Gap Resonator. *Rev. Sci. Instrum.* **57**, 1095-1099.

Hwang, P.T.R. and Pusey, W.C. (1973) U.S. Patent 3,740,641.

Hyde, J.S. (1995). Electron Paramagnetic Resonance in *Handbook of Microwave Technology, Volume 2,* Ishii, T. K., ed., pp. 365-402, Academic Press, New York.

Hyde, J.S. (1965a). Gyromagnetic Resonance Spectroscopy, U.S. Patent 3,197,692.

Hyde, J.S. (1965b). ENDOR Of Free Radicals In Solution. *J. Chem Phys.* **43**, 1806-1818.

Hyde, J.S. (1966). Microwave Cavity Resonator, U.S. Patent 3,250,985.

Hyde, J.S. (1975). EPR Spectrometer Resonant Cavity, U.S. Patent 3,878,454.

Hyde, J.S. (1987). Narrow Cavity Low Cost EPR Spectrometer, U.S. Patent. 3,931,569.

Hyde, J.S. (1998a). Multiquantum EPR. in *Foundations of Modern EPR*, G.R. Eaton, S.S. Eaton and K.M. Salikhov, eds., pp. 741-757, World Scientific, New York.

Hyde, J.S. (1998b). EPR at Varian: 1954-1974. in *Foundations of Modern EPR*, G.R. Eaton, S.S. Eaton and K.M. Salikhov, eds., pp. 695-716, World Scientific, New York.

Hyde, J.S. and Froncisz, W. (1989). Loop Gap Resonators. In *Advanced EPR: Applications in Biology and Biochemistry*, A.J. Hoff, ed., pp. 277-306, Elsevier, Amsterdam.

Hyde, J.S., Froncisz, W. and Kusumi, A. (1982). Dispersion Electron Spin Resonance with the Loop-Gap Resonator. *Rev. Sci. Instrum.* **53**, 1934-1937.

Hyde, J.S., Newton, M.A., Strangeway, R.A., Camenisch, T.G. and Froncisz, W. (1991). Electron Paramagnetic Resonance Q-Band Bridge with GaAs Field-Effect Transistor Signal Amplifier and Low-Noise Gunn Diode Oscillator. *Rev. Sci. Instrum.* **62**, 2969-2975.

Hyde, J.S., Mett R.R. and Anderson, J.R. (2002). Cavities With Axially Uniform Fields For Use In Electron Paramagnetic Resonance. III. Re-Entrant Geometries. *Rev. Sci. Instrum.* **73**, pp. 4003-4009.

Korber, M.R., Teuthorn, D. and Lockie, D. (2002). YIGs Tune High-Speed Millimeter-Wave Radios. *Microwaves & RF*, pp. 92-100 (July).

Lesniewski, P. and Hyde, J.S. (1990). Phase Noise Reduction of a 19 GHz Varactor-Tuned Gunn Oscillator for Electron Paramagnetic Resonance Spectroscopy. *Rev. Sci. Instrum.* **61**, 2248-2250.

Mchaourab, H.S., Christidis, T.C., Froncisz, W., Sczaniecki, P.B. and Hyde, J.S. (1991). Multiple-Quantum Electron-Electron Double Resonance. *J. Magn. Reson.* **92**, 429-433.

Mchaourab, H.S., Christidis, T.C. and Hyde, J.S. (1993). Continuous Wave Multiquantum Electron Paramagnetic Resonance Spectroscopy. IV. Multiquantum Electron-Nuclear Double Resonance. *J. Chem. Phys.* **99**, 4975-4985.

Mett, R.R., Froncisz, W. and Hyde, J.S. (2001). Axially Uniform Resonant Cavity Modes For Potential Use In Electron Paramagnetic Resonance Spectroscopy. *Rev. Sci. Instrum.* **72**, 4188-4200.

Nelson, F.A. and Baker, G.A. (1967). Gyromagnetic Resonance Apparatus Utilizing Two Sample Signal Comparison, U.S. Patent 3,348,136.

Newton, M.E. and Hyde, J.S. (1991). ENDOR at S-Band (2-4 GHz) Microwave Frequencies. *J. Magn. Reson.*, **95**, 80-87.

Ono, M. and Uchida, M. (1998). Analytical Method For Evaluating Flavor Stability Of Fermented Alcoholic Beverages Using Electron Spin Resonance, U.S. Patent 5,811,305.

Pace, M.D., Christidis, T.C. and Hyde, J.S. (1993). S-Band ENDOR of Hyperfine Interactions of ^{1}H and ^{14}N Nuclei in Trinitrophenylmethylnitroxide. *J. Magn. Reson.*, **102A**, 101-104.

Pfenninger, S., Forrer, J. and Schwieger, A. (1988). Bridged Loop-Gap Resonator: A Resonant Structure For Pulsed ESR Transparent To High-Frequency Radiation. *Rev. Sci. Instrum.* **59**, 752-760.

Pfenninger, S., Froncisz, W. and Hyde, J.S. (1995). Noise Analysis of EPR Spectrometers with Cryogenic Microwave Preamplifiers. *J. Magn. Reson.* **113A**, 32-39.

Piasecki, W., Froncisz, W. and Hyde, J.S. (1996). Bimodal Loop-Gap Resonator. *Rev. Sci. Instrum.* **67**, 1896-1904.

Piette, L.H., Ludwig, P., Adams, R.N. (1962). EPR And Electrochemistry. Studies Of Electrochemically Generated Radical Ions In Aqueous Solution, *Anal. Chem.*, **34**, 916.

Pusey III, W.C. (1973). How To Evaluate Potential Gas And Oil Source Rocks. *World Oil,* April

Rempel, R.C., Ward, C.E., Sullivan, R.T., St. Clair, M.W., Weaver, H.E. (1964). Gyromagnetic Resonance Method and Apparatus, U.S. Patent 3,122,703.

Rinard, G.A., Quine, R.W., Song, R., Eaton, G.R., Eaton, S.S. (1999). Absolute EPR Spin Echo and Noise Intensities. *J. Magn. Reson.* **140,** 69-83.

Robins, R.J. (1982). *Phase Noise in Signal Sources*, Peregrinus Ltd. on behalf of the Institute of Electrical Engineers, Herts, England.

Saraceno, A.J., Fanale, D.I. and Coggeshall, N.D. (1961). An Electron Paramagnetic Resonance Investigation of Vanadium in Petroleum Oils. *Anal. Chem* **33**, 500.

Saraceno, A.J. (1963). Determination of Vanadium Content of Hydrocarbon Oils by Electron Paramagnetic Resonance Spectrometry, U.S. Patent 3,087,888 (reissued 1967).

Schwieger, A. and Jeschke, G. (2001). *Principles Of Pulse Electron Paramagnetic Resonance,* Oxford University Press, New York.

Strangeway, R.A., Mchaourab, H.S., Luglio, J., Froncisz, W. and Hyde, J.S. (1995). A General Purpose Multiquantum Electron Paramagnetic Resonance Spectrometer. *Rev. Sci. Instrum.* **66,** 4516-4528.

Symons, M.C.R. (1995). Whole Body Electron Spin Resonance Imaging Spectrometer, In Bioradicals Detected By ESR Spectroscopy, H. Ohya-Nishiguchi and L. Packer, eds., pp. 93-102, Birkhäuser Verlag, Basel, Switzerland.

Uchida, M. and Ono, M. (1996). Improvement For Oxidative Flavor Stability Of Beer – Role Of OH-Radical In Beer. *J. Am. Soc. Brew. Chem* **54,** 198-204.

Uchida, M., Suga, S. and Ono, M. (1996). Improvement For Oxidative Flavour Stability Of Beer – Rapid Prediction Method For Beer Flavour Stability By Electron Spin Resonance Spectroscopy. *J. Am. Soc. Brew. Chem* 54, 205-211.

Uchida, M. and Ono, M. (2000). Technological Approach To Improve Beer Flavor Stability: Analysis Of The Effect Of Brewing Processes On Beer Flavor Stability By The Electron Spin Resonance Method. *J. Am. Soc. Brew. Chem.* **58,** 8-13.

Varian Associates Technical Information Bulletin: Signal Amplitudes in Electron Paramagnetic Resonance, Fall 1965, pp. 10-13.

Chapter 14

Prognosis

Sandra S. Eaton and Gareth R. Eaton
Department of Chemistry and Biochemistry, University of Denver, Denver, Colorado 80208

Abstract: Enhancements are foreseen in EPR instrumentation and application of EPR to biological systems. The need for educating future researchers about the use of EPR is emphasized.

Each of the authors in this two-volume perspective on biomedical EPR was asked to look into the crystal ball and divine the future of the topic surveyed in that chapter. Here, we attempt to look a bit more broadly. We have made previous attempts at prognoses, some of them specific enough to be confirmed by subsequent events (Eaton and Eaton, 1988a,b, 1993, 1995; Eaton, Eaton and Salikhov, 1998). This section is expressed in the context of biomedical applications, but most of what is said here applies equally well to applications in the study of non-biological materials and characterization of "small" chemical species.

Although EPR spectra are often presented and interpreted as if the only important observables are g-value and hyperfine, details of line shape and relaxation times and the magnetic field/microwave frequency dependence of these parameters provide increased insight into the system under study. EPR imaging reveals the power of including spatial dimensions along with the spectral dimension, and time-domain measurements add the dimension of time, for the spins (spin relaxation times; spin polarization) and for the kinetics of the system under study. Within time-domain EPR the observables include electron spin echoes, free induction decays, CW-detected recovery from power saturation, and other measures of spins such as coherent Raman beats. There are many characteristic times to consider in spin systems, and there are many readouts of non-linear spin responses, including saturation-transfer. Only a few of these have been explored so far - no doubt many more will be reported even before this book is in print.

The environmental factors that influence the spins from one perspective are problems to be controlled, but also each becomes a topic that can be studied via the spins. These include temperature, viscosity, O_2 and other paramagnetic species, electrochemical potential, pH, time (kinetics) and so on. Many of these topics are discussed in the chapters in these two books, and new applications converting problems to insights can be expected in the near future.

The table lists a few of the over 100 defined EPR measurements. One such list is in a 1997 issue of the EPR Newsletter (Eaton and Eaton), and on the authors' web site. Were one to put a check in each box for which there is at least one published experiment, there would still be many empty boxes. It is still true that at least 99% of all EPR spectra, whether for routine analysis or for research, in materials or in biomedical areas, are linear, CW X-band spectra, and most use a rectangular TE_{102} cavity resonator. However, for many problems one will achieve better insight into the sample using pulsed, time-domain, or non-linear CW EPR spectroscopy, and often frequencies other than X-band will be optimum, and we predict trends in this direction. Each type of experiment may become the crucial way to look at a particular problem.

Table 1. The EPR Perspective

		1D	2D	3D	4D
CW					
	Linear				
	Multi-frequency (MHz to THz)				
	In-vivo				
	Multi-resonance (ELDOR, ENDOR, TRIPLE)				
	Field-swept, frequency-swept				
	Time-resolved (transient) EPR				
	Non-linear (ST-EPR, saturated, rapid passage)				
	Microwave amplitude modulation				
	ODMR, etc.				
Pulsed					
	Multi-frequency				
	ESE (T_1, T_2, T_m)				
	Pulsed ENDOR				
	FT-EPR				
	Saturation recovery				
	Pulsed ELDOR				
	Pulsed magnetic field				
Multi-quantum					
	Multi-frequency				
	LODESR, DOMESR, etc.				

Combining EPR with other physical methods provides yet a further dimension from which to view biological systems. In this concept we mean more than just hyphenated techniques such as liquid-chromatography-EPR or laser-excitation-EPR, as important as these are. We mean conceptually (and maybe graphically) overlaying MRI images and EPR images, or comparing molecular dynamics with macroscopic flow patterns. Analogous extensions might combine first-principles calculations and multidimensional EPR parameter measurements.

National EPR Centers, funded by the National Institutes of Health in the USA, and comparable Centers in Europe, provide access to specialized equipment and cutting-edge methodology, especially for biomedical applications. In late-2003, the U.S. Centers include the National Biomedical ESR Center in Milwaukee, which was discussed in the introduction to these volumes, the EPR Center for the Study of Viable Systems (Dartmouth), the Center for Imaging In Vivo Physiology (Chicago, Denver, and Baltimore), and the Advanced ESR Technology Center (ACERT) (Cornell). These Centers are funded to develop the technological advances that will define the future of biomedical ESR, and to make these tools available to researchers. Major EPR labs in the USA and other countries also help with specialized measurements. For example, the National High Magnetic Field Laboratory in Tallahassee, Florida, provides access to EPR at much higher magnetic fields and frequencies than are available elsewhere in the USA, and thus makes it possible to explore properties of spin systems hitherto unattainable. The most useful instrumentation and methodology developed in research labs will find its way into future generations of commercial EPR spectrometers. At present, the primary constraint on expansion of commercial EPR instrumentation is the limited size of the market. As the research community recognizes the importance of new techniques the capability will be come available. This process is well illustrated by the recent availability of pulsed ELDOR instrumentation and software to implement the DEER method for measuring distances between electron spins (see *Biol. Magn. Reson.* **19**, 2000). One can anticipate the commercial availability of application-specific modules as accessories to EPR spectrometers, and the development of specialized spectrometers for measurements not supported by the standard spectrometers, such as the recently-released transient response spectrometer (Bruker E500-T).

The chapters in these two volumes already give a glimpse of the future, since they are written by researchers working to create better ways of learning about biological systems with EPR. There will be increasing use of multiple microwave frequencies, especially much higher and much lower than X-band, and of two-frequency (ELDOR) methods. The slow-scan, analog filter, lock-in detection methods that are almost universal in CW EPR will be supplanted by many kinds of direct detection, direct digitization, and post-processing methods. Increasing ability to study aqueous samples will bring EPR methods closer to physiological conditions. Faster electronic components will reduce the dead time in pulse experiments and increase the digitization rates in CW and pulse experiments.

Resonators are the heart of the EPR spectrometer. The development of the loop-gap-resonator (LGR) was important in its own right, but arguably its most important contribution was to free us from the constraints of thinking about cavities. Now, a wide variety of resonant structures are being developed to optimize S/N for special samples, ranging from flat radiation dosimeter film to animals. Anything that the electrical engineer can design to resonate at the appropriate frequency potentially could be an EPR resonator for some application. Computer programs such as the High Frequency Structure Simulator (HFSS, Ansoft Corporation) will help refine intuitive approaches to nearly optimized resonator designs. As Jim Hyde pointed out in Volume 24 chapter 7, there are many types of resonators available for experiments at X-band, but no where near that variety at lower or higher frequencies. This is a challenge and an opportunity. Multifrequency EPR will grow in proportion to the availability of new resonators designed to facilitate special experiments.

Almost all EPR investigations, especially those of biological systems, ultimately are limited by signal-to-noise of available instrumentation, because the spin concentrations are inherently low, or the exploration of parameters requires a study of concentration dependence extending to very low spin concentrations. Some ways to approach the thermal noise limit of spectrometers are being explored. For example, using a low-noise preamplifier cooled to the same temperature as the sample helps reduce noise by the predictable amount (Rinard et al., 1999). A crossed-loop resonator makes it possible to place an amplifier right at the signal source even in pulsed EPR (Rinard et al. 1996a,b). The importance of minimizing noise as well as maximizing signal will result in the design of improved low-noise microwave sources, resonators specific to the sample type, as described by Hyde in this volume, and optimization of the location of the first stage amplifier as mentioned above. The result, which we term an "application-specific module" will not have the general applicability of the TE_{102} rectangular X-band cavity that has so dominated EPR in the past. However, building tools for specific problems will open new horizons for EPR.

Even magnets may not look the same in the future. At high microwave frequencies, superconducting magnets are required because iron-core electromagnets cannot achieve the required high magnetic field. Development effort for high-field EPR is mostly in national magnet development facilities, such as in France (e.g., Isber et al., 2001), Japan (e.g., Motokawa 2000), and the USA (National High Magnetic Field Laboratory, Tallahassee, Florida) (e.g., Hassan et al., 2000). EPR at frequencies lower than L-band can use air-core magnets (e.g., Rinard et al., 2002). Some potentially useful resonators can be small enough that application-specific spectrometer systems can be envisioned with very small, easily portable, permanent magnets or electromagnets.

There are numerous vendors of accessories and supplies essential for various aspects of EPR measurements, including Oxford Instruments (cryostats), Cryo Industries (cryostats), Wilmad (sample tubes, quartz Dewars, standard samples), Oxis (spin traps), Molecular Specialties (loop gap resonators), Research Specialties (accessories and service), Summit Technology (small spectrometers, accessories), Scientific Software Services (data acquisition), Cambridge Isotope Labs (labeled compounds), and CPI (klystrons). Addresses and up-to-date information can be accessed via the EPR Newsletter.

Few EPR spectrometers have the degree of automation that has become common in other spectroscopy areas. Bruker has a dosimetry spectrometer with an automatic sample changer, and software to prepare reports. The latest pulsed EPR spectrometers are able to execute long (including overnight) multipulse experiments without operator intervention. The most recent cryogenic temperature control systems also can be programmed and can control the cryogen valve in addition to the heater. There is one research lab that has highly automated ENDOR spectrometers. Except for these examples, EPR has not been automated to a very significant degree. In the future, one can anticipate being able to run EPR spectra from a remote site if someone puts the sample into the spectrometer and sets up what ever environmental control (e.g., temperature, atmosphere over the sample) is needed for the experiment. Some applications of EPR will see broad use only if multiple samples can be studied with sufficient automation that high throughput can be attained.

Obtaining an EPR spectrum may be the easy part of EPR spectroscopy. Interpretation of the results often is the hard part. By this we mean not just the interpretation in terms of what species is present, or what the line shape or T_2 means about this species, but rather what the EPR reveals about the biological system. One aspect of this is to educate scientists who have expertise in both the spectroscopy and the biology, so that each can be interpreted without falling into traps. It is toward this end that these books have tried to give an overview, with references to other books and reviews to help researchers make the transition to modern biomedical ESR.

Software is available, some commercially and some from individual labs (easily locatable via the EPR Society software exchange), for simulation of many spin systems. The understanding of many spin systems is now at the stage that a full simulation of the experimental line shape is a necessary step in interpreting an EPR spectrum. There will always remain important problems for which the key is to understand the spin system, so simulation is at the edge of the state of the art. For example, for many high-spin Fe(III) systems there is little information about ZFS terms, so one does not even know which transitions should be included in the simulation. At the other extreme, for S = 1/2 organic radicals in fluid solution one should be able to fit spectra within experimental error if the radical has been correctly identified. Future directions include combining spectral interpretation as outlined above with quantum mechanical and molecular dynamics descriptions of the biological system.

The application of EPR to biological systems has become sophisticated enough, with a large arsenal of tools, each available in at least a few labs, that the main problems are now biological. That is, the EPR spectroscopy is discriminating enough that it becomes increasingly important to have a very well-defined biological system, or one will focus in great detail on an impurity, or on an ill-poised pH or redox condition. Putting "dirty" ill-defined samples into the spectrometer will lead to "dirty" ill-defined ideas about what the EPR spectra mean. The spectroscopy can give very well-defined results for whatever sample happens to be put into the resonator.

As we said in the Preface, there are spins everywhere, and recognition of the importance of studying them in biological systems will increase. EPR is uniquely suited to this study. Instrumentation, methodology, software for analysis and simulation will develop in concert, simultaneously optimizing particular experiments as the horizons expand within a multi-frequency, multidimensional milieu. As electronic components with the needed capabilities become available, data acquisition will, for example, move toward direct digitization of signals so that multiple harmonics can be extracted from the raw data as outlined by Hyde in the chapter on future trends. We also envision that the EPR response region between slow-scan CW and pulsed EPR, a CW region where relaxation times affect the signal even in the absence of saturation, will increasingly be exploited. Finally, computers will be used increasingly to create new types of graphic displays to communicate multidimensional data sets to the biomedical researchers.

The focus in these two volumes has been on methodology and instrumentation. Advances in theory, and in computational methods, are also needed. Beyond these, there is an even more important element - education. There needs to be greater focus on educating the next generation of scientists who will use the powerful methods described in these volumes.

That task, also, requires ingenuity and stands as a challenge to universities world-wide.

We will end this attempt at prognosis with the same hope that we ended Foundations of Modern EPR: "The most confident prediction we can make is the most exciting discoveries and enhancements will be ones we have not thought of."

1. REFERENCES

Eaton, G. R., and Eaton, S. S. (1988a) The Future of EPR Instrumentation, *Spectroscopy* **3**, 34-36.

Eaton, G. R., and Eaton, S. S. (1988b) Workshop on the Future of EPR (ESR) Instrumentation - Denver, Colorado, August 7, 1987, *Bull. Magn. Reson.* **10**, 3-21.

Eaton, G. R., and Eaton, S. S. (1993) The Future of Electron Paramagnetic Resonance Spectroscopy, *Spectroscopy*, **8** 20-27.

Eaton, G. R., and Eaton, S. S. (1995) The Future of Electron Paramagnetic Resonance, *Bull. Magn. Reson.* **16**, 149-192.

Eaton, G. R., and Eaton, S. S. (1997) EPR Methodologies - Ways of Looking at Electron Spins. *EPR Newsletter* **9** (1), 15-18.

Eaton, G. R., Eaton, S. S., and Salikhov, K. M. (1998) The Next 50 Years, in *Foundations of Modern EPR*, Eaton, G. R., Eaton, S. S., and Salikhov, K. M., eds., World Scientific, Singapore, pp. 792-793.

Hassan, A. K., Pardi, L. A., Krzystek, J., Sienkiewicz, A., Goy, P., Rohrer, M., and Brunel, L.-C. (2000) Ultrawide Band Multifrequency High-Field EMR Technique: A Methodology for Increasing Spectroscopic Information. *J. Magn. Reson.* **142**, 300-312.

Isber, S., Christidis, T., Tabbal, M., Charar, S., and Goiran, M. (2001) High-Frequency EPR of Co^{2+} in CdSe. *Physica B* **293**, 304-307.

Motokawa, M. (2000) Electron Spin Resonance of Magnetic Materials in High Fields and High Frequencies. *Appl. Magn. Reson.* **19**, 77-91.

Rinard, G. A., Quine, R. W., Ghim, B. T., Eaton, S. S., and Eaton, G. R. (1996a) Easily Tunable Crossed-Loop (Bimodal) EPR Resonator, *J. Magn. Reson. A* **122**, 50-57.

Rinard, G. A., Quine, R. W., Ghim, B. T., Eaton, S. S., and Eaton, G. R. (1996b) Dispersion and Superheterodyne EPR Using a Bimodal Resonator, *J. Magn. Reson. A* **122**, 58-63.

Rinard, G. A., Quine, R. W., Song, R., Eaton, G. R,. and Eaton, S. S. (1999) Absolute EPR Spin Echo and Noise Intensities, *J. Magn. Reson.* **140**, 69-83.

Rinard, G. A., Quine, R. W., Eaton, S. S., Eaton, G. R., Barth, E. D., Pelizzari, C. A., and Halpern, H. H. (2002) Magnet and Gradient Coil System for Low-Field EPR Imaging, *Magn. Reson. Engineer.* **15**, 51-58.

Rinard, G. A., Quine, R. W., Eaton, G. R., and Eaton, S. S. (2002) 250 MHz Crossed Loop Resonator for Pulsed Electron Paramagnetic Resonance, *Magn. Reson. Engineer.* **15**, 37-46.

Contents of Previous Volumes

VOLUME 5

VOLUME 6

VOLUME 13

VOLUME 19

VOLUME 21

VOLUME 22

Index

GPSR Compliance

The European Union's (EU) General Product Safety Regulation (GPSR) is a set of rules that requires consumer products to be safe and our obligations to ensure this.

If you have any concerns about our products, you can contact us on ProductSafety@springernature.com

In case Publisher is established outside the EU, the EU authorized representative is:

Springer Nature Customer Service Center GmbH
Europaplatz 3
69115 Heidelberg, Germany

Batch number: 10739606

Printed by Printforce, the Netherlands